ANIMAL MODELLING IN SURGICAL RESEARCH

ANIMAL MODELLING IN SURGICAL RESEARCH

Edited by

Bengt Jeppsson

Malmö University Hospital, Lund University, Sweden

harwood academic publishers
Australia • Canada • China • France • Germany • India
Japan • Luxembourg • Malaysia • The Netherlands • Russia
Singapore • Switzerland • Thailand

Amsteldijk 166
1st Floor
1079 LH Amsterdam
The Netherlands

British Library Cataloguing in Publication Data

Animal modelling in surgical research
1. Animal models in research 2. Surgery - Research
I. Jeppsson, Bengt
610. 7'24

ISBN: 90-5702-213-3

Contents

PREFACE

An animal model is 'a living organism with an inherited, naturally acquired or induced pathological process that in one or more respects closely resembles the same phenomenon in man' (Wesslers, National Institute of Health, 1976). This definition of an animal model of human disease is useful. If animal models are used, they should be appropriate for the situation that is to be studied. There is probably no ideal model for any disease and the disease itself may be variable and may have many facets, so that more than one animal model may be required.

During a workshop on 'Needs for a New Animal Model', Leader and Padgett (*American Journal of Pathology*, 1981) listed nine criteria for a good animal model:

1. It should accurately reproduce the disease or lesions under study.
2. It should be available for multiple investigations. Sharing of animals and data among instructions has been an important factor in many researchers' successes. It allows monitoring of the scientific validity of observations and stimulates further investigations.
3. It should be exportable.
4. If a disease under study is genetic, the species should be polytocous – producing multiple young at each birth.
5. The animal should be large enough for multiple biopsy samples.
6. It should fit into the available facilities of most laboratories. The accelerating cost of any changes in animal housing and care standards makes this criterion particularly relevant.
7. The animal should be easily handled by most investigators. However, convenience should not be the determining factor on the selection of the model.
8. It should be available in multiple species.
9. The animal used in a model should survive long enough to be usable.

Numerous animal models have been presented during the years, but it is difficult to say which is most appropriate for the study of a particular disease or effect of therapy in a human population. This book therefore contains animal models for the study of human disease conditions. It is however not only a description of the models but also an academic discussion on their relevance for studying specific questions in human illnesses.

Bengt Jeppsson

CONTRIBUTORS

Bo Ahrén
Department of Medicine
Malmö University Hospital
Malmö, Sweden

R. Aldini
Istituto di Scienze Cliniche
Università di Bologna
Bologna, Italy

Aamer Ar 'Rajab
Department of Surgery
Lund University
Lund, Sweden

H. G. Beger
Department of General Surgery
University of Ulm
Ulm, Germany

Günther Bernatzky
University of Salzburg
Naturwissenschatfliche Fakultät
Central Animal Faculty
Salzburg, Austria

Marcus Büchler
Department of Visceral and Transplantation Surgery
University of Bern
Bern, Switzerland

Bernard de Hemptinne
Dienst voor Levertransplantatie
Digestive Heelkunde
Universitair Ziekenhuis Gent
Gent, Belgium

Jens Christian Djurhuus
Institute of Experimental Clinical Research
University of Aarhus
Aarhus N, Denmark

Laureano Fernández-Cruz
Experimental Microsurgery Unit
Department of Surgery, Hospital Clinic
University of Barcelona,
Barcelona, Spain

Karl-Erik Giercksky
Department of Surgery
The Norwegian Radium Hospital and
Institute of Cancer Research
Oslo, Norway

Finn Gottrup
Copenhagen Wound Healing Centre
Bispebjerg University Hospital
Copenhagen, Denmark

Claus Hammer
Institute für Chirurgische Universität
Ludwig Maximilians Universität
München 70, Germany

Gottfried Harbauer
Department of Experimental Surgery
Sarland University
Homburg (Saar)
Germany

Rosemary Hickman
Department of Surgery
University of Cape Town Medical School
Cape Town, South Africa

Bengt Jeppsson
Department of Surgery
Malmö University Hospital
Malmö, Sweden

P. J. Klopper
Department of Surgical Research
Academic Hospital
University of Amsterdam
Amsterdam, The Netherlands

Antonio Lopez-Pena
Dienst voor Levertransplantatie
Digestive Heelkunde
Universitair Ziekenhuis Gent
Gent, Belgium

Runar Lundblad
Department of Surgery
The Norwegian Radium Hospital and
Institute of Cancer Research
Oslo, Norway

M. Montagnani
Dipartimento di Medicina
Interna e Gastroenterology
Universitá di Bologna
Bologna, Italy

José Ramon Morandeira
Servicio de Biomedicina y Biomateriales
Universidad de Zaragoza
Facultad de Medicina
Zaragoza, Spain

Eckhard Nagel
Klinik für Abdominal-und Transplantationschirurgie
Der Medizinische Hochschule in Hannover
Hannover
Germany

Bernard Nordlinger
Laboratory of Experimental Hepatic Surgery and
Digestive Surgery Unit
Hôpital Saint-Antoine
Paris, France

Yves Panis
Digestive Surgery Unit
Hôpital Lariboisière
Paris
France

Rudolf Pichlmayr
Klinik für Abdominal-und Transplantationschirurgie
Der Medizinischen Hochschule in Hannover
Hannover, Germany

Ib Rasmussen
Department of Surgery
University Hospital
Uppsala, Sweden

M. Teresa Ribera
Experimental Gastroenterology Laboratory
School of Digestive Pathology
Hospital de la Santa Cruz Y San Pablo
Autonomous University of Barcelona
Barcelona, Spain

Gerhard Rümenapf
Department of Surgery
University Hospital
Erlangen, Germany

Angelika Schmiedl
Department of Surgery
University Hospital
Erlangen, Germany

Michael H. Schoenberg
Department of Surgery
University of Ulm
Ulm, Germany

Paul Otto Schwille
Division of Experimental Surgery
University Hospital
Erlangen, Germany

Hans Stødkilde-Jørgensen
The MR Research Center
Skejby Hospital
Skejbygaardsvej
Aarhus, Denmark

Per Svendsen
Biomedical Laboratory
Odense University
Odense M, Denmark

Eduardo M. Targarona
Department of Surgery
Hospital Clinic
Barcelona, Spain

Nicholas L. Tilney
Surgical Research Laboratory
Harvard Medical School
Brigham and Women's Hospital
Boston, Mass, USA

G. Tufveson
Department of Transplantation Surgery
University Hospital
Uppsala, Sweden

Giovanni Ussia
Dipartimento di Discipline Chirurgische
Anestesiologiche e die Trapianti
"A. Valsalva"
Università di Bologna
Bologna, Italy

Jan Hein van Dierendonck
Department of Pathology
University of Leiden
Leiden, The Netherlands

Henk-Jan van Slooten
Department of Pathology
University of Leiden
Leiden, The Netherlands

Hans-Werner Waclawiczek
Department of Surgery
Salzburg Hospital
Salzburg, Austria

David Whitley
Surgical Research Laboratory
Harvard Medical School
Brigham and Women's Hospital
Boston, Mass, USA

PART I

GENERAL ASPECTS

PART I

1. Comparative Biology of Animals and Man in Surgical Research

Per Svendsen[1] and Finn Gottrup[2]

[1]*Biomedical Laboratory, Odense University, 21, Winslow Parken, DK-5000 Odense, Denmark*
[2]*Copenhagen Wound Healing Centre, Bispebjerg University Hospital, DK-2400 Copenhagen, Denmark*

Advances in Medical Research Resulting from Experiments with Animals

The first known use of animals in medical research is attributed to Galen of Pergamon (129–200 AD). He removed or damaged organs of conscious pigs and monkeys in order to study their function. He studied the results of damage to the spinal cord, perforation of the thoracic wall, and ligation of arteries.

It was not until the beginning of the nineteenth century, however, that systematic experimental research using animals evolved. In 1822 – before anesthesia had been invented – the French physiologist Francois Magendie published results of his experiments with dogs, showing that transection of the dorsal roots of the spinal cord caused loss of sensation, while section of the ventral roots caused loss of muscular contraction. These experiments formed the basis for understanding the complicated function of the nervous system, and are still considered among the most important discoveries in the history of biological science.

Claude Bernard, who was a student of Magendie's, published a book in 1865 called *Introduction to the Study of Experimental Medicine*, which was to become the foundation of modern physiological research. Bernard is known for his experiments on the arrow poison curare, which became the basis for understanding neuro-muscular nerve impulse conduction.

In the last part of the nineteenth century the Russian scientist Ivan Pavlov conducted his famous experiments on digestive physiology in dogs, leading to the understanding of the cooperation between physiological function and external stimuli, known as conditioned reflexes.

Other pioneers in modern experimental medicine were Louis Pasteur, Rudolph Wirchow, Emil Behring and Robert Koch, who examined infectious diseases in animals as models for diseases in man.[52]

In the twentieth century, experimental medicine has expanded dramatically, and the use of laboratory animals has increased rapidly. The most remarkable achievements have been the prevention and treatment of infectious diseases through the development of antisera, vaccines, sulphonamides and antibiotics. Knowledge of endocrine glands and the function of hormones in metabolism and reproduction are the results of animal experimentation. The most important discovery was probably that of Banting and Best on the role of the pancreas in the development of diabetes, and the use of insulin in the treatment of this disease. Dogs were used as models in these studies.

During the past fifty years the development of new drugs has been extensive. The treatment of psychiatric diseases has been dramatically changed. New anesthetics have made progress possible in surgery, cytotoxic drugs have improved the treatment of cancer, and immunosuppressive compounds have made it possible to perform organ transplantations.

In the nineteenth century surgery was the last resort, used only when all other treatments had been tried. Progress in surgery since then can be attributed to many factors, one of them being experimental operations performed on animals.[38] This chapter will focus on some of the basic preconditions for experimental research in surgery.

The Animal Model

An animal model may be defined as a living organism with an inherited, naturally acquired or induced pathological process that closely resembles the same condition in the target species.[26]

Experimental models for the study of insulin-dependent diabetes mellitus are good examples: mice in which diabetes is provoked by multiple injections of streptozotozin or by infection with encephalomyocarditis (EMC-M) virus are examples of induced models. The BB rat and the non-obese diabetic (NOD) mouse, are spontaneous or genetically determined models; they both develop diabetes with glucosuria, hypoglycemia, ketonuria and hypoinsulinemia.

In most species, complete surgical extirpation of the pancreas is extremely difficult because the gland extends so widely into the mesentery, and the blood supplies to the gland and the duodenum are difficult to separate. Surgical removal also removes the exocrine pancreatic function and the gastro-intestinal hormones secreted by the islet cells.

A number of preconditions must be met to ensure the animal model is relevant. The first consideration is the choice of *animal species*. In the past dogs and cats have frequently been used as models in surgical research. They were often chosen because they were easily available and easy to handle rather than as a result of an anatomical or physiological comparison between the animal and man. Growing public concern about the use of companion animals in scientific research has forced scientists to accept other animal species for their projects, and to consider the choice of species from a scientific point of view.

Many diseases are common to both man and animals, and spontaneous clinical cases in domestic or laboratory animals can be very useful as models for human conditions. Most mammalian species are largely anatomically similar in terms of cardiovascular, respiratory and nervous systems, whereas there are marked differences in the anatomy of their digestive and reproductive systems. The anatomical differences are frequently accompanied by physiological and metabolic differences. In herbivores (plant-eating animals), parts of the gastro-intestinal tract have developed into specialized compartments for cellulose digestion, for instance the rumen–reticulum of ruminants or the coecum–colon of equidae. The digestive process in these species differs from that of humans in that it is primarily microbial rather than enzymatic, and the products of carbohydrate digestion are volatile fatty acids rather than glucose.

The second consideration when using animals as experimental models is the influence of external factors on the homeostasis. Physical factors such as temperature, humidity, noise and light intensity, and chemical factors such as feed composition and pesticide residues may affect metabolism by stimulation of metabolic enzymes, and thus cause misinterpretation of experimental results.

How the animals are handled and the experimental procedure used are other factors that may affect experimental results. Painful procedures necessitate anesthesia, but anesthesia itself may affect the cardiovascular and respiratory systems and influence metabolism, and consequently lead to misinterpretation of experimental results.

Experimental failures are often caused by infectious diseases in laboratory animals, and biological factors such as pathogens, parasites or latent infections may affect the animals and influence the experimental results. Great efforts have been made to prevent the spread of pathogens among laboratory animals.

Comparative Anatomy and Physiology

The Musculo–Skeletal System

The musculo-skeletal system is composed of the skeleton, joints, ligaments, muscles and tendons. Generally the skeletons of man and animals are quite similar. Major differences are seen in the teeth, the spine and the distal parts of the limbs. The joints show great anatomical similarities, but are differently positioned because animals are quadrupedal.

The *dental formula* of mice and rats is: I: 1/1, C: 0/0, P: 0/0, M: 3/3 = 16. The teeth of the guinea-pig are slightly different as one premolar tooth is present in both upper and lower jaw. The dental formula is: I: 1/1, C: 0/0, P: 1/1, M: 3/3 = 20. In the rabbit a second upper incisor is present behind the first, and the dental formula is: I: 2/1, C: 0/0, P: 3/2, M: 3/3 = 28. Owing to the absence of enamel on the lingual surface of the incisors, a sharp cutting edge is maintained on the incisors of these species. The pulp cavity of the incisors remains open, and growth continues throughout the animal's lifetime. The canine and premolar teeth are absent, and caudal to the incisors is a space known as the diastema, into which folds of cheek mucosa extend, thus separating the gnawing apparatus from the chewing apparatus.

The dental formula for the permanent teeth of the pig is: I: 3/3, C: 1/1, P: 4/4, M: 3/3 = 44. The incisors are flattened and curved. The crowns have no distinct neck. The enamel covering is more extensive on the labial surface than on the lingual surface. The canine teeth of the male are well developed and project out of the mouth; in the sow they are smaller. The cheek teeth have complex tuberculate crowns with a distinct neck and round, pointed roots. The first premolar has two roots, the other premolars have three or four roots. The molars all have four roots.

The dental formula for the permanent teeth of the dog is I: 3/3, C: 1/1, P: 4/4, M: 2/3 = 42. In some brachycephalic breeds the number of cheek teeth may be reduced. The incisors are placed vertically. The labial surface is convex and the lingual surface slightly concave. The roots are narrow. The canine teeth are large, conical and curved, with a 2–3 cm long root. The upper fourth premolar and lower first molar teeth, the carnassial teeth, are much larger than the remaining cheek teeth.

The number of teeth in *non-human primates* is usually thirty two with a dental formula of I: 2/2, C: 1/1, P: 2/2, M: 3/3. Some species variations occur.

The *vertebral column* of mice and rats is composed of 7 cervical, 12–14 thoracic, 5–6 lumbar, 4 sacral and 27–30 caudal vertebrae. In the guinea-pig, the os sacrum is reduced to 3 vertebrae and the cauda to 7 vertebrae. In the rabbit, the vertebral formula is C7, T12, L7, S4 and C16. Some breeds, however, have 13 thoracic vertebrae. The vertebral formula of the pig is C7, T14–15, L6–7, S4 and C20–23, and of the dog C7, T13, L7, S3 and C20–23.

The appendicular skeleton includes the paired pectoral and pelvic girdles, and the bones of the limbs. The pectoral girdle consists of the scapula and the clavicle. The latter is present in small rodents, guinea-pigs and rabbits. In the dog the clavicle is a small irregular bone or cartilage embedded in the brachiocephalicus muscle in front of the shoulder joint. It forms no articulation with the rest of the skeleton. In the pig the clavicle is absent.

The *pelvic girdle* is composed of the paired os coxae, which articulate with the transverse processes of the os sacrum. Os coxae comprise the ilium, ischium and pubis bones.

The *long bones* of the limbs are similar in all species, whereas the number of digits varies. Small rodents, rabbits and dogs have five digits, pigs have four, and guinea-pigs have four on the front legs and three on the hind legs.

A number of diseases of the musculo–skeletal system, including dental diseases (23, 47, 61), osteomyelitis, osteoarthritis (1, 57), and scoliosis (36), and surgical techniques such as dislocations (64), tendon healing (20), and prosthetic joint surgery (9) have been studied experimentally using animal models. A wide range of animal species have been used: most frequently, however, the monkey, the dog, and the rabbit (Table 1.1).

The Nervous System

In all species the nervous system consists of the central nervous system (brain and spinal cord) and the peripheral nervous system (cranial and spinal nerves). The anatomical structures are qualitatively

Table 1.1. Commonly used animal models in the study of diseases of the skeletal system

	Animal model				
Condition	*Rat*	*Rabbit*	*Dog*	*Pig*	*Monkey*
Teeth					
Alveolar wound healing	++		+		+
Tooth transplantation			+		++
Columna					
Scoliosis		+			+
Bones					
Osteomyelitis	+	+	+		
Joints					
Osteoarthritis	+	++	++		+
Caput femoris					
Avascular necrosis		+	++	+	
Coxa vara		+	+	++	
Congenital dislocation	+	+	++	+	
Prosthetic joints			++		
Tendons					
Wound healing	+	+	+		

(++) frequently cited
(+) occasionally cited
( ) rarely cited

similar but quantitatively different. A detailed description of these systems is outside the scope of this chapter, and so only a simplified description in relation to relevant differences will be given.

The *central nervous system* consists of the grey matter (nerve cell-bodies and mostly non-myelinated fibres), and the white matter (mostly myelinated fibres) with neuroglial cells occurring throughout. Macroscopically and topographically, the central nervous system can be divided into the forebrain (the cerebrum and diencephalon), midbrain, hindbrain (the pons, medulla oblongata and cerebellum), and the spinal cord. The midbrain, pons and medulla oblongata collectively form the brain stem. The size, shape and structure of the central nervous system vary between the different species of laboratory animals.

The forebrain consists of the cerebrum and diencephalon. The cerebrum consists of the paired hemispheres which are partly separated by a sagittal fissure crossed by a massive commissure, the corpus callosum. A frontal, parietal, temporal, limbic and occipital lobe can be described. The diencephalon is situated between the cerebral hemispheres and the brain stem. It consists of the thalami, hypothalamus and epithalamus. The olfactory and optic cranial nerves originate in the forebrain.

The midbrain or mesencephalon is situated between the forebrain and the hindbrain. It consists of the tectum, the cerebral penducles and the cerebellar penducle. The occulo–motor and trochlear cranial nerves originate in the midbrain.

The hindbrain or rhombencephalon consists of the pons, medulla and cerebellum. The cerebellum, which is located dorsal to the pons and medulla, consists of two hemispheres and a midline vermis. The trigeminal, abducent, facial and vestibulocochlear nerves originate in the pons, and the glossopharyngeal, vagal, cranial accessory and hypoglossal cranial nerves originate in the medulla oblongata.

The *spinal cord* continues rostrally with the medulla oblongata and ends caudally as the conus medullaris. Cervical and lumbar enlargements of the spinal cord are associated with limb innervations. The dorsal (sensory) and ventral (motor) roots of the spinal nerves originate in the spinal cord. The lumbar, sacral and coccygeal nerve roots extend beyond the spinal cord as the cauda equina.

The *peripheral nervous system* consists of the cranial and spinal nerves and their ramifications. Peripheral nerves comprise fasciculi of myelinated and non-myelinated fibres, enclosing three sheaths of connective tissue: the epineurium, the perineurium and the endoneurium.

All species, including man, have 12 cranial nerves, whereas the number of spinal nerves varies between species according to the number of vertebrae. In man the number of spinal nerves is 31, whereas the horse has 42, the cow 37, the pig 33, the dog 36–37, the rabbit 37, the guinea-pig 32–36, the rat 34, the mouse 33 and the hamster 32–33. The peripheral nerves innervate somatic and visceral structures and carry efferent (motoric) and afferent (sensoric) impulses. The somatic efferent fibres pass directly from the central nervous system to the skeletal muscles, whereas the autonomic (visceral) efferent fibres pass from the central nervous system to smooth muscle and glands. These efferent fibres are preganglionic, with synapses in peripheral ganglia. The somatic and autonomic afferent fibres are sensory neurons passing from peripheral receptors to the spinal dorsal root ganglia or similar ganglia of certain cranial nerves.

The *autonomic nervous system* has sympathetic and parasympathetic sections. The sympathetic nerves originate in the spinal thoracolumbal region (T1 to L2), whereas the parasympathetic nerves originate in the cranial and sacral nerves (cranial nerves III, VII, IX and X, and S2–S4).

Several animal models of *nervous system disorders* such as increased intracranial pressure and spinal cord injury,[15] are described in the literature. The most commonly used species are dogs, cats, and non-human primates (Table 1.2).

The Eye

The orbit of the guinea-pig, sheep, goat and non-human primates is completely surrounded by bone as in humans and is termed 'closed'. Other laboratory animal species have an 'open' orbit, which is surrounded by bone except on the lateral side, where bone is replaced by the orbital ligament. The position of the eye in the skull reflects the natural behavior of the animal species. Carnivores such as cats and dogs need binocular vision to hunt their prey and consequently have their eyes placed on the front of the head, whereas herbivores have their eyes placed on the sides of the head to provide a maxi-

Table 1.2. Commonly used animal models in the study of surgical diseases in the nervous sytem

	Animal model				
Condition	*Rat*	*Rabbit*	*Dog*	*Pig*	*Monkey*
Hydrocephalus		+		+ +	+
Epilepsy	+		+ +		+
Spinal cord injury		+	+ +	+	

(+ +) frequently cited
(+) occasionally cited
( ) rarely cited

mum panoramic view: rabbits, ruminants and equids, are typical examples The extraocular muscles are the four rectus muscles (dorsalis, ventralis, lateralis and medialis) and the two oblique muscles (dorsalis and ventralis). The retractor oculi is not present in all species, but is well developed in herbivores. The upper eyelid is opened by the levator palpebral superioris. In the rabbit a musculus depressor palpebral inferioris is present to open the lower eyelid. The cat has a thin smooth muscle used in the retraction of the nictitating membrane. This structure is a specialized eyelid that protects the eye. It is well developed in birds, ungulates and carnivores, but it is poorly developed in rodents, rabbits and primates.

Animals have three types of *ocular gland*. The lacrimal gland is located behind the orbital rim dorsolateral to the eyeball, the Hardian gland is found nasal to the eyeball, and the nictitans gland is found within the nictitating membrane in the species in which this structure is present.

The size of the *cornea* correlates with the behavior of the species. Nocturnal animals usually have large corneas (30–35% of the eyeball) relative to the eyeball to allow more light to strike the retina. In animals such as rabbits and cats the cornea occupies about 30% and in man approximately 17% of the eyeball. As in man the cornea of animals is a laminar structure. Bowman's membrane, however, is frequently absent.

The shape of the pupils differs between species. Herbivores have horizontal and oval pupils, which provide a wide field of vision, whereas most carnivores and primates have round pupils. The cat is an exception as it has vertically split pupils. Rodents and rabbits have round pupils.

The size of the *lens* varies greatly. Nocturnal animals such as rats and mice have very large spherical lenses, occupying most of the eyeball. In diurnal animals, and in man the lens only occupies about 10% of the eyeball volume.

Nocturnal and arrhythmic animals have a specialized layer of the *choroid* known as the tapetum lucidum, which increases the amount of light striking the retina by reflection. The *retina* of the common laboratory animals consists of the same ten layers as in man, although there is some difference in the layer of rods and cones. The fovea is absent in nocturnal animals, but always present in birds, apes and some monkey species. Most animals have both rods and cones, and rods are more numerous than cones. Nocturnal animals have mainly rods and diurnal animals have large numbers of cones.

The blood supply to the retina can be holangiotic as in man, where the retina receives blood directly from main arteries or from the cilio–retinal plexus. This vascular pattern is found in non-human primates, cats, dogs, sheep, cows and pigs. In rabbits only a small part of the retinal blood supply is provided with obvious blood vessels. This pattern is called merangiotic. If the fundus is completely devoid of blood vessels, as in the guinea-pig and birds, the pattern is known as anangiotic. In horses the blood vessels are limited to the peripapillary region; this is known as the paurangiotic pattern. In rabbits the blood–aqueous barrier is labile, and invasive surgery induces protein leakage into the aqueous epithelium from the ciliary epithelium, miosis, increased intraocular pressure and anterior uveal hyperemia.

A number of *eye diseases* and methods of surgical treatment of eye diseases have been studied experimentally in laboratory animals. Refractive errors have been experimentally produced in rabbits, cats, domestic fowl and non-human primates.[60,79] Corneal epithelial healing has been studied in rabbits, cats, dogs and primates.[7,29] Corneal transplantation has been performed experimentally in cats,[4] and transplantation immunology has been studied in inbred mice and rats.[35,66] Radical keratotomy has been performed and studied in rabbits[10] and monkeys,[6] experimental glaucoma in cats and monkeys,[55,56,58] cataract and lens extraction in monkeys and rabbits,[45] and retinal detachment in rabbits and monkeys,[44] and in cats,[2] (Table 1.3).

The Cardiovascular System

The anatomy of the human heart is very similar to that of other mammals. In the rat, however, there is an extra coronary myocardial blood supply. The atria receive a major input from the subclavian and the internal mammary arteries, and the rat's heart is more like that of fish than of other mammals. The heart of the rabbit differs from that of other mammalian species in that the right atrioventricular valve is formed by two cusps rather than three.

Table 1.3. Commonly used animal models in the study of diseases of the eye

	Animal model				
Condition	*Rat*	*Rabbit*	*Dog*	*Pig*	*Monkey*
Refractive errors			+	+	+
Corneal healing			++		+
Corneal transplant	+	+		++	
Radical keratotomy					++
Glaucoma				+	+
Cataract			+		+
Retinal detachment			++	+	+

(++) frequently cited
(+) occasionally cited
( ) rarely cited

The *peripheral circulatory system* does not differ significantly in the animal species commonly used as laboratory animals. Access to the veins is simple in all species except the guinea-pig. In mice and rats superficial veins are located in the tail, and in rabbits along the ear margin. In dogs and cats the cephalic vein on the plantar side of the front leg and the vena saphena on the lateral aspect of the hock are superficial and easily accessible. Vein puncture in the pig is performed in the cranial vena cava with a long needle introduced just in front of the manubrium sterni at an angle of 45°. Percutaneous arterial puncture can be performed in the rabbit using the central artery of the ear. Cardiovascular data for the different species of laboratory animals are shown in Table 1.4.

Experimental studies on *cardiovascular diseases* using animal models are common. Diseases of the heart valves have been experimentally produced primarily in dogs[27,59,63] and to some extent in sheep.[70] The dog[43] and the pig have frequently been used as models in the study of myocardial infarction. Experimental cardiac failure has been produced in dogs,[28,62,67] calves and sheep (Table 1.5).

The Respiratory System

The upper respiratory system of the commonly used laboratory animals presents no unusual anatomical features, however the lower respiratory organs show some anatomical differences between species. In mice and rats the left lung forms a single lobe, whereas the right lung is divided into four lobes: the apical, cardiac and diaphragmatic are located in a row, and the intermediate lobe is located against the diaphragm. In the guinea-pig and the rabbit the left lung comprises three lobes, while the right lung is similar to that of small rodents. In the dog the right lung is approximately 25% larger than the left, divided into four lobes by deep fissures. The left lung is divided into three lobes: the apical, cardiac and diaphragmatic. In the pig the right lung has four lobes, and the left lung has two lobes, as the apical and the cardiac lobes are not separated by an interlobular fissure as seen in other species. Respiratory data for the commonly used laboratory animals are presented in Table 1.6.

Animal models of pulmonary edema, pulmonary atelectasis and surgical lung resection are described in the available literature. Most commonly dogs, sheep, pigs, and rabbits have been used[8,16,53,69,75] (Table 1.7).

The Digestive System

The anatomy of the digestive system shows great variation between the commonly used species of laboratory animals and also between these species and humans.[68]

In the oesophagus, two major differences appear. The first relates to the musculature of the esophagus and is thus of interest in studies of esophageal motility. The second difference relates to the charac-

Table 1.4. Cardiovascular data for common laboratory species

Animal (body weight)	*Heart rate* (beats/min)	*Blood pressure* (mm/Hg)	*Blood volume* (ml/kg)	*Haemoglobin* (g/100ml)	*Haematocrit* (%)	*Exsangguination* (ml/kg)	*Safe sample* (ml/kg)
Mouse (30g)	570	113/81	80	15(10–20)	40(35–45)	35	7
Rat (300g)	365 (300–430)	116/90	50–54	15(12–18)	40(35–45)	20	5
Guineapig (600g)	260	90/55	75	14(11–17)	40(30–42)	35	7
Rabbit (3kg)	220	110/80	70	12(8–15)	40(30–50)	35	7
Cat (3kg)	110–140	120–75	75	11(18–14)	40(30–50)	35	7
Dog (15kg)	100–130	112/56	90	15(11–18)	45(38–53)	45	9
Minpig (30kg)	83	97	65	11(10–13)	40(30–50)	25	6
Sheep (40kg)	70(70–80)	114/90	60	11(10–12)	34(29–38)	25	5
Goat (30kg)	80(70–125)	120/84	70	11(18–14)	34(29–38)	30	7
Cow (400kg)	60–70	130/90	60	11(8–14)	37(33–48)	30	7

Table 1.5. Common animal models in the study of cardiovascular diseases

Condition	*Animal model* *Rat*	*Guinea-pig*	*Dog*	*Cat*	*Pig*	*Lamb*	*Calf*
Heart valves							
Mitral stenosis			+			+	
Mitral insufficiens			+				
Pulmon. stenosis			+	+		+	
Tricusp. insufficiens			+				
Aortic insufficiens			+				
Aortic stenosis			+				
Coarctation of aorta			+		+	+	
Left ventricular hypertrophy	+	+	+		+		
Myocardium							
Infarction			+		+		
Heart failure			+			+	+

(+ +) frequently cited
(+) occasionally cited
( ) rarely cited

Table 1.6. Respiratory data for common laboratory animal species

Animal (*body weight*)	*Respiratory rate* (breaths/min)	*Tidal volume* (ml)	*Minute volume* (ml/min)	*Blood* pCO_2 (mm Hg)	*Blood pH*
Mouse (30g)	100–250	0.15	24	50	7.40
Rat (300g)	70–150	1.6	220	42	7.35
Guineapig (600g)	90–150	1.0–4.0	100–400	40	7.35
Rabbit (3kg)	35–60	19–25	800–1,200	40	7.35
Cat (3kg)	20–30	12–18	300–400	36	7.35
Dog (15kg)	30	200–250	3,500–4,500	38	7.36
Minipig (30kg)	16–18	250–280	4,500	40	7.43
Sheep (4kg)	12–30	280–330	5,500	38	7.40

Table 1.7. Commonly used animal models in the study of pulmonary diseases

Condition	*Animal model* *Rabbit*	*Dog*	*Pig*	*Sheep*
Pulmonary edema	+	+	+	+
Pulmonary atelectasis		+	+	
Lung resection	+	+		

(+ +) frequently cited
(+) occasionally cited
( ) rarely cited

teristics of the esophageal mucosa, and is important in the study of mucosal injury. In man the esophageal musculature is striated in the upper one-third and smooth in the lower two-thirds. The muscles are arranged with an inner circular and an outer longitudinal layer. Sphincters are located at the pharyngo-oesophageal and the gastro–esophageal junctions. The esophageal mucosa is a non-keratinized squamous cell epithelium, which protects against ingesta and gastric reflux. Except for a few mucous-secreting glands the human esophagus produces no secretions.

In rodents, dogs and pigs the striated musculature appears along almost the whole length of the esophagus, whereas the musculature of the cat is more comparable to that of man with the distal one-third being smooth. In animals adapted to a diet of coarse plant material the esophageal mucosa is a keratinized stratified squamous mucosa, whereas the epithelium of carnivores is less keratinized. Like humans, rabbits have few mucous-secreting glands, whereas these glands are frequent in the dog and absent in the cat. For these reasons, the cat and the rabbit seem to be the best models of human esophageal injury.

Mice, rats, and hamsters have simple stomachs with clearly demarcated cardiac and pyloric regions. The cardiac region has a stratified squamous mucosa without glands, whereas the pyloric region is anatomically similar to that of man and other monogastric animals. The stomach of the rabbit and the guinea-pig lacks the keratinized cardiac region. The stomach of the dog and the cat is of the simple monogastric type divided into cardiac, fundic and pyloric regions. The stomach of the pig is relatively large and its major curvature extends caudally to a point midway between the xiphoid cartilage and the umbilicus. The left part of the stomach is large and round, whereas the right part is small and bends sharply to form the pylorus. The pig's stomach can be divided into four regions: the esophageal region, which lacks glands, and is found around the cardia; the cardiac gland region, which extends to the middle of the stomach; the fundic region; and the pyloric region.

In the ruminants (cow, sheep and goat) the esophageal region has developed into the forestomachs, which occupy nearly three-quarters of the abdominal cavity. The forestomachs comprise the rumen, reticulum and omasum. The fourth compartment of the ruminant stomach, the abomasum, is an elongated sac situated mainly on the abdominal floor. Physiologically, the rumen acts as a fermentation vat, in which carbohydrates are digested to volatile fatty acids by microbial digestion. This process produces large amounts of methane and carbon dioxide, which are removed by eructation. When placed in lateral or dorsal recumbency, for instance during general anesthesia, the eructation reflex is inhibited and the animal may develop ruminal tympany and suffocation. The abomasum resembles the true stomach of the monogastric animal. It is divided into an anterior blind sac, the fundus, the abomasal body and the pyloric part. The physiological function of the abomasum is mainly protein digestion, especially denaturation of the microbial protein entering the abomasum from the rumen.

In small rodents, guinea-pigs and rabbits, the *duodenum*, *jejunum* and *ileum* correspond to the same structures found in man. In carnivores, especially in the cat, the wall of the small intestine is relatively thick. In the pig the small intestine has a fan-like arrangement of the mesenteric vessels with the vascular arcades forming in the muscular mucosa rather than in the mesentery as in other mammals.

Mice and rats have a simple large intestine consisting of the *coecum*, *colon dextrum*, *colon transversum*, *colon sinistrum* and *rectum*. The coecal wall has lymphoid aggregates, but lacks the vermiform appendix. In the guinea-pig and the hamster the coecum is relatively large and lacks the appendix; the colon is similar to that of other rodents. The rabbit has a large colon. Near the ileocoecal valve is an accumulation of lymphoid tissue known as the sacculus rotundus, and a vermiform appendix comprising lymphoid tissue is located at the apex of the coecum. The carnivores have a relatively small coecum without an appendix. The colon and rectum are similar to those of man. The pig, like man, is a true omnivore. The large colon, however, differs anatomically. The coecum and right and transverse colon are combined into a spiral structure called the spiral colon. The vermiform appendix is absent in the pig.

The *liver* of the mouse is divided into four lobes joined dorsally. The large median lobe is subdivided into a right and a left lobule. The right lateral lobe is divided horizontally into anterior and posterior portions. The caudal lobe consists of two lobules dorsal and ventral to the esophagus. The left lateral lobe is undivided. The gall-bladder is located in the bifurcation of the median lobe. The cystic and hepatic ducts unite to form the common bile duct, which receives one or more pancreatic ducts before entering the duodenum.The rat liver is lobulated like that of the mouse; however, it lacks the gall-bladder, and the rat is unable to concentrate and store bile. The ductus choledocus is formed by the hepatic ducts of the hepatic lobes. It enters the duodenum 2–3 cm from the pylorus.The liver of the rabbit is divided into four lobes, with an irregular subdivision. The gall-bladder is situated in a deep depression on the visceral surface of the right anterior lobe. The hepatic ducts unite with the cystic duct to form the ductus choledocus, which enters the duodenum immediately posterior to the pylorus. There is no connection to the pancreatic duct.The dog liver is divided into five lobes, of which the left lateral is the largest. The right central lobe is second in size. It is divided by a deep fossa in which the gall-bladder is located. The common bile duct is formed by the cystic duct and the hepatic ducts and opens into the duodenum 5–8 cm from the pylorus. The minor pancreatic duct usually opens into the common bile duct. The pig liver is divided into four lobes, right lateral, right central, left lateral and left central. On the dorsal part of the right lateral lobe is the caudate lobe. The gall bladder is located in a fossa on the right central lobe. The cystic duct joins the hepatic duct at an acute angle immediately after the emergence of the latter from the portal fissure. The ductus choledocus enters the duodenum at the papilla duodeni 2–5 cm from the pylorus. The vena cava caudalis is entirely embedded in the liver.

The exocrine function of the *pancreas* has been studied most thoroughly in the dog. The canine pancreas is suspended in the duodenal mesentery and is freely mobile. Two, or occasionally three, ducts enter the duodenum directly, and the minor pancreatic duct enters the common bile duct. In cats and non-human primates the pancreatic duct enters the common bile duct as in man. The pig's pancreas is located retroperitoneally in the same way as the human pancreas. The single pancreatic duct enters the duodenum near the caudal lobe of the pancreas. Cannulation or ligation of the pancreatic duct is a relatively simple procedure in the pig, and this species is excellent for the study of exocrine pancreatic function. The rabbit pancreas extends widely into the mesentery. The single pancreatic duct opens into the duodenum and is well separated from the opening of the bile duct. The pancreatic duct of the rabbit can be cannulated using duodenotomy.

Multiple animal models have been developed to study *diseases* of the digestive system, including esophageal stricture,[25,33,50] gastric reflux,[34,39] gastric and duodenal ulcers, pancreatitis, portal hypertension,[37] gall stones, intestinal resection,[5,11,73] intestinal obstruction,[54] intestinal anastomosis,[12,31] ulcerative colitis[3,17,42] and organ transplantations.[22,30,48,74] All the usual species of laboratory animals have been used (Table 1.8).

The Urogenital System

The urogenital system shows some anatomical and physiological differences between the different laboratory animal species, and considerable differences between animals and man.

The urinary organs consist of the kidneys, ureters, vesica and urethras. The *kidneys* of mice, rats and guinea-pigs are paired bean-shaped organs located retroperitoneally against the dorsal abdominal wall. The right kidney is located more cranially and is larger than the left. The kidney is dorso-ventrally flattened with a concave medial border. Blood vessels and ureter join the kidney at the hilus located at the medial border. Transection of the kidney reveals the cortex and the medulla. The cortex follows the shape of the convex border, and the medulla is located against the inner concave surface of the cortex. The rodent kidney is unipapillated with a single papilla and calyx, surrounded by the renal pelvis. The *ureters* extend from the kidneys,

Table 1.8. Commonly used animal models in the study of gastro-intestinal diseases

	Animal model						
Condition	*Mouse*	*Rat*	*Rabbit*	*Guinea-pig*	*Dog*	*Cat*	*Pig*
Oesophagus							
Stricture		+			+	+	
Reflux		+	++		+	+	
Stomacon							
Vagotomy		+			++		
Acid secretion		+			++		
Resection		+			++		+
Endocrine factor		+			++		+
Small intestine							
Obstruction		++	+		+		+
Resection		++			+		+
Anastomosis		+	+		+		+
Large intestine							
Ulceractive colitis	+	+	+	++			
Pancreas							
Transplantation		+				++	
Liver							
Port. hypertens		++	+		+		+
Transplantation		+					++

(++) frequently cited
(+) occasionally cited
( ) rarely cited

dorsal to the uterine horns or the vas deferens, to the vesica, where they enter the dorsal wall of the bladder neck separately. The *vesica* is triangular in shape with the apex directed caudally. The caudal portion of the vesica narrows into the urethra. In the female the *urethra* is a short tube which extends from the bladder to the external orifice of the clitoris just anterior to the vaginal orifice. In the male the urethra passes into the penis. The urinary system of the rabbit differs little from that of the small rodents, except that in the female the urethra opens into the vagina.

The *kidneys* of the pig are bean-shaped. They are located ventral to the transverse processes of the first four lumbar vertebrae, with the left kidney slightly cranial to the right one. The hilus is situated at the centre of the medial border, and the pelvis is divided into two calyces majores, which are subdivided into eight to twelve calyces minores, each of which contains a papilla. The *urinary bladder* of the pig is relatively large and located in the abdominal cavity. The dorsal surface is covered by the peritoneum, whereas the peritoneum does not extend so far caudally on the ventral surface. The urethra is 7–8 cm long. The urethral orifice is found in the vagina, bounded on each side by a thick fold which extends caudally.

The *kidneys* af the dog are relatively large, with the right kidney situated opposite the first three lumbar vertebrae and the cranial pole in the renal impression of the liver. The left kidney is loosely attached by the peritoneum and subject to some variation in position. With an empty stomach, the left kidney is positioned from the second to the fourth lumbar vertebra. When the stomach is full, the left kidney is displaced the length of one vertebra caudally. The renal papilla is divided by dorso-ventral crests. The empty *bladder* is entirely in the pelvic cavity, whereas the distended bladder is mainly in the abdominal cavity. The urinary bladder is almost completely covered by the peritoneum. The *urethra* opens into the caudal end of the vagina.

The *female genital organs* consist of the ovaries, oviduct, uterus and vagina. In mice, rats and guineapigs the *ovaries* are located at the caudo-lateral poles of the kidneys. In the adult animal the ovary appears as a mass of follicles and corpora lutea embedded in fatty tissue. The *oviduct* is a long, narrow, coiled tube connecting the ovarian bursa with the bicornuate uterus. The *uterus* is a Y-shaped tubular structure formed by the two cornua and the corpus. The cranial part of the corpus uteri consists of two cavities separated by a median septum. The caudal part is undivided and forms the neck of the cervix. The *vagina* of mice and rats is short, thick and muscular. The urethra opens near the tip of the clitoris, within the shallow depression of the clitoral fossa. The vagina of the guineapig has a U-shaped orifice, and is separated from the anus by the perineal sac, which contains the perineal glands. A thin membrane covers the vaginal orifice during anoestrus.

In the rabbit the uterus is completely divided into two cavities, which open independently into the vagina.

The *ovaries* of the juvenile sow are situated at or near the lateral margin of the pelvic inlet. The surface of the ovaries has an irregular lobulated appearance due to projections of large follicles and corpora lutea. The mature follicle is 7–8 mm, and the corpus luteum is 12–15 mm in diametes. The *oviducts* are 12–15 cm long. The abdominal end forms an ampulla with a large opening, and the uterine end opens into the uterine cornua. The body of the *uterus* is only about 5 cm long, whereas the two cornua are 120–150 cm long and arranged in numerous coils. The cervix is approximately 10 cm long and directly continued by the *vagina* without forming any intravaginal projection. The vagina of the adult sow is 10–12 cm long, narrow and with a thick muscular coat. The *urethra* opens into the vagina.

The *ovaries* of the dog are oval-shaped, about 2 cm long and located 1–2 cm from the caudal end of the corresponding kidney. The oviducts are 5–8 cm long. The corpus uteri is short (2–3 cm), whereas the uterine cornua are 12–15 cm long and narrow. The cervix is short with a thick muscular coat. Ventrally it forms a cylindrical projection which lies in a depression of the vaginal wall. The *vagina* is relatively long with a thick muscular wall.

The *male genital organs* consist of the testes, excretory ducts, accessory glands, urethra and penis. In mice and rats the *testes* lie in two separate thin-walled scrotal sacs. The inguinal canal remains open throughout life, and the testes are often in the retracted position in the abdominal cavity. The excretory ducts include the rete testis, the efferent ducts, the epididymis and the ductus deferens. The *epididymis* consists of the proximal caput epididymis, the tubular corpus and the distal cauda epididymis. The cauda continues in the *vas deferens*, which joins with blood vessels and nerves to form the *spermatic cord* and continues through the inguinal canal and crosses the ureter to join the urethra at the ampulla. The male *urethra* extends from the urinary bladder to the external orifice of the penis. The *penis* lies within the loose *prepuce*. It contains three masses of erectile tissue, the corpus cavernosum urethrae and two corpora cavernosa penis. A small bone, the *os priapi*, is located on the ventral wall of the penis. Associated with the male genital organs are six paired glands. The largest are the *vesicular glands*. Each gland has a wide duct that enters the urethra together with the vas deferens near the neck of the bladder. The *coagulating glands* are attached to the *vesicular glands*. The folded mucosa is visible through the thin walls. Each gland has two ducts entering the dorsal wall of the collum vesica cranial to the entrance of the vesicular glands. The *ampullary glands* are branched tubular glands with many short ducts that open directly into the ampulla. The two pairs of *prostate glands* (dorsal and ventral) open with several ducts into the lateral and ventral walls of the urethra. The *bulbourethral glands* are located lateral to the junction of the membraneous urethra and the penis. Finally there are two *preputial glands* situated in the subcutaneous tissue near the end of the penis. Their ducts open into the lateral wall of the preputial cavity.

The genitals of the male guineapig differ somewhat from those of rats and mice. The paired *ductus deferentes* are 40–60 mm long and divided into a coiled epididymal portion and an uncoiled distal portion. The diametre of the duct remains unchanged and the ampulla is lacking. The ducts pass through the inguinal canals into the peritoneal cavity dorsal to the urinary bladder and ventral to the ureter. They enter the urethra separately ventral to the *vesicular gland* and dorsal to the bladder. The *accessory glands* are the paired *vesicular glands*, the *coagulating gland*, the ventral and dorsal lobes of the *prostate gland*, and the *bulbourethral gland*. The vesicular gland consists of two elongated vermiform tubes located dorsal to the vesica. They converge medially and enter the urethra immediately dorsocaudal to the ductus deferentes. The coagulating gland is a paired lobulated organ situated laterodorsal to the vesicular gland, craniodorsal to the dorsal lobe of the prostate and lateral to the urethra. Each gland has a single duct that opens into the urethra craniolateral to the opening of the ductus deferens and the vesicular gland. The secretion of the coagulating gland forms the vaginal plug after coitus. The prostate consists of two pairs of lobes joined by the transverse isthmus. The ducts from each lobe enter the urethra caudolateral to the entrance of the vesicular gland. The paired bulbourethral glands are small oval masses, each connected with the urethra by a single duct. The *penis* of the guineapig extends cranially for about 1 cm, then turns caudally 2–3 cm. The glans penis is cylindrical and is covered by keratinized scales. The urethral opening is slightly ventral to the tip of the penis, which contains a penis bone. Caudoventral to the urethral opening is a pouch which contains two horny styles. This pouch is everted during erection so that the styles project externally.

In the male rabbit the *ductus deferentes* enter the ventral wall of the *vesicular seminalis* dorsal to the vesica. The *vesicular gland* and the *prostate gland* are located in the dorsal wall of the seminal vesicle, which they cover. These glands are much smaller in size than in rodents and guineapig. The small paired *bulbourethral glands* are located in the dorsal wall of the urethra immediately caudal to the prostate gland. Two paired *glandulae perinealis* are located subcutaneously dorsolateral to the penis. They empty their secretion into the hairless inguinal spaces.

The *testicles* of the boar are large. They are situated in the scrotum with the long axis directed upward and backwards with the cauda epididymis the highest. The *spermatic cord* is 20–25 cm long in the adult animal. The ductus deferens opens into the urethra without forming a distinct ampulla. There are three accessory glands. The *vesicular gland* is large, and the two lobes extend into the abdominal cavity. They cover the caudal part of the bladder, the ureters, the ductus deferentes, the *prostate gland* and the *bulbourethral gland*. Several ducts unite to form a small excretory duct which terminates on the colliculus seminalis of the urethra. The prostate consists of the body, which is about 2.5 cm wide, and the pars disseminata. The bulbourethral gland is large, cylindrical in shape and located on either side of the caudal two-thirds of the pelvic urethra. The lobes are approximately 12 cm long and 3 cm wide. Each gland has a large excretory duct which enters the urethra at the ischial arch. The accessory glands are reduced in size in animals which have been castrated early in life. The *penis* of the adult boar measures about 45–50 cm in length. It forms an S-shaped curve, the sigmoid flexure, when retracted. The external part of the penis has no glans, and is shaped as a spiral. The urethral orifice is located ventrolaterally.

The male dog lacks both the vesicular gland and the bulbourethral gland. The prostate is relatively large and located near the cranial border of the pelvis, surrounding the neck of the bladder. The gland is often enlarged in older animals. The *penis* has two distinct corpora cavarnosa and a relatively long penis bone. The glans penis is cylindrical in shape. Behind the glans is an enlargement known as the *bulbus glandis*, which is made up of erectile tissue. The dog's *urethra* is relatively long and partly enclosed in the prostate.

Diseases in the urinary and reproductive systems have been studied in animal models. Kidney diseases such as renal hypertension,[49] acute and chronic renal failure,[19,51,65] and hydronephrosis[14,46] are frequently studied in mice, rabbits, dogs and pigs. Kidney transplantations are performed in rats, rabbits, dogs, and pigs.[18,24,32,41] Most of the surgical research on the male reproductive system has been performed on rodents,[21,72] rabbits,[13,71] and dogs.[40] Reproductive physiology, *in vitro* fertilization and surgical sterilization techniques have been studied in rodents and rabbits. Congenital abnormalities, especially those caused by toxic substances, are studied in animal models. Rats and rabbits are frequently used. Surgery on fetuses, for instance in the study of intrauterine growth retardation, is commonly performed in sheep, guinea-pigs and monkeys (Table 1.9).

Factors Affecting the Relevance of the Animal Model

Genetics and Animal Breeding

If all animals of a species were identical, only two animals would be required for an experiment: one for the experimental group and one for the control group. However, apart from identical twins, no two individuals are genetically identical, and if a recorded difference between the experimental group and the control group is to be statistically significant, the two groups must contain a certain number of individuals. The greater the similarity between the individuals, the smaller the acceptable size of the groups.

When breeding laboratory animals two methods are used: inbreeding and outbreeding: *Inbreeding* aims at homozygocity and thus individual similarity in genotype. Inbreeding is achieved by crossing closely related individuals, most often brother and sister. The degree of inbreeding is expressed as the inbreeding coefficient, which is an expression of the percentage reduction in heterozygotic loci in the population. It is calculated using the equation: $F_n = 1 - (1 - \Delta F)^n$ where ΔF is the increase of the inbreeding coefficient per generation (19% for brother–sister crossing) and n is the number of generations. Approximately twenty generations of continuous brother–sister breeding are necessary to produce an inbred strain. Owing to spontaneous mutations, genetic drift and selection for fertility, the individuals of an inbred strain are not completely identical. The inbreeding coefficient approaches 98%.

In practice, an inbred strain is produced from one brother–sister pair. From their offspring, new breeding pairs are selected. Some of these pairs may fail to reproduce due to inbreeding depression, whereas others may have offspring. New pairs are selected, and after a few generations most lethal genes have been eliminated. The most fertile animals are selected, and all other animals are discarded. After twenty successive generations of brother–sister breeding, the strain is, by definition, an inbred strain.

The advantage of the inbred animal is the genetic similarity between individuals. The homozygocity, however, causes a weak constitution and makes the animals very dependent upon the environment. For certain experiments it is therefore common to use the F_1 generation produced by crossing two inbred strains. These animals are heterozygous, and therefore more vigorous, but are still genetically similar.

Another advantage of inbred animals is that they can be kept in very small numbers. Two to three breeding pairs are sufficient to maintain the strain over periods when animals are not used for experiments.

Control of genetic contamination of inbred strains is an important speciality of laboratory animal science. Skin transplantation tests, skeleton morphology and biochemical parameters are used for this purpose.

Inbred strains of animals are only possible in species with a short reproduction time like mice and rats. Species with a longer gestation period are pro-

Table 1.9. Animal models in the study of diseases in the urogenital system

	Animal model							
Condition	*Mouse*	*Rat*	*Guineapig*	*Rabbit*	*Dog*	*Pig*	*Sheep Goat*	*Monkey*
Renal hypertension		+		+	+	+	+	
Acute renal failure				++				
Chronic renal failure	++							
Hydronephrosis		+			+	+		
Peritoneal dialysis				+	+			
Kidney transplantation		+		+	+	+		
Cryptorchidism	+	++	+	+		+		
Hydrosalpinx				++				
Fetal surgery			+				++	+

(++) frequently cited
(+) occasionally cited
( ) rarely cited

duced by a method known as *outbreeding*. This method secures the genetic stability in a population of animals from generation to generation rather than in individual animals. Rabbits and guinea-pigs are produced in this way, as well as several mouse and rat stocks. To avoid inbreeding, the animals are bred randomly, and the colony must consist of a large number of breeding animals.

Surrounding Temperature

A homeothermic animal can maintain a constant body temperature when subjected to external temperatures within a range specific for each species. Within this temperature range is a smaller range, known as the thermoneutral zone, where body temperature is regulated exclusively by variations in the flow of heat from the core of the body to the surface, these variations are known as conduction and convection of heat.

If the surrounding temperature falls below the thermoneutral zone, the body temperature is regulated by an increased metabolic rate and consequent heat production. Above the thermoneutral zone, body temperature is maintained by evaporative heat loss.

Laboratory animals are usually kept under conditions where the room temperature varies between 16–26°C. This temperature, however, is well below the thermoneutral zone, and consequently changes in the surrounding temperature cause changes in the metabolic rate of the animals.

An example may illustrate the importance of this phenomenon. LD_{50} for chlorpromazine in mice, (the dose in mg/kg body weight that causes death in 50% of the animals), varies considerably with the surrounding temperature. At 13°C, LD_{50} is 12 mg/kg, whereas at 28°C the drug appears much less toxic with an LD_{50} of 350 mg/kg. At a room temperature of 38°C toxicity is again significantly higher.

Light–Dark Cycle

The intensity of visible light affects the organism via the photoreceptors in the retina, the pineal body in the brain and the endocrine system. The estrous cycle is affected by the light–dark cycle. In rats, for example, the cycle lasts four days with a light-dark cycle of 12 : 12 hours. At 16 hours of light the cycle is 5 days and at 22 hours the cycle becomes irregular. Alterations in light intensity may also affect the enzyme activity in the liver, and alter the metabolism of drugs and toxic compounds.

Noise

The effect of noise on the homeostasis of laboratory animals is well documented. One example that illustrates this is the experiment where rabbits were fed a fat-rich diet, and the level of cholesterol was measured over a period of several months. One group of animals, exposed to a constant noise level of 102 decibels showed significantly higher cholesterol levels than control animals living in a silent environment.

Food

The nutritional requirements of laboratory animals are well defined, and commercially available diets are usually adequate. Long-term storage under inappropriate conditions may, however, damage essential vitamins and proteins. As metabolic processes are enzymatic in nature, vitamin and protein deficiencies may alter the rate of metabolism in laboratory animals, and thus affect experimental results.

Chemical Residues

Chemical substances, either naturally present or introduced into the environment, may affect enzyme systems and metabolism of the laboratory animals. An example is cage bedding made from spruce or pine wood, which contains an aromatic compound known as alpha-pinene, which was found to stimulate the cytochrome P-450 enzyme system in the liver, causing increased metabolism of drugs and toxic compounds. Similar alterations in metabolism may result from residues of insecticides or herbicides in animal feed.

Reference

1. Adams, M. E. and Billingham, M. E. J. (1982). Animal models in degenerative joint disease. *Curr Top Pathol.*, **71**, 265–297
2. Anderson, D. H., Stern, W. H., Fisher, S. K., Erickson, P. A. and Borgula, G. A. (1983). Retinal detachment in the cat: The pigment epithelial–photoreceptor interface. *Invest Ophthalmol. Vis. Sci.*, **24** 906–926.
3. Aner, M. R. and Cohen, B. J. (1976). Animal model of human disease. Ulcerative colitis. Animal model: Ulcerative colitis induced in guinea pigs with degraded carrageenan. *Am. J. Pathol.*, **84**, 431–434.
4. Bahn, C. F., MacCallum, D. V., Lovet, E. J., Varani, J., Meyer, R. F., Lillie, J. H. Martonyi (1983). Corneal homograft rejection in the cat *Cornea*, **2**, 9–25
5. Bahr, R. and Flash, A. (1978). Morphological and functional adaptation after massive resection of the small intestine: Experiments using minipigs of the Gottingen strain. *Prog Pediatr. Surg.*, **12**, 107–141.
6. Binder, P. S., Zavala, E. Y., Baumgartner, S., Deg, J. K. and Grossman, R. K. (1984). Refractive keratoplasty. Keratophakia in a nonhuman primate. *Arch Ophthalmol.*, **102**, 1671–1677
7. Bito, L. Z. (1984). Species differences in the response of the eye to irritation and trauma: a hypothesis of divergence in ocular defence mechanisms, and the choice of experimental animals for eye research. *Exp Eye Res.*, **39**, 807–829.

8. Bland, R., Damling, R., Selinger, S. and Staub, N. (1977). Effects of alveolar hypoxia on lung fluid and protein transport in unanesthetized sheep. *Circ. Res.*, **40**, 269–274.
9. Chen, P. Q., Turner, T. M., Ronnigen, H., Galantem, J., Urban, R. and Rostoker, W. (1983). A canina cementless total hip prosthesis model. *Clin. Orthop.*, **176**, 24–33.
10. Cherry, P. M. H. and Britton, R. (1984). Experimental radical keratectomy in rabbits for the correction of aphakia. *Cornea*, **3**, 61–64.
11. Clatworthy, H. W., Saleeby, R. and Lovingood, C. (1952). Extensive small bowel resection in young dongs: Its effect on growth and development. *Surgery*, **32**, 341–351.
12. Crowling, R. H. and Booth, C. C. (1967). Structural and functional changes following small intestinal resection in the rat. *Clin. Sci.*, **32**, 139–149.
13. Cummins, J. M. and Glover, T. D. (1970). Artificial cryptorchidism and fertility in the rabbit. *J. Reprod. Fertil.*, **23**, 423–433.
14. Durhuus, J. C., Nerstørm, B., Gyrd-Hansen, N. and Rask-Andersen, H. (1976). Experimental hydronephrosis An electrophysiologic investigation before and after release of obstruction. *Acta. Chir. Scand.*, **472**, 17–28.
15. Eidelberg, E., Staten, E., Watkins, J. C., McGraw, D. and McFadden, C. (1976). A model of spinal cord injury. *Surg. neurol.*, **6**, 35–38.
16. Enjeti, S., Terry, P. R., Menkes, H. A. and Traystman, R. J. (1982). Mechanical factors and the regulation of perfusion through atelectatic lungs in pigs. *J. Appl. Physiol.*, **52**, 647–654.
17. Fath, R. B. Deschner, E. E. Winawer, S. J. and Dworkin, B. M. (1984). Degraded carrageenan-induced colotis in CFl mice. A clinical, histopathological and kinetic analysis. *Digestion*, **29**, 197–203.
18. Fisher, B. and Lee, S. (1965). Microvascular surgical techniques in research, with special referene to renal transplantation in the rat. *Surgery*, **58**, 904.
19. Gagnon, R. F. and Dugid, W. P. A. (1983). A reproducible model for chronic renal failure in the mouse. *Urol Res.*, **11**, 11–14.
20. Gelberman, R. H., Van de Berg, J. S., Lundborg, G. N. and Akeson, W. H. (1983). Flexor tendon healing and restoration of the gliding surface. *J. Bone. Joint. Surg.*, **65A**, 70–79.
21. Goldberg, E. and Hawtrey, C. (1968). The effect of experimental cryptorchidism on the isoenzymes of lactic dehydrogenase in mouse testes. *J. Expl. Zool*, **167**, 411–418.
22. Grace, D. M., Feindel, C. M. S., Harris, K. A., Inculet, R. I. and Wallace, A. C., Stiller, C. R. (1920). Immune response to transplanted pancreas in pigs. *Transplant Proc.*, **14**, 649–655.
23. Guglielmotti, M. B. and Cabrini, R. L. (1985). Alveolar would healing and ridge remodeling after tooth extraction in the rat A histologic, radiographic, and histometric study. *J. Oral Maxillofac Surg.*, **43**, 359–364.
24. Guttman, R. D., Lindquist, R. R., Parker, R. M., Carpenter, C. B. and Merrill, J. P. (1967). Renal transplantation in the inbred rat. 1. Morphologic, immunologic and functional alterations during acute rejection. *Transplantation*, **5**, 668–681.
25. Haller, J. A. and Bachman, B. A. (1964). The comparative effect of current therapy on experimental caustic burns of the esophagus. *Pediatrics*, **34**, 236–245.
26. Hau, J., Andersen, L. L. I., Rye Nielsen, B. and Poulsen, O. M. (1989). Laboratory animal models. Scand *J. Lab. Anim. Sci.*, **16**, [1], 7–9.
27. Hawe, A., Tsakiris, A. G., McGroon, D. C. and Rastelli, G. C. (1970). Experimental production of chronic graded mitral stenosis. *J. Thoroc. Cardiovasc. Surg.*, **60**, 559–564.
28. Higgins, C. B., Pavelec, R. and Vatner, S. F. (1973). Modified techniques for produciton of experimental right-sided congestive heart failure. *Cardiovasc Res.*, **7**, 870–874.
29. Hirst, L. W., Kenyon, K. E., Fogle, J. R., Hanninen, L. and Stark, W. J. (1981). Comparative studies of corneal surface injury in the monkey and rabbit. *Arch. Opthalmol.*, **99**, 1066–1073.
30. Kamada, N. and Calne, R. Y. (1983). A surgical experience with five hundred thirty liver transplants in the rat. *Surgery*, **93**, 64–69.
31. Kilam, S.K. Jackson, F. L. and Williams, H. T. G. (1975). Bacterial peritonitis and the bursting strength of intestinal anastomosis. *Can. J. Surg.*, **18**, 372–375.
32. Kirman, R. L., Colvin, R. B., Flye, M.W., Leigth, G. S., Rosenberg, S. A., Williams, G. M. and Sachs, D. H. (1979). Transplantation in miniature swine. VI. Factors influencing survival of the renal allografts. *Transplantation*, **28**, 18–23.
33. Knowx, W. G., Scott, J. R., Zintel, H. A., Guthrie, R. and McCabe, R. E. (1967). Bouginage and steroids used singly or in combination in experimental corrosive esophagitis. *Ann. Surg.*, **166**, 930–941.
34. Kranendouk, S. E., Westbrook, D. L., Obertop, H., Tencate, C. J. W. and van Huoten, H. (1979). Experimental reflux esophagitis in the rat. *Eur. Surg. Res.*, suppl 2:11.
35. Lang Rf, Riekhof FT, Steinmuller, D. (1975). Interlamallar corneal grafts in rats. Effects of histocompatibility. *Arch. Ophthalmol.*, **93**, 349–353.
36. Langenskiold, A. and Michelsson, J. E. (1961). Experimental progressive scoliosis in the rabbit. *J. Bone. Joint. Surg.*, **43B** 116–120.
37. Laufman, H., Furr, W. E., Ross, A., Craig, R. L. and Bernhard, V. (1952). occlusion of the portal vein in experimental ascits. *Arch. Surg.*, **65**, 886–893.
38. Larmouth, J. (1954). The surgeon's debt to animal experiments. *Conquest*, **134**, 3–10.
39. Lillemoe, K. D., Johnson, L. F. and Harmon, J. W. (1983). Alkaline esophagitis: A comparison of the ability of components of gastroduodenal contents to injure the rabbit esophagus. *Gastroenterology*, **85**, 621–628.
40. Linardh, G., Duchek, M., Nelson, C.-E. and Winblad, B. (1974). Vasovasostomy nuture technique. An experimental study inthe dog. *urol. Res.*, **2**, 43–47.
41. Lowry, R.P., Blackburn, J. and Marghesco, D. (1985). Lymphocyte subsets and lymphotoxin in rat renal allograft rejection. *Transplant Proc.*, **17**, 1919–1921.
42. MacPherson, B. and Pfeiffer, C. J. (1976). Experimental Colitis. *Digestion*, **14**, 424–452.

43. Marcus, E., Kaz, L. N., Pick, R. and Stamler, J. (1958). The production of myocardial infarction, chronic coronary insufficiency and chronic coronary heart disease in the dog. *Acta Cardio*, **13**, 190–198.
44. Marmor, M. F., Martin, L. J. and Tharpe, S. (1980). Osmotically induced retinal detachment in rabbits and primates. *Invest Ophthalmol. Vis. Sci.*, **19**, 491–493.
45. Mirate, D. J., Morgan, T. R., Green, K. and Hull, D. S. (1981). A simple technique for extracapsular lens extraction in the rabbit. *Curr. Eye Res.*, **1**, 491–493.
46. Naber, K. G. and Madsen, P. O. (1974). Renal function in chronic hydronephrosis with and without infection and the role of the lymphation. an experimental study in the dog. *urol. Res.*, **2** 1–9.
47. Nasjleti, C. E., Caffesse, R. G., Castelli, W. A. and Hoke, J. A. (1975). Healing after tooth re implatation in monkeys. *Oral Surg.*, **39**, 361–375.
48. Nolan, M.S., Lindsey, N. J., Savas, C. P., Herold, A., Beck, S., Slater, D. D. and Fox, M. (1983). Pancreatic transplantation in the rat. Long-term study following different methods of management of exocrine drainage. *Transplantation*, **36**, 26–33.
49. Olson, J. L. O. and Swindle, M. M. (1988). Renal hypertension. In: *Experimental Surgery and Physiology: Induced Animal Models of Human Disease*. Swindle M.M. & Adams R.J. (eds). Baltimore Williams and Wilkins.
50. Orlamdo, R. C., Powell, D. W. and Carney, C. N. (1981). Pathophysiology of acute acid injury in rabbit esophageal epithelium. *J. Clin. Invest.*, **68**, 286–293.
51. Ormrod, D. and Miller, T. (1980). Experimental uremia. Description of a model producing varying degrees of stable uremia. Nephron **26**, 249–254.
52. Paton, W. (1984). *Man and Mouse: Animals in Medical Research*. Oxford: Oxford University Press.
53. Pavlin, J. and Cheney, J. W. (1979). Unilateral Pulmonary edema in rabbits after re-expansion of collapsed lung. *J. Appl. Physiol.* **46**, 31–35.
54. Prihoda, M. Flatt, A. and Summers, R. W. (1984). Mechanisms of motility changes during acute intestinal obstruction in the dog. *Am. J. Physiol.*, **247**, G37–G42.
55. Quigley, H. A. and Addicks, E. M. (1980). Chronic experimental glaucoma in primates. I. Production of elevated intraocular pressure by anterior chamber injection of autologous ghost red blood cells. *Invest. Ophthalmol. Vis. Sci.*, **19**, 126–136.
56. Quingley, H. A. and Addicks, E. M. (1980). Chronic experimental glaucoma in primates. II. Effect of extended intraocular pressure elevation on optic nerve head and axonal transport. *Invest. Ophthalmol. Vis. Sci.*, **19**, 137–152.
57. Radin, E.L., Martin, R. B., Burr, D. B., Caterson, B., Boyd, R. D. and Godwin, C. (1984). Effect of mechanical loading on the tissues of the rabbit knee. *J. Orthop. Res.*, **2**, 221–234.
58. Radius, R. L. and Bade, B. (1981). Pressure-induced optic nerve axonal transport interruption in cat eyes. *Arch. Ophthalmol.*, **99**, 2163–2165.
59. Raju, S., Cibulaski, A. and Hendrix, M. B., (1980). Experimental induction of intrinsic valve stenosis. *Am. Surg.*, **46** 485–493.
60. Raviola, E. and Wiesel, T. N. (1985). An animal model of mypia. *N. Engl. J. Med.*, **312**, 1609–1615.
61. Riviere, G. R. (1981). Tooth retransplantation among matched and mismatched rhesus monkeys. *J. Oral. Surg.*, **39**, 331–335.
62. Roshe, J. and Morrow, A. G. (1955). The experimental production of g raded aortic insufficiency. *Surg. Gynecol. Obstet*, **101**, 305–310.
63. Sink, J. D., Attarian, D. E., Chitwood, W. R., Hill, R.C., Pellom, G. L. and Wechsler, A. S. (1980). An improved technique for producing ventricular hypertrophy with a subcoronary valvular aortic stenosis model *Ann Thorac Surg.*, **30**, 285–290.
64. Smith, W. S., Coleman, C. R., Olix, M. L. and Slager, R.F. (1963). Etiology of congenital dislocation of the hip. An experimental approach to the problem using young dogs. *J. Bone. Joint. Surg.*, **45A**, 491–500.
65. Solez, K. Acute renal failure. (1982). *Pathology of the Kidney* Heptinstall RH (ed). Boston: *Little, Brown and Co.*, **2**, 1069–1148.
66. Streilein, J.W., McCulley, J. and Niederkorn, J. Y. (1982). Heterotropic corneal grafting in mice: a new appraoch to the study of corneal alloimmunity. *Invest Ophthalmol. Vis. Sci.*, **23** 489–500.
67. Su-Fan, A., Brum, J. M., Kaye, M. P. and Bove, A. A. (1982). A new technique for producing pure aortic stenosis in animals. *Am. J. Physiol.*, **246**, H296–H301.
68. Swindle, M. M. (1988). Comparative anatomy of the gastrointestinal tract. In: *Experimental Surgery and Physiology: Induced Animal Models of Human Disease* Swindle, M. M. and Adams, R. J. (eds). Baltimore: Williams and Wilkins.
69. Sylvester, J. T., Mitzner, W., Ngeow, Y. and Permutt, S. (1983). Hypoxic constriction of alveolar and extra-alveolar vessels in isolated pig lungs. *J. Appl. Physiol.*, **54**, 1660–1666.
70. Taylor, D. E. M. and Whamond, J. S. (1975). A method of producing graded stenosis of the aortic and mitral valve in sheep for fluid dynamic studies. *J. Physiol.*, **244**, 16–17.
71. Vasques, G., Oberti, C., Boeckx, W., Winston, R. M. L. and Borsens, I. A. (1981). The evaluation of experimentally produced hydrosalpinges in rabbits. *Fert Steril.*, **35**, 342–348.
72. Walsh, P. C. and Swerdloff, R. S. (1973). Experimental cryptrochidism. Effect of serum LH and FSH in the rat. *Urol Res.*, **1**, 22–26.
73. Weser, E. and Hernandez, M. H. (1971). Studies of small bowel adaptation after intetinal resection in the rat. Gastroenterology **60**, 69–75.
74. Woodel, E. S., Hunter, G. C., Eisele, P., Talken, L. and Ward, R. E. (1985). Orthotopic porcine liver transplantation: operative technique. *J. Surg. Res.*, **39** 483–488 .
75. Yee, N. M. and Hyatt, R. E. (1983). Effect of left pneumonectomy on lung mechanics in rabbits. *J Appl Physiol.*, **54**, 1612–1617.
76. Yinon, U. (1984). Myopia induction in animals following alteration of the visual input during development: a review. *Curr. Eye. Res.*, **3**, 677–690.

PART I

2. Anesthesia, Pain Control and Analgesia in Experimental Animals

Günther Bernatzky[1], Rudolf Likar[2] and Hans-Werner Waclawiczek[3]

[1]*University Salzburg, Naturwissenschaftliche Fakultät, Central Animal Facility (Head: Univ.-Prof. Dr. Alois Lametschwandtner) Hellbrunnerstr. 34, A-5020 Salzburg, Austria*
[2]*Hospital Klagenfurt, Dep. for Anästhesiology and Intensivmedicine, A-9026 Klagenfurt, Austria*
[3]*Hospital Salzburg, Dep. for Surgery, Müllnerhauptstr. 48, A-5020 Salzburg, Austria*

Introduction

Pain research in animals over the past two decades has led to some discussions among animal rights movement groups.[64] It has also led to new, improved methods for controlling and alleviating acute and chronic pain.[17,44,88] At the same time, animal models of chronic pain[83] will be necessary in order to understand the mechanism of chronic pain which is one of the most costly health problems in some countries in the world.[26,27,63,88,98]

Pain literature is filled with hundreds of different techniques and modifications of techniques for measuring responses to noxious stimulation in animal subjects. These techniques range from the measurement of simple somatic and autonomic reflexes to the analysis of complex learned behavior.[62] Painful stimuli do not only elicit sensations of pain, they also elicit emotional or affective reactions, such as anxiety or arousal, defensive and autonomic reflexes, and behavior directed at terminating or avoiding stimulation.[7,90]

There are many similarities between animals and humans in anatomical and chemical pathways of pain perception. There is strong evidence that the neuropeptide Substance P (SP) is involved in pain.[36,40,84] Receptors for opiates are distributed over the central nervous system in regions belonging to pain perception.[13,48,57]

In 1985, the Council for International Organizations of Medical Sciences (CIOMS) issued "International Guiding Principles for Biomedical Research Involving Animals."[41] In Article VI and Article VII research studies in which pain or distress may play a role are given: Investigators should assume that procedures that would cause pain in humans cause pain in other vertebrate species, although more needs to be known about the perception of pain in animals (Article VI). Procedures with animals that may cause more than momentary or minimal pain or distress should be performed with appropriate sedation, analgesia, or anesthesia in accordance with accepted veterinary practice. Surgical or other painful procedures should not be performed on unanesthetized animals paralysed by chemical agents (Article VII).

Pain, Stress and Distress

Pain is closely related with fear, suffering and distress; this relationship can hardly be separated. Pain, suffering and distress are subjective phenomena. With the present state of knowledge, it is often possible to recognize these states but less easy to define them. Pain may be defined: "Pain is a basically unpleasant sensation referred to the body which represents the suffering induced by the psychic perception of real, threatened, or phantasied injury."[19] Pain may be defined as discomfort resulting from injury or disease. Distress results from pain, anxiety or fear. Pain may also be psychosomatic, the product of emotional distress.[68,69]

Fear and pain activate quite different kinds of behavior[12] and both are associated with fundamentally different motivational systems. Fear activates species-specific defensive behaviors, such as freezing, flight or fighting, that minimize encounters with pain-producing stimuli. In contrast pain appears to be conducive to the types of behaviors that provide for healing and recuperation (e.g. rest, grooming, licking the affected area and sleep). It is possible that fear can inhibit pain under some circumstances: For example, a deer that has been wounded by hunters may flee, showing defensive behavior that is different from that of uninjured behavior animals. But once the deer is out of danger, pain-related recuperative behaviors predominate.[12] Suffering and distress describe unpleasant emotions which people would normally prefer to avoid. Stress is a physiological response and is mainly assessed by a biochemical reaction of the body. Such feelings may be a consequence of severe pain or lasting harm.[60]

Pain and stress may also introduce unwanted variables into an experiment and complicate the results obtained. Analgesia and anesthesia may also jeopardise scientific results.[26,34,85,93] Pain can be induced by mechanical, electrical, thermal or chemical stimuli.[51]

It is difficult to understand exactly what pain is and how it can be assessed.[49] It is generally recognized that assessment of pain is subjective; man knows what his pain is and infers that the same sensations occur in animals. On the other hand, terms used for pain in animals, such as prolonged, severe, enduring, inferring, discomfort, suffering or distress are also subjective. So, "Let the animal tell by certain signs what it feels and not what man thinks it should feel."

There are two general kinds of animal experimentation in which pain may occur:[4]

Firstly, studies that investigate the nature of pain itself and the anatomical, behavioral, chemical, pharmacological and physiological mechanisms that are responsible for it. Secondly, a larger group of those studies in which pain is but a byproduct of the procedures used.[47]

Examples of laboratory pain and distress may be divided into four categories:[55] see Table 2.1.

Painassessment

The animal may not be able to say that it is in pain, but there are many clinical signs (see, for example, Tables 2.4 and 2.5) to guide the observer that are unlikely to be misleading. The key signs of pain for the six major species of laboratory animal are listed in Table 2.2.[50]

The perception of pain is largely subjective: it is best described as an awareness of discomfort resulting from injury, disease, or emotional distress and evidenced by biological or behavioral changes.[4] There are also some guidelines to the recognition of pain, distress and discomfort in experimental animals during experiments and in the animal facility:[34,61]

The degree of pain or distress in animals during experiments may be classified in several ways. Firstly, pain may be classified according to the methods that are applied; the experimental method can be rated on a scale. This approach is, however, quite inexact and the results are very variable. For example, people who had long experience of a particular animal experiment, which others described as "very cruel", rated such experiments as less severe.[34] There are some experiments that the public sees as very gruesome but that others consider to be less so. Secondly is classification by means of clinical or other signs shown by animals. This is a better method. Some authors estimate many characteristics by quantification or by semiquantitative scores such as body weight conditions and the food and water intake of the animals. For instance, each characteristic is quantified, they are summarized, and the final number gives some basis for the assessment of pain, distress and discomfort in experimental ani-

Table 2.1. Examples of laboratory pain and distress

1. *Experiments which should not be performed on unanesthetized animals*
Decelerations, microwaves, irridations, heatstroke, rapid decompression, peritonitis (infection), toxicity, electric shock and other pain experiments, prey killings
2. *Experiments which should require analgesics*
Burns, poisonings, irradiations, woundings and tumors
3. *Impermissible pain and distress*
Sleep deprivation, electric shock avoidance and punishment, electric shock preference, electroconvulsive shock, inescapable electric shock for "learned helplessness", non electric shock punishment, drug addiction to observe behavior, induced psychosis, maternal deprivation, heatstroke on behavior, microwaves on behavior, irradiation on behavior, chemical warfare weapons on behavior, cigarettes, toxicity testing by the LD 50, blindings, forced fighting
4. *Livestock and horses*
Acceleration, addictive drugs on behavior, burns, woundings, electric shock avoidance and punishment, toxicity testing of known poisonous substances

mals. From polyarthritis experiments Gärtner[34] proposed that the decline of body weight in adult rats should be one of the guide characteristics: if the decline is less than 5%, the animal suffers mildly, if the decline is less than 6–25% the animal suffers moderately and above 25% the animal suffers severely.

A practicable method of classification according to clinical or other signs shown by the animals was proposed by Morton and Griffith[61] who estimated many characteristics by quantification or by semiquantitative scores: The following tables[61] provide details of signs based on the experience of animal technicians, animal nurses, research workers and veterinary surgeons who have worked with animals for many years (see Tables 2.3 and 2.7).

Table 2.2. Key signs of pain for the six major species of laboratory animal[50]

Rat:	Vocalization, struggling, licking/guarding, weight loss, piloerection, hunched position, hypothermia
Mouse:	Reflex withdrawal, biting response, piloerection, hunched back, sunken eyes and abdomen, dehydration, weight loss
Guinea-pig:	Struggling, withdrawal, vocalization, animals do not resist restraint or capture, staring, coat, unresponsive
Hamster:	Weight loss, increased aggression or depression, extended sleep periods
Gerbil:	Hunched appearance, weight loss, shock syndrome
Rabbit:	Unresponsive, failure to eat or drink, faces back of cage, vocalizes when handled

Table 2.3. Species-specific signs of behavior indicating pain, distress or discomfort in experimental animals (from 61)

Species	*Posture*	*Vocalizing*	*Temperament*	*Locomotion*	*Other*
Rat*	Persistent dormouse posture	Squeals on handling or pressure on affected area	May become more docile or aggressive		Abdominal writhing in mice; eats bedding and neonates
Rabbit	Looks anxious, faces back of cage (hiding posture)	Piercing squeal	Kicks and scratches or "dozey"		No spillage of food or water, eats neonates
Guinea-pig		Urgent repetitive squealing	Rarely vicious: usually quiet; terrified, agitated	Drags back legs	No spillage of food or water
Dog	Anxious glances; seeks cold surfaces; tail between legs; "hangdog" look	Howls; distinctive bark	Aggression or cringing and extreme submissiveness; runs away		Penile protrusion; frequent urination
Cat	"tucked in" limbs; hunched head and neck	Distinctive cry or hissing and spitting	Ears flattened; fear of being handled; may cringe		
Monkey	Head forward; arms across body	Screams	Facial grimace		

*Many signs in rats may also be seen in mice

Table 2.4. Common clinical signs indicating pain, distress or discomfort in experimental animals (from 61)

System	*Signs*
Cardiovascular	Heart rate altered; pulse quality affected; peripheral circulation decreased; blue and cold extremities (ears, paws)
Respiratory	Abnormal breathing pattern: rate and depth altered, labored, panting; nasal discharge
Digestive	Bodyweight loss or poor growth; feces altered in volume, color or consistency (e.g. black with blood; pale, lack of bile pigments; undigested food; diarrhoea/constipation); vomiting; jaundice; salivation
Nervous and musculoskeletal (locomotory)	Twitching, fitting, tremors, convulsions, paralysis, pupils dilated; shivering, hyperesthesia, reflexes sluggish or absent; unsteady gait, lameness, muscle flaccidity, rigidity or weakness; protecting affected area (e.g. bordering abdomen) or reluctant to move a limb (e.g. arthritis)
Miscellaneous	Any abnormal swelling, protrusion (hernia, rupture) or abnormal discharges from natural orifices; raised body temperature; dehydration; sunken eyes, skin tents; urine specific gravity increase, decrease in volume

Table 2.5. Species specific clinical signs indicating pain, distress or discomfort in experimental animals (from 61)

Species	*Cardiovascular*	*Respiratory*	*Other*
Rat	Dark claws and feet, eyes bulge and pale	Shallow rapid breathing; grunting on expiration	Red staining around eyes and nose; cyanosis, congestion and jaundice in mucous membranes or non-pigmented and non-hairy areas. Square tail (dehydration).
Rabbit		As rat	White discharge from eyes, nose and on inside of fore paws; cyanosis, congestion, and jaundice in mucous membranes, or non-pigmented and non-hairy areas
Guinea-pig		As rat	Cyanosis, congestion and jaundice in mucous membranes or non-pigmented and non-hairy areas
Dog		As rat; salivation and panting	As guinea-pig; raised body temperature; increase in specific gravity of urine and decrease in volume; sweaty paws, pupils dilate, eyes glazed
Cat		As dog	As dog; circumal gland discharge, third eyelid may protrude
Monkey		As dog	As dog

Not all signs may be present at one time and no single sign can indicate the degree of pain (Tab. 2.6).

Any score of 3 is potentially serious and one should always look for valid alternative explanations. For instance, the animal may not have come round fully from an anaesthetic, or the surgery may have directly affected the parameter being measured.[61]

Table 2.6. Relationship between signs and degree of pain, distress and discomfort (from 61)

	Normal (0)	*Mild* (1)	*Moderate* (2)	*Severe* (3/4)
Appearance	Coat loses sheen, hair loss, starcy, harsh…	…	…	…
		Failure to groom, soiled perineum…	…	…
	Discharge from eyes and nose…	…	…	…
	Eyelids partly closed…	…	…	…
			Eyes sunken and glazed…	…
		Hunched up look…	…	…
		Respiration labored, abnormal panting…	…	…
				Grunting before expiration; grating teeth…
Food/water intake	Reduced…	…	…	Zero (prolonged)..
	Faecal/urine output reduced…	…	…	Zero…
Behavior	Away from cage mates, isolated…	…	…	Unaware of extraneous activities or bullying from mates…
		Self mutilation…	…	…
		Restlessness, reluctant to move, recumbent…	…	…
		Change in temperament…	…	…
		Squealing, howling, etc. especially when provoked…	…	…
Clinical signs	Strong pulse…	…	Weak pulse…	…
Cardiovascular		Cardiac rate increased or decreased…	…	…
	Abnormal peripheral circulation…	…	…	…
		Pneumonia, pleurisy…	…	…
Digestive		Altered fecal volume, colour, consistency…	…	…
		Abnormal salivation…	…	…
		Vomiting (high frequency)…	…	…
				Boarded abdomen as in peritonitis…
Nervous			Lameness and arthritis…	…
(musculoskeletal)		Twitching…	…	Convulsions…

Table 2.7. Possible interpretation of total scores from an overall assessment of an experimental animal (from 61)

Total score	*Overall assessment*
0–4	Normal
5–9	Monitor carefully; should consider the use of analgesics and sedatives
10–14	Ample evidence of suffering; some form of relief must be seriously considered; should be under regular observation; seek expert advice; consider termination
15–20	Relief should be given, unless the animal is comatose. Is it a worthwhile experimental animal because it is likely to be physiologically abnormal? There is ample evidence of severe pain. If likely to endure, terminate the experiment

Suffering cannot be measured biochemically adequately by analysing a compound in the blood or by recording electroencephalogram patterns. Studies on blood levels of free fatty acid, glucagon, catecholamines and corticosteroids, Bradykinin and Substance P, have been carried out however, to try to associate biochemical variation with pain. At present, there are not enough biochemical markers for pain; we are left solely with clinical assessment (See Table 2.4).

For any experiment in which an animal is likely to suffer pain, federal law in most countries now requires that a veterinarian (or similar person) should be consulted in the planning of the experiment and that anesthetics, tranquilizers and analgesics should be used, except for when they would interfere with the purpose of the experiment. These are the principal tools for the reduction of experimental pain and distress.

In what ways, however, can a research protocol be modified to minimize animal pain, stress and distress?[67] We will consider this question next.

Analgesia and Anesthesia: Reducing Pain and Distress

Pain is relieved with analgesics or anesthetics; distress is eased with tranquilizers.

There are different ways of reducing pain and distress:

- by providing relief with drugs
- by changing the procedures so that less pain or distress is produced
- by using a less sensitive species.[4]

One should aim to protect animals from unnecessary pain by methods appropriate to the aim of the experiment. An analgesic is an agent that relieves pain without causing loss of consciousness. On the other hand, an anesthetic is an agent that causes loss of the sensation of pain, usually without loss of consciousness. A tranquilizer is an agent that quiets, calms, and reduces anxiety and tension with some alteration of the level of consciousness and without effecting analgesia. Tranquilizers are particularly useful in reducing distress and resistance to confinement.[4]

Animal's pain and distress can be minimized by pharmacological treatment. Several sites, within the central nervous system as well as the periphery, may be sensitive to analgesic drugs. Pharmacological treatment can be categorized as analgesics, including narcotic and non-narcotic drugs; sedatives, including muscle relaxants and tranquilizers; and anti-inflammatory drugs. Little information is available on recommended doses.[95] Most treatment is largely anti-inflammatory in nature:[88] The steroidal and non-steroidal anti-inflammatory agents have their primary pain-relieving actions peripherally. Other pharmacological studies have shown the existence of functionally distinct subpopulations of opioid receptors (mu, kappa, and sigma) that are all blocked by the antagonist naloxone, but that differ in regard to agonist specificity.[43,45,54,56,70,78] Good results have been obtained by using intrathecal[8,9] or by peridural application of drugs.[35]

It is well known that the brain produces its own painkillers. With this knowledge, scientists have tried to develop non-addictive painkillers.[80] For example electrodes can be implanted in specific parts of the brain of persons suffering from chronic pain.[6]

Analgesics may not affect the research but those that are anti-inflammatory or have central effects may be unacceptable and an alternative method of controlling pain will have to be found. The management of pre- and post-operative pain and distress in experimental animals is important.[26] There are some demands which should be observed:

- choice of the anesthetic agent in relation to the animal species and the nature of experiment
- species-dependent variation in response to anesthetics
- interaction of anesthetics with experimental results.

General anesthesia refers to a reversible generalized loss of sensory perception. This is accompanied by a sleep-like state which is induced by drugs or by physical means. **Local anesthesia** refers to reversible loss of sensation in a discrete region of the body. **Analgesia** denotes block of pain perception with or without retention of other sensory modalities.

Drugs which produce **general anesthesia** are thought to do so by affecting synaptic transmission in the central nervous system.[42] This effect of general anesthesia is based on analgesia, muscle relaxation and inhibition of reflexes. For problematic patients, such as dogs, weakened by disease or experimental procedures, general anesthesia in connection with surgical trauma poses a considerable risk. A general anesthetic is preferable to a local one for surgical manipulations as it suppresses both pain and fear.[46]

Local anesthetics, on the other hand, produce their effect by blocking action potentials in axons, or their genesis at nerve endings. The use of local anesthetics requires postsurgical care, because anesthetized surfaces are particularly liable to accidental and self-inflicted damage.[71]

Both forms of anesthesia have widespread effects in the body. For example, they affect the respiratory response to CO_2, thermoregulation and blood-pressure control. In turn, these changes will affect anesthetic action.[26,27] Hypothermia, for instance, potentiates the action of general anesthetics. There is much interest in acupuncture as a means of producing both general and local anesthesia.[96]

In 1983 a committee[22] issued "Ethical Guidelines for Investigation of Experimental Pain" in Conscious Animals which proclaimed that studies of pain in animals paralyzed with a neuromuscular blocking agent should not be performed without a general anesthetic or an appropriate surgical procedure that eliminates sensory awareness.[97]

Muscle relaxants or paralytic drugs (e.g. succinylcholine or other curariform drugs) are not anesthetics. They must not be used alone for surgical restraint, although they can be used in conjunction with drugs known to produce adequate analgesia.[65]

Svendsen and Carter[82] have tested five anesthetic regimes on blood pressure, heart rate, respiratory frequency, blood gas tensions and acid-base balance in laboratory rats during two hours of analgesia. The most stable respiratory and cardiovascular status was seen during anesthesia with fentanyl-fluanisone: 0.08mg/kg fentanyl i.m.; +4.0mg/kg fluanisone i.m., +0.064mg/kg atropin sulf. s.c, +3.4mg/kg diazepan. Pentobarbitone (89mg/kg i.p) gave the next best results. Anesthesia with ketamine conveyed no further advantage and etorphine-acepromazine is contra-indicated from a physiological point of view.[82]

Studies of Skolleborg *et al.*[79] indicate that midazolam-fentanyl-fluanisone (3.0ml/kg: 1.25mg/ ml midazolam, 0.05mg/ml fentanyl and 2.5mg/ml fluanison) anesthesia is preferable to pentobarbital because tissue perfusion was generally better preserved. With sodium pentobarbitone anesthetic highly significant strain differences were found in sleeping time in mice.[87] Sleeping time on different test days also varied, indicating that environmental factors were affecting the results. Lovell[52,53] has observed that age, dose level and fasting before the test, bedding material and temperature significantly altered this pentobarbitone sleeping time. Erhardt *et al.*[20,21] have observed that ketamine-xylazine (and ketamine prolongation) seems to be the most reliable method for long term anesthesia (18–24 h immobilization) in the male and female Wistar rat. They also proclaim that halothane/N_2O in 30% oxygen is the most reliable and the safest form of anesthesia, in rabbits although severe hypotension may be present, requiring supportive therapy.

Fentanyl/fluanisone tranquilization may interfere with the development of the Shwartzman reaction in Chinchilla rabbits.[62] The Shwartzman reaction in rabbits is one of the more common tests used to establish the endotoxic activity of bacterial cell-wall lipopolysaccharides.[59]

For animal experimental studies for example on pigs, an anesthetic method should be used which guarantees adequate analgesia and muscle relaxation together with a high degree of stability of hemodynamic, respiratory and metabolic functions for the duration of the intended surgical interventions.[74] After premedication with an i.m. injection of 2 mg/kg azaperone and 0.08 mg/kg atropine sulphate, anesthesia can be induced with 8mg/kg metomidate i.v. After tracheal intubation, pigs should be supplied with a nitrous oxide/oxygen mixture in the ratio 3:2. Relaxation can be carried out with 0.2mg/kg hexacarbacholine, and 0.03mg/kg buprenorphine is then to be administered.[74] Buprenorphine is suitable as an analgesic component in combinations used for anesthetising a pig. The hemodynamic system remained largely uninfluenced by operation-mediated sympathoadrenergic stimulation and homeostasis in terms of metabolic parameters remained steady.

West and Green (1987) have undertaken a trial of the sedative effects of a novel Benzodiazepine Climazolam in Sprague-Dawley and Lewis strain rats of both sexes at the recommended dosage of 0.25 mg/kg i.v. Climazolam sedated the rats but failed to inhibit the bighting reflex or responses to painful stimuli.

The Management of Post-Operative Pain is also Important

Post-surgical care should include observation of the animal to ensure uneventful recovery from anesthesia and surgery.[65] Equipment and supply items that can be helpful in intensive care include heating pods, vaporizers, vacuum equipment, respirator, cardiac monitor, and oxygen.

Pain-relieving drugs should be administered to animals after surgery whenever this would not interfere with the behavior under study; animals should be carefully monitored so that any complications that develop may be treated.[81,91] Transection of the spine or brain stem is recommended for surgical experiments when possible.[46] Where post-surgical study is considered necessary (e.g. in cardiology), intensive post-operative control of pain is necessary for controlling the animal under general anesthesia until death.[11]

There is much scope for better post-procedural analgesia.[26,27] Although this is routinely given to dogs, primates and farm animals, it is seldom given to small rodents, which account for over 80% of animals used in the U.K. Moreover, rodent analgesics have not been adequately explored.[5,39] The analgesic selected for this purpose is buprenorphine (Temgesic) as it offers the important advantage of possessing a long duration of action (6–14h) in a wide range of species. Immediately following a surgical procedure, animals received buprenorphine at 8–14h intervals for at least 48 hours at the dose rates[25] listed in Table 2.8.

Immediately following a surgical procedure, rodents and rabbits should also receive buprenorphine at 8–14h, following anesthesia with a combination of fentanyl/fluanisone and midazolam.[26]

There are some other methods by which pain or stress from the laboratory can be reduced to a level without the loss of scientific results. For example, observations of pigs have indicated that providing

Table 2.8. Doses for buprenorphine (Temgesic) for different species; application by i.m., s.c. or i.v[25]

Dogs	0.01–0.02 mg/kg
Rabbits	0.02–0.05 mg/kg
Primates	0.01 mg/kg
Rats	0.1–0.5 mg/kg

them with toys and positive human contact reduces their excitability.[37] A calm animal can be trained to cooperate during restraint and handling. This makes handling less stressful for both the animals and their handlers. The quality of the research data derived from experimentation with tamed, calmed pigs will be improved, because the physiological alterations associated with stress and excitement will be minimized. More research is needed on the use of toys and other means of environmental enrich and, if possible, data derived.

Other methods of reducing pain in animals include:

- New types of instruments (microinstrumentation), as they can lead to the more refined or reduced use of live animals
- Use of small samples for analysis by better analytical instrumentation[1,2]
- *In vitro* experiments (cell culture, isolated cell nerves, isolated rat skin)[99]
- *In vivo* measurements using fiber optics[73]
- Other minimally invasive techniques in animal research such as immunoszintigraphy, amniocentesis and use of laser surgery.
- The production of target-specific monoclonal antibodies.

Application of Drugs

Drugs are injected either intraperitoneally,[8,9] intravenously, intramuscularly or intracranially. If the drug is to be given repeatedly, the injection procedure can be stressful and may cause discomfort, so capsules of porous rubber, and small implantable minipumps are now available to deliver drugs taken over a long period.[18]

Drugs used to control pain have been administered in essentially two ways: either systematically or directly into the nervous system. Systemic administration is the more common form.

Small needles should be substituted for large ones. Smaller doses can be used and tests can be terminated at the earliest feasible time. Sometimes smaller doses will actually result in increased sensitivity of the test.[32]

Flecknell *et al*[30] have described a procedure by which lignocaine-prilocaine local anaesthetics (EMLA-cream) may contribute to pain-free venepuncture in laboratory animals. Application of EMLA cream enabled percutaneous insertion of catheters into the cephalic vein in dogs and cats and the marginal ear vein in rabbits without causing any detectable pain or discomfort.

Administration of drugs directly into the nervous system in pain research has involved injecting drug solutions either into the cerebrospinal fluid ventricular system or directly into brain tissue.[94] The principal difference between these two forms of administration is that when drugs are injected intraventricularly, their sites of action in the nervous system may be diffuse and unspecifiable in contrast to the relatively well-localized effects of drug solutions when injected directly into brain tissue. The action in the nervous system may be diffuse and unspecific. Even more precise is iontophoretic application of drugs, whereby small electric currents in a micropipette are used to deposit extremely small amounts of a drug in the vicinity of a single neuron.[44]

Conclusion

Research workers have a responsibility to minimize the pain which may be produced by procedures that they carry out on experimental animals.[26,27] Buprenorphine (Temgesic) offers the important advantage of possessing a long duration of action in most animals. For animal experimental studies (for example on pigs), an anesthetic method should be used which guarantees adequate analgesia and muscle relaxation with a high degree of stability of hemodynamic, respiratory and metabolic functions for the duration of the intended surgical interventions.

The problem of pain should be avoided through preventive measures and good management. Clearly the avoidance of painful injury would be the ideal. Recognition of clinical signs of pain or distress requires careful clinical observation and education of animal-care personnel.

We have to be careful to enowe most experiments today should not involve pain. Most animals used in experiments should not suffer pain, and the degree of pain that is inflicted during some experiments has to be greatly reduced through the establishment of rules for the humane conduct of experiments and the development of new types of drugs.

There are great ethical responsibilities for everybody working in this field.[16,22,76] The ethical training of the young scientists is therefore important in order to instil awareness of this large ethical responsibility when dealing with experimental animals.

Reference

1. Abelson, P. H. (1984). Analytical instruments. *Science*, **226**, 249.

2. Abelson, P. H. (1985). Instrumentation. *Science*, **230**, 245..
3. AMA White Paper, (1988). Use of animals in biomedical research – the challenge and response. *American Medical Association*, 36 Seiten, USA...
4. Alternatives to animal use in research, testing and education. (1988). Office of Technology Assessment, Congress of the United States. ISBN 0-8247-7977-0. Marcel Dekker , Inc. 270 Madison Avenue, New York: Marcel Dekker Inc., pp. 442.
5. Baumberger, A. (1983). Schmerzbekämpfung beim Tier: Ein Bericht ber die praktische Anwendung von Analgetika bei Versuchstieren. *Schweiz. Arch. Tierheil.*, **125**, 17–19.
6. Bennett, T. L. (1982). *Introduction to physiological psychology*. Monterey, CA: Cole Brooks.
7. Benton, D. (1988). The role of opiate mechanisms in social relationship. The Psychopharmacology of Addiction (Ed) M. Lader. Oxford University Press.
8. Bernatzky, G., Doi T. and Jurna, I. (1983). Effects of intrathecally administered pentobarbital and naloxone on the activity evoked in ascending axons of the rat spinal cord by stimulation of afferent A and C fibers. *Naunyn-Schmiedeberg's Arch. of Pharmacol.*, **323**, 211–216.
9. Bernatzky, G. and Jurna, I. (1986). Intrathecal injection of codeine, buprenorphine, tilidine, tramadol and nefopam depresses the tail-flick response in rats. *Eur. J. Pharmacol.*, **120**, 75–80.
10. Beyeler, K. L., Adler, R. H. and Hürny, Ch. (1991). The IASP - Classification of pain and the bio-psycho-sieval model of medicine. *Eur. J. Pain* , **12(2)**, 54–57.
11. Bloom, S. (1984). Alternative method for heart attack studies reduces animal pain. *Sci. Center Newsletter*, **6(2)**, 4.
12. Bolles, R. C. and Fanselow, M. S. (1980). A perceptual defensive recuperative model of fear and pain. *Behav. Brain Sci.*, **3**, 291–323.
13. Chang, K., Cooper, B., Hazum, E. and Cuatrecasas, P. (1979). Multiple opiate receptors: different regional distribution in the brain and differential binding of opiates and opioid peptides. *Mol. Pharmacol.*, **16**, 91–104.
14. Curl, J. L. (1988). Ketamine – xylazine anaesthesia in the Djungarian hamster (*Phodopus sungorus*). *Laboratory Animals*, **22**, 309–312.
15. Editorial (1980). Ethical standards for investigations of exp. pain in animal. *Pain*, **9**, 141–143.
16. European Communities Biologists Association (1988). Competence of biologists for experiments on animals. Report of the Workshop at Amsterdam, the Netherlands, March 7–9 Chapter 3 The European Convention: Consequences for biological research (Gaertner K.) p. 14–20.
17. Dubner, R. (1987). Research on pain mechanism in animals. *Journal of the American Medical Association*, **191(10)**, 1273–1276.
18. Ellison, G, Nielsen, E. B. and Lyon M. (1981). Animal models of psychosis hallucinatory behaviours in monkeys during the late stage of continuous amphetamine intoxication. *J. Psychiatr. Res.*, **16**, 13–22.
19. Engel, G. L. (1969). Pain. Chapter 3. In C. M. McBryde (Ed.) *Signs and Symptoms: Applied Physiology and Clinical Interpretation*. 5th ed. J. B. Lippincott.
20. Erhardt, W., Rechl, H., Bauer, M. and Bülmel, G. (1985). Short anaesthesia in the rabbit with Propanidid or Azaperon – Alfentanil – Etomidat. *Z. Versuchstierkunde Proceeding*, **27**, 82–83.
21. Erhardt, W., Scherer, M., Greiner, C. and Bülmel, G. (1985). Methods of long-term anaesthesia in the rat. *Z. Versuchstierkunde Proceeding*, **27**, 84.
22. Editorial (1980). Ethical standards for investigations of experimental pain in animals. *Pain*, **9**, 141–143..
23. Fenwick, D. C. and Blackshaw, J. K. (1989). Carbon dioxide as a short-term restraint anaesthetic in rats with subclinical respiratory disease. *Laboratory Animals*, **23(4)** 220–228.
24. Field, K. J. and Lang, C. M. (1988). Hazards of urethane (ethyl carbamate): a review of the literature. *Laboratory Animals*, **22**, 255–262.
25. Flecknell, P. A. (1984). The relief of pain in laboratory animals. *Laboratory animals*, **18(2)**, 143–146.
26. Flecknell, P. A. (1985). Post-operative pain relief in experimental animals. *Z. Versuchstierkunde Proceeding*, **27**, 84–85
27. Flecknell, P. A. (1985). The management of post-operative pain and distress in experimental animals. *Animal Technology*, **36(2)**, 97–103.
28. Flecknell, P. A., Liles, J. H. and Wootton, R. (1989). Reversal of fentanyl / fluanisone neuroleptanalgesia in the rabbit using mixed agonist / antagonist opioids. *Laboratory Animals*, **23**, 147–155.
29. Flecknell, P. A., Hooper, T. L., Fetherstony, G., Locke, T. J. and McGregor, C. G. A. (1989). Long-term anaesthesia with alfentanil and midazolam for lung transplantation in the dog. *Laboratory Animals*, **23**, 278–284.
30. Flecknell, P. A., Liles, J. H. and Williamson, H. A. (1990). The use of lignocaine – prilocaine local anaesthetic cream for pain-free venepuncture in laboratory animals. *Laboratory Animals*, **24**, 142–146.
31. Flecknell, P. A. and Liles, J. H. (1991). The effects of surgical procedures, halothane anaesthesia and nalbuphine on locomotor activity and food and water consumption in rats. *Laboratory Animals*, **25**, 50–60.
32. Freeberg, F.E., Griffith, J.F. and Bruce, R.D. (1984). Correlation of animal test method with human experience for household products. *J. Toxicol. Cut. Ocular Toxicol.*, **1**, 53–64.
33. Gamse, R., (1988). Ohne Schmerz für das Tier. *Alternativen zum Tierversuch*, 208–213, Lembeck F.(Ed) New York: Thieme Verlag Stuttgart.
34. Gärtner, K. (1985). Handling and experimental procedure of laboratory animals as stress induction. *Z. Versuchstierkunde Proceeding*, **27**, 57–119.
35. Gerlach, K., Bonath, K. and Schaller, G. (1985). Peridural long-term and short-term anaesthesia – haemodynamically non depressive methods, suitable for experimental procedures in the dog. *Z. Versuchstierkunde Proceeding*, **27**, 87.
36. Giles, L. G. F. and Harvey, A. R. (1987). Immunohistochemial demonstration of nociceptors in the capsule and synovial folds of human zygapophyseal joints. *British Journal of Rheumatology*, **26**, 362–364.

37. Grandin, T. (1986). Minimizing stress in pig handling. *Laboratory Animals*, **15(3)**, 15–20..
38. Guerreiro, D. and Page, C. P. (1987). The effect of neuroleptanalgesia on some cardiorespiratory variables in the rabbit. *Laboratory Animals*, **21**, 205–209.
39. Hampson, J. E. and Silcock, S. R. (1985). Pain in laboratory animals – the case for refinement. *Z. Versuchstierkunde Proceeding*, **27**, 66.
40. Henry, J L. (1982). Relation of substance P to pain transmission: neurophysiological evidence. *In Ciba Foundation Symposium 91. Substance P in the nervous system.* Proter, R., O'Conner, M. (Ed) London: Pitman Co
41. Howard-Jones, N. (1985). A Cioms Ethical code for animal experimentation, *WHO Chronicle*, **39**, 51–56.
42. Hughes, H. C. (1981). Anaesthesia of laboratory animals. *Laboratory Animals*, **10(5)**, 40–56.
43. Iwamoto, E. (1981). Locomotor activity and antinociception after putative mu, kappa and sigma opioid receptors agonists in the rat: influence of dopaminergic agonists and antagonists. *The Journal of Pharmacology and Experimental T herapeutics*, **217**, 451–460.
44. Jurna, J. (1981). Die Bedeutung des Tierversuches für die Schmerzforschung und die Entwicklung schmerzdämpfender Arzneimittel. pp. 82–84. Tierversuche-K. H. (Ed) Sontag Edition Cantor Antendorf.
45. Kavaliers, M. and Hirst, M. (1986). Food hoarding and ingestion in the deer mouse: selective responses to mu and kappa opiate agonists. Pharmacology, *Biochemistry and Behaviour*, **25**, 543–548.
46. Kerr, F. W. L. (1982). The investigators responsibilities in research using animals. Scientific *Perspectives on Animal Welfare*. W. J. Dodds and F. B. Orlans (Eds) New York: Academic Press.
47. King, F. A. (1984). The concept of pain and its reduction in animal research. Paper presented at the 92nd Annual Meeting of the American Pschological Association, Toronto, August 25.
48. Kosterlitz, H. (1983). Opioid peptides and pain – an update. *In Advances in Pain Research and T herapy*, vol. 5. (Eds.) J. Bonica *et al*. New York: Raven Press, 199–208.
49. Laininger, M. (1990). Kann mit einfachen Verhaltenstests das Wohlbefinden von Labortieren beurteilt werden ? – Verhaltensuntersuchung am Beispiel der Ratte, Inaug. Diss., Berlin.
50. LASA working party Assessment and control severity. Section IV (1990). Recognition and assessment of the signs of pain and distress in laboratory rodents and rabbits. 115–117.
51. Lineberry, C. G. (1981). Laboratory animals in pain research, Methods of Animal Experimentation. W. I. Gay (Ed) New York: Academic Press.
52. Lovell, D. P. (1986). Variation in pentobarbitone sleeping time in mice. 1. Strain and sex differences. *Laboratory Animals*, **20**, 85–90.
53. Lovell, D. P. (1986). Variation in barbiturate sleeping time in mice. 3. Strain and environmental interactions. *Laboratory Animals*, **20**, 85–90.
54. Mansour, A., Khachaturian, H., LewisM., Akil H. and Watson, S. (1988). Anatomy of CNS opioid receptors. *Trends in Neurosciences*, **11**, 308–314.
55. Marion, J. R. *et al* (1986). Animals Legal Defense Fund, S 19.
56. Martin, W. (1967). Opioid antagonist. *Pharmacological Review*, **19**, 463–521.
57. Millan, M. J., Czonkowkski, A., Pilcher, W. T., Almeida, O. F. X., Millan, M. H., Colpaert, F. C. and Herz, A. (1987). A model of chronic pain in the rat: functional correlates of alterations in the activity of opioid systems. *The Journal of Neuroscience*, **7(1)**, 77–87.
58. Milligan, S. R. and Edwards, C. (1988). The effects of different anaesthetic agents on the estimation of uterine vascular permeability in mice. *Laboratory Animals*, **22**, 343–346.
59. Milner, K. C., Rudbach, J. A. and Ribi, E. (1971). General characteristics. In *Microbial Toxins*, Vol. IV, *Bacterial Endotoxins* (Eds). G. Weinbaum, S. Kadis and S. J. Ajl. New York: Academic Press.
60. Moberg, G. P. (Ed) (1985). Animal Stress. American Physiological Society, pp. 324.
61. Morton, D. B. and Griffiths, P. H. M. (1985). Guidelines on the recognition of pain, distress and discomfort in experimental animals and an hypothesis for assessment. *The Veterinary Record*, April 20: 431–436.
62. Naess, V., Fosse, R. T. and Hofstad, T. (1986). A possible inhibitory effect on the Shwartzmann reaction in the rabbit caused by fentanyl - fluanisone tranquillization. *Laboratory Animals*, **20**, 304–306.
63. National Institute and Drug Abuse (1981). New approaches to treatment of chronic pain. A review of multidisciplinary pain clinics and pain centers: Research Monograph Series No. 36 pp. 198.
64. Nicoll, Ch. S. and Russell, Sh. M. (1989). Animal research vs. animal rights. *News* **3**, 1668–1671.
65. NIH Guide for the Care and Use of Laboratory Animals. National Institutes of Health Publications No. 86–23 (revised, 1985), Bethesda, MD 20892.
66. Oliverio, A., Castellano, C. and Puglisi-Allegra, S. (1984). Psychobiology of opioids. *International Review of Neurobiology*, **25**, 277–355.
67. Orlans, F. B. (1987). Research protocol review for animal welfare. *Invest. Radiol.*, **22**, 253–258.
68. Panksepp, J., Herman, B. H., Vilberg, T., Bishop, P. and DeEskinazi, F. G. (1980). Endogenous opioids and social behavior. *Neuroscience & Biobehavioral Reviews*, 473–487.
69. Panksepp, J. (1986). The psychobiology of prosocial behaviors: separation distress, play and altruism. In Altruism and Aggression: *Biological and Social Origins*, C. Zahn-Waxler, E.M. Cummings and R. Iannotti (Eds.) Cambridge: Cambridge University Press: pp. 19–57
70. Panksepp, J., Lensing, P. and Bernatzky, G. (1989). Delta and kappa opiate receptor control of separation distress. 19th Annual Meeting Society for Neuroscience, Phoenix, Arizona, October 29. November 3.
71. Paton, W., (1984). *Man and Mouse: Animals in Medical Research*. New York: Oxford University Press.
72. Peeters, M. E., Gil, D., Teske, E., Eyzenbach, V., Brom, W. E., Lumeij and DeVries H. W. (1988). Four methods for general anaesthesia in the rabbit: a comparative study. *L aboratory Animals*, **22**, 355–360.
73. Peterson, J. I. and Vurek, G. G. (1984). Fiber-optic sensors for biomedical applications. *Science*, **224**, 123–127.

74. Pfenninger, E., Gürnert, A. and Siegler, W. (1985). Anaesthesia of the domestic pig with buprenorphine-oxygen-nitrous oxide. *Z. Versuchstierkunde Proceeding*, **27**, 108–109.
75. Reid, W. D., Davies, C., Pare, P. D. and Pardy, R. L. (1989). An effective combination of anaesthetics for 6-h experimentation in the golden Syrian hamster. *Laboratory Animals*, **23**, 156–162.
76. Rollin, B. E. (1985/86). Animal Pain, Advances in Animal Welfare Science. (Ed) Fox, M. W. and Mickley, L. D. The Humane Society of the United States. pp. 91–106.
77. Sales, G. D., Wilson, K. J., Spencer, K. E. V. and Milligan, S. R. (1988). Environmental ultrasound in laboratories and animal houses: a possible cause for concern in the welfare and use of laboratory animals. *Laboratory Animals*, **22**, 369–375.
78. Shippenberg, T. and Herz, A. (1987). Place preference conditioning reveals the involvement of D-1 dopamine receptors in the motivational properties of mu and kappa opioid agonists. *Brain Research*, **436**, 169–172.
79. Skolleborg, K. C., Grönbeck, J. E., Grong, K., Abyholm, F. E. and Lekven, J. (1990). Distribution of cardiac output during pentobarbital versus midazolam/fentanyl/fluanisone anaesthesia in the rat. *Laboratory Animals*, **24**, 221–227.
80. Snyder, S. (1977). Opiate receptors and internal opiates. *Scientific American*, **236**, 44–58.
81. Stevens, C. G. (1975). Human Perspectives in National Research Council/National Academy of Sciences, The Future of Animals, Cells, Models, and Systems in Research, Development, Education, and Testing. Washington, DC. National Academy of Sciences.
82. Svendsen, P. and Carter, A. (1985). Influence of five anaesthetic regimens on blood pressure, heart rate, respiratory frequency, blood gas tensions and acid-base balance in laboratory rats during two hours of analgesia. *Z. Versuchstierkunde Proceeding*, **27**, 114.
83. Sweet, W. H. (1981). Animal models of chronic pain: their possible validation from human experience with posterior rhizotomy and congenital analgesia (Part I of the Second John J. Bonica Lecture). *Pain*, **10**, 275–295.
84. Takahashi, T. and Otsuka, M. (1975). Regional distribution of substance P in the spinal cord and nerve roots of the cat and the effect of dorsal root section. *Brain Res.*, **87**, 1–11.
85. Ursin, H. and Murison, R. C. C. (1985). Psychological stress factors – impact on endocrinology and immunology. *Z. Versuchstierkunde Proceeding*, **27**, 62.
86. Uvarov, D. O. (1985). Research with animals: requirement, responsibility, welfare. *Laboratory Animals*, **19**, 51–75.
87. Vieregge, Th. and Hackbarth, H. (1985). Barbiturate sleeping time and antipyrine elimination for pharmacogenetical differentiation of 10 inbred strains of rats. *Z. Versuchstierkunde Proceeding*, **27**, 115;
88. Von Voigtlander, Ph. F. (1982). Pharmacological alteration of pain: the discovery and evaluation of analgestics in animals In *Central Analgestics* (Ed) Lednicer, D., Adria Laboratories. A Wiley-Interscience Publication. New York, Chichester, Brisbane, Toronto, Singapore: John Wiley & Sons. pp 51–79.
89. Voipio, H.-M, Pekonen, H. and Nevalainen, T. O. (1985). A method for oral administration of tablets to swine. *Z. Versuchstierkunde Proceeding*, **27**, 115–116.
90. Wall P. D. (1979). On the relation of injury to pain. *Pain*, **6**, 253–264.
91. Warren, A. G. (1985). Some considerations in the care of small experimental animals before, during and following surgical operations. *Z. Versuchstierkunde Proceeding*, **27**, 117.
92. West, C. D. and Green, C. J. (1987). The sedative effects of climazolam and climazolam with fentanyl-fluanisone in rats (*Rattus norvegicus*). Lovell, D. P. (1986). *Laboratory Animals*, **21**, 143–148.
93. Wiesenfeld-Hallin, Z., (1985). The effects of stress and the environment on nociception, pain reaction and the development of tolerance to opiates in laboratory animals. *Z. Versuchstierkunde Proceeding*, **27**, 62–63.
94. Wootton, R., Cross, G., Wood, S. and West, C. D. (1988). An analgesiometry system for use in rabbits with some preliminary data on the effects of buprenorphine and lofentanil. *Laboratory Animals*, **22**, 217–222.
95. Wright, E. M., Marcell, K. L. and Woodson, J. F. (1985). Animal pain: evaluation and control. *Laboratory Animals*, **14(4)**, 20–35.
96. Yoxall, A. T. (1981). Pain in small animals – its recognition and control. *ICAR News*, **XXV(1)**, 16–24.
97. Zimmermann, M. (1983). Ethical guidelines for investigations of experimental pain in conscious animals. *Pain*, **16**, 109–110.
98. Zimmermann, M. (1986). Ethical consideration in relation to pain in animal experimentation The Ethics of Animal Experimentation: Proceedings of the Second CFN Symposium held in Stockholm Sweden. *Acta Physiologica Scandinavia*, **128** (Suppl. 554), 221–233.
99. Zimmermann, M. (1988). Schmerzforschung ohne Qual. In *Schuppan/Hardegg, Tierschutz durch Alternativen*, Fischer Verlag, 59–65..

PART II

WOUND HEALING, TISSUE REPAIR, BIOMATERIALS

PART II

3. Physiology and Pathophysiology of Wound Healing *In Vivo* Models

Finn Gottrup

Copenhagen Wound Healing Center, Bispebjerg University Hospital, DK-2400 Copenhagen, Denmark

Introduction

The primary goal of the wound healing process after injury of the tissue is the restoration of continuity between the wound edges and/or repair of the tissue defect. This takes place in mammals, most often regardless the type of original tissue, by bridging the gap with a scar of collagenous connective tissue. Regeneration of the original tissue can, however, be found in epithelia and endothelia tissues and has also been demonstrated in kidney, liver, striated muscle as well as in the peripheral nervous tissue. The healing by scar formation can thus not immediately be taken for granted in design of an experimental model.

The multifactural nature of acute as well as chronic wounds make the interpretation of the result of clinical trials difficult. In order to allow a more detailed study of the repair process experimental models are needed. Especially in surgical research animal models have been fundamental in order to improve our knowledge and understanding of the wound healing process and many human lives have been saved using therapeutic techniques learned from these models. They can simplify a complex problem, but is has to emphasized that models are not substitutes for a hypothesis or creative thinking. Surgical research has also be based on in vitro investigations studying the cellular and biochemical facets of the biological responses in the wound healing process. These in vitro models have, however, not yet been able to imitate the complex biological processes like wound healing or inflammation. Consequently animal models still have to be used frequently in this field of surgical research.

What Model to Use?

Basically two types of model can be used to clarify a problem: theoretical and practical. In the type of research described in this chapter mainly practical animal models have been used. A practical animal model has been defined as "a living organism with an inherited, naturally acquired or induced pathological process that in one or more respects closely resembles the same phenomenon in man."[1] An animal model may vary from a one-cell protozoan to chimpanzees depending of the purpose of the study. Thinking in terms of animal models for diseases no ideal model probably is available. The disease itself may be variable and may have so many facets that one animal model may not be enough for a fully description on the disease. In order to find the most appropriate model available many influencing factors have to be taken in consideration. Nine *criteria for a good animal model* have been listed by Leader and Padgett:[2] 1. Accurately reproduce the disease or lesion; 2. Availability for multiple investigations; 3. Exportable; 4. If genetic disease the species should be polytocous; 5. Allow multiple biopsy samples; 6. Fit into animal facilities; 7. Should be easily handled; 8. Available in multiple species; 9. Should survive long enough to be usable. Also the *genetic of the animal population* has to be considered in order to ensure maximum comparability of the results.[1] Different types of populations can be considered: 1. Randomly-mating populations; 2. Outbred strains; 3. Inbred strains; 4. F1 hybrids; Coisogenic and congenic strains.

When the above mentioned considerations have been made the range of models have to be thought.

Most surgical experimental work uses mammalian preparations. The models may then be: *In vitro studies; Isolated organ studies* or *Whole animal studies*. The choice of mammalian model has to be done ranging from mouse to chimpanzees. The final choice of model is influenced by some important practical factors[3] as: 1. The space and physical facilities required for animal work; 2. Laboratory facilities; 3. Technical expertise; 4. Financial resources; 5. Time and 6. Ethical and legal factors.

Which Model in Wound Healing Research

In wound healing research the above mentioned considerations has to be thought before choosing a model. It has to be decided whether is shall be a *in vitro* (cell cultures or organ cultures) or *in vivo* (isolated organ or whole animal) model. This decision should be based on which type of parameters are relevant to describe the raised problem. The advantages of *in vitro* models in wound healing research are that it allows the direct examination of the effect of an environmental change or substance on the tissue without influence of other tissue components. The major disadvantages is the difficulty in extrapolation of results to the wounded tissue in an otherwise normal organism. The advantages of an *in vivo* models in wound healing research are that wounded tissue is similar to wounds found in the clinical practice and in case of skin experimental wounds can be made in human subjects. The disadvantages are that direct examination of single tissue components is difficult, and in case of human skin wounds only small, clean wounds can be produced and even this may be an ethical problem. In wound healing research the following **parameters** are most often investigated: I. *In vivo*: Biomechanical and biochemical studies of the healing tissue, observation of healing macroscopically, measurements of different wound healing markers, cellular response and immunological responses; II. *In vitro*: Cultures of cells (fibroblasts, keratinocytes, macrophage, epidermal or endothelial cells etc) and observation of activity in these cultures by monitoring different types of collagen metabolites, aminoacids, glycosaminoglycans etc. Or following the modulation of cultures by cloning of cells or adding types of wound fluid or tissue factors. Organ cultures of the skin can also be used. Migration of epidermis can provide information on the first phase of wound healing and the system is easily used for studying the influence of "internal" as well as "external" factors.

All types of tissue injury and degeneration are followed by a wound healing process. This book aim at presenting animal models in surgical research for the study of human disease states. For this reason this chapter will primarily focus on *in vivo* animal models based on a surgical trauma. Many of the above mentioned criteria for good animal models and population can be simplified in surgical wound healing research because of the biological process which is the background for most wound healing is similar for almost all types of tissue or species used for this type of research.

Before choosing a model two factors are important: 1. The problem itself and 2. the degree of simplification which will be acceptable. In surgical wound healing research point 2. is a minor problem because of the similarity of the healing process as earlier mentioned. The problem, however, has to be defined in order to find which types of wound model should be used, which type of animal would be most appropriate and which parameters should be used for evaluation of the problem.

In vivo wound healing models can primarily be divided into *1. Artificial wound models and 2. Wound similar to the surgical trauma*. While 1. can be used only for evaluation of granulation tissue and its component 2. can be a model of all types of human surgical wounds in all types of tissue. The problem which is raised must decide which type of model is most suitable for solving this problem. In Table 3.I types of models in surgical wound healing research are shown.

Artificial Models for Wound Healing

These models have primarily been based on implanted foreign material in the tissue investigated most often in subcutis. The originally used materials were subcutaneous implanted *sponges of polyvinyl cellulose sponges*[4,5] and stainless steel wire mesh chamber.[6,7] *The wire mesh chambers* are also called "Shilling - Hunt chambers." These chambers are still used and placed subcutaneously it rapidly fills with

Table 3.1. Types of experimental models in surgical wound healing research

In Vitro		
Cell culture	–	fibroblast macrophages epidermal endothelial
Organ culture		
In Vivo		
Artificial Models	–	sponges wire mesh chambers porous other types of tubes combined
Surgical Wound Models	–	closed open

wound fluid and connective tissue matrix. Fluid and tissue inside this cylinders can be measured by histological, histochemical and biochemical analysis. Furthermore, the fluid can be aspirated daily and this way uptain a temporary relationship of wound factors cells and respiratory gas tension. Various compound or agents can also be placed inside the chamber.

More recently new devices have been developed especially for wound healing studies in man. In 1976 Viljanto reported the use of a tube structure based on the experienced gained with viscose cellular sponge.[8] The device called *'cellstic'* was inserted as a drain and removed within days of surgery. The experience with this model was gained by animal studies but in 1991 the wound healing of more than 1700 patients has been investigated by this method. In 1982 a *miniature porous Polytetra fluorethylene teflon tube (PTFE) (Gore-Tex^R)* tubing was presented by Goodson and Hunt.[9] It was inserted percutaneously and placed in the subcutaneous tissue. The devise is a strong, porous material allowing in growth of connective tissue which can be evaluated by histology and measurements of hydroxyprolinee accumulation. Using this material a lesser inflammatory reaction compaired to the older described implants is found. In order to replace the very expensive PTFE model (Gore-Tex^R) with less expensive and increased ingrow a new material has been investigated. This material (Impra^R) with a pore size of 90–120 μm has been found improving the PTFE method in relation to higher ingrow of hydroxyproline and lesser variability in a rat model.[10, 11] Presently this new material is tested in human studies, where the intra- and inter personal variability have been investigated in detail [10] and related to the surgical trauma.[12]

In 1986 a *polyvinyl alcohol sponge incased in a perfurated tubing of silicone (PVA implant)* has been developed by Diegelmann *et al.*[13] This can be placed percutaneously in animals and humans using a 12-gauge needle and trochar. This devise is normally pulled out after 14 days of healing and has been described more gently to remove than the PTFE tube. In our laboratorium, however, we have had no problems removing the Impra tube after 15 days in rat or 10 days in man. Recently the PVA implant model has been compared to the Impra model. [14] It was concluded that the Impra model accumulates 2.5 times more hydroxyproline than the PVA model, for which reason the Impra model is preferable when quantifying the collagen deposition in humans. Both methods can, however, be used in smaller animals where the collagen accumulation is higher than in man.

Wound Models Similar to the Surgical Trauma

This type of wound healing models fulfill almost all the mentioned criteria for being a good animal model: It can accurately reproduce the lesion; It is available for multiple investigations and for multiple investigators and the most often the animals used also fulfill the criteria.

The lesion can in most cases be performed the same way as in human clinical work and the lesion can be made the same way each time by different investigators. The species or types of animals is not critical because the basic tissue developed in all soft wounds is similar in all types of mammals. The animal chosen has to be a much related to the human surgery as possible but in wound healing research the model in many respects be chosen also according to the facility of the laboratory. The most important factors for choosing these types of models seems to be *where* and *how* the wound healing process should be evaluated.

Where to Evaluate the Wound Healing Process

In clinical practice all types of organs can be involved in surgery. This means that especially the *size of the involved organ* has to be taken into account when the animal model has to be chosen. The surgical procedure in the animal model has to be as similar as possible to the human clinic in order to give reliable and relevant information. On the other hand, the animal model used has to be as small as possible in order to fit into available physical and laboratory facilities. Other important factors are: Financial resources, technical expertise and ethical and legal factors are as directing against as small a animal model as possible. The small size of organs in small models may, however, in some studies be overcomed by using stereoscopic operation microscopes allowing a sufficient surgical technique.

How to Evaluate the Wound Healing Process

The other important determinant of an animal models in wound healing research is the **parameters** which will be used for evaluating the healing process. From a functional quantitative point of view the *strength of the wound* is the most important property in surgical practice. The time interval from injury to a healed scar is strong enough to resist the forces occurring in the tissue under physiological conditions, is of major importance for the surgeon. Interest has therefore been focused on the biomechanical properties of healing wounds. Mechanical testing of tissues can be done *in-vivo* and *in-vitro*. Beside these parameters also the *biochemical properties* have been investigated, primarily collagen, elastin and glycosamines. These parameters have primarily been used in all types of surgical wounds to study the physiology of the wound, the influence of pathopysiological factors and different types of surgical techniques on the healing process.

In an open surgical wound (or for that matter also a chronic ulcer) measurements of *surface area* has been the most used parameter for quantitative evaluation of wound healing. Reduction of wound areas has been used as a measure of progress of healing. The parameters allow an evaluation of the influence of intrinsic as well as extrinsic factors as dressings, local pharmacal etc. *Other parameters* for evaluation of the wound healing process *in vivo* have been the use of wound healing markers as tissue oxygen tension, fibrin activity, tissue ultrastructure, cellular response (receptors of adhesions) and immunological response (mononuclear antibodies and different types of growth factors).

Decision on type of model has to take the mentioned parameters in account for which reason each parameter need a more detailed description.

I. Biomechnical Parameters and the Wound Healing Process

Biomechanical testing involve methodological problems universal for testing of tissues. From a biological point of view *in-vivo* and *in-situ* testing should provide the most relevant and reliable information. Two techniques have normally been used for this type of biomechanical testing: The blow-out technique and the tear-apart technique.

The *blow-out technique* has been used for measuring strength of healing wounds in hollow organs as well as in abdominal wall, with the abdominal cavity being the "hollow" organ.[15] This method measure the bursting strength in the testing situation approximating the physiological distribution of forces in the organ wall. The argue for this measurement has been its relation to the clinical situation. The problem with this type of measurements is that the hollow organs in the body, except from blood vessels and other ducts, are not cylindrical tubes. In case of transverse wounds, as an anastomosis, as well as an longitudinal wound will always disturb the geometric configuration during testing. This means that the wall pressure tension according to La Place's law cannot be accurately calculated. Another drawback for this type of measurements is that after a certain healing time the rupture in these types of tissue occurs outside the incision line, and the acutally strength of the wound is not measured. This method, therefore, only allows a rough quantitative comparison between results obtained from different investigators. This parameter may, however, be usable early in the healing process where other methods are not suitable. The *tear-apart technique* has been used for studying skin wounds. Hooks are pieced through the skin on either side of the linear wound and they are forced apart.[16, 17] The resulting force, however, is not only dependable on the mechanical properties of the wound but also on those of the tissue surrounding the hooks. This is probably less important during early wound healing. Instead of hooks in the skin stubs have been glued on the skin.[18]

The biomechanical properties can also be evaluated by *in vitro* testing. This can be done using strips or membrane specimens containing the wound. The *strip technique* is based on a strip fixed in two clamps and with the perpendicular wound placed in between.[16, 17] This technique enables the recording of a number of parameters besides breaking and/or tensile strength. This enables more detailed information and comparison of results from different investigators. One of the disadvantages of the method is that meticulous preparation may be required with possible damage of the wound from specimen manipulation. This technique is the best method for quantifying wound strength, but is less usable in early wound healing. The *membrane technique* could in theory avoid some of the drawbacks of the strip method. This method test a circular specimen containing the wound in the middle.[19] The specimen is mounted as a seal on a chamber in which the pressure increases until the specimen ruptures. This method yields a more detailed description of the strength. The tissue membrane is, however, inhomogeneous and has no spherical expansion. It is doubtful whether the more detailed data recorded compared to the blow-out method allow more comprehensive interpretations.

It can be concluded that biomechanical measurements of healing wounds the *in vitro* test is considered more accurate and reproducible than *in vivo* and *in situ* tests. In early wound healing, however, the *in vivo* and *in situ* test may be more useful because of lesser damage to the wound during the preparation procedure. These circumstances have to be taken in consideration before a model is chosen.

II. Biochemical Parameters and the Wound Healing Process

The principle component of non-specific scar tissue is fibrous collagen, which provide the main part of the mechnical strength. The scar tissue consist of collagen fibers with small amounts of ground substances. Therefore investigations of collagen synthesis, content and distribution has been the most used biochemical parameter in surgical as well as other types of wound healing research. Recently also measurements in wound fluid of collagen metabolites (especially PICP and PIIINP)[20, 21] have been introduced as important parameters for the collagen synthesis. However, also elastin, glycos mainoglycans, lymphokinons and fibrinolytic activity have been investigated. This technique has also been used in wound fluid from mastectomy patient wound fluid.[22] The biochemical assessment of the wound healing process is based on measuring on accepted procedures identical for all laboratories, and comparisons can most often be made without problems. Biochemical factors can for the mentioned reasons be used in most wound healing models if the facilities of measuring the parameters are available. Recently also the microdialysis technique has been introduced in wound healing research. This method allows dynamic measurement of different compunds involved in the wound healing process (lactate, glucose etc.) and compounds or drugs influencing the healing process (antibiotics, antiseptics etc.). These measurements can be performed in different types of tissue in animals and in man primarily in subcutaneous tissue close to a wound. The concentration of antibiotics has been described in blood and wound fluid but not dynamically in tissue. During this procedure it is possible to investigate whether the antibiotic concentration in tissue is higher than the MIC value of the infecting bacteria. This has been studied for gentamycin [23] and for iodine.[24]

III. Surface Area and the Wound Healing Process

Studying skin wound healing direct tracing using sterile transparent sheets has been used.[25] or color photography.[26] This type of measurement, however, involves methodological problems based on by the fact that biological surfaces are not flat and should not be allowed to come into contact with the measuring equipment in order to avoid contamination. This means that neither direct tracing nor photography are ideal measures of the surface area of the healing wound. The introduction of sterophotogrammetry or computer-linked stereo camera[27] have made it possible to measure healing non-invasively, allowing accurate, reproducible measurements of irregular defects in three dimensions. The procedure, although, is to some extend time consuming, requires some training and the importance of measuring volume has still to be proven of clinical value. For surface measurements in man this method, however, seems most appropriate, while in animal research the traditional methods under certain conditions may be usable if the above mentioned limitations are observed.

IV. Other Parameters and the Wound Healing Process

Other parameters have been used to quantify the wound healing process or parts of it. To quantify peripheral tissue perfusion, in order to optimise wound healing and resistance to infection tissue oxygen tension has been measure.[28] Angiogenesis has been measured using a immunohistological staining procedure,[29] the ultrastructure of capillaries and surroundings by electron microscopy and different components in wound fluid by different immunological procedures. Using these procedures require some expertise which has to be learned before using these procedures in an animal model.

Criteria for Choosing a Model in Wound Healing Research

In Table 3.2 the criteria for choosing a research model in surgical wound healing are summarized. First the *problem have to be defined*: Are the problem the effect of an environment change or substance on cell types an 0)*in vitro* model 0) may be used. The primary advantage of this type of model is that no animals are needed for the study which from an ethical point of view always is to prefer.

Furthermore, this type of studies allow investigation of single effects on each types of cells without influence from other tissue components. The disadvantages, however, is the difficulty in extrapolation of results to the wounded tissue in an otherwise normal organism. For the purpose of this book the **in vivo model** in surgical research will be primarily discussed. The first question will then be: Can the problem be solved using an artificial wound or do it need to be a wound similar to the one made in surgery? If the problem is a basic question related to development of wound healing components as collagen etc, cells, bacteria or influencing factors in the wound fluid an *artificial wound model* may be usable. The advantages of an artificial wound are that this wound allow us to investigate other wise hidden information in a close surgical wound. The information of an artificial or surogate wound can not be determit by blood samples and this model allow us to say something about the healing properties of the connective tissue after a specific surgical procedure. This is especially important in human wounds where this type of model presently is the only

Table 3.2. Important criteria for choosing a model in surgical wound healing research

The model should:

1. Reproduce the surgical trauma
2. Be available and suitable for surgery
3. Be suitable for other types of monitoring
4. Be large enough for surgery, biopsy samples, and chronic cannulation
5. Suitable for anaesthesia in longer periods
6. Be available in more than one species
7. Fit into practical factors (described in the text)

available. Using this type of model different part of the wound healing process can be investigated separately. The model allow investigation of the inflammation reaction, differential count of wound cells, timing of wound healing speed and matrix deposion. The disadvantages of this type of model is the lacking comparability of the model to wound in surgical practice. Furthermore, the question of the significance of the implant has been raised: Does the implant enhance the inflammatory reaction?; Does the implant limit in growth of healing tissue?; Is the matrix depacit in the implanted typical for the wounds? and is the separation of the connective tissue inside the sponges from the surrounding connective tissue a problem? These draw backs, however, are partly eliminated in newer devises and for several studies in surgical wound healing research this model is the only available (human studies etc.).

If the question is related directly to the configuration of the wound before or after surgery (e.g. healing of damaged soft tissue or bone, surgical wounds made in different types of tissue, an anastomosis in a hollow organ etc) *a model similar to the surgical trauma* has to be used. The advantages of this types of wound are the similarity of the animal model to wounds in the surgical clinic. Furthermore, the influence of materials and procedures can be evaluated in clinical related wounds and in case of surface wounds be applied directly to the wounded sites. The disadvantages of these types of models is the complexity of involved tissue and tissue interaction which may result in lost information of the single processes in the wound healing process. Furthermore, most of these models only allow information of the uncomplicated clean wound healing and models for complicated clinical wounds may be difficult to prepare.

Finally the *involved parameters have to be defined.* Which type of parameters are of especial interest for this study: 1. The directly measured healing parameters as wound strength, the build up of collagen or other compounds or the wound surface; 2. The parameters important for an optimal wound healing capacity as angiogenesis, tissue perfusion and tissue oxygen tension or 3. Components in the wound tissue or fluid related to the healing or immunological processes.

Finally, the choice of a model will be influenced of *practical factors*: 1. Space and physical facilities required for animal work; 2. Laboratory facilities; 3. Technical expertise; 4. Financial resources; 5. Time and 6. Ethical and legal factors. These factors are important and it has in each situation to be evaluated whether a less optimal model have to be used or the study planned must be changed or cancelled.

Conclusions

Our present knowledge of the physiology and pathophysiology of the wound healing process in surgical wounds is primarily based on experimental research. No single model, however, full fill all the purpose of describing the complex nature of the wound healing process. In order to allow detailed studies of the single component in the wound healing process different types of experimental models have to be used. Each model, however, may be used with full recognition of the advantages and disadvantages of the model. In *vitro models* have the ethical advantages of no use of animals. The results on the contrary are difficult to extrapulate to the wound healing process in a normal organisms. In *vivo models* involve animal and human experiments, but allow us to follow up the healing process sequencely and measure the different components.

The interest of investigating the wound healing process has increased dramatically the last 10–20 years. In the future new models experimentally and especially in human research will increase our understanding of the components of the wound healing process leading to improve outcome of surgery and patient care.

Reference

1. Isselhard, W. H. and Kusche, J. (1986). Animal experimentation. In *Principles and Practice of Research; Strategies for Surgical Investigators*, edited by H. Troidl, W. O. Spitzer, B. McPeek and M. F. McKneally, pp. 149–161. Berlin, Springer-Verlag.
2. Leader, R. W. and Padgett, G. A. (1981). The genesis and validation of animal models. *American Journal of Pathology*, **101**, 11–17.
3. Calman, K. C. (1989). The choise of an experimental model. In *Principles of Surgical Research*, edited by R. T. Mathie, K. M. Taylor and J. S. Calnan, pp. 17–25. London, Butterworths.
4. Woessner, J. F. and Boucek, R. J. (1961). Connective tissue development in subcutaneous implanted polyvinyl sponge. *Archieves of Biochemic*, **93**, 85–94.
5. Viljanto, J. (1964). Biochemical basis of tensile strength in wound healing. *Acta chirurgica Scandinavica*, Supplementum 333.
6. Schilling, J. A., Joel, W. and Shurby, H. M. (1959). Wound Healing: A comparative study of the Histochemical changes in granulation tissue contained in stainless steel wire mesh and polyvinyl sponges clynders. *Surgery*, **46**, 702–710.
7. Hunt, T. K., Twomey, P., Zederfeldt, B. and Duuphy, J. E. (1967). Respiratory gas tensions and pH in healing wounds. *American Journal of Surgery* **114**, 302–307.

8. Viljanto, J. (1976). Cellstic: A device for wound healing studies in man. Description of the method. *Journal of Surgical Research*, **20**, 115–119.
9. Goodson, W. H. and Hunt, T:K. (1982). Development of a new miniature method for the study of wound healing in human subjects. *Journal of Surgical Research*, **33**, 394–401.
10. Jørgensen, L. N, Kallehave, F., Karlsmark, T., Vejlsgaard, G. L. and Gottrup, F. (1994). Evaluation of the wound healing potential in human beings from the subcutaneous insertion of expanded polytetrafluorethylene tubes. A methodologic study. *Wound Repair and Regeneration*, **2**, 20–30.
11. Wicke, C., Halliday, B. J., Scheuenstuhl, Forbee, E. F. and Hunt, T. K. (1995). Examination of expanded polytetrafluorethylene wound healing models. *Wound Repair and Regeneration* **3**, 284–291
12. Jørgensen, L. N., Kallehave, F., Karlsmark, T. and Gottrup, F. (1996). Reduced collagen accumulation after major surgery. *British Journal of Surgery*, **83**, 1591–1594.
13. Diegelmann, R. F., Lindblad, W. J. and Cohen, I. K. (1986). A, subcutaneous implant for wound healing studies in humans. *Journal of Surgical Research*, **40**, 229–237.
14. Jørgensen, L. N., Olsen, L., Kallehave, F., Karlsmark, T., Diegelmann, Cohen, I. K. and Gottrup, F. (1995). The wound healing process in surgical patients evaluated by the expanded polytetrafluoroethylene and the polyvinyl alcohol sponge: a comparison with special reference to intra-patient variability. *Wound Repair and Regeneration*, **3**, 527–532.
15. Gottrup, F. (1996). Experimental tissue trauma and healing. In *Essentials of Experimental Surgery: Gastroenterology*. Edited by S. L. Jensen, H. Gregersen, F. G. Moody and MH Amiri, New York, Harmond Academic Publishers, Chap **34**, pp. 1–11.
16. Fogdestam, I. and Gottrup, F. (1980). Biomechanical methods in wound-healing research with special reference to skin and gastrointestinal trat. In *Biology of Collagen*, edited by A. Viidik and J. Vuust, pp. 363–371. London, Academic Press.
17. Viidik, A. and Gottrup, F. (1986). Mechanics of healing soft tissue wounds. In *Frontiers in Biomechnics*, edited by G. W. Schmid-Schönbein, S.L-Y. Woo and B. W. Zweifach, pp. 263–279. New York, Springer-Verlag.
18. Wijn, P. (1980). The Aliniar viscoelastic properties of human skin in vivo for small deformations. *Thesis*, Katholieke Universiteit Nijmegen, Nejmegen, The Netherlands.
19. van Noort, R., Black, M. M., Greaney, M. G. and Irvin, T. (1978). A new *in vitro* method for the measurement of mechanical strength of abdominal wounds in laboratory animals. *Engineering in Medicine*, **7**, 217–221.
20. Haukipuro, K., Risteli, L., Kairahuoma, M. I. and Risteli, J. (1987). Aminoterminal propeptide of type III procollagen in healing wounds in Humans. *Annals of Surgery*, **206**, 752–56.
21. Haukipuro, K., Melkko, J., Risteli, L., Kairaluoma, M.I. and Risteli, J. (1991). Synthesis of type I collagen in Healing wounds in Humans. *Annals of Surgery*, **213**, 75–80.
22. Kaarsen, L., Mollerup, C., Karlsmark, T., Hjøjer, A., Vejlsgaard, G. L. and Gottrup, F. (1993). Glycosaminoglycaner in wound fluid from mastectomized patients *Wound Repair and Regeneration*, **1**: 109.
23. Lorentzen, H., Kallehave, F., Kolmos, H. J., Knigge, U., Bülow, J. and Gottrup, F. (1996). Gentamicin concentrations in human subcutaneous tissue. *Antimicrobial Agents and Chemotherapy*, **40**, 1783–1789.
24. Lorentzen, H. and Gottrup, F. (1995). A microdialysis study on acute wounds in a pig model – preliminary results. In: *Iodine and wound physiology* Information Transfer, Cambridge, pp. 9–14.
25. Gowland Hopkins, N. F. and Jamieson, C. W. (1983). Antibiotic concentration in exudate in venous ulcers: The prediction of ulcer healing rate. *British Journal of Surgery*, **70**, 532–534.
26. Myers, M. B. and Cherry, G. (1971). Zink and the healing of chronic leg ulcers. *American Journal of Surgery*, **120**, 77–81.
27. Bulstode, C.J.K., Goode, A.W. and Scott, P.J. (1986). Stereo photogrammetry for measuring rates of cutaneous healing: A comparison with conventional techniques. *Clinical Science*, **71**, 437–443.
28. Gottrup, F. (1992). Measurement and evaluation of tissue perfusion in surgery: In order to optimising wound healing and resistance to infection. In *International surgical practice*, edited by D. Leaper and F. J. Branicki. Oxford University Press, Oxford, pp. 15–39.
29. Pickworth, J. J. and De Sousa, N. (1988). Differential wound angiogenesis: Quantitation by immunohistological staining for Factor VI-3-related antigen. In *Beyond occlusion: Wound care proceedings*, edited by T. J. Ryan. pp. 19–24. London, Royal Society of Medical Service.

PART II

4. Biomaterials

P. J. Klopper

Professor of Experimental Surgery, (emeritus), Academic Medical Centre, University of Amsterdam, Amsterdam, The Netherlands

Introduction

The use of foreign material in surgery is contradictory for, according to the classic rule of wound treatment, all foreign material had to be removed to enhance normal healing. In modern surgery large amounts of foreign material are introduced.

Implantation as such is a surgical trauma followed by an inflammatory reaction.[164–166] Moreover, foreign material can influence this process by its chemical and physical properties, its form and by the eventual change in structure, caused by the biological environment and mechanical factors.[1–5]

The introduction of foreign material to close wounds has been performed for ages; the choice was dependent on geography, for example silk in China or linen in the European area. Experience with larger amounts of foreign material in wounds was obtained in the treatment of fractures. Cerclage with metal wires, possibly already known in Arabic medicine, received renewed attention in French literature of the eighteenth century.

In the first years of the nineteenth century more became known about the behavior of different metals in the body and the effect of galvanic corrosion. The principles of Lister stimulated a more extended use of metals and at the start of this century several ingenious methods of internal fixation were introduced.[6]

The development of metallurgy played an important role in the safer use of metal alloys. The interpretation of clinical and experimental observations was enhanced by the standardization of these materials by institutes of standardization in different countries.[7,8]

The advancement of technology in the industrial period of the nineteenth century enabled availability of new materials and at the end of that century the first attempts at clinical implantation were published. Severe complications were mentioned after implantation of celluloid, artificial horn (caseineformaldehyde) and bakelite. More synthetic materials followed between 1910 and 1930, such as polyvinylchloride, cellulose-acetate and poly-acrylates.[8] The term "macromolecule" was introduced in 1926 and from the 1930's up until now a large number of macromolecular materials became available.

In the 1930's surgical experience was obtained with synthetic sutures.[7,9] After the Second World War history repeated itself and many disadvantageous results were observed after implantation of material developed for industrial use.

More recent development is marked by the availability of materials produced to use inside the human body. Development and testing of biomaterials has become a multi-disciplinary science and it has stimulated contacts between medical and technical disciplines.[10]

This cooperation will remain important also in the future. The surgeon has to define the properties of the material wanted for specific application. It is not an easy task, for behavior in the human body is difficult to foretell and knowledge of the mechanical forces and the activity of the tissue in a specific region is often limited. Implants and the presence of bacteria is often followed by extrusion and the surfaces of some materials possess a tendency to adsorb bacteria.[11]

The future will also be marked by increasing legal regulations.[12] Since the 1970's strict rules have been introduced in the USA (FDA, Food and Drug Administration). In Europe regulations were made by local governments. In view of the European unity regulation of medical devices is introduced across the whole community.

For surgical research it means that experiments have to be planned and performed according to existing or new standards. At the same time, due to the overwhelming quantity of available materials, it is almost impossible to possess exact knowledge of the devices used in surgical practice.[13,14] The decision of whether to introduce devices is the responsibility of the surgeon and surgical experimentation remains indicated.

Definitions

The term "biomaterial" became known in the 1970's, when the number of materials developed for use as implants grew. In view of the increasing use of implants and regulations for applications, definitions were formulated by international organizations and at a conference of the European Society for Biomaterials in 1986.[15]

Biomaterials are non-viable materials used in medical device, intended to interact with biological systems.[8] These devices are defined by the International Standard Organization (ISO) as "an instrument, apparatus, implement material or other article used singly or in combination (together with any accessory thereto), which is intended by the manufacturer to be used in humans for:

a) Diagnosis, prevention, monitoring, treatment or alleviation of disease or injury;
b) Investigation, replacement or modification of the anatomy or a physiological process;
c) Control of contraception

and that does not achieve its principal intended action by pharmacological means, but which may be assisted in its functioning by such means. Hereafter, use of the term medical device will include dental devices."

The devices are categorized by the nature of the interaction and depending on the definition. The standard materials used by surgeons are in class II or III of the ASTM or as externally communicating devices in direct or indirect contact with blood (extra-corporeal circulation, transfusion etc.), natural channels (urinary, intestinal pathways) or internal devices.

Not only is the way of application of interest but also the duration of interaction, which varies from limited (less than twenty four hours) to permanent contact (more than thirty days).

Biomaterials can be tested by a number of methods, concerning the properties of the material used and the way in which the final product will be applied. General tests are those on cytotoxicity, mutagenicity, carcinogenicity and reproductive and developmental toxicity.[16]

Knowledge of the acute and systemic toxicity and, in particular for implants, of the subchronic and chronic toxicity is necessary. In accordance with tests on drugs pyrogenicity has to be absent. The application stipulates whether tests on irritation of the skin, eyes or mucous membranes, sensitization, hemocompatibility, hemolysis or stone formation have to be included.[17,18]

Mechanical properties have to be known, for it makes no sense to start experiments in animals if the material is weak but strong forces can be expected. Some of this information can be obtained by laboratory testing. In view of the worldwide wish for limitation of animal experiments it can be expected that the use of cell or tissue cultures will be strongly advised.[167] Discussion continues, however, as to how far the results of these tests contribute to real judgements when the aim is implantation.

Most laboratory research therefore has to be followed by implantation in animals, where factors such as different types of inflammatory cells, circulation and mechanical load are present.

In most standards, subcutaneous, intramuscular and intraosseous implantation are advised to be performed in mice, guinea pigs, rabbits or other appropriate animals.[19] In relation to the length of observation short-term (one to two weeks) and long-term experiments (90 days to 104 weeks) with intervals are distinguished. Sometimes the producer of the material can give more information.

In summary, various tests give information about mechanical strength, physical properties and chemical stability of the materials and reaction of the tissue in view of short- or long-term implantation.

Sterilization

Sterilization by Heat

All implants have to be sterile; materials that can be treated in an autoclave are favored for practical reasons, thermostability will then be necessary. For this reason, a number of materials have to be excluded.

Gamma Sterilization

Irradiation is a method much used by manufacturers. To apply this type of sterilization specialized knowledge is necessary, as several polymers (natural and synthetic) lose mechanical strength or show degradation after irradiation. Repeated sterilization, (and by other methods as well), can cause increased aging, accompanied by loss of quality.

Gas Sterilization

Ethylene oxide irradiation is a rather easy method, but its use has been discussed for several reasons.

Some materials used in medical devices can adsorb this gas or its derivatives and implantation can cause toxic effects especially in individuals with a low body-weight or who are in bad general condition. Personnel of the sterilization unit can be exposed to this gas; in some countries central sterilization facilities are installed to guarantee better safety measures.

Other methods of sterilization are in progress but are not included in normal hospital facilities.

For surgical experiments biomaterials must be tested after sterilization with a method appropriate for clinical use.

Materials

Biomaterials can be distinguished according to firstly the type of material: **metals** (alloys or in pure form), **ceramics and polymers** (natural or synthetic); proteins, polysaccharides and semi-synthetic polymers are also used, secondly adsorbability and thirdly the possibility of ingrowth.

The type of material is an important factor; however, surface and form are also involved in the tissue reaction following implantation. Treatment of the surface plays an important role in orthopedic devices. Finishing can prevent corrosion; the properties of stainless steel are determined by the presence of an oxide-layer.[20] Damage of this layer can contribute to early corrosion, enhanced by the fact that in the early stages after implantation a distinct hypoxic zone can be determined around the implant.[21,22] This finding accentuates the importance of the "no-touch" technique to prevent mechanical damage of the implant.

An irregular surface attracts more inflammatory cells. In some tests the presence of a standardized groove is prescribed,[23] and at the same time it prevents movement of the sample.

Porous metals were developed to obtain a better fixation. Synthetic fabrics fulfil the same task in soft tissues. For ages natural polymers were used as sutures.

Metals

Metals are widely used and assure high mechanical properties in the fixation of skeletal fractures;the stiff material can act as a stressprotector of the treated bone and late inflammatory reactions cannot be excluded.[24,25] For this reason a second operation is often necessary to remove the material. Patients are sometimes sensitized by ions present in the alloy, like nickel and chromium. The action of some of the ions on an eventual malignancy after a long period of implantation is still in discussion.[26]

Experience with metals covers a long period and has contributed considerably to the knowledge of material behavior in the body. In view of their mechanical qualities metals still have the advantage -in comparison to polymers- of being able to obtain a high mechanical performance can be obtained with a relatively small mass.

In 1810 it had already become known that galvanic corrosion occurred after implantation. Extended research conducted on the galvanic order of rank for metals of different alloys in biological environment. Simultaneous implantation of different metals is followed by formation of a galvanic element. The resulting current leads to corrosion of the metal and resorption of bone.[7]

Other types of corrosion became known: for example, those caused by mechanical tension and by damage of the polished surface. Corrosion has also been demonstrated at the contact site between screw and plate; this finding accentuated that both should be made of the same alloy.

Surgical technique plays an important role in osteosynthesis with screws, for drilling can evoke high temperatures, resulting in bone necrosis. The screw becomes surrounded by fibrous tissue and in a few weeks the initial strength of the fixation is lost, as demonstrated by the pull-out method.

Metallosis can be seen macroscopically as a slightly blue to yellow-stained fibrous layer around the device in the case of Fe or Ni oxide. Other oxides are colorless (Co, Mo and Cr) but can be visualized by histological staining or more advanced techniques, such as electronmicroscopy with x-rays and immunochemical methods.[27,168–170] The presence of ions in the surrounding tissue can cause cell-death, accompanied by fibrous tissue formation and calcification.

The alloys of stainless steel are divided according to the amount of chromium and nickel present. The first stainless steel available was V2A (Cr/Ni: 18/8), but it was soon been replaced by an alloy with molybdenum (Cr:16 – 18, Ni:10 – 14, Mo:2 – 3). The material became well known as V4A or Stellite.

Later on, experience was obtained with cobalt-chromium alloys, known as Vitallium, Vinertia or Wrought Vitallium. For a long period the cobalt-based alloys were judged as highly bioinert, but no real contact with bone could be observed. An interface consisting of a layer of fibrous tissue was always present.[28]

In the last decade, experience was obtained with titanium alloys and pure titanium. Swedish dentists showed that the titanium oxide layer acts as a ceramic material and real osteo-integrity could be demonstrated and no interface between bone and metal could be detected.[20,29,171–173]

Recently metals coated with hydroxyapatite were studied and came in trade. The fixation between bone and implant is enhanced. In discussion are still the maximal distance between device and bone that can be bridged by newly formed bone and the behavior of the coating after a longer period of implantation.[174,175] The hydroxyapatite coating contributes to the increased use of cementless implants of joint prostheses. In the near future continued research on metals with so-called internal memory can be expected.[30] This material can be formed in the wanted shape at a certain temperature. After implantation in a more suitable form, body temperature evokes the pre-formed shape.

Ceramics

The technique of ceramics production goes back to prehistoric times; the first use as a biomaterial goes back to the last part of the nineteenth century.

Investigations were performed to judge the behavior of calcium salts in the bone. Natural hydroxyapatite (ivory) became popular in reconstructive surgery between 1910 and 1920. It is interesting that casual reports are present, mentioning extruded material after a period of almost fifty years.

Ceramics have a high resistance to compression as the tensile strength is low. In dental surgery, ceramic implants reinforced with metal are applied. Hydroxyapatite cones were implanted to prevent the disappearance of the alveolar ridge by atrophy of the mandible but atrophy was not inhibited.

Ceramics are sintered at temperatures between 1000°C and 1500°C to get a polycrystalline structure. With special techniques a monocrystalline structure can be obtained which is related to that of precious stones. Ceramics are widely applied in industry: for example damage of metal surfaces in aircraft engines is prevented by ceramic coatings. Orthopedic implants can be coated with the plasma spray technique.[31,32]

Ceramics can be subdivided according to the composition of the material.

Silicate Ceramics. Bioglass, made by producers of lenses is well-known. The material is completely inert and is not adsorbed. After implantation the surface is covered by natural hydroxyapatite.[33–36]

Aluminum Ceramics. These are used in artificial joints:the development of other metal oxides such as zirconia is in progress.[37]

Carbon. This material has been developed for the heatshields of space carriages; the use of carbonimplants has been studied for various applications, such as urostoma and nerve electrodes.[38]

Carbon can be obtained as a fiber and as glassy or vitreous material. Graphite was already being used in the 1950's as a coating combined with heparin at the inner surface of vascular prostheses. Nowadays polymers or proteins are more suitable.[39]

Much experience with carbon fibers was obtained in the replacement of ligaments.[40,41] One of the disadvantages was fragmentation of the material with displacement of particles to the synovial membrane and the lymph nodes. In veterinary surgery favorable results were obtained in the replacement of tendons, the material being completely covered with natural collagen. Carbon fibers can also be used as a reinforcement of polymers.

The use of solid material as a nerve electrode or as cystostoma remained limited.[42–44] The subcutaneous use as a spacer in reconstructive surgery resulted in complaints regarding color.

Calcium Phosphate Ceramics. Hydroxyapatite is the main ceramic substance in natural bone and numerous technological and biological publications are concerned with the application of dense and porous forms hydroxyapatite. The basic material can be obtained from industry and is used to whiten baker's flour. The powder can be moulded in form and by sintering, a dense or more porous form can be obtained. Temperature affects the properties of the final product. The material is available as block or as granulate. In the body hydroxyapatite takes years to degrade or it is not adsorbed at all; it is bioinert.

One of the properties of this ceramic is, that after implantation no interface can be found between bone and hydroxyapatite. Implantation of calcium triphosphate (TCP) is followed by replacement of the material by an equal volume of bone in a few months.[45–50] Based on this observation, recent investigations give attention to the biphasic calcium ceramics: mixtures of calcium-triphosphate and hydroxyapatite.[31] Beside hydroxyapatite, other forms are also present in normal bone: octacalcium phosphate, brushite and calciumpyrophosphate. Formation of bone after resorption of TCP takes place on or in the nucleus of hydroxyapatite. The question as to whether calcium phosphate ceramics possess bone-inductive or bone-conductive activity is still discussed.[51] It seems, however, that specific proteins present in living bone are more active.[52] Up until now, the action of trace elements present in these ceramics is not clear.

Hydroxyapatite has been investigated in different fields of surgical research. Applications in otolaryngology, orthopedic, neurologic and dental surgery can be found and the material is bioinert.[53,54] It can be expected that composites of calcium ceramics will be a main subject in the 1990's, particularly in combination with adsorbable polymers or as coating on metals.[32]

Polymers

Natural Polymers

A number of natural polymers is easily degradable. This property is of interest in suturing, as fixation of the wound edges by connective tissue takes place within a week.

For a long time the only adsorbable material available was catgut, obtained from the processed submucosal or subserosal layer of the intestines of cattle or sheep. The name is still a mystery. According to some authors, its origin is from the name of an Arabic musical instrument. It is certain that fabrication of catgut for string instruments is a very old art: in the Middle Ages the material was already being used for suturing.[55]

The catgut can be treated in several ways, resulting in differences in mechanical strength and rate of resorption. Up until now plain or chromic catgut has been available, but it is being replaced by synthetic adsorbable polymers in most countries. One of the reasons is that, in extended studies in animals, mild to severe inflammatory reaction was seen around catgut sutures.[57–61] Degradation takes place by lysozymatic enzymes formed by inflammatory cells, in contrast to the modern materials, which are hydrolyzed. The mechanical strength of catgut may be variable because natural material is not homogeneous.[62]

Collagen from other origins, derived from tendons of different species for example bull tendons, has been used as suture material.[63] If resorption within a few days is necessary, catgut is still a good choice. Other natural polymers, such as linen and silk, were used for a long time as non-absorbable sutures. Silk in particular has been extensively studied; in large clinical series it was found that this material gave rise to a considerable inflammatory reaction often followed by thread fistula.

Processing of natural polymers can diminish the antigenicity and in the last decade a revival of the use of some materials has taken place.[64]

Collagen is on the market as processed natural material or in reconstituted form. Reconstituted collagen is made from animal collagen, mostly from tendons.[63] The material is mechanically and chemically processed, formed to sponges, powder, liquid or films and used as hemostatic or expander in plastic surgery.[65–67]

Natural collagen has the advantage of possessing a three-dimensional structure, facilitating ingrowth of cells and blood vessels.[41,68] Pigskin has been used in the treatment of burns and processed human umbilical veins or processed veins of the lower extremity have been used as vascular prostheses.[68] Recently arterial canine grafts preserved for fourteen days in UW solution were advocated.[173]

Processed dermal collagen of sheep was introduced to keep the shape of nipple valves,[69] to fill defects of the abdominal and thoracic wall,[70,71] to reconstruct ligaments, nerves and as wound dressing.[177–181]

Collagen from cattle is treated according to techniques of the leather industry and is tanned with glutaraldehyde or diisocyanate. The tanning procedure evokes cross-linking of the collagen, resulting in a higher strength, slower degradation and a decreased or absent antigenicity. A disadvantage can be that the tanning chemical can leach from the material. The untanned material degrades more rapidly and can evoke an antigenic reaction.[72]

In cardiovascular surgery experience has been obtained with tanned heartvalves of pigs,[73] and in vascular surgery collagen was used as an vascular prosthesis.[73–76] Examples are the bovine artery, the reinforced human umbilical vein and, recently, reinforced connective tissue stimulated in the sheep and processed with glutaraldehyde after excision.

As is the case for all absorbable fixation materials, an ideal situation is reached when degradation takes place simultaneously with the repair of connective tissue of the host. At the same time the increasing mechanical force contributes to the alignment of the newly formed collagen fibers.[77]

Synthetic Absorbable Polymers

Since the 1970's catgut has been almost completely replaced by synthetic materials, mostly copolymers of glycolic and lactic acid (PGA and PLA). First an American product (Dexon) became available, later on a suture of PGA mixed with polyglactin 910 (Vicryl). In the 1980's PDS (Polydiaxonone) was introduced as suture material or fabric.[78–89]

Lactic acid is derived from molasses or starch. The mechanical strength of l-lactate polymers increases with molecular weight.[90] In pure form polylactate is, however, slowly degradable and crumbling takes place. The particles can be present for a long period (up to years), giving rise to an inflammatory reaction and encapsulation. Particularly when a large amount of material is present in the form of staples, rods or plates, this effect may be enhanced.[91,92] This reaction is not seen around suture material with a limited mass combined with a short or moderate rate of degradation. It has been shown clinically that if slowly degrading sutures are used, the frequency of thread fistulae does not differ from that of the older non-absorbable materials.

Synthetic Polymers

After the Second World War a number of new synthetic polymers became available which were often directly implanted in patients. This empirical way of treatment was sometimes followed by remarkably good results; the development of eye lenses of methylacrylate can be given as an example. But the same material used in the hip joint gave dissatisfying results, for the particles produced by wear gave rise to severe synovitis accompanied by loss of function.[93]

New materials were implanted as vessel prostheses and stimulated a new field of research which is still ongoing.[94,95] The main problem was that only bulk materials were available; manufactured by large chemical firms and intended for industrial use. Smaller firms processed these products into medical devices and the amount and type of additives needed for processing were variable or secret. Moreover, materials of the same chemical structure are often on the market under different names and, leading to even more complexity, polymers of the same chemical group can differ considerably in behavior in biological situations. For example some polyurethanes are bioinert and others are toxic or carcinogenic. It can be stated as a rule that most additives and monomers are toxic.

Polymers are made from monomers by chemical processes which require catalysts and only a few synthetic materials are completely polymerized. A number of polymers need additives to obtain the suitable property for medical application; others are colored for practical or cosmetic reasons.

Phthalic acid is added to PVC to obtain flexible tubes. This chemical can leach from the material and long exposure to blood can result in intoxication; necrotic colitis has been described in infants.

Staining improves the aspect and can be necessary to distinguish (suture) material in the wound. Most dyes are, however, toxic and can influence inflammatory reactions.

Polymers can be subdivided into:

Thermoplastics. Organic polymers which soften reversibly above a certain temperature.

Thermosetting resins. A higher degree of cross-linking is present without softening over a certain temperature, but degradation can occur instead.

Elastomers. Synthetic rubbers.

After orientation with the use of laboratory tests, the polymer has to be tested at the place of use and for a longer period of time.

The main changes in the structure of polymers are: degradation by body fluid and living cells in combination with mechanical load. This can lead to changes in the surface structure of the polymer or to changes in the material itself.

The main problems to be expected are related to the experience obtained with metals. The entrance of leachables (monomers, catalysts, stains and additives) into the living tissue has to be controlled and assessed for systemic and local action and for carcinogenicity. Low concentrations of these products complicate judgement, in particular because the conditions in a culture are different from those in a well-vascularized region of the body.

The problem of malignant degeneration was recognized at an early stage, but convincing reports in humans are very scarce, while in animal experiments some species are more susceptible than others. Implantation has been limited to older people with a lower life-expectancy. Registration of the behavior of implants in patients over a long period of time is made difficult, factors such as mobility and the limited interest of pathologists. The fact that the number of devices made from new materials is always on the increase contributes to limited knowledge after long-term implantation. At the acceptor's death it is seldom possible to obtain the implant and the high costs of specialized research prevent a more thorough investigation, even if certain arrangements have been made by codicil. The patient's age will remain an important factor in the indication to implantation; a tendency to lower the age limit can be found, although absolute certainty on the behavior of biomaterials cannot be given. Studies on animals can only partially contribute to our understanding as there is no consensus on the type of species to be used in relation to applications in man, and long-term implantation in larger animals is expensive.

The following synthetic polymers (see Fig. 4.1) are frequently used in surgery.

Cellulose Derivatives. These are semisynthetic materials used as hemostaticum or in combination with acrylate. Tape has been advised in the closure of skin wounds.[96]

Acryl Polymers. Polymers of acrylic acid are widely used. Polymethylmethacrylates (PMMA), well-known as plexiglass or perspex, are used as prostheses and bone-cement. The latter application gave rise to a extensive literature on the toxic action of monomers and the influence of heat on the bone.

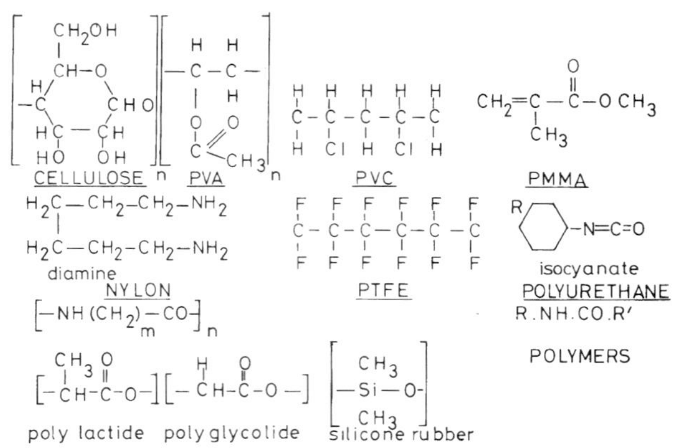

Figure 4.1. Formula of various polymers.

Acrylic hydrogel (PolyHEMA Hydron) has been used as an expander in joints.[97,98] Cyanoacrylates were investigated on a large scale as glues, but the disadvantage is that after glueing a certain amount of foreign material remains present in the wound.[28] New biodegradable materials are presented.[178]

Polyamide. (Nylon-6). One of the older polymers, still used as suture material in microsurgery.

Polyvinylchloride. (PVC). PVC is one of the most discussed synthetics and is in use on a large scale as tubing. The added phthalic acid can be toxic and the amount that is permitted to be extracted is regulated in some pharmacopoeias.

Polyvinylalcohol. (PVA-Ivalon). In the past PVA was extensively studied after implantation in the heart. The findings showed the importance of dynamic loading on the behavior of implanted synthetics. The forces present in the arterial system degraded the material in a short time, while it remained unchanged in the venous circulation.

Polyethylene. Depending on the method of fabrication,two types can be obtained: high- (supra high molecular) and low-density polyethylene. The high-density form is widely used as cups in artificial hips.The particles seen in patients after a long period of implantation are part of current research.[178] The other type is well-known as a fabric and frequently used in the repair of fascial defects.[86,99,100]

Polyethylene Terephthalate. (Dacron). This material is available as a fabric that is suitable for the fabrication of vascular prostheses. Investigations on this material led to the worldwide use of synthetic vascular prostheses. To diminish thrombogenicity seeding with endothelial cells has been performed.[11,101–103] The final result of implantation is, however, dependent of the underlying disease[182]

Polypropylene. The use of this polymer as a monofil suture material in vascular sur-gery is well known. Extensive experience has been obtained in the use as a mesh to reinforce defects in abdominal and thoracic wall. [104–108,183–186]

Polytetrafluorethylene. (PTFE-Teflon). Polymers of fluorocarbon are available in firm and in porous forms (expanded PTFE Goretex).[109] The bioinert nature is well known and it is often used as a control in comparison with materials that are being investigated.

Goretex has been studied in vascular surgery and showed reasonable results in relation to patency. In general the reconstruction of small vessels is, however, limited. In vascular research the species influences the results of implantation. In rats most materials maintain patency. Although dogs are often used for testing prostheses. Good results in dogs will not guarantee success in other animals or man for endothelial repair is rapid in contrast to that in man.

PTFE is in competition with other fabrics that can also be used for the closure of abdominal defects.[100,108,110]

Epoxy-resins. This resin is used in some orthopedic implants and as a pacemaker case. It is known for its bioinert properties.

Polycarbonates. This material has been investigated as plates in fracture treatment. It is used in hospitals to make a mould to fix the patient during radiation therapy.[111]

Polyurethanes. This group of polymers have good qualities regarding compliance and strength, permitting its use in artificial valves. It is a good example of how synthe-tics which belong to the same chemical group van behave differently as testing of a high number of samples showed that only a few were suitable for medical use.[112] Carcinogenicity of this polymer is currently being discussed.

Films and sponges of polyurethane with a high water vapor permeability are widely used as wound dressing.[112] Other polymers, covered by a thin layer of polyurethane to obtain the desired permeability, are on the market.

Elastomers. (Synthetic rubbers – Polydimethylsiloxane – Silicone rubbers). This polymer is widely known by the trade-name 'Silastic' and can be obtained in different forms, as blocks, fluid, rods and membranes. The material is bioinert and used on a large scale in reconstructive surgery.[114] The reaction of surrounding tissue is dependent on the form. In its use as a mamma-prosthesis, intense formation of connective tissue can occur; leakage of the gel inside the prosthesis is held responsible.[115] The use of this prosthesis has been forbidden recently in the US.

The oil form was used in joints, skin and as an expander but gave intense inflammatory reactions.

Natural rubber has been used for catheters and drains for a considerable time. Its use in gloves resulted in complications, the talc, or later on, starch powder produced complicated by inflammation (granuloma) if the peritoneal cavity was contaminated, simulating acute peritonitis.[187] The rubber itself could evoke sensitization through the action of the sulfuric components necessary for vulcanization.[116] Recently new gloves were introduced which are made from a polymer of natural rubber (latex) and alginate.

New developments in polymer chemistry do not only concern the introduction of new materials;

changes in physical structure have also led to improvements. In some polymers linear arrangement of monomer molecules considerably improved mechanical strength.

Consideration of the desired application remains important; new strong fibers -such as Kevlar- possess a high tensile strength, but knotting the fibre results in a considerable loss of strength.

Modern development is also marked by the introduction of composites. Polymers can be reinforced by the combination of fibers or fabric. Fibers used are made of carbon, glass or synthetic polymers. Composites of absorbable biomaterials have been investigated for uses as vessels, tubes, meniscus and plates.[104,188,189] When a non-absorbable component remains present, unfavorable results have to be excluded. The introduction of lactate polymers started with suture material: nowadays, combinations of polylactates, polyurethanes, natural collagen, cellulose derivatives and ceramics are in trial. Composites enable the ideal situation -degradation of material in balance with repair of host tissue- ton be reached; with this advantage they will mark the development in next future.

Methodology

In surgical research, interest is centered on implants, which are defined as any non-active medical device intended to be totally or partially introduced into the human body or a natural orifice or to replace an epithelial surface, the surface of the eye or of a tooth, by medical or surgical intervention and which is intended to be left after the procedure. The European community has introduced norms which are valid in the whole community. Medical devices have to be marked with CE, derived from CEN (Comité Européen de Normalisation). Before introduction on the market the product has to be tested by a so called 'notified body' present in all countries of the community. Eventual supplementary testing can only take place in recognized laboratories. If authorization has been obtained the device can be brought into the market in all states of the community.

Chemical and Biological Tests

When a new material will be used in the development of an implant, biological tests have to be performed to assess and control eventual hazardous properties of the material as mentioned above. This can also be necessary when new samples of material of well-known structure are used.

In the ISO norms, two conditions are mentioned: tests on the effect of released substances (extraction) of the device material and the effect on cells directly exposed to the device material. If sterilization is required, the device has to be subjected to tests which mimic the circumstances in clinical practice. Although leachables can be detected by biological methods, determination of the concentration of leaking chemicals after extraction can also be performed by chemical analysis (staining methods, gas-chromatography, spectroscopy).[117,118]

Non-standardized tests are also mentioned in the surgical literature, often developed for specific uses. Materials which are intended for use in the intestinal tract and bile ducts can be tested under laboratory conditions with artificial fluids based on the normal composition of the juice. For materials which are intended for use in the bloodstream a whole series of tests has been developed to get information on the compatibility with blood (animal blood can be easily obtained from a slaughterhouse). The procedure can be complemented with *in vitro* of a circuit (pump, tubing and oxygenator) if the material will be used in an extracorporeal circuit. Systemic anti-coagulation can prevent clotting and streamlining of the interior side of the circuit can prevent or inhibit damage to the corpuscles. New developments on the possibility of coating different circuit elements have been published recently. This a subject of research that began years ago. Since the development of hemodialysis, there has been considerable progress towards the properties of an ideal system.The increasing use of cardiac assist in infants and adults accentuates this line of research.[190]

The reliability of the results can be augmented by the planning of projects supported by technical universities and hematologists. In the development of vascular prostheses generally accepted tests on permeability for water or blood are available.[119] Knowledge of the permeability for water and gas is necessary in the development of materials which will be used as wound dressings, and also in materials which are present in devices for oxygenation.

In several countries series of tests were developed and published as standards according to ISO,CEN, DIN, ANSI, ASTM, USP XXII or the European Pharmacopeia. In view of the tendency to minimize the use of laboratory animals an awareness of different methods advised for biological testing is recommended.

Mechanical Testing

Mechanical testing is a valuable method to demonstrate changes of the material or device after implantation as the non-implanted sample can be used as control.

The standard loading modes are : measurement of tensile strength, compression, bending (three or four point-bending) and torsion. Shear forces are of importance in material that will be exposed to friction and in vascular work.[117,120–122] (see Fig. 4.2).

With a standardized surface, information on material changes can be obtained. The degree of im-

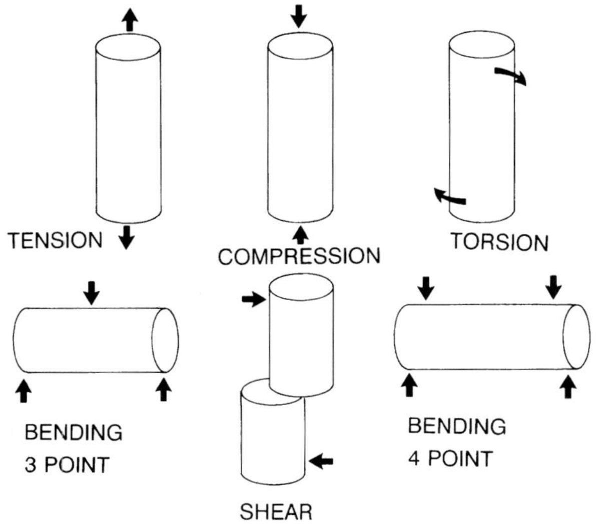

Figure 4.2. Various types of loading of materials.

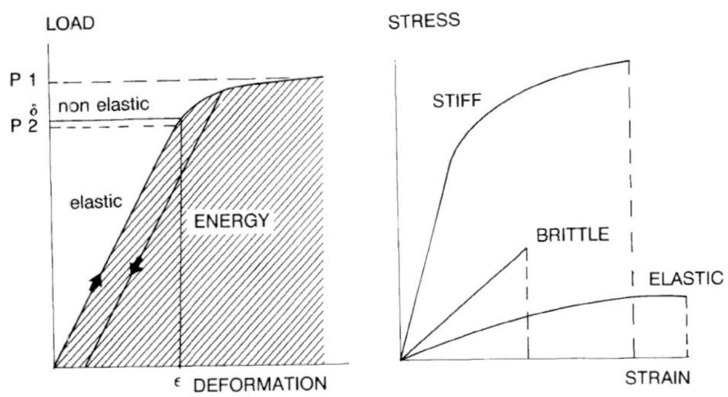

Figure 4.3. Stress-strain curve.
left: P1-P2 non elastic (plastic) deformation. e Calculation of Young's modulus (Stress divided by strain).
right: Stress-strain relation in different type of materials.

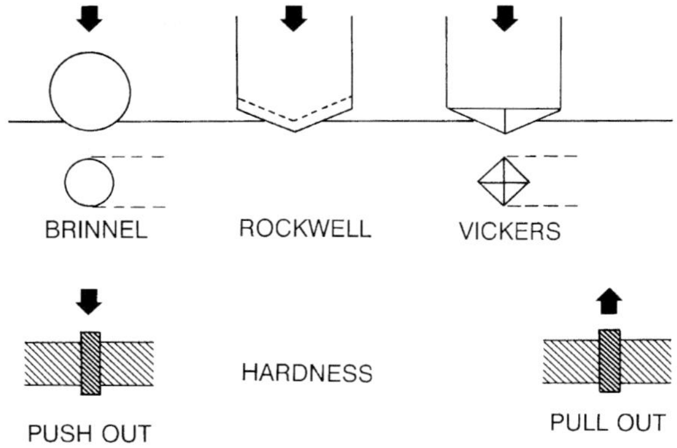

Figure 4.4. Three types of hardness tests.

pression will give information about the hardness of the material before and after implantation. The methods of Brinell, Rockwell, Vickers and others can be used. Small instruments to test hardness are available for easy comparison of properties of the surface before and after implantation.[123,124] Most tensiometers also have equipment to determine hardness. In all tests the impression made by a load given at a certain time is measured (see Fig. 4.4).

All mechanical tests determine the relation between load resulting in stress of the material and deformation (strain-impression, bending or lengthening). Most biological materials are viscoelastic.[125] The first part of the stress strain curve is linear and shows that the material resumes its original dimension after unloading. If loading is continued plastic deformation starting at the bending point and failure follows.

If the material is brittle, the non-elastic region of the graph is small, whereas ductile material is marked by a long period of deformation. The relation between stress and strain at the top of the linear line of the graph gives the module of elasticity of the tested material (Young's modulus).[126] The total area under the graph is a measure of the storage of energy (see Fig. 4.3).

Loading can be performed in relation to time (static and dynamic loading). Loading of bone at low speed presents simple fractures, whilst at high speed communitive fractures are the result. This reaction is dependent on the dissipation of the stored energy.[127,128]

Several types of model are used to analyze the mechanical relation between implant and bone: experimental or computer models for example the finite element method. The relation between the upper part of the femur and hip-prostheses is a particular subject for current research.[129]

Tensile strength is often accentuated in the judgement of suture material. However, the tensile strength of a thread as such is a relative value in contrast to the resistance of knot and thread against loading, the so called 'loop holding capacity'.[130–132]

Fixation of the thread in the tensiometer is rather complicated and breaking is often seen at the fixation point. Loading of a loop is easier and gives information on the combination of knot and thread, as used in surgical practice. The knot can break or slip (more than 2 mm of lengthening) and breakage often occurs in the knot, for knotting causes damage to the thread, especially in monofil material.[133,134]

Implantation Tests

Orientation on biological and mechanical properties is followed by testing of the material in laboratory animals; guinea pigs, mice and rats are mentioned in the literature but for surgical implants other appropriate animals can be used. The device can be implanted subcutaneously, in muscle or in bone and controls of a bioinert material such as PTFE can facilitate the interpretation of histological results.

The device should always be tested at the site and under the conditions of intended use. To get appropriate information short-term observation (two weeks) and long-term observation (twelve weeks minimum) are required (see Fig. 4.5).

Interpretation of histological reaction in the slides is classified as none or minimal, moderate and severe . This interpretation is variable, as standardization of tissue reaction is still in progress.[187]. In the near future computerized interpretation to obtain comparable results may be a possible development. The judgement of the reaction with immunochemical methods is rarely used in research on the properties of polymers, in contrast to metals where staining methods have already been known for a long period.

The following definitions are in use: tensile strength of the material expressed in Newton or Pascal per mm^2 of surface.

Loop holding capacity: the force needed to break a tied loop or provoke slippage over 2 mm of length. Knot efficiency can be expressed in the quotient between breaking strength of the knotted suture and breaking strength of the thread.[135–142] Other factors in the judgement of suture material are the handling, determined by stiffness of the thread, and the friction coefficient; the latter to a large extent defines the damage to the tissue.[137,143,144]

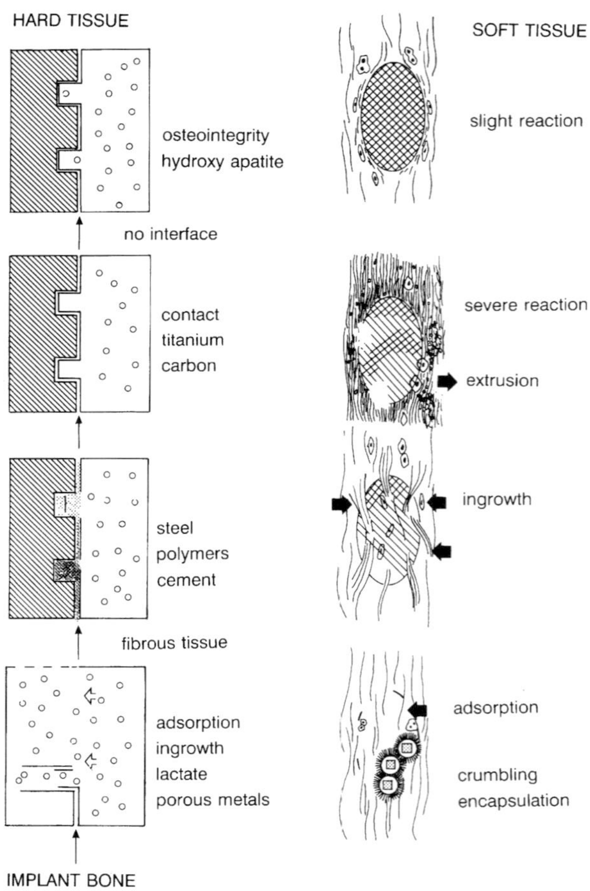

Figure 4.5. Scheme of inflammatory reaction of hard and soft tissue after implantation.

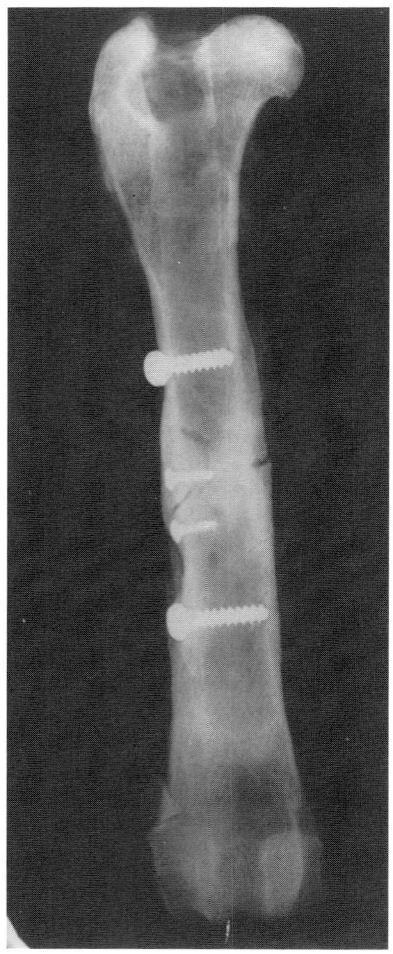

Figure 4.6. Canine femur after intramedullary fixation with a polyester rod. Small fragments fixed with screws in the rod.

X-Ray Investigation

In osseous implants X-ray investigation can provide a general review of the eventual changes after implantation. Technical roentgenfilm and cassettes without screen can contribute to images of high quality (see Fig. 4.6).

In samples of hard tissue and in samples of vessels filled with a radiopaque solution such as barium or silver microradiography is a valuable method. Optimal results are obtained with an X-ray diffraction tube with a small focal spot. The voltage required to obtain the wavelength varies from 5 KV to 10 KV, density of rays is expressed in mA.[145] Exposure time is dependent on thickness of the sample, current and type of film, it varies from 10 to 45 minutes. Images with a high resolution are obtained with high resolution films or plates. By light microscopy magnification up to 300 times is possible. Two methods are mentioned, contact- and projection microradiography. By the first pictures with high contrast are obtained, quality can be increased by the use of vacuumplates.

Microradiography is a simple method to judge formation of bone around an implant and the vascular reaction around biomaterial encapsulated in soft tissue.[146,147] Documentation can be made with an apparatus for photomacroscopy, with the advantage over a microscope that the diaphragm facilitates focusing.[148] X-ray microanalysis can be used for detection of metals and salts in tissue[149–151] (see Fig. 4.7).

Microscopy

Lightmicroscopy is still the first choice for imaging[191]. Polarizing and interference contrast microscopy contributes to the easy detection of foreign material in the sample, most polymers and the collagen in tissue is birefringent (see Fig. 4.8). All implanted samples have to be compared with materials implanted at the same time.[152] In some standards for implantation indications are given. Recently the use of microwaves decreased the time of fixation.[153,154]

Reference samples known for their bio-inertness include teflon, polyethylene and polydimethyl siloxane. Most samples can be treated for routine paraffin sectioning to obtain a section of 5 to 7 μm. Hard tissue can be embedded in methylmethacrylate and sectioned with the aid of a glass-knife or by a sawing-microtome. For quantitative analysis sections of 1 μm can be necessary.[155,156] The sections are primarily stained with conventional methods and for special purposes specific stains can be used.[152,157,158]

The amount of tissue reaction has to be compared with reference to the thickness of scar, and the

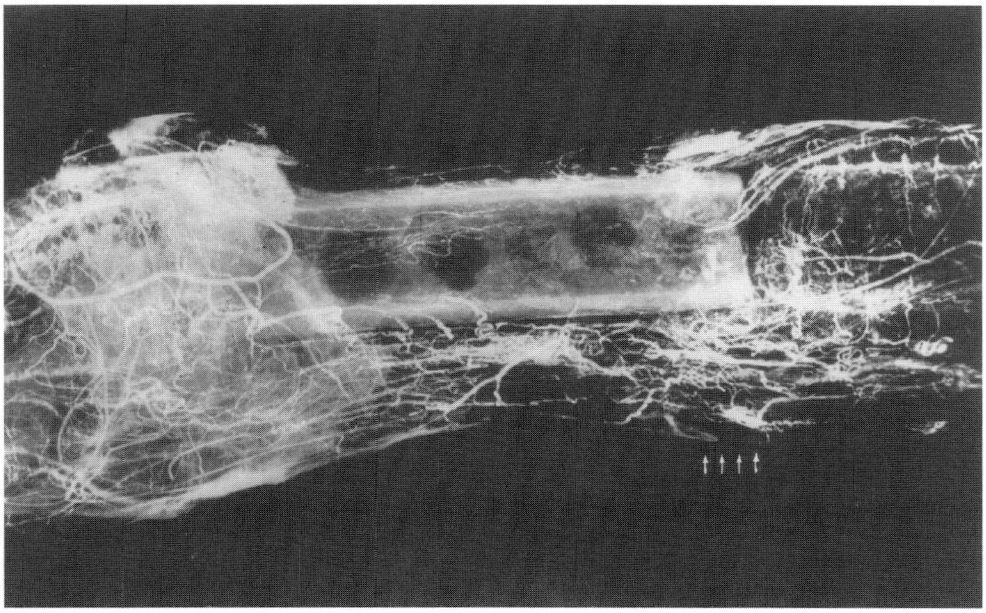

Figure 4.7. Microradioangiography after implantation of a prosthesis in the trachea; at the arrows granulation tissue.

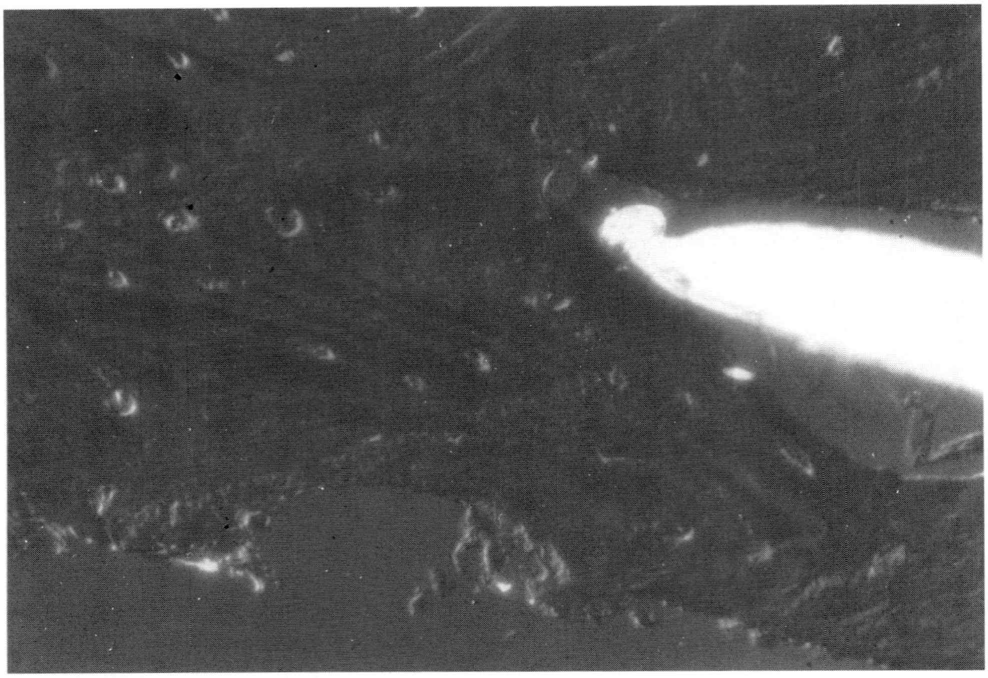

Figure 4.8. Birefringence of a small particle of a polymer in bone.

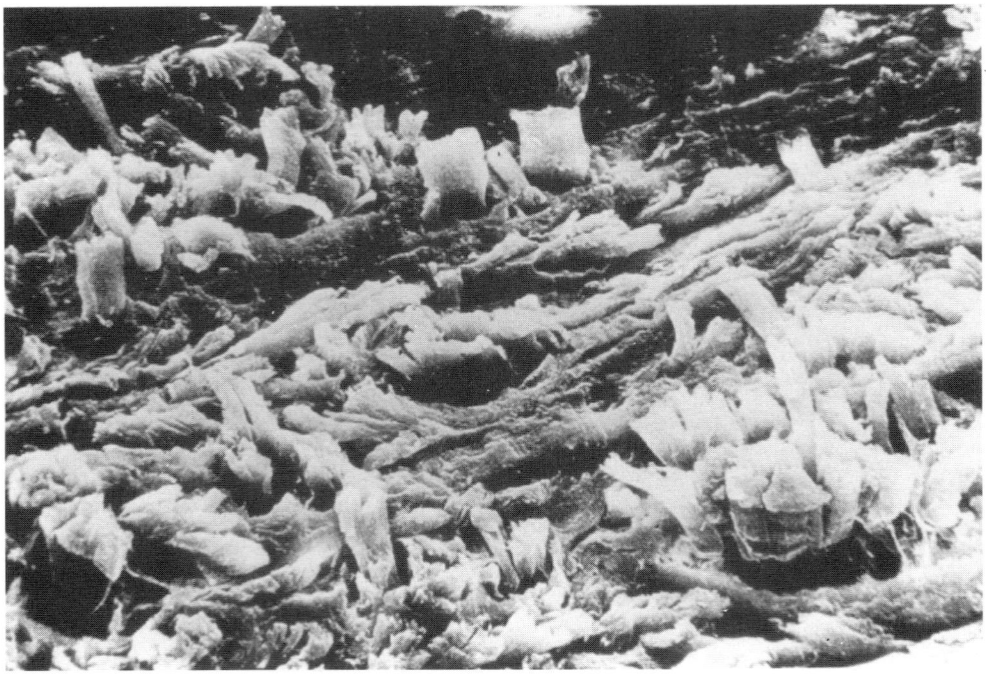

Figure 4.9. Scanning electron microscopy of surface of a natural polymer (dermal collagen).

presence of inflammatory cells or particles.[27,159] The results can be made objective by a simple score list. In the near future further standardization may be obtained using systems for quantitative digital image analysis. The latter will contribute to a more standardized interpretation of tissue reaction.[160,161,192]

Scanning Electron Microscopy (SEM)

As mentioned in the chapter on intestinal anastomoses (Chapter 6), SEM is indicated for evaluation of the surface of either tissue or material.[193] In particular, surface changes of the implanted material can be judged for degradation or wear in comparison to a non-implanted specimen. Failure of threads after knotting is also easily detected[162,163] (see Fig. 4.9).

Transmission Electron Microscopy

This method is time-consuming and requires a great experience. It can be of value in detecting particles inside macrophages and, with advanced instruments, analysis of the particles is possible.

Conclusions

In modern surgery implantation of polymers has taken a high profile in view of the desire to ameliorate the quality of live.

Recently the re-introduction of natural materials as implant material has evoked interest. In the near future the use of composites to increase mechanical strength and to regulate biodegradation will be a subject for research. The increasing number of rules for application of medical devices made from biomaterials has stimulated experimental investigation in order to obtain safe implants for clinical use. Careful measurement of the experimental and clinical data will be an important factor.[194]

References

1. Zimmerman, L. M. and Levine, R. (1957). *Physiologic principles of surgery*. Baltimore: Saunders Co.
2. Grillo, H. C. (1964). Research in wound healing. *In Research methods in surgery*, edited by W. F. Balinger II, pp. 235–255. Boston: Little Brown and Company.
3. Hunt, T. K. and Dunphy, J. E. (1974). Woundhealing. *In Scientific Foundations of Surgery*, edited by C. Wells, J. Kyle and J. E. Dunphy, 2nd edn, pp. 713–718. London: Heinemann.
4. Ledingham, I. McA. and MacKay, C. (1978). *Jamieson & Kay's Textbook of surgical physiology*. Edinburgh: Churchill Livingstone.
5. Peacock, E. E. (1984). *Woundrepair*. 3d edn. Philadelphia: Saunders Cy.
6. Lambotte, A. (1913). *Chirurgie operatoire des fractures*. Paris: Masson & Cie.
7. Contzen, H., Straumann, F. and Paschke, E. (1967). *Grundlagen der Alloplastik mit Metallen und Kunststoffen*. Stuttgart: Thieme.
8. Williams, D. F. and Roaf, R. (1973). Implants in surgery, London: Saunders.
9. Nockeman, P. F. (1980). *Die chirurgische Naht*. 3e Aufl. Stuttgart: Thieme.
10. DePutter, C. , Lange, G. L. de, Groot, K. de and Lee A. J. C. (1988). *Implant materials in biofunction. Advances in Biomaterials*. Amsterdam: Elsevier.
11. Harris, J. M. and Martin, L. F. (1987). An in vitro study of the properties influencing Staphylococcus epidermidis adhesion to prosthetic vascular graft material. *Annals of Surgery*, **206**, 612–620.
12. Daniels, A. U. (1989). Applied biomaterials standards report. *Journal Biomedical Materials Research*, **23**, 9–11.
13. Christel, P. (1989). The regulation of orthopaedic devices in France. *Journal Biomedical Materials Research*, **23**, 3–11.
14. Sevastianov, V. I. and Tseytlina, E. A. (1991). Biomedical research in the USSR. *Journal of Biomedical Materials Research*, **25**, 255–265.
15. Williams, D. F. (1988). Consensus and definitions in biomaterials. In *Implant materials in biofunction*. Advances in Biomaterials, editors C. de Putter, G. L. de Lange, K. de Groot and A. J. C. Lee, **8**, 11–16. Amsterdam: Elsevier Science Publishers.
16. Rosdy, M. and Claus L.-C. (1990). Cytotoxity testing of wound dressings using normal human keratinocytes in culture. *Journal of Biomedical Materials Research*, **24**, 363–377.
17. Mendenhall, H. V. (1986). Principal considerations for experimental surgical procedures. In *Handbook of biomaterials evaluation*, edited by A. F. von Recum, pp. 265–275. New York: Macmillan.
18. Edlich, R. F., Rodeheaver, G. T. and Thacker, J. G. (1987). Considerations in the choice of sutures for wound closure of the genitourinary tract. *Journal of Urology*, **137**, 373–379.
19. Jansen, J. A., Waerden, J. P. C. M. van der, Lubbe, H. B. M. van der and Groot K. de (1990). Tissue response to percutaneous implants in rabbits. *Journal of Biomedical Materials Research*, **24**, 295–307.
20. Davies, J. E., Lowenburg, B. and Shiga, A. (1990). The bone-titanium interface in vitro. *Journal of Biomedical Materials Research*, **24**, 1289–1306.
21. Silver, I. A. (1980). The physiology of wound healing. In *Wound healing and Wound infection. Theory and surgical practice*, edited by T. K. Hunt, pp. 11–31. New York: Appleton Century-Crofts.
22. Bouet, T., Schmitt, M., Desuzinges, C. and Eloy R. (1990). Quantitative in vivo studies of hyperemia in the course of the tissue response to biomateral implantation. *Journal of Biomedical Materials Research*, **24**, 1439–1461.
23. Chehroudi, B., Gould, T. R. L. and Brunette, D. M. (1990). Titanium-coated micromachined grooves of different dimensions affect epithelial and connective-tissue cells differently in vivo. *Journal of Biomedical Materials Research*, **24**, 1203–1219.
24. Tonino, A. J., Davidson, C. L., Klopper, P. J. and Linclau, L. A. (1976). Protection from stress in bone and its effects. *Journal of Bone and Joint Surgery*, **58B**, 107–113.

25. Klopper, P. J. and Tonino, A. J. (1980). Internal fixation of femur fractures with plastic rods-an experimental study. In *Current concepts of internal fixations of fractures*, edited by H. K. Uhthoff, pp. 416–422. Berlin: Springer Verlag.
26. Blumenthal, N. C. and Cosma, V. (1989). Inhibition of apatite formation by titanium and vanadium ions. *Journal Biomedical Materials Research*, **23**, 13–22.
27. Reuling, N., Wisser, W., Jung, A. and Denschlag, H. O. (1990). Release and detection of dental corrosion products in vivo: Development of an experimental model in rabbits. *Journal of Biomedical Materials Research*, **24**, 979–991.
28. Henck, L. L. (1980). The interfacial behavior of biomaterials. *Journal Biomedical Materials Research*, **14**, 803–811.
29. McCutchen, J. W., Collier, J. P. and Mayor, M. B. (1990). Osseointegration of titanium implants in total hip arthroplasty. *Clinical Orthopaedics and Related Research*, **261**, 114–125.
30. Baumgart, F., Bensmann, G. and Haasters, J. (1980). Memory alloys-New material for implantation in orthopedic surgery. In *Current concepts of internal fixation of fractures*, edited by H.K. Uhthoff, Part I and II. pp. 122–135. Berlin: Springer Verlag.
31. Ducheyne, P., Henck, L. L., Kajan, II A., Martens, M., Bursens, A. and Mulier, J. C. (1986). Effect of hydroxyapatite impregnation on skeletal bonding of porous coated implants. *Journal Biomedical Materials Research*, **14**, 225–237.
32. Groot, K. de, Geesink, R. G. T., Klein, C. A. P. T., Serekian, P. (1987). Plasma-sprayed coatings of hydroxyl-apatite. *Journal of Biomedical Materials Research*, **21**, 1375–1382.
33. Griss, P., Greenspan, D. C., Heimke, G., Krempien, B., Buchinger, R., Hench, L. L. and Jentschura, G. (1976). Evaluation of a bioglasscoated Al2O3 total hip prosthesis in sheep. *Journal of Biomedical Materials Research*, **7**, 511–518.
34. Gibbons, D. F. (1980). Biocompatibility of Macor glass ceramic. *Journal Biomedical Materials Research*, **14**, 177–180.
35. Gross, U., Brandes, J., Strunz, V., Bab, I. and Sela, J. (1981). The ultrastructure of the interface between a glass ceramic and bone. *Journal of Biomedical Materials Research*, **15**, 291–305.
36. Ono, K., Yamamuro, T., Nakamura, T. and Kokubo, T. (1990). Mechanical properties of bone after implantation of apatite-wollastonite containing glass ceramic-fibrin mixture. *Journal of Biomedical Materials Research*, **24**, 47–63.
37. Bertoluzza, A., Simoni, R., Tinti, A., Morocutti, M., Ottani, V. and Ruggeri, A. (1991). Calcium phosphate materials containing alumina: Raman spectroscopical, histological and ultrastructural study. *Journal of Biomedical Materials Research*, **25**, 23–38.
38. Kenner, G. H., Williams, W. S., Lovell, J. E. and Eatherly, W. P. (1975). Two-year biocompatibility study of ORNL graphite. *Journal Biomedical Materials Research*, **6**, 67–72.
39. Hennink, W. E., Dost, L., Feijen, J. and Kim, S. W. (1983). Interaction of albumin heparin conjugate preadsorbed surface with blood. *Transactions American Society Artificial Internal organs*, **29**, 200–205.
40. Jenkins, D. H. R. (1978). The repair of cruciate ligaments with flexible carbon fibre. *Journal of Bone and Joint Surgery*, **60B**, 520–522.
41. Woo, S. L.-Y. and Sites, T. J. (1988). Current advantages on the study of the biomechanical properties of tendons and ligaments In: *Collagen Vol II. Biochemistry and Biomechanic*, edited by M. E. Nimni, pp. 223–242. Boca Raton Florida: CRC Press Inc.
42. Mooney, V., Hartman, D. B., McNeal, D., Benson, J. (1974). The use of pure carbon for permanent percutaneous electrical connector systems. *Archives of Surgery*, **108**, 148–153.
43. Kingma, Y. J. (1977). Measurements on glassy-carbon electrodes. *Medical Biological Engineering and Computers*, **15**, 327–332.
44. Foster, I. W., Ralis, Z. A., McKibbin, B. and Jenkins, D. H. R. (1978). Biological reaction to carbon fiber implants: The formation and structure of a carbon-induced "Neotendon". *Clinical orthopaedics and related research*, **131**, 299–307.
45. Redja, A. B. V., Peelen, J. G. J. and Groot, K. de (1977). Calciumphosphate as a bone substitute. *Journal of Bioengineering*, **1**, 93–99.
46. Denissen, H., Mangano, C. and Venini, G. (1985). *Hydroxylapatite implants*. Padua: Piccin.
47. Verburg, A. D., Klopper, P. J., Hooff, A. van den, Marti, R. K. and Oechsner, P. E. (1988). The healing of biologic and synthetic bone implants. *Archives of orthopaedic and traumatic surgery*, **107**, 293–300.
48. Hollinger, J. O., Schmitz, J. P., Mizgala, J. W. and Hassler, C. (1989). An evaluation of two configurations of tricalcium phosphate for treating craniotomies. *Journal Biomedical Materials Research*, **23**, 17–29.
49. Lange, G. L. de, De Putter, C. and Wijs, F. L. J. A. de (1990). Histological and ultrastructural appearance of the hydroxyapatite-bone interface. *Journal of Biomedical Materials Research*, **24**, 829–845.
50. Daculsi, G., Passuti, N., Martin, S., Deudon, C., Legeros, R. Z. and Raher, S. (1990). Macroporous calcium phosphate ceramic for long bone surgery in humans and dogs. Clinical and histological study. *Journal of Biomedical Materials Research*, **24**, 379–396.
51. Ohgushi, H., Okumura, M., Tamai, S., Shors, E. C. and Caplan, A. I. (1990). Marrow cell induced osteogenesis in porous hydroxyapatite and tricalcium phosphate: A comparative histomorphometric study of ectopic bone formation. *Journal of Biomedical Materials Research*, **24**, 1563–1570.
52. Damien, C. J., Parsons, J. R., Benedict, J. J. and Weisman, D. J. (1990). Investigation of a hydroxyapatite and calcium sulfate composite supplemented with an osteoinductive factor. *Journal of Biomedical Materials Research*, **24**, 773–787.
53. Blitterswijk, C. A. van, Hesseling, S. C., Grote, J. J., Koerten, H. K. and de Groot, K. de (1990). The biocompatibility of hydroxyapatite ceramic. A study of retrieved human middle ear implants. *Journal of Biomedical Materials Research*, **24**, 433–453.
54. Hollinger, J. O., Schmitz, J. P., Mark, D. E. and Seyfer, A. E. (1990). Osseous wound healing with

xenogeneic bone implants with a biodegradable carrier. *Surgery*, **107**, 50–54.

55. Fandre, A. (1944). *Le catgut, les ligatures et les sutures chirurgicales a travers les ages*. Paris: Masson & Cie.
56. Lawrie, P., Angus, G. E. and Reese, A. J. M. (1960). The absorption of surgical catgut II. The influence of size. *British Journal Surgery*, **47**, 551–555.
57. Postlethwait, R. W. (1968). Tissue reaction to surgical sutures. In *Repair and regeneration. The scientific basis for surgical practice*, edited by J. E. Dunphy and W. van Winkle Jr, pp. 263–286. New York: McGraw-Hill Book Company.
58. Laufman, H. (1977). Is catgut obsolete? *Surgery Gynecology & Obstetrics*, **45**, 587–588.
59. Laufman, H. (1977). Synthetic absorbable sutures. *Collective review SGO*, **145**, 597–608.
60. Capperauld, I. and Bucknall, T. E. (1984). Sutures and dressings. In *Wound healing for surgeons*, edited by T. E. Bucknall and H. Ellis, pp. 75–93. London: Balliere Tindall London.
61. Peper, W. A, Alloz, D. R., Peterson, E. L. and Miagilligan, Jr. D. J. (1986). Pigskin as a topical hemo-stat in arterial, liver and splenic injuries. *Surgery*, **99**, 557–563.
62. Postlethwait, R. W. (1975). A new synthetic absorbable suture. *Surgery Gynecology & Obstetrics*, **140**, 377–380.
63. Klopper, P. J. (1986). Collagen in surgical research. *European Surgical Research*, **18**, 218–223.
64. Heidemann, E. (1988). The chemistry of tanning In *Collagen Vol III. Biotechnology*, edited by M. E. Nimni, pp. 39–6. Boca Ranton Florida: CRC Press Inc.
65. Gruber, U. F. (1969). *Blood replacement*. Berlin: Springer Verlag.
66. Blair, S. D., Backhouse, C. M., Harper, R., Matthews, J. and McCollum, C. N. (1988). Comparison of absorbable materials for surgical haemostasis. *British Journal Surgery*, **75**, 969–971.
67. Miller, E. J. (1988). Collagen Types: Structure Distribution and functions. In *Collagen Vol I. Biochemistry*, edited by M. E. Nimni, pp. 139–156. Boca Raton Florida: CRC Press Inc.
68. Calligaro, K. D., Friedell, M. L., Rollins, D. L., Semrow, M. and Buchbinder, D. (1991). A comparitive review of in situ versus reversed vein grafts in the 1980s. *Surgery, Gynecology & Obstetrics*, **172**, 247–252.
69. Meijer, D. W. (1992). The continent ileal reservoir. *Eur. J. Surg.*; suppl. 568.
70. Wal, A. C. van der, Out, N. J. M., Hooff, A., van den and Klopper, P. J. (1988). Reconstruction of the abdominal wall with processed sheep dermal collagen. In *Implant materials in biofunction*. Advances in Biomaterials, edited by C. de Putter, G. L. de Lange, K. de Groot and A. J. C. Lee, Vol. 8. Amsterdam: Elsevier Science Publishers.
71. Rudolphy, V. J., Tukkie, R. and Klopper, P. J. (1991). Chest wall reconstruction with degradable processed sheep dermal collagen, an experimental study in dogs. *Annals of Thoracic Surgery*, **52**, 821–825.
72. Gulik, Th. van and Klopper P. J. (1987). Studies on the shrinkage temperature and mechanical properties of gamma irradiated sheepskin splits. *Journal Society of Leather Technologists and Chemists*, **71**, 75–79.
73. Cohn, L. H., Collins, J. J., Disesa, V. J., Couper, G. S., Peigh, P. S., Kowalker, W. and Albred, E. (1989). Fifteen-year experience with 1678 Hancock porcine biopres-thetic heart valve replacements. *Annals of Surgery*, **210**, 435–443.
74. Palmar, J. C., Sibbitt, R. R., Tio, F. O., Reuter, S. R., Peters, J. E. and Garcia, F. (1986). Expandable intraluminal vascular grafts: a feasibility study. *Surgery*, **99**, 199–205.
75. Nojori, C., Noishiki, Y. and Koyanagi, H. (1987). Aorta-coronary bypass grafting with heparinized vascular grafts in dogs. A preliminary study. *Journal Thoracic and Cardiovascular Surgery*, **93**, 867–877.
76. Nishida, H., Grooters, R. K., Soltanzadeh, H., Thiemen, K. C. and Schneider, R. F. (1991). Clinical alternative bypass conduits and methods for surgical coronary revascularization. *Surgery, Gynecology & Obstetrics*, **172**, 161–174.
77. Ramshaw, J. A. M., Peters D. E., Werkmeister, J. A. and Ketharanathan, V. (1989). Collagen organization in mandrel-grown vascular grafts. *Journal of Biomedical Materials Research*, **23**, 649–660.
78. Varma, S, Ferguson, H. L., Breen, H. and Lump, W. V. (1974). Comparison of seven suture materials in infected wounds: an experimental study, *Journal Surgical Research*, **17**, 165–170.
79. Pavan, A., Bosio, M. and Longo, T. (1979). A comparitive study of polyglycolic acid and catgut as suture materials. Histomorphology and mechanical properties. *Journal Biomedical Materials Research*, **13**, 477–496.
80. Ippisch, A., Wriedt-Luebbe, I., Duspiva, W. and Bluemel, G. (1980). Mikrochi-rurgische Nerven-und Gefaessnaehte mit resorbierbarem Nahtmaterial. *Zeitschrift für Plastische Chirurgie*, **4**, 1–14.
81. Ray, J. A., Doddi, N., Regula, D., Willimans, J. A. and Melveger, A. (1981). Polydioxanone (PDS), a novel monofilament synthetic absorbable suture. *Surgery Gynecology & Obstetrics*, **153**, 497–507.
82. Albers, W., Geldmacher, J. and W. Beyer, W. (1982). Sehnennaehte mit einem neuen monofilen synthetischen resorbierbaren Fadenmaterial (PDS-faden der Staerke 6–0). *Der Chirurg*, **53**, 168–171.
83. Schaefer, C. J., Colombani, P. M. and Geelhoed, G. W. (1982). Absorbable ligating clips. *Surgery Gynecology & Obstetrics*, **154**, 513–516.
84. Bucknall, T. E., Teare, L. and Ellis, H. (1983). The choice of a suture to close abdominal incisions. *European Surgical Research*, **15**, 59–66.
85. Russell, P. J. (1985). Resorbable materials and composites. New concepts in orthopedic biomaterials. *Orthopedics*, **8**, 907–915.
86. Browning, A. (1986). The effect of annealing treatments on the tensile properties and hydrolytic degradative properties of polyglycolic acid sutures. *Journal of Biomedical Materials Research*, **20**, 613–632.
87. Sanz, L. E., Patterson, J. A., Kamath, R., Willet, G. W., Ahmend, S. W. and Butterfield, A. B. (1988). Comparison of Maxon suture with Vicryl, chromic catgut and PDS sutures in fascial closure in rats. *Obstetrics and Gynaecology*, **71**, 418–422.
88. Robinson, P. H. (1989). Microarterial grafting into carotid artery of the rabbit. Some considerations con-

cerning species dependant thrombogenicity. *British Journal of Plastic Surgery*, **42**, 59–64.

89. Schroeder, H. P. von, Kwan, M., Amiel, D. and Coutts, R. D. (1991). The use of polylactic acid matrix and periosteal grafts for the reconstruction of rabbit knee articular defects. *Journal of Biomedical Materials Research*, **25**, 329–339.
90. Patterson-Brown, S., Cheslyn-Curtis, S., Biglin, J., Dye, J., Easmon, C. S. F. and Dudley, H. A. F. (1987). Suture materials in contaminated wounds: a detailed comparison of a new suture with those currently in use. *British Journal Surgery*, **74**, 734–735.
91. Clarke-Pearson, D. L. and Creasman, W. T. (1985). A clinical evaluation of absorbable polydiaxonone ligating clips in abdominal and pelvic operations. *Surgery Gynecology & Obstetrics*, **161**, 250–252.
92. Schakenraad, J. M., Hardonk, M. J., Feijen J., Molenaar, I. and Nieuwenhuis, P. (1990). Enzymatic activity toward poly(L-lactic acid) implants *Journal of Biomedical Materials Research*, **24**, 529–545.
93. Dowson, D. and Wright, V. (1976). Wear characteristics of prosthetic materials. In *Biocompatibility of implant materials*, edited by D. Williams, pp. 15–20. London: Sector publishing ltd.
94. Park, J. B. (1979). *Biomaterials: An introduction*. New York: Plenum Press.
95. Chegini, N., Metz, S. A. and Masterson, B. J. (1990). Tissue reactivity and degradation patterns of absorbable vascular ligating clips implanted in peritoneum and rectus fascia. *Journal of Biomedical Materials Research*, **24**, 929–937.
96. Danto, L. A., Albertazzi, V. J., Eliott, E. E. and Love, F. G. (1987). The use of paper tape in wound closures. *Surgery Gynecology & Obstetrics*, **164**, 389–390.
97. Hastings, G. W. (1976). Adhesives and tissues. In *Biocompatibility of implant materials*, edited by D. Williams, pp. 144–151. London: Sector Publishing Limited.
98. Kon, M. and Visser, A. C. de (1981). A poly (HEMA) sponge for restoration of articular defects. *Plastic and Reconstructive Surgery*, **67**, 289–294.
99. Brown, G. L., Richardson, J. D., Malangoni, M. A., Tobin, G. R., Ackerman, D. and Polk, Jr., H. C. (1985). Comparison of prosthetic material for abdominal wall reconstruction in the presence of contamination and infection. *Annals of Surgery*, **202**, 705–711.
100. Chu, C. C., and Welek, L. (1985). Characterization of morphologic and mechanical properties of surgical mesh fabrics. *Journal of Biomedical Materials Research*, **19**, 903–916.
101. Sheppard, A. S., Eldrup-Jorgensen, J., Keough, E. M., Foxhall, T. F., Ramberg, K., Connoly, R. J., Mackey, W. C., Gavris, V., Auger, K. R., Libby ., O'Donell, T. F. and Callow, A. D. (1986). Endothelial cell seeding of small caliber synthetic grafts in the baboon. *Surgery*, **99**, 318–326.
102. Wachem, P. B. van, Vreriks, C. M., Beugeling, T., Feijen, J., Bantjes, A., Detmers, J. P. and Aken, W. G. van (1987). Adhesion of cultured human endothelial cells into methacrylate polymers with varying surface wettability and charge. *Biomaterials*, **8**, 323–328.
103. Koeveker, G. B., Graham, L. M., Burkel, W. E., Sell, R., Wakefield, T. W., Dietrich, K. and Stanley, J. C. (1991). Extracellular matrix preparation of expanded polytetrafluoroethylene grafts seeded with endothelial cells. Influence on early platelet disposition, cellular growth, and luminal prostacyclin release. *Surgery*, **109**, 313–319.
104. Lei, B., van der, Bartels, H. J., Nieuwenhuis, P. and Wildevuur, C. H. R. (1985). Microporous, compliant, biodegradable vascular grafts for the regeneration of the arterial wall in rat abdominal aorta. *Surgery*, **98**, 955–963.
105. Adloff, M. and Arnaud, J. P. (1987). Surgical management of large incisional hernias by an intraperitoneal mersilene mesh and an aponeurotic graft. *Surgery Gynecology & Obstetrics*, **165**, 204–206.
106. Pourdeyhimi, B. (1989). Porosity of surgical mesh fabrics. New Technology. *Journal Biomedical Materials Research*, **23**, 145–152.
107. Zilla, P., Fasol, R., Preiss, P., Kadletz, N., Deutsch, M., Ichima, H., Tsangkaris, G. and Groscurth, P. (1989). Use of fibringlue as a substrate for in vitro endothelialization of PTFE vascular grafts. *Surgery*, **105**, 515–522.
108. Wants, G. F. (1991). Incisional hernioplasty with mersilene. *Surgery Gynecology & Obstetrics*, **172**, 129–137.
109. Bauer, J. J., Salky, B. A., Gelernt, I. M. and Kreel, I. (1987). Repair of large abdominal wall defects with expanded polytetrafluorethylene PTFE. *Annals of Surgery*, **206**, 765–769.
110. Tyrell, J., Silbermann, H., Chandrasoma, P., Niland, J. and Shill, J. (1989). Absorbable versus permanent mesh in abdominal operations. *Surgery Gynecology & Obstetrics*, **168**, 227–232.
111. Katz, A. R., Mukherjee, D. P., Kaganov, A. L. and Gordon, S. (1985). A new synthetic monofilament absorbable suture made from polytrimethylene carbonate. *Surgery Gynecology & Obstetrics*, **161**, 213–222.
112. Belisle, J., Maier, S. K., and Tucker, J. A. (1990). Compositional analysis of Biomer. *Journal of Biomedical Materials Research*, **24**, 1585–1598.
113. Jonkman, M. F., Bruin, P., Hoeksma, E. A., Nieuwenhuis, P., Klasen, H. J., Pennings, A. J. and Molenaar, I. (1988). A clot inducing wound covering with high vapour permeability: enhancing effects on epidermal wound healing in partial thickness wound in guinea pigs. *Surgery*, **104**, 537–545.
114. Ahn, S. T., Monafo, W. W. and Mustoe, T. A. (1989). Topical silicone gel: A new treatment for hypertrophic scars. *Surgery*, **106**, 781–787.
115. Beranek, J. T. (1990). Why does topical silicone gel improve hypertrophic scars. A hypothesis. *Surgery*, **108**, 122–123.
116. Ellis, H. A. (1990). The hazards of surgical glove dusting powders. Collective review. *Surgery Gynecology & Obstetrics*, **171**, 521–527.
117. Roberts, G., McCormak, H., Ketharanathan, V., Macleish, D. G., Field, P. L. and Milme, P. Y. (1989). The role of physical and chemical characteristics in assessing the performance of a new biological vascular graft. *Journal Biomedical Materials Research*, **23**, 442–450.
118. Megerman, J., Reddy, E., L'Italien, G. J., Warnock, D. F. and Abbott, W. M. (1991). A laboratory model

to quantitate the resistance of collagen vascular grafts to biodegradation. *Journal of Biomedical Materials Research*, **25**, 295–313.

119. Zikria, B. A., King, T. C., Stanford, J. and Freeman, H. P. (1989). A biophysical approach to capillary permeability. *Surgery*, **105**, 625–631.
120. Sentissi J. M., Ramberg, K., O'Donell, T. F., Connolly, R. J. and Callow, A. (1986). The effect of flow on vascular endothelial cells grown in tissue culture on polytetrafluorethylene grafts. *Surgery*, **99**, 337–343.
121. Abidin, M. R., Towler, M. A., Rodeheaver, G. T., Thacker, J. G., Cantrell, R. W. and Edlich, R. F. (1989). Biomechanics of curved surgical needle bending. *Journal Biomedical Materials Research*, **23**, 129–143.
122. Szivek, J. A. and Yapp, R. A. (1989). A testing technique allowing cyclic application of axial bending and torque loads to fracture plates to examine screw loosening. *Journal Biomedical Materials Research*, **23**, 105–116.
123. Evans, G. P., Behiri, J. C. and Bonfield, W. (1988). Microhardness, Young's Modulus and mineral content in bone and a bone analogue. In Implant *Materials in Biofunction. Advances in Biomaterials*, edited by C. de Putter, G. L. de Lange, K. de Groot and A. J. C. Lee, **8**, 311–315. Amsterdam: Elsevier Science Publishers.
124. Campbell, J. A., Temple, W. J., Frank, C. B. and Hudicroft, S. A. (1989). A biomechanical study of suture pull out in linea alba. *Surgery*, **106**, 888–892.
125. Yamada, H. (1970). *Strength of biological materials.* Baltimore: Williams and Wilkins.
126. Frankel, V. H. and Nordin, M. (1980). *Basic biomechanics of the skeletal system*. Philadelphia: Lea and Febiger.
127. Visser, J. D. (1980). Dynamic strength of surgical suture materials. In *Evaluation of biomaterials*, edited by G. D. Winter, J. L. Leray and K. de Groot. London: Wiley and Sons.
128. Sauter, A. J. M., and Klopper, P. J. (1983). Fat embolism after static and dynamic load. *Acta. Orthopaedia Scandinavica*, **54**, 94–100.
129. Huiskes, R. (1991). Biomechanics of artifical joint fixation. In: '*Basic Orthopaedic mechanics*'' A Textbook for orthopaedic residents and graduate students specializing in biomechanics, edited by V. C. Mow and W. C. Hayes, Chapter 8. New York: Raven Press.
130. Hermann, J. B. (1971). Changes in tensile strength and knot security of surgical sutures in vivo. *Archives of Surgery*, **106**, 707–710.
131. Nilsson, T. (1982). Effect of strain rate on tensile strength and strain of surgical suture materials *Scandinavian Journal of Plastic Surgery*, **16**, 97–100.
132. Ashley, C. W. (1988). *The Ashley book of knots*. London: Faber and Faber.
133. Rijssel, E. J. C. van, Trimbos, J. B. and Booster, M. H. (1990). Mechanical performance of square knots and sliding knots in surgery: a comparative study. *American Journal Obstetrics and Gynecology*, **162**, 93–97.
134. Brouwers, J. E., Oosting, H., Haas, D.de and Klopper, P. J. (1991). Dynamic loading surgical knots. *Surgery, Gynecology & Obstetrics* (in press).
135. Taylor, F. W. (1938). Surgical knots. *Annals of Surgery*, **107**, 458–468.
136. Holmlund, D. E. W. (1974). Knot properties of surgical suture materials. A model study. *Acta Chirurgica Scandinavica*, **140**, 355–362.
137. Holmlund, D. E. W. (1976). Physical properties of surgical suture materials: stress-strain relationship, stress-relaxation and irreversible elongation. *Annals of Surgery*, **184**, 189–194.
138. Tera, H. and Aberg, C. (1976). Tensile strength of twelve types of knot employed in surgery, using different suture material. *Acta Chirurgica Scandinavica*, **142**, 1–7.
139. Tera, H. and Aberg, C. (1977). Strength of knots in surgery in relation to type of knot, type of suture material and dimension of suture thread. *Acta. Chirurgica Scandinavica*, **143**, 75–83.
140. Thacker, J. G., Rodeheaver, G. T., Kurtz, L., Edgerton, M., Edlich R. F. (1977). Mechanical performance of sutures in surgery. *American Journal Surgery*, **133**, 713–715.
141. Trimbos, J. B. and Klopper, P. J. (1985). Knot security of synthetic absorbable suture material, a comparison of polyglycolic acid and polyglactin-910. *European Journal Obstetrics Gynecology and Reproduction Biology*, **19**, 183–190.
142. Trimbos, J. B. (1986). Security of various knots commonly used in surgical practice. *Obstetrics and Gynaecology*, **68**, 425–430.
143. Trimbos, J. B., Rijssel, E. J. C. van and Klopper, P. J. (1986). Performance of sliding knots in monofilament and multifilament suture material. *Obstetrics and Gynaecology*, **68**, 425–430.
144. Chu, C. C. and Kizil, Z. (1989). Quantitative evaluation of stiffness of commercial suture materials. *Surgery Gynecology & Obstetrics*, **168**, 233–238.
145. Nuffield, E. W. (1966). X-ray diffraction methods. New York: John Wiley & Sons.
146. Rhinelander, F. W. and Baragry, R. A. (1962). Microangiography in bone healing. I. Undisplaced closed fractures. *Journal of Bone and Joint Surgery*, **44**, 1273–1298.
147. Rhinelander, F. W., Phillips, R. S., Steel, W. M. and Beer, J. C. (1968). Microangiography in bone healing. II. Displaced closed fractures. *Journal of Bone and Joint Surgery*, **50**, 643–662.
148. Blickman, J. R., Recourt, A. and Klopper, P. J. (1958). Microradiographical investigation of the vascular pattern in the liver of rats. *Journal Belge de Radiologie*, **41**, 452–465.
149. Engstroem, A. (1949). Microradiography. *Acta Radiology*, **31**, 503–521.
150. Lafay, F. (1961). Les techniques d'examen microradiographique. *La Presse Medicale*, **69 (11)**, 512–514.
151. Brodsky, B., Tanaka, S. and Eikenberry, E. F. (1988). X-ray diffraction as a tool for studying collagen structure. *In Collagen Vol I. Biochemistry*, edited by M. E. Nimni, pp. 95–112. Bocaranton Florida: CRC Press Inc.
152. Woodward, S. C. and Salthouse, T. N. (1986). The tissue response to implants and its evaluation by light microscopy. In *Handbook of biomaterials evaluation*, edited by

A. F. von Recum, pp. 364–378. New York: Macmillam Cy.
153. Boon, M. E. and Kok, L. P. (1987). *Microwave cookbook of Pathology: The art of microscopic visualization* Leiden: Coulomb Press Leyden.
154. Ruijgrok, J. M., Boon, M. E. and Wijn, de J. R. (1990). The effect of heating by microwave irradiation and by conventional heating on the aldehyde concentration in aqueous glutaraldehyde solutions. *Histochemical Journal*, **22**, 389–393.
155. Salthouse, T. N. (1986). Histochemical and quantitative microscopy In *Handbook of Biomaterials Evaluation*, edited by A. F. von Recum, pp. 398–407. New York: Macmillan.
156. Sandberg, M., Aro H., Multimaki, P., Aho, H. and Vuorio, E. (1989). In situ localization of collagen production by chondrocytes and osteoblasts in fracture callus. *Journal of Bone and Joint Surgery*, **71A**, 69–77.
157. Luna, L. G. (1968). Manual of histologic staining methods of the armed forces institute of pathology, 3d edn. New York: McGraw-Hill Cy.
158. Bancroft, J. D. and Stevens, A. (1990). Theory and practice of histological techniques, 3d edn. Edinburgh: Churchill Livingstone.
159. Goodman, S. B., Fornasier, V. L., Lee, J. and Kei, J. (1990). The histological effects of implantation of different sizes of polyethylene particles in the rabbit tibia. *Journal of Biomedical Materials Research*, **24**, 517–524.
160. Turner, E., Lawrence, W. H. and Autian, J. (1973). Subacute toxicity testing of biomaterials using histopathological evaluation of rabbit muscle tissue. *Journal of Biomaterial Research*, **7**, 39–58.
161. Rahn, B. A., Geret, V., Capaul, C., Lardi, M. and Solothunnann, B. (1982). Morphometric evaluation of tissue reaction to implants using low cost digitizing techniques. In: *Clinical Application of biomaterial*, edited by A. J. C. Lee, T. Albrektsson and P. Branemark. New York: John Wiley and Sons.
162. Chu, C. C. and Campbell, N. D. (1982). Scanning electron microscopic study of the hydrolytic degradation of poly(glycolic acid). suture. *Journal Biomedical Research*, **16**, 417–430.
163. Dociu, N. and Bilge, F. H. (1986). Evaluation by scanning microscopy. In: *Handbook of biomaterials evaluation*, edited by A. F. von Recum, pp. 390–397. New York: MacMillan.
164. Cohen I. K., Diegelmann R. F. and Lindblad, W. J. (1992). *Woundhealing Biochemical and clinical aspects* Philadelphia W. B. Saunders Company..
165. Luyn, M. J. A., Wachem, P. B., Leta, R. and Blauw, E. H. (1994). Modulation of the tissue reaction to biomaterials. *J. Mat. Sc. Mat. Med.*, **5**, 671–678.
166. Cenni, E., Ciapetti, G., Cervellati, M., Cavedagna, D., Falsane, G., Gamberini, S., and Pizzoferrato, A. (1996). Activation of the plasma coagulation system induced by some biomaterials. *J. Biomed. Mat. Res.*, **31**, 145–148.
167. Luyn, M. J. A. van Wachem, P. B. van Nieuwenhuis, P. and Jonkman M. F. (1992). Cytotoxicity testing of wound dressings using methylcellulose cell culture. *Biomaterials*, **13**, 267–275.
168. Bruijn, J. D., Seelen, J. L., Veldhuizen, R. W., van Raaij, J. J. A. M., Koerten, H. K. and Bernosky, F. P. (1996). Loosening of cementless threaded acetabular cups in relation to particulate debris.A radiographic and histologic study of 20 revision cases. *Acta Orthop. Scand.* in press.
169. Bruijn, D. J. (1996). The cementless Mecring acetabular prosthesis; clinical studies and failure analysis. *Ph.D. Thesis*, University of Amsterdam.
170. Hirakawa, K, Bauer, T. W., Stulberg, B. M. and Wilde, A. H. (1996). Comparison and quatitation of wear debris of failed total hip and knee arthroplasty. *J. Biomed. Mat. Res.*, **31**, 257–263.
171. Santavirta, S., Konttinen, Y. T., Hoikka, V. and Eskola, A. (1991). Immunopathological response to loose cementless acetabular components. *J. Bone Joint Surg.*, **73-B**, 38–44.
172. Tracana, R. B., Sousa, J. P. and Carvalho, G. S. (1994). Mild inflammatory response to stainless steel corrosion products *J. Mat. Sc. Mat. Med.*, **5**, 596–600.
173. Torgersen, S. and Gjerdet, N. R. (1994). Retrieval study of stainless steel and titanium miniplates and screws used in maxillofacial surgery, *J. Mat. Sc. Mat. Med.*, **5**, 256–262.
174. Furlong R. J., and Osborn, J. F. (1991). Fixation of hip prostheses by hydroxyapatite ceramic coatings *J. Bone Joint. Surg.*, **73-B**, 741–746.
175. Dhert, W. J. A., Thomsen, P., Klein, C. P. A. T., Groot, K. de, Rozing, P. M. and Ericson, L. E. (1994). Fluorapatite-coated implants in experimental arthritis:the response of rabbit trabecular bone. *J. Mat. Sc. Mat. Med.*, **5**, 59–66.
176. Vischjager, T.M., van Gulik, Pfaffendorf, M., Zwieten, P.A. van, Kromhout, J. G., Klopper, P. J. and Jacobs, M. J. H. M. (1995). Morphology and function of dog arterial grafts preserved in UW solution. *Eur. J. Vasc. Endovasc. Surg.*, **10**, 431–439.
177. Casagranda, F., Werkmeister, J. A., Ramshaw, J. A. M. and Ellender, G. (1994). Evaluation of alternative glutaraldehyde stabilization strategies for collagenous biomaterials. *J. Mat. Sc. Mat. Med.*, **5**, 332–337.
178. Oei, T. S., Klopper, P. J., Spaas, J. A. J. and Buma, P. (1995). Reconstruction of the flexor tendon sheath An experimental study in rabbits. *J. Hand Surg.*, **21b**, **1**, 72–83.
179. Heijke, G. C. M., Klopper, P. J. and Dutrieux, R. P. (1993). Vein graft conduits versus conventional suturing in peripheral nerve reconstructions *Microsurgery*, **14**, 584–588.
180. Wachem, P. B., Luyn, M. J. A., Olde Damink, L. H. H., Dijkstra, P. J., Feijen, J. and Nieuwenhuis, P.(1992). Tissue interactions with dermal sheep collagen implants: A transmission electron microscopical evaluation *Cells and Mat.*, **1**, 251–263.
181. Olde Damink, L. (1993). Structure and properties of crosslinked dermal sheep collagen. *PhD. Thesis.* Universiy of Twente, Enschede.
182. Ross, R. (1996). Cell biology of atherosclerosis. *Ann. Rev. Physiol.*, **57**, 791–804.
183. Fansler, R. F., Taheri, P., Cullinane, C., Sabates, B. and Flint, L. M. F. (1995). Polypropylene mesh clo-

sure of the complicated abdominal wound. *Am. J. Surg.*, **170**, 15–18.

184. Robbins, A. W. and Rutkow, I. M. (1993). The mesh-plug hernioplasty. *Surg. Clin. N. America*, **73**, 501–512.
185. Arregui, M. E., Navarette, Davis C. J., Castro, D. and Nagan, R. F. (1994). Laparoscopic inguinal herniorrhaphy. *Surg. Clin. N. America*, **73**, 513–527.
186. Shaghayegh, A. -W, Choe, E. U., LaBarre, J. J., Flint, L. M. and Ferrara, J. J. (1996). Evaluation of a novel synthetic material for closure of large abdominal wall defects. *Surgery*, **119**, 141–145.
187. Luijendijk, R. W., Lange, D. C. D. de, Wouters, C. C. A. P, Hop, W. C. J., Diron, J. J., Pailler, J. L., Comprodon, B. R., Holindahl. L., Geldorp, H. J. van and Jeekel, J. (1996). Foreign material in postoperative adhesions. *Ann. Surg.*, **223**, 242–248.
188. Dunnen, W. F. A den, Stokroos, I., Blaauw, E. H., Holwerda, A., Pennings, A. J., Robinson, P. H., and Schakenraad, J. M. (1996). Light microscopic and electron-microscopic evaluation of short-term nerve regeneration using a biodegradable poly (DL-lactide ε caprolacton). nerve guide. *J. Biomed Mat. Res.*, **31**, 105–115.
189. Otani, Y., Tabata, Y. and Ikada, Y. (1996). A new biological glue from gelatin and poly(L-glutamic acid *J. Biomed. Mat. Res.*, **31**, 157–166.
190. Gründeman, P. F., Tukkie, R., Moulijn, A. C., Rudolphy, V. J., Jong, J. W. de, Frederiks, W. M., Gijsbers, G. H. M., Marx, F. and Klopper, P. J. (1993). Short-term biventricular criculatory support in the regionally stunnedpigmyocardium. *Thorac. Cardiovasc. Surgeon*, **41**, 290–296.
191. Farkas, D. L. *et al* (1993). Multimode light microscopy and the dynamics of molecules, cells, and tissues *Ann. Rev. Physiol.*, **55**, 785–817.
192. Ungersböck, A., Schlegel, U. and Rahn, B. A. (1994). A method for quantitative histomorphometric evaluation of soft tissue to implants. *J. Mat. Sc. Mat. Med.*, **5**, 557–562.
193. Steinberg, P. J. and Hodde, K. C. (1990). The morphology of synovial lining of various structures in several species as observed with scanning electron microscopy. *Scanning Microscopy*, **4**, 987–1019.
194. Wright, J. G., McLeod, R. S. Lassing Al. Walters, B. C., Hu, X. (1996). Measurement in surgical clinical research. *Surgery*, **119**, 241–244.

PART III

GASTROENTEROLOGY

PART III

5. Intestinal Anastomoses

P. J. Klopper

Professor of experimental Surgery (emeritus), Academic Medical Centre, University of Amsterdam, Amsterdam, The Netherlands

Injuries of the intestinal wall has attracted the interest of surgeons since prehistoric times and surveying this remarkable history some distinct periods can be memorized.

In an early period chiefly casuistic information was published.[1–3] In the last part of the eighteenth and first half of the nineteenth century the base of the main suture techniques was investigated experimentally and clinically. Thirdly, the period following the work of Lister in which abdominal surgery as such was started and the numerous techniques were explored concerning safety and rapidity and finally the period up to now, in which mechanical suturing was developed on a large scale and more systemic comparison of the effect of the different techniques took place.[4–6] Some new developments are still in progress, but although mortality following colonic anastomoses has decreased considerably, morbidity still remains at a high level, justifying continued research.[7,8]

At the time of Hippocrates a difference was already noted between the longitudinal lesion and the transverse lesion of the intestine, patients with the first had some chance of survival. Celsus indicated that suturing of small intestines never led to healing, while wounds in the colon could be treated.

Arabian literature revealed the first mechanical suturing of intestines by the jaws of large ants, while in the Middle Ages in the famous Italian surgical school of Salerno, the first endoprosthesis was developed by placing the trachea of an animal in the lumen of the intestine. This early experience also showed that only longitudinal lesions, closed with the technique of the furriers, gave the patient some chance of survival.

In a later stage several authors showed the importance of the serosal layer, for if an intestinal lesion was covered by a neighboring intestine, closure was enhanced. In the eighteenth century some surgeons advised use of this technique for the risk was smaller than with suturing.

Surgical history before the time of Lister shows that the subject of intestinal suturing was studied at large scale and the most ingenious methods were developed to get reasonable results. Already in this period everting and inverting techniques were compared, and the question of whether a continuous or a interrupted suture had to be used formed part of the discussion.

Besides publications on clinical observations, experiments in animals were also performed. Ligation of the duodenum resulted in repair of the passage in about two weeks. Resections were also studied. A student in Heidelberg showed that a dog could survive pyloric resection, but the period of observation was limited by the action of antivivisectionists.

More than seventy years later Pean performed the same intervention in man. In this period the development of mechanical instruments also started, varying from rings with wooden pins-acting in the same way as the instrument of Nakayama for small vessel anastomosis a hundred years later-to endoprostheses of bone or pieces of potato.[9]

All trials to suture the intestinal wall were influenced by infection, which accompanied laparotomy in the pre-Listerian period by high mortality.[3]

All early efforts made under difficult circumstances were of importance in the post-listerian period. The finding that antiseptical measures prevented postoperative infection spread in a few years through the whole of Europe.

Between 1870 and 1880 Billroth and his coworkers performed a series of experiments in dogs that formed the base for gastric surgery, still a subject of actual research.

Most anastomoses were made according to the technique of Lembert and Albert, based on the finding that adaptation of the serosal layer by interrupted seromuscular sutures prevented leakage, a large number of variations has since been introduced .

Not only prevention of leakage of the anastomosis in the postoperative period was the aim of the different techniques, a second important factor was the duration of the operation, for the technical capacity of the surgeon was considerably influenced by the effectiveness of anesthesia. Experiments were performed to compare the length of time of the different suture techniques.

Mechanical suturing was studied by several methods and based on experiments in animals. Murphy[10] in 1892 introduced the so-called Murphy-button permitting a rapid and easy connection between two parts of the intestine by an ideal inverting adaptation. His name is well-known, but literature of that time reveals that a large number of variations can be found.[11] The advantage of the method was that end-to-end anastomoses were made, a type of repair that was also the preference of most authors in later periods.

Parallel to the development of suture techniques numerous types of intestinal clamps were developed to prevent contamination of the peritoneal cavity by intestinal contents during the intervention. The importance of the use of non-traumatic instruments was accentuated, to prevent impaired healing of the anastomosis. In 1909 an engineer in Budapest, Humer Hueltl, developed an instrument using metal staples for distal gastrectomy (von Petz-clamp).[12–14]

After World War Two an institute for the development of mechanical stapling instruments in different fields of surgery was founded in Moscow. In the sixties these stapling instruments were further developed by American, Canadian and Japanese investigators.[15–18]

Several firms are currently bringing highly developed instruments onto the market. Discussion concerning costs and the limited experience in hand-laid anastomoses is not yet closed.

In spite of the technological development morbidity is still rather high. In colorectal surgery 13% clinically diagnosed leaks are mentioned and it is even greater when dehiscence of the anastomosis is detected radiologically.

Methodology

The main factors involved in normal healing of anastomoses determine the parameters of value in research of this subject. In view of technique some rules can be a guideline.[19] The anastomosis of the gut has to be air- and water-tight; careful apposition enhances primary healing and an optimal blood supply has to be preserved.

Mechanical Parameters

In the first postoperative period (up to a few days) the mechanical strength is -dependent on the type of suturing performed- slightly increased by the formation of fibrin between the layers of the divided intestine and the serosal layer if present.[20] A few days later formation of collagen starts in the submucosal layer and the serosal layer. In the following period strength is dependent on the repair of connective tissue in the submucosal layer, the eventual formation of myogenic elements and strength of the sutures. Careful knotting is of course essential, in particular if slide knots with monofil material are used (see Fig. 5.1).

The strength of the suture material decreases gradually; from a few days if catgut has been used to several weeks in the case of synthetic absorbable sutures. Most non-adsorbable sutures keep the initial values but some degrade.[21,22]

Distension can also affect the property of suture material. If high loading by distension is present, the lengthening of sutures can lead to plastic deformation of the material. It means that after unloading the sutures are loosely present in the wound.

When non-absorbable sutures have been used, migration of suture material can occur. It is dependent on the site of suturing. Most material will be extruded into the lumen or remains present at the outer surface of the intestine. Migration causes complications if metal staples or non-absorbable suture material is used in the urinary tract. Formation of stones has been observed inside the lumen of the bladder or in the intestine used in deviation procedures.[23]

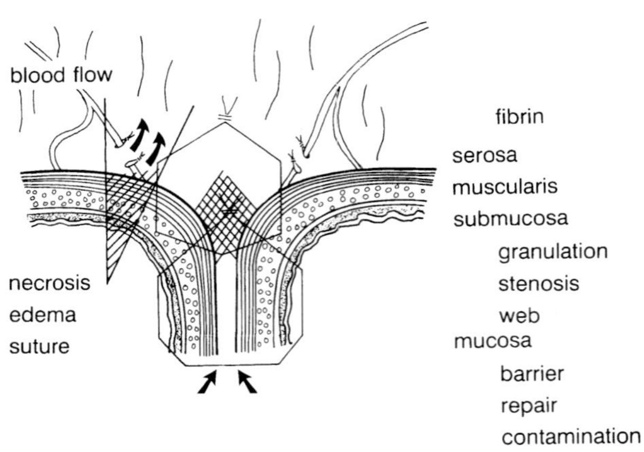

Figure 5.1. Factors influencing the healing of an (inverting) anastomosis.

Registration of Forces

Bursting Pressure

To study the strength of anastomoses determination of bursting pressure is a frequently used parameter. The obtained values give insight in the mechanical properties of the anastomosis in the first week after intervention,[174] in particular on those of the collagen. The way of loading is, however, not physiologic.

Little is known about the direction of forces present in the sutured anastomosis. In anastomoses at the border of the abdomen and thorax considerable forces in axial direction can occur. Transverse force can arise if distension is present.

Pressure is measured by injecting fluid or gas into the lumen until leakage is observed, detected by using a colored fluid or placing the gut under fluid. If volume and pressure are registered with an X–Y recorder compliance can be measured.[24–29]

Measurements with a Tensometer

Stress-strain graphs can be obtained by loading standardized strips into a tensometer.[30] The estimation of thickness in biological samples is difficult,[31,32] some authors have correlated force with the collagen content of the strip.[33–35]

Registration in Vivo

The presence of peristalsis shortly after suturing determines the load of the anastomosis.[20,36,37] Registration can be made by electromyography or by registration of forces with waterproof strain gauges.[32,38–40] Recently pressure measurements above and beneath a rectal anastomosis have been published, indicating that pressure as such is unlikely to cause leakage, the distal values being slightly higher.[41]

Ultrasonography can play a role in detecting the passage of intestinal fluid through an anastomosis as was shown in studies of the passage through the pyloric channel. Post-operative measurements of peristalsis have been made by continuous recording of sounds, the disadvantage is that no correlation between sound and propulsion can be made.

Surface vibration analysis was recently introduced as a new non-invasive technique.

Radiology is frequently used postoperatively to assess anastomotic integrity. One to two weeks after the intervention a watersoluble contrast medium is given into the rectum.

The main biological factors involved in healing of anastomoses are the mucosa, the vascularization, the condition of the connective tissue, the muscular and serosal layers and the intestinal contents, chemically as well as bacteriologically.

Morphology

Mucosa

Light microscopy remains the classical method[175] and the result is dependent on the manner of fixation and staining.[42] Polarization or interference microscopy can be a valuable addition, in particular for judging connective tissue. Histochemistry of mucus formed by the mucous cells after gastrotomy has been used in stomach studies.[26,43] Cutting of paraffin embedded histological samples can be problematic if metal sutures are present. Embedding in methylmethacrylate followed by cutting with a sawmicrotome -as is done in the preparation of skeletal bone- can be recommended.[22] New marking techniques with monoclonal antibodies directed to specific components of the cell, like keratin in mesothelial cells, facilitate identification.[44]

More advanced methods became available, scanning electron microscopy is of interest for surgical research.[176] A rapid and undisturbed repair of epithelium in skin, mucosa or endothelium prevents entrance of exogenous agents into the underlying layers of the damaged organ. The amount of inflammation in these layers plays a role in the final function of the gut.

Repair of skin in burns has been studied intensively, data on repair of intestinal or urinary mucosa are rather scarce.[45,46] Investigation of the bile duct showed that two processes are involved, migration and mitosis of epithelial cells.[47–50] Migration takes place in the first twenty four to forty eight hours, but is strongly influenced by the condition of the underlying structures.[51] If the underlying structure is intact defects are covered in a few days. Repair after complete division of the intestinal wall is always followed by an inflammatory reaction between the two divided ends and shows a parallel with the processes seen after similar conditions in the skin. Only when the granulation tissue reaches a certain stage will covering of the defect be possible,[52] also because at both sides of the wound an increasing amount of new epithelial cells has been formed. The advantage of SEM over light microscopy is that the thin layer of new cells covering a granulating defect can be more easily demonstrated[53–59] (see Figs. 5.2, 5.3).

A diminished function of the mucosa and the mucosal barrier influences the general condition after abdominal surgery.[177–180] The acute changes following stress or administration of drugs like potassium-chloride or salicylic acid are well known. Damage of the mucosa in combination with a changed bacterial flora is one of the factors in portal encephalopathy.[60] Loss of function of the mucosal barrier after thermal injury and wound sepsis was demonstrated in rats.[61,62] Specific factors present in the plasma enhance this effect.

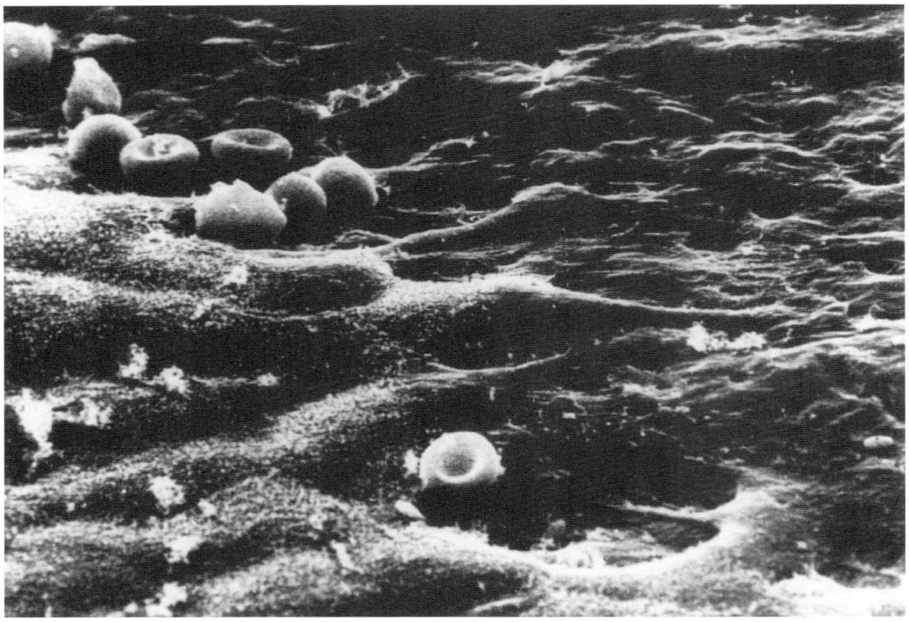

Figure 5.2. Migration of epithelial cells in a superficial defect of the mucosa of the common bile duct of the rabbit.

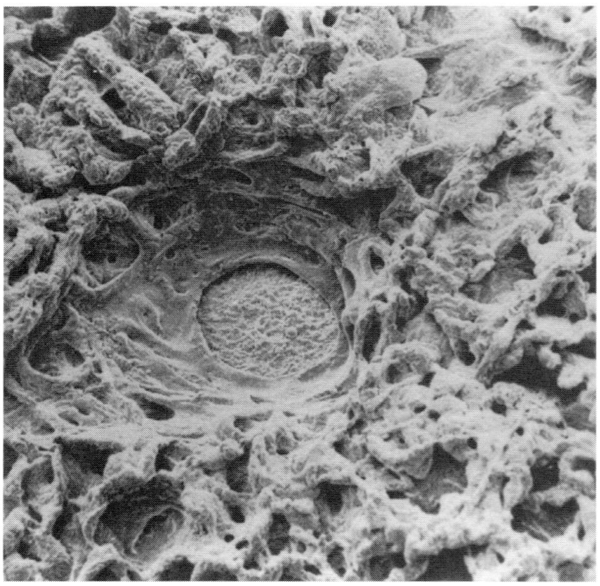

Figure 5.3. Blockade of migration of epithelial cells after a superficial defect in gastric mucosa of the rat.

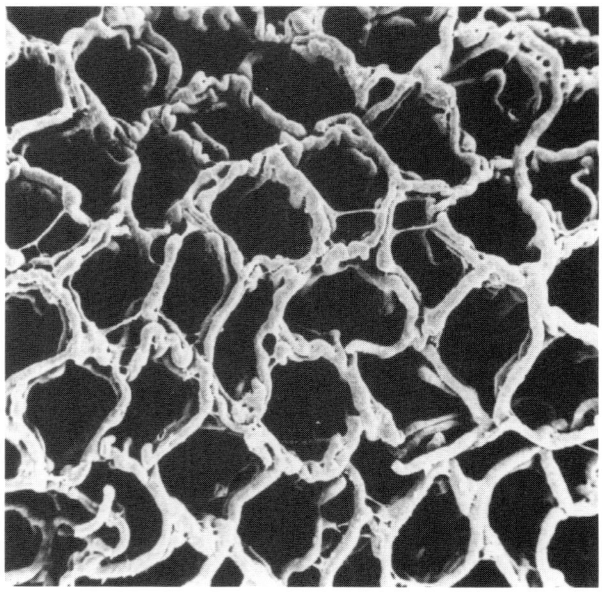

Figure 5.4. Corrosion cast of gastric mucosal vascularisation in the rat following administration of cortisone.

The mucosal layer of the gastrointestinal tract is a dynamic unit, loss of cells is continuously followed by renewal of cells. Investigation is based on the relation between production of new cells, loss of cells,the total amount of cells and the properties of the mucus[181,182] (see Fig. 5.5).

Proliferation can be studied by determination of the mitotic- and DNA-synthesis index, the latter with autoradiography following injection of labelled precursors of DNA.[63–65] Regeneration of cells and formation of the barrier are influenced by growth factors,[183,184] prostaglandins and leucotrines.[66]

Some authors mean that the coming years will be marked by a highly developed pharmacology of wound healing.[185] In skin wounds some externally administrated factors enhanced the healing process.[186–189]

The observation that healing in the fetus can be scarless[190] has made that also drugs like hyaluronic acid can be used in the wounds of adults.[191–194]

In the interpretation of the processes that take place at the borders of the wound, electron microscopy can be of value to differentiate the cells involved and the ultrastructure of the smooth muscle.[67,195]

It is not excluded that after many years the border of two different types of epithelium can lead to malignant degeneration.[68–70] The application of cultured intestinal cells can contribute to a better understanding of this process.[196]

Vascularization

The condition of vascularization is crucial for normal healing as a limited blood supply enhances dehiscence in colorectal surgery. The blood supply is affected by surgical technique; the condition in both ends of the wall will be disturbed by clamps, ligation or coagulation of vessels, necessary to expose the intestine.[47,71–75]

Numerous methods were introduced to determine peroperative intestinal viability. Laser Doppler flowmetry to determine the chance of undisturbed healing appears to be valuable information.[197,198]

Introduction of suture material causes small areas of less vital or necrotic tissue to form which contributes to an increased inflammatory reaction and to forming of new vessels. Angiogenesis and the factors involved are better understood.[199,200]

The type of material can also influence this healing process.[76–78] The much-used catgut could give rise to microabscess, while braided suture threads have a saw-effect, enhancing the introduction of small tissue particles and bacteria during the passage through the tissue. Also enclosure of foreign material present in the intestinal contents cannot be excluded, the chance of infection is less if careful preoperative lavage of the gut is performed.[79,80]

Microradioangiography

To obtain radio-opacity of the vessels, the circulation has to be filled with a diluted suspension of barium under physiological pressure to prevent extravascularization.[73]

The samples are cut in slices with a thickness of 0.5 to 1 mm and X-ray contactprints are made on high-resolution photographic material. Disadvantages are that a crystallographic X-ray tube has to be used and that the method can be used only at post mortem. An advantage is that histological slides can still be made (see Figs. 5.6 and 5.7).

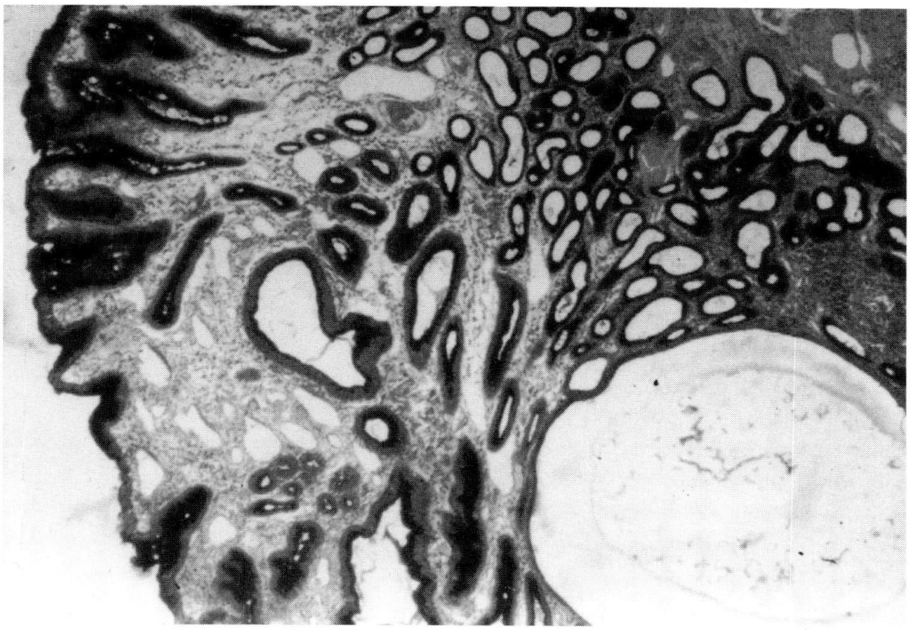

Figure 5.5. Dilatation of gastric glands after gastrojenustomy in the dog.

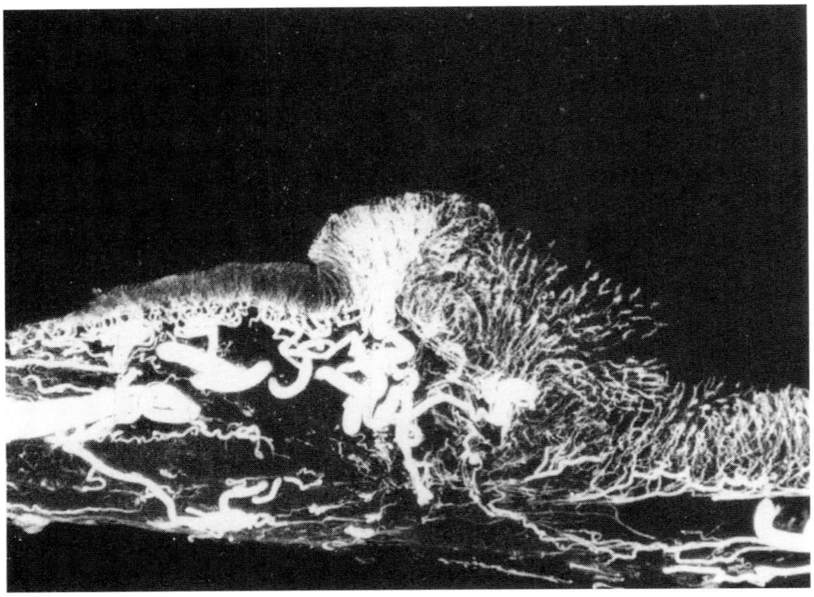

Figure 5.6. Microradioangiogram of gastrojejunual anastomosis in the dog. Right: Gastric corpus; Left: Jejunum. (one week after intervention).

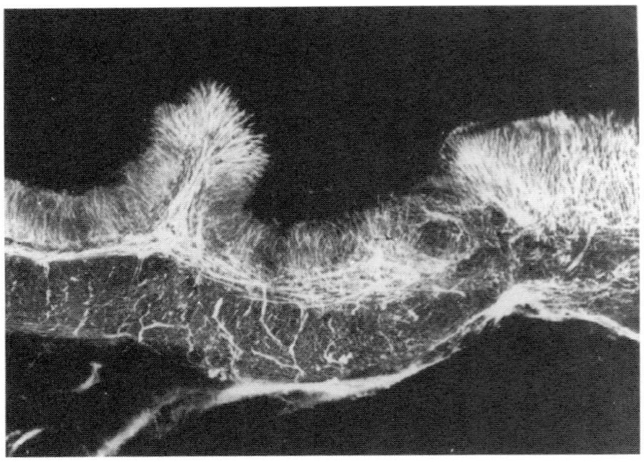

Figure 5.7. Microradioangiogram of gastrojejunal anastomosis. Right: Gastric corpus; left: Jejunum. (one year after the intervention)

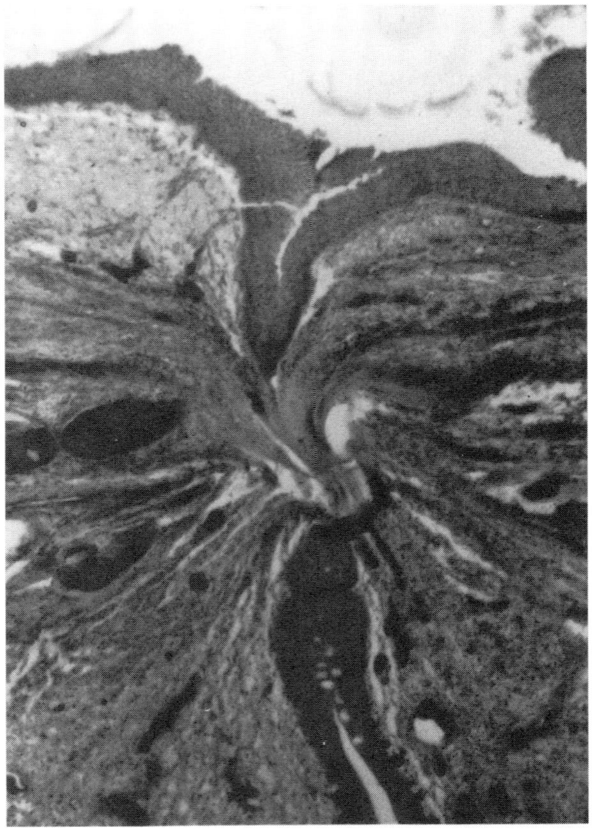

Figure 5.8. Histology of ureteral anastomosis in the dog (48 hours). The epithelial gap is already closed. The compression of the muscles by the staple is visible.

Corrosion Technique

Several authors have used the corrosion technique after filling the vessels with silicon rubber or methylmethacrylate.[81] Casts of the latter can be studied with light or scanning electron microscopy[82–84] (see Fig. 5.4).

In vivo vital microscopy and recently measurement of flow with a laser flowprobe have been performed.[43] Dyes and labelled colloids were used to judge viability of the intestinal wall and lymphatic innervation of the wall. Omentoplasty became actual since the use of microsurgery in reconstructive surgery.[85] It also stimulated research on the omentum to cover defects in the intestinal wall.[201–203]

Submucosa

The submucosal layer of the intestine became very well-known by the work of Halstedt,[86] defining this layer as the skeleton of the intestine.

The layer forms also the main mechanical factor in the first days following operation, for the sutures are placed through this structure. In recent work the meaning of this layer became again of interest for its important role in inflammation and collagen metabolism.[204,205]

Trauma by laparotomy and the intestinal wall is always followed by changes in the properties of the mucosa and collagen metabolism.[87–91]

The structure of the collagen fibers can be studied with lightmicroscopy, but does not give insight in the mechanical properties following resorption in the first stage of woundhealing and remodelling in a later period.[92–96]

Polarization- or interference microscopy facilitates identification of the fibers. Degrading collagen looses its birefringence and the process of remodelling can be followed (see Fig. 5.8).

Biochemical Methods

Determination of metabolites of collagen metabolism (proline and hydroxyproline) is used as a parameter.[63] In experiments performed in rats and rabbits a distinct degradation of collagen has been demonstrated in the colon and the small intestine.[97] The collagenolytic activity is increased three- to four-fold in the first 12 hours after the intervention.[98] In the ileum the changes seem to be limited to the anastomotic region, in the colon also the proximal part is affected. It is not certain whether these changes are related to species, for in particular in rodents a high collagenolytic activity is present.[26,99,100]

Recently attention has been drawn to the fact that in the skin during the process of woundhealing the original fraction of collagen specific for this tissue is changed and repair takes place in the stage of remodelling. Determination of the fraction can be made with immunochemical methods, but they are specific for the species.

In skin wounds collagen synthesis could be followed by analysis of the wound fluid.[101,102]

Collagen content of the submucosal layer can be judged by determination of hydroxyproline and proline by chemical or radioactive methods.[63,103] The general condition can influence collagen metabolism in the wall.[54] Healing is impaired by infection, sometimes followed by intraperitoneal abscess formation. Plasminogen activators from the tissue are involved in this outcome.[104,105]

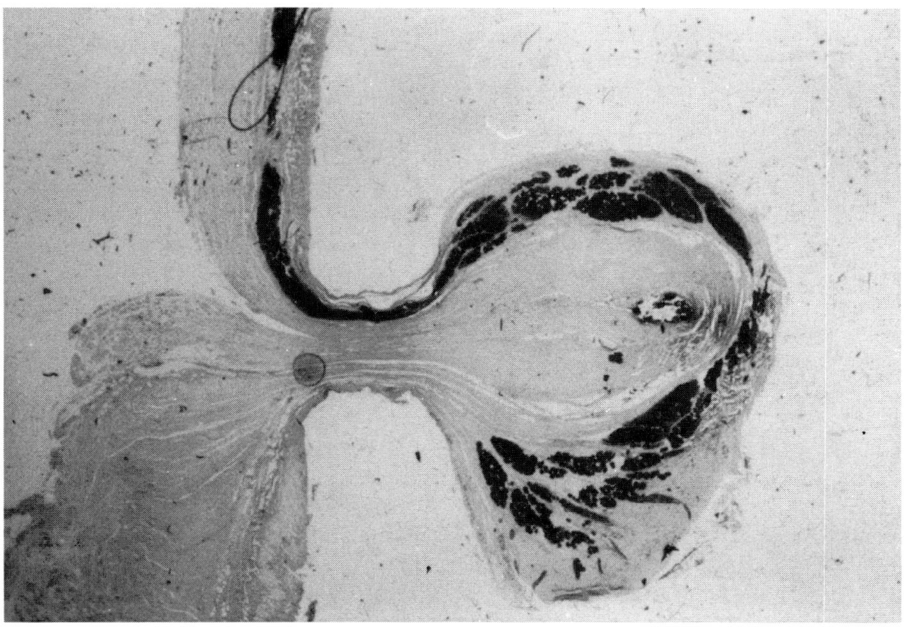

Figure 5.9. Microscopic view of the compression of the canine esophagus 24 hours after introduction of a button-type device. The dark-stained tissue: glands situated at the junction.

Muscular Layers

The disturbed function of local and distant muscular activity was studied intensively in view of postoperative ileus. Clinically microphones were placed on the abdominal wall, but the relation between peristalsis as such and effective propulsion forms a point of discussion.[106]

Recently developed ultrasonographic techniques are worthwhile to investigate in postoperative determination of effective propulsion in combination with pressure measurements or electromyography.[107–109]

In experimental work registration of the electrical activity of the gastrointestinal tract by means of intraluminal or serosal electrodes has been studied in several parts of the tract, in the distal part of the common bile duct[110] and after functional end-to-end anastomoses.

Correlation between peristalsis and electrical activity is difficult in view of the time-consuming interpretation of the registered amplitudes.[111–113] Computer-assisted analysis of both parameters is in progress. A combination of electrodes and strain-gauge force transducers sutured to the serosa has the advantage that the action of the circular or the longitudinal layer can be studied separately combining potential and mechanical action[114] (see Fig. 5.10).

Much experience has been obtained by the use of gastric electromyography and areas with a pacemaker activity have been demonstrated. Resection of intestine is followed by a decrease in activity beneath the anastomosis. In the rectum the activity was more pronounced in the distal part, suggesting the loss of an inhibitory mechanism.

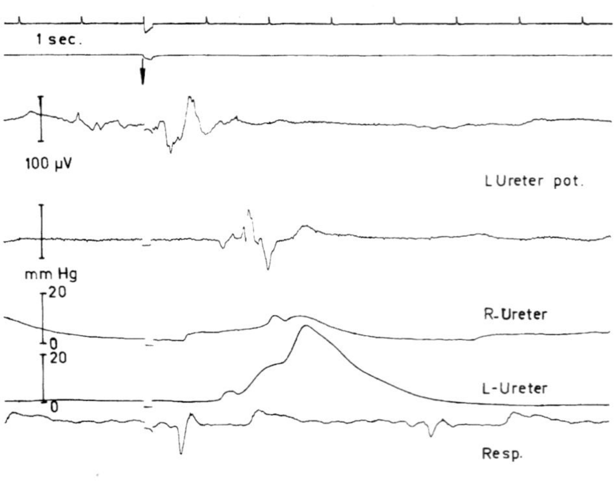

Figure 5.10. Simultaneous measurement of electrical activity, pressure of the right and left ureter and respiration in the dog.

The intestine has a basic electrical rhythm (*BER*), in the stomach the frequency is about three waves per minute. In the proximal small intestine the activity is higher and decreases in distal direction, varying from 12 to 6 waves/min.[111] In the colon a slow wave pattern is present, defined as electrical control activity (*ECA*).[115] During contraction superposed spike potentials occur. During fasting spike potentials are seen intermittently, sometimes only once per hour.

In the sixties it was found that in the fasting dog migrating bursts of spiking could be observed every 90 to 120 minutes. During this activity four phases can be distinguished: **phase I** no spike potentials visible, **phase II** intermittent spike activity, **phase III** strong activity parallel with peristalsis (Migrating Motor Complex – *MMC*) and **phase IV** a short period with decreasing frequency of spikes.

Myoelectrical activity can be measured by intraluminal electrodes or by electrodes fixed on the seromuscular layer. In man measurements have been made by this method in the postoperative period up to one month.[116] It showed that electrical activity was present shortly (twelve to twenty four hours) after the intervention with only small changes in frequency. Comparison of electrical activity of the functional end-to-end anastomosis to a classical end-to-end anastomosis in dogs showed that even after two years only half of the migrating myoelectric complex (MMC) crossed the first-named anastomosis. Clinical experience revealed that side-to-side anastomoses constructed by mechanical instruments, lead to bacterial overgrowth, in contrast to the manual side-to-side technique.

Research on electrical activity of the intestine stimulated the use of electrical pacemakers, but the effect in restoring intestinal activity in the postoperative period was not well proven.

All muscular activity is controlled by the action of **neurogenic** elements in the intestinal wall;[117] histochemistry of these elements and neuropeptide forming cells has been made extensively but literature concerning changes in the structure after surgical intervention is scarce.[118,119] The enteric neural system can be visualized by general stains and stains for enzymes like acetylcholinesterase and catecholamines. Peptides are demonstrated with specific immunohistochemical methods.

In monkeys interruption of the vagal innervation of the colon is followed by a distinct decrease in contractions. The effect of denervation by surgical intervention is a main subject in sphincter surgery.[120]

Serosa

In the history of the intestinal layer the strain after safe-suturing played an important role. In the classical view the continuous suture going through all layers made a water-tight closure, while the following interrupted seromuscular sutures gave adaption of the serosal layers at both ends of the wall, resulting in closure of the split in only a few hours. It was in accordance with the observation that a perforation of the intestine could be treated by covering by adjacent intestines. Formation of fibrin followed by resorption and ingrowth of new connective tissue contributed to mechanical reliability of the new-laid anastomosis. Parts of the gut without a serosal layer show a higher frequency of complications.[206]

Serosa has a tendency to cover the gap in everted intestinal anastomoses in a few days, but this process does not prevent the higher frequency of adhesions seen after the everted closure technique [121–129] (see Fig. 5.11).

To obtain safe intestinal anastomoses the following factors are of importance:

1) The adaptation of both divided walls has to be uniform, either inverting or everting. In one layer anastomoses end-to-end adaption is an ideal situation, which is difficult to obtain (see Figs. 5.12, 5.13).

2) The mechanical force of the suture line has to be sufficient, at least up to the period in which new reliable collagen is formed. In most anastomoses mechanical forces are limited, in the esophago-jejunostomy or esophago-gastrostomy they are more pronounced. Experimental experience showed that coverage of the suture-layer by collagen can act as a splint while at the same time preventing leakage.

Coverage of a sutured layer with non-porous material like silicon rubber or polyethylene, however,

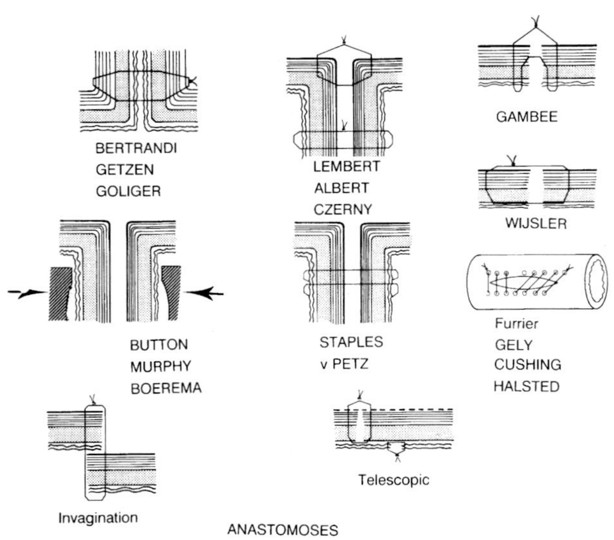

Figure 5.11. Various types of anastomosing techniques.

caused leakage, indicating the important role of the serosa.

3) Vascularization of both divided walls has to be optimal, the presence of suture material gives local non-vascularization, depending on the number of sutures; the force used during knotting contributes to the amount of necrosis. The amount of foreign material (sutures and particles) can influence the extent of inflammation. The histological image obtained after use of certain types of catgut indicated formation of microabscesses. Most sutures are extruded after a certain time, but seromuscular sutures laid during a gastrotomy in dogs could be found without reaction up to 10 years.

4) Careful adaption of the mucosa can enhance early repair of epithelium followed by closure of the entrance for harmful agents and also limits inflammation in the underlying layers. The type of suture material, monofil or braided, can be of influence as was shown in the skin. Braided material can spread bacteria and small particles of tissue by a sawing effect (see Fig. 5.19).

5) Undisturbed repair of the submucosa limits the amount of connective tissue and prevents the formation of strictures. In older people the condition of the tissue can influence the anastomosis, the tissue being more frail and easily damaged. Experiments on healing in older and young animals did not reveal a distinct difference.[207,208].

Hemorrhage can occur in the lumen or abdominal cavity followed by circulation disorders and bleeding inside the suture influences normal healing.

Mechanical Suturing Methods

In the last decade mechanical suturing has been widely used in patients and research projects. Referring to the factors before mentioned, the following advantages are found.

The methods can save time during operation but in some studies this gain was, however, limited. Experience with the instruments is necessary but, in a recent publication the number of complications seen after anastomoses made by experienced surgeons or residents in trainee was equal, in contrast to anastomoses made by hand (see Figs. 5.17, 5.18).

Most experience in Europe and the USA was obtained in colorectal surgery, in Japan more esophago-jejunal anastomoses were made. The adaptation is uniform, but can be influenced by incorrect use of the purse string. Vascularization is scarcely influenced by the form and placement of the staples. In a number of publications mucosal bleeding is one of the complications.[130–134]

Tissue reaction around the inert metal is minimal and the presence of metal probably influences conduction of electrical activity.[135] It can be expected

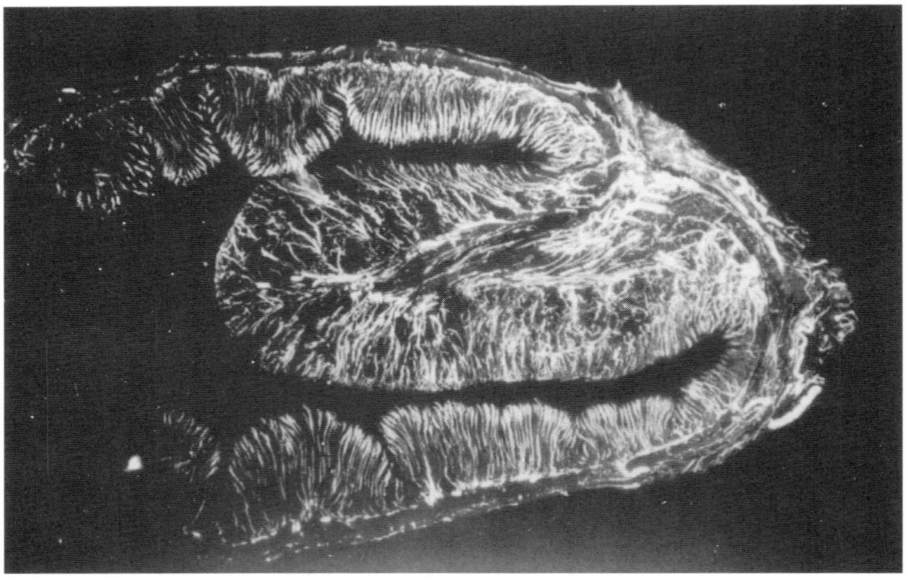

Figure 5.12

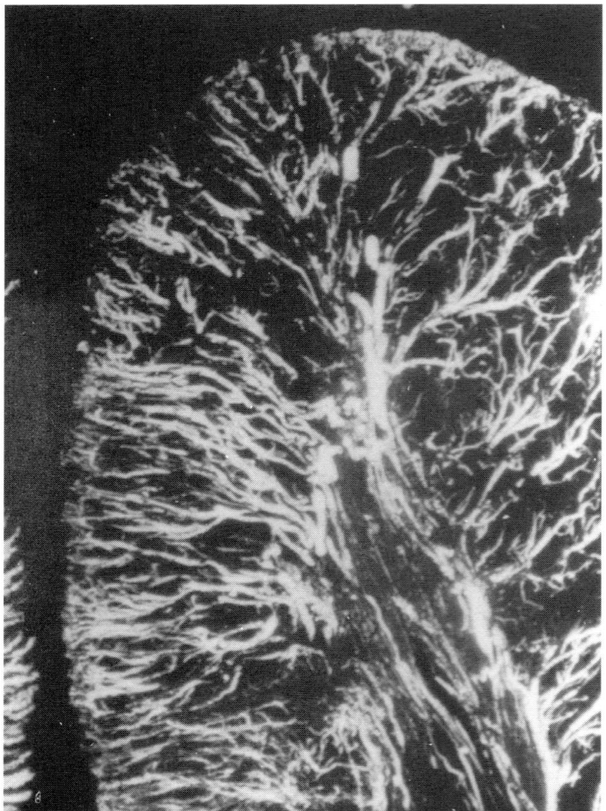

Figure 5.13 Microradiogram of closed duodenal stump (right), with enlargement at the left.

that most materials migrate into the lumen. In urological literature and in experiments stone-formation on staples used in deviation procedures has been described (see Fig. 5.15). Discussion about cost in relation to hospital stay is not yet closed.

Leakage after stapled anastomosing can still occur and was seen up to thirty and sixty days.[136] Damage of the vascularization can be one of the causes (see Figs. 5.14, 5.16).

Narrowing of anastomosis is divided into three groups[137–139]; the membranous narrowing caused by a web comparable with the old term "Wultz", granular stricture and cicatricial narrowing, all influenced by the mentioned factors, with percentages varying from 8 to 36%.[14,45] The new trend in surgery -endoscopic surgery- appears to have an indication in the treatment of strictures following intestinal anastomoses.[80]

With modern stapling instruments inverting anastomoses are made. Narrowing of the lumen is one of the issues in the discussion on the everting and inverting type of anastomoses.[140–142] The discussion concerning the "Nahtwulst" began in the past century[143] and the formation of webs sometimes formed by narrowing due to granulation is described in modern literature. Experiments in growing animals showed that the ridge formed by the staples disappeared after some weeks. After twelve weeks the lumen showed a normal diameter, independent of the method of anastomosing.[144–147] In view of the repair of epithelium on the ridge that is formed by the staples, scanning electron microscopy of the area revealed that in rabbits a defect was still

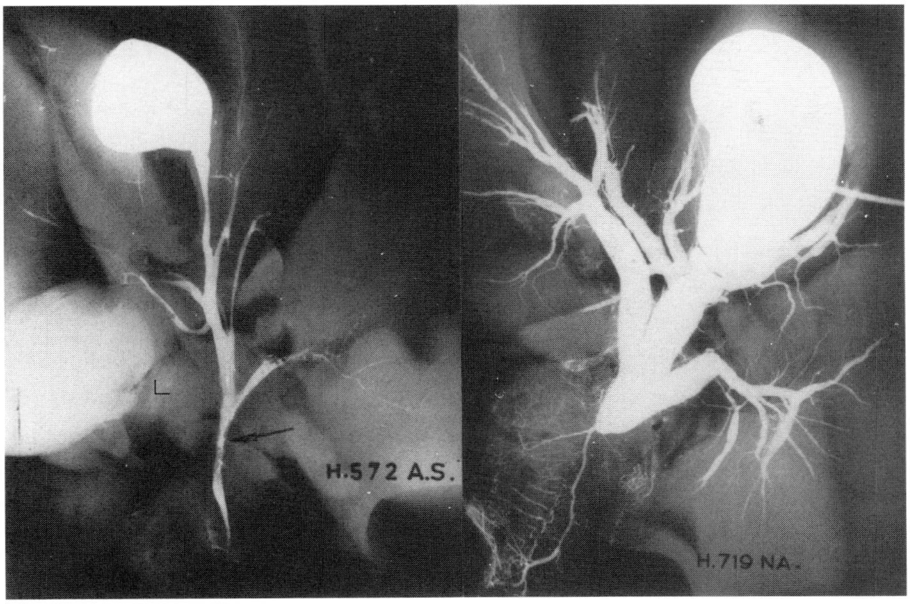

Figure 5.14. Anastomosis of the common bile duct, right with Nakayama's instrument for small vessels, left with Codman's vascular stapler.

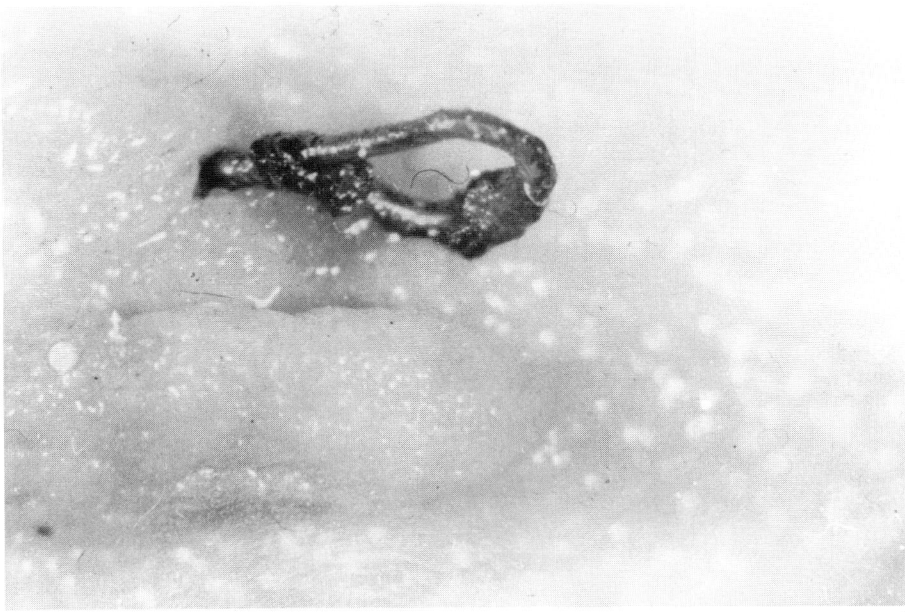

Figure 5.15. Formation of a stone on a metal staple.

present at three to four weeks after the intervention.[94,148–153]

Stapling instruments with adsorbable material are available, but the mass of the staples is rather high[154,155] and the period of degradation is long. The material crumbles and is encapsulated. It means that two unfavorable factors are present: the tissue reaction is more pronounced and foreign material remains present. Little is known of the effect of increased local concentration of polylactic acid in a wound area.[156] The presence of foreign material in a wound, one of the stimuli that can increase inflammatory reaction, is of importance in the development of instruments anastomosing without sutures.

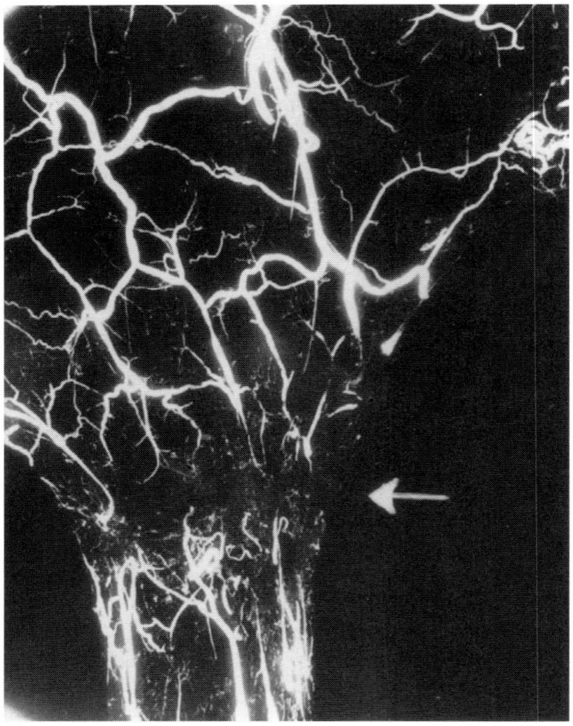

Figure 5.16. Avascular zone in bile duct Stricture.

Mechanical Anastomosing without Sutures

In 1894 Murphy introduced his button after experimentation in animals and stated that the application of his instrument limited the duration of the operation and enhanced an undisturbed healing of the anastomosis.[10] The apparatus acted in the same way as an instrument with silver rings of Denans from 1826.

Several authors confirmed Murphy's results and introduced variations on the original device, including the use of degradable material. The application of the button-type instruments remained, however, limited (see Fig 5.20).

In the treatment of gastric cancer with total gastric resection Boerema used a button driven by a spring to perform the esophago-jejunostomy (see Fig 5.9). The whole procedure could be made via the abdomen.[157,158]

Later on experience was also obtained in emergency treatment of esophageal varices. This indication was based on the blocking of the connection between portal and caval systems. Stapling guns are also used for this indication. The results are, however, largely dependent on the cause of cirrhosis.[159,160] In alcoholics mortality remains high. Jansen[161,162] developed a similar device acting with magnets.

The instruments were used in a certain period, widespread use has never been reached. One of the reasons was that the button had to be driven out by natural means or physically removed and that the passage of intestinal contents was obstructed. The two parts of the device compress the tissue between both halves, followed by necrosis and loosening. Histologically the healing was undisturbed and often complete repair of muscular layers was found, however, strictures could not be excluded.[163,164]

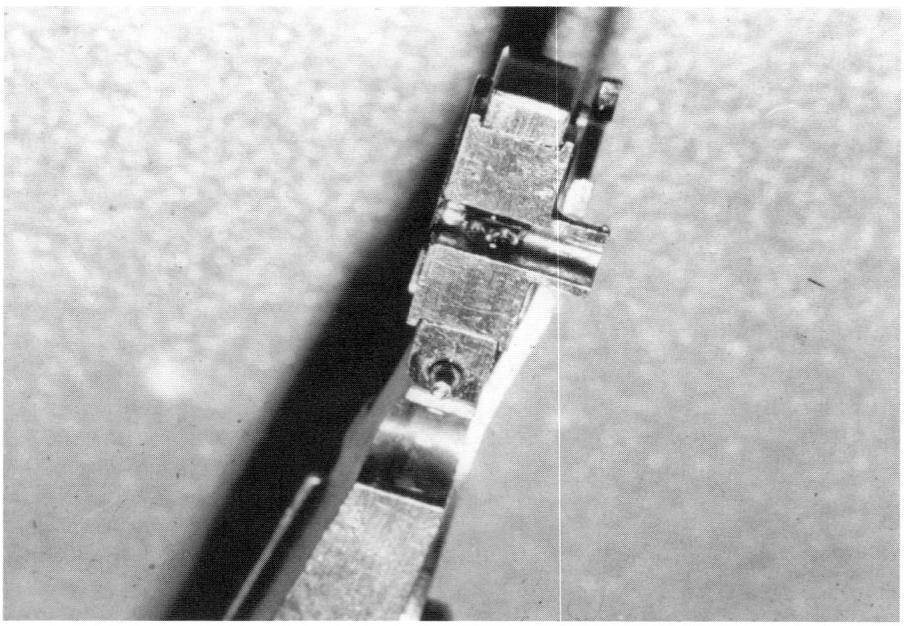

Figure 5.17. Detail of the American vascular stapling instrument (Codman).

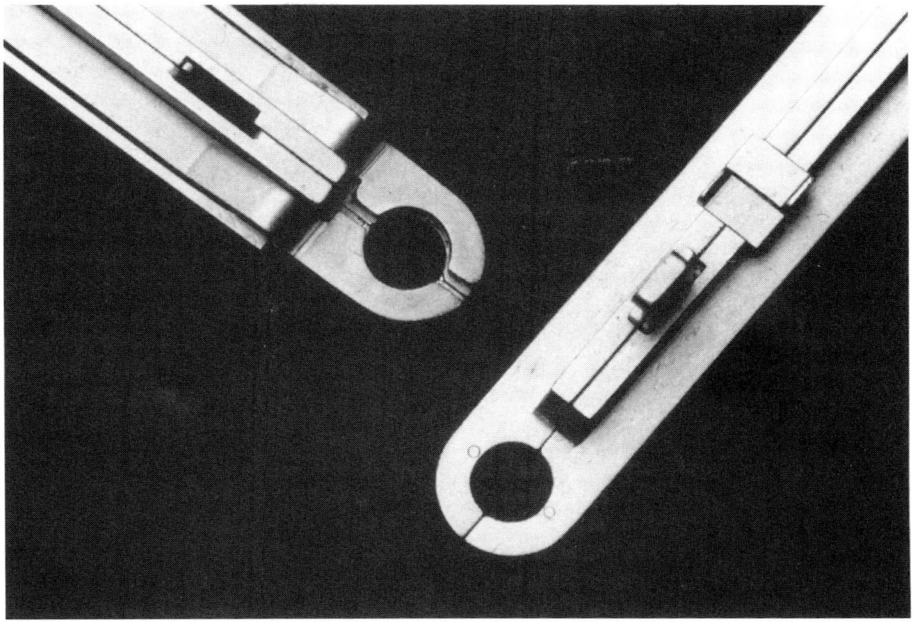

Figure 5.18. Detail of the Canadian vascular stapling instrument.

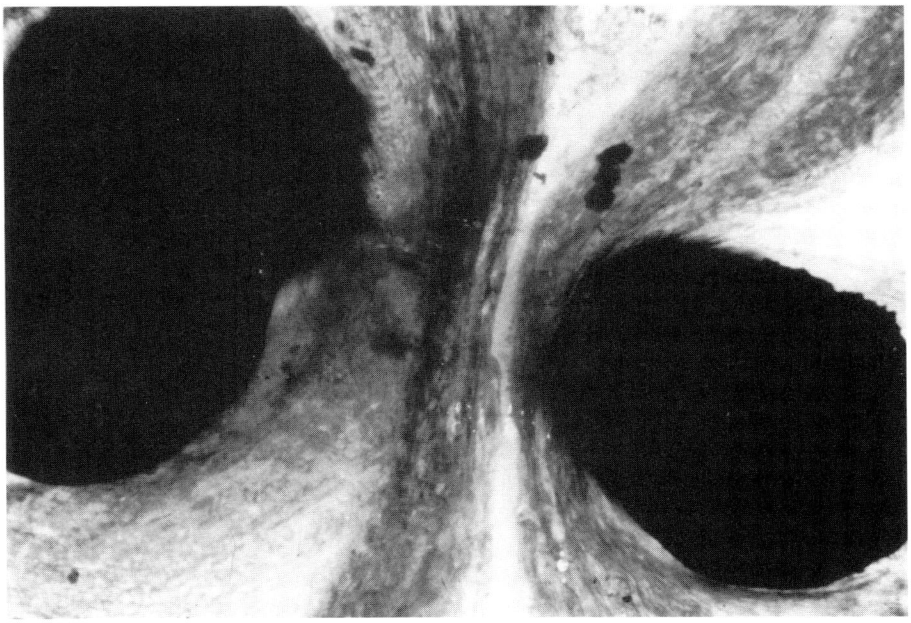

Figure 5.19. Detail of metal staple (USSC) in intestinal anastomosis. The two black dots are the staple cut with a sawmicrotome. A few vessels are visible. (small black dots). No signs of inflammation (three months after closure of duodenum).

The same principle was introduced by other investigators, but this time the button is made of casein or, recently, of a synthetic polymer fragmenting after one to two weeks. (Biofragmentable anastomosing ring-BAR).[165,166] The construction of the devices is comparable to the older types.[13,167] After

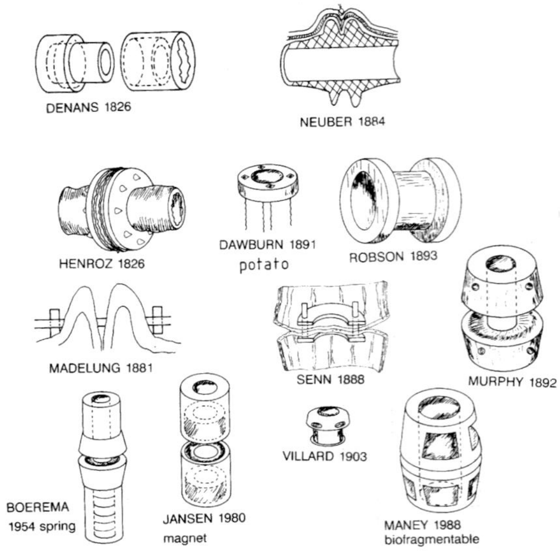

Figure 5.20. Mechanical anastomosing devices.

one year observation in dogs, the healing of the anastomoses proved to be uncomplicated, with a limited amount of growth of connective tissue. Since then several hundreds of patients have been treated successfully with this device.[168–172] In contrast to the Murphy type buttons necrosis of the intestinal wall is limited. The device disintegrates and healing continues without the presence of suture material.

In conclusion anastomosing of organs with a function of transport has been since hundreds of years. Numerous experiments contributed to a decrease in mortality. In spite of the development of ingenious techniques and careful surgery, the mean morbidity in colorectal surgery is still high.

Continued experimental work, standardization of clinical trials and new surgical techniques can play a role in lowering this morbidity in favor of the patient and the cost of health care. The development in future will be increasingly marked by the question if the new method gives real progress or is just a gimmick.[209]

Reference

1. Wangensteen, O. H. (1955). *Intestinal obstruction.* 3rd ed. Springfield: Thomas.
2. Wangensteen, O. H. (1969). Historical aspects of the management of acute intestinal obstruction. *Surgery*, **65**, 363–383.
3. Wangensteen, O. H. and Wangensteen, S. D. (1978). *The rise of surgery. From empiric craft to scientific disciplin.* Folkestone: Dawson.
4. Trimpi, H. D., Khubchandani, I. T., Sheets, J. A. and Stasik, J. J. (1977). Advances in intestinal anastomosis: Experimental study and an analysis of 984 patients. *Diseases Colon & Rectum*, **20**, 107–112.
5. Goligher, J. C. (1980). *Surgery of the anus, rectum and colon* 4th edn. London: Bailliere-Tindall.
6. Oz, M. C. and Forde, K. A. (1990). Endoscopic alternatives in the management of colonic strictures. *Surgery*, **108**, 513–519.
7. Castiglioni, A. (1931). *Histoire de la medecine* Paris: Payot.
8. Haeger, K. (1988). *The illustrated history of surgery.* Gothenburg: AB Nordbok.
9. Nachlas, M. M., Younis, M. T., Roda, C. P. and Wityk, J. J. (1972). Gastro-intestinal motility studies as a guide to postoperative management. *Annals of Surgery*, **175**, 510–522.
10. Murphy, J. B. (1892). Cholecysto-intestinal, gastro-intestinal anastomosis and approximation without structures (original research). *Medical Records*, **42**, 665–676.
11. Hocking, M. P., Carlson, R. G., Courington, K. R. and Bland, K. I. (1990). Altered motility and bacterial flora after functional end-to-end anastomosis. *Surgery*, **108**, 384–391.
12. Petz, A. von (1924). Zur Technik der Magenresektion. Ein neuer Magen-Darmnahtapparat. *Zentralblatt fuer Chirurgie*, **51**, 179–183.
13. Fraser, I. (1982). An historical perspective on mechanical aids in intestinal anastomosis. *Surgery Gynecology & Obstetrics*, **155**, 566–574.
14. Steichen, F. M. and Ravitch, M. M. (1984). *Stapling in surgery.* Chicago: Year book medical publishers Inc.
15. Nakayama, K., Tamiya, T., Yamamoto, K. and Akimoto, S. (1962). A simple new apparatus for small vessel anastomosis (free autograft of the sigmoid included). *Surgery*, **52**, 918–922.
16. Beattie, W. G., Brown, F. N., Devitt, J. E. and Vogelfanger, I. J. (1965). Experimental reconstruction of the cervical esophagus with a free enteric graft. *Canadian Journal of Surgery*, **8**, 221–228.
17. Guthy, E. and Brendel, W. (1969). Stapling devices and their use in surgery. *Progress Surgery*, **7**, 56–113.
18. Irvine, A. H., Leahy, C. F. and Vogelfanger, I. J. (1970). Management of the ureter in renal transplantation. *British Journal of Urology*, **42**, 402–406.
19. Foster, M. E. and Leaper, D. J. (1984). The alimentary tract. In Woundhealing for surgeons, edited by T. E. Bucknall and H. Ellis, pp. 161–179. London: Bailliere-Tindall.
20. Arey, L. B. (1936). Woundhealing. *Physiological Review*, **16**, 327–406.
21. Hess, J. L., McCurnin, D. M., Riley, M. G. and Koehler, K. J. (1981). Pilot study for comparison of chromic catgut suture and mechanically applied staples in enteroanastomoses. *Journal of the American Animal Hospital Association*, **17**, 409–414.
22. Gross, U. M., Powers, D. L. and L. Clemence-Benson L. (1986). General aspects of hard tissue processing pp. 439–460 In *Handbook of biomaterials evaluation*, edited by A. F. von Recum. New York: MacMillan.
23. Bergman, S. M., Sears, H. F. and Javadpour, N. (1978). Complication with mechanical stapling device in creation of ileoconduit. *Urology*, **12**, 71–73.

24. Attisha, R. P. and Smith, A. N. (1969). Pressure activity of the colon and rectum in diverticular disease before and after sigmoid myotomy. *British Journal Surgery*, **56**, 891–894.
25. Winkle Jr., W. van and Hastings, J. C. (1972) Considerations in the choice of suture material for various tissues. *Surgery Gynecology & Obstetrics*, **135**, 113–126.
26. Klopper, P. J. (1973). Some functional and morphological aspects of the healing of a gastrotomy. *Archivum Chirurgicum Neerlandicum*, **25**, 160–171.
27. Kron, I. L., Harman, P. K. and Nolan, S. P. (1984). The measurement of intra-abdominal pressure as a criterion for abdominal re-exploration. *Annals of Surgery*, **199**, 28–30.
28. Bemelman, W. A., Verburg, J., Brummelkamp, W. H. and Klopper, P. J. (1988). A physical model of the intrathoracic stomach. *American Journal of Physiology*, **254**, G168–G175.
29. Bemelman, W. A., Brummelkamp, W. H., Lubbers, M. J., van Royen, E. A. and Klopper, P. J. (1990). Postoperative intermittent positive pressure ventilation decreases duodenogastric reflux after intrathoracic reflux after intrathoracic esophagogastrostomy. *Surgical Research Communications*, **8**, 29–38.
30. Schuurkes, J. A. J., van der Schee, E. J., Grashuis, J. L. and Charbon, G. A. (1978). Calibration of strain gauge force transducers and quantification of gastric-motility. In *Gastrointestinal motility in health and disease*, edited by H. L. Duthie pp. 647–654. Lancaster: MTP.
31. Cronin, K., Jackson, D. S. and Dunphy, J. J. E. (1968). Changing bursting strength and collagen contents of the healing colon. *Surgery Gynecology & Obstetrics*, **126**, 747–753.
32. Daly, J. M., Vars, H. M. and Dudrich, S. J. (1972). Effects of protein depletion on strength of colonic anastomoses. *Surgery Gynecology & Obstetrics*, **134**, 15–21.
33. Posthlethwait, R. W. and Durham, N. C. (1975). Rate of breaking strength loss of absorbable sutures in the stomach. *Surgery*, **78**, 531–533.
34. Tera, H. and Aberg, C. (1976). Tissue holding power to a single suture in different parts of the alimentary tract. *Acta Chirurgica Scandinavia*, **142**, 343–348.
35. Thacker, J. G., Rodeheaver, G., Kurtz, L., Edgerton, M. T. and Edlich, R. T. (1977). Mechanical performance of sutures in surgery. *American Journal Surgery*, **133**, 713–715.
36. Wilbur, B. G. and Kelly, K. A. (1973). Effect of proximal gastric, complete gastric and truncal vagotomy on canine gastric electric activity, motility and emptying. *Annals of Surgery*, **178**, 295–303.
37. Waldhausen, J. H. T., Shaffrey, M. E., Skenderis, B. S., Jones R. S. and Schirmer, B. D. (1990). Gastrointestinal myoelectric and clinical patterns of recovery after laparotomy. *Annals of Surgery*, **211**, 777–785.
38. Price, P. B. (1948). Stress, strain and suture. *Annals of Surgery*, **128**, 408–421.
39. Fleisher, D. R. (1970). On the measurement of intestinal tonus. *Gastroenterology*, **58**, 685–691.
40. Nimmo, W. S. (1984). Effect of anaesthesia on gastric motility and emptying. *British Journal Anaesthesiology*, **56**, 29–35.
41. Burkitt, D. S. and Donovan, I. A. (1990). Intraluminal pressure adjacent to left colonic anastomoses. *British Journal Surgery*, **77**, 1288–1290.
42. Luna, L. G. (1968). *Manual of histologic staining methods of the armed forces institute of pathology*, 3rd edn. New York: McGraw-Hill.
43. Pearse, A. G. E. (1968). *Histochemistry*. Theoretical and applied. 3d edn. London: J. & A. Churchill Ltd.
44. Modlin, I. M., Sank A. and Albert D. (1981). Current aspects of gut hormones. *Journal of Surgical Research*, **30**, 602–618.
45. Peacock, E. E. and van Winkle, W. (1970). *Surgery and biology of wound repair*. Philadelphia: W. B. Saunders Comp.
46. Jennings Jr., W. D., Vaughan, B. L., Moretz, W. H., Rhode, C. M. and Collins, J. G. (1977). The mucosal factor in intestinal anastomoses. *American Surgeon*, **43**, 55–59.
47. Ravitch, M. M. (1975). Observations on the healing of wounds of the intestine. *Surgery*, **77**, 665–673.
48. Ravitch, M. M. (1977). Observations on the healing of wounds of the intestines. *Surgery*, **77**, 665–673.
49. Hattum, A. H. van, James J., Klopper, P. J. and Muller, J. H. (1979). A model for the study of epithelial migration in wound healing. *Virchows Archiv. Cellular Pathology*, **30**, 221–230.
50. Hattum, A. H. van, James J., Klopper, P. J. and Muller, J. H. (1981). Healing of an superficial lesion in the common bile duct of the rabbit. *Netherlands Journal of Surgery*, **33**, 25–31.
51. Schroeder, J. J. M., James, J., Muller, J. H., Everts V. and Klopper, P. J. (1976). The epithelium in healing experimental standard lesions in the gastric mucosa of the rat. *Digestion*, **15**, 397–410.
52. Notigher, F. (1978). Tierexperimentelle Studie mit verschiedenen Nahtmaterialien an der Kolonanastomose der Ratte. *Helvetica Chirurgica Acta*, **45**, 171–173.
53. Brown, R. A., Chin, C. J., Scott, H. J. and Gurd, F. N. (1970). Ultrastructural changes in the canine ileal mucosa cell after mesenteric arterial occlusion. *Archives of Surgery*, **101**, 290–297.
54. Dougerthy, S. H. (1984). Role of enterococcus in intraabdominal sepsis. *American Journal Surgery*, **148**, 308–311.
55. Miller, T. A. (1988). Gastroduodenal mucosal defense: Factors responsible for the ability of the stomach and duodenum to resist injury. *Surgery*, **103**, 389–397.
56. Stein, H. J., Esplugues, J., Brendan, J. R., Whittle, J. R., Bauerfeind, P., Hinder, R. A. and Blum, A. L. (1989). Direct cytotoxic effect of oxygen radicals on the gastric mucosa. *Surgery*, **106**, 318–324.
57. Thompson, J. S., Saxena, S. K., Geraton, C., Schultz, G. and Sharp, J. G. (1989). The effect of the route of delivery of urogastrone on intestinal regeneration. *Surgery*, **106**, 45–51.
58. Jones, W. G., Minei, J. P., Barber, A. E., Rayburn, J. L., Fahey, T. J. and Shires, G. T. (1990). Bacterial translocation and intestinal atrophy after thermal injury

and burn wound sepsis. *Annals of Surgery*, **211**, 399–405.
59. Mooney, D.P., Reilly, M. O. and Gamelli, R. L. (1990). Tumor necrosis factor and wound healing. *Annals of Surgery*, **211**, 124–129.
60. Barie, P. S., Christou, N. V., Dellinger, E. P., Rout, W. R., Stone, H. H. and Waymack, J. P. (1980). Pathogenicity of the enterococcus in surgical infections. *Annals of Surgery*, **212**, 155–159.
61. Carter, E. A., Tompkins, R. G., Schiffrin, E. and Burke, J. F. (1990). Cutaneous thermal injury alters macro-molecular permeability of rat small intestine. *Surgery*, **107**, 335–341.
62. Houston, K. A., McRitchie, D. I. and Rotstein, O. D. (1990). Tissue plasminogen activator reverses the deleterious effect of infection on colonic wound healing. *Annals of Surgery*, **211**, 130–135.
63. Wirl, G. (1977). Extractable collagenase and carcinogenesis of mouse skin. *Connective Tissue Research*, **5**, 171–178.
64. Willems, G. (1979). Cell population kinetics in the mucosa of the gastrointestinal tract. In *Scientific basis of gastroenterology*, edited by H. L. Duthie and K. G. Wormsley. Edinburgh: Churchill Livingstone.
65. Willems, G. (1981). Cellular kinetics and gastric mucosal disease. In *Magen und Magenkrankheiten*, edited by W. Domschke und K. G. Wormsley, pp. 195–219. Stuttgart: Thieme.
66. Pihan, G. and Szabo, S.(1988). Microvascular injury and the role of leukotrienes and prostaglandines in acute mucosal damage and protection. In *Prostaglandins and leukotrienes in gastro-intestinal diseases*, edited by W. Domsche and H. G. Dammann, pp. 23–29. Berlin: Springer Verlag.
67. Gabella, G. (1988). Structure of smooth muscle pp. 33–45. In *An illustrated guide to gastrointestinal motility*, edited by D. Kumar and S. Gustavsson. Chichester: Wiley and Sons.
68. Stewart, R. (1973). Influence of malignant cells on healing of colonic anastomoses: experimental observations. *Proceedings Royal College Medicine*, **66**, 1089–1091.
69. Gadacz, T. R., McFadden, D. W., Gabrielson, E. W., Ullah, A. and Berman, J. J. (1990). Adenocarcinoma of the ileostomy: The latent risk of cancer after colectomy for ulcerative colitis and familial polyposis. *Surgery*, **107**, 698–703.
70. Lohmuller, J. L., Pemberton, J. H., Dozois, R. R., Ilstrup D. and van Heerden J. (1990). Pouchitis and extraintestinal manifestations of inflammatory bowel disease after ileal-anal pouch anastomosis. *Annals of Surgery*, **211**, 622–629.
71. Benjamin, H. B. and Becker, A. B. (1959). A vascular study of the small intestines. *Surgery Gynecology & Obstetrics*, **108**, 134–140.
72. Delaney, J. (1963). Collateral bloodflow to a devascularized segment of small intestine. *Surgery Gynecology & Obstetrics*, **116**, 494–496.
73. Abramowitz, H. B. and McAlister, W. H. (1969). A comparative study of small bowel anastomoses by angiography and microangiography. *Surgery*, **66**, 564–569.
74. Schrock, T. R., Deveney, C. W. and Dunphy, J. E. (1973). Factors contributing to leakage of colonic anastomoses. *Annals of Surgery*, **177**, 513–518.
75. Oestmann, J., Langenbruch, K., Meier, H. J. *et al.* (1985). Nachweis von Insuffizienzen enteroenteraler Anastomosen nach Gastrektomien durch Untersuchungen met wasserloeslichen Kontrastmitteln-Zeitpunkt und Untersuchungsqualitaet. *Roentgen Blaetter*, **38**, 356–358.
76. Deveney, K. E. and Way, L. W. (1977). Effect of different absorbable sutures on healing of gastrointestinal anastomoses. *American Journal Surgery*, **133**, 86–94.
77. Forrester, J. C. (1980). Sutures and woundrepair pp. 194–207. In *Wound healing and wound infection*, edited by T. K. Hunt. New York: Appleton-Century Crofts.
78. Novell, J. R. and Lewis, A. A. M. (1990). Peroperative observation of marginal artery bleeding: a predictor of anastomotic leakage. *British Journal Surgery*, **77**, 137–138.
79. Manz, C. W., LaTendresse, C. and Sako, Y. (1970). The detrimental effects of drains on colonic anastomoses: an experimental study. *Diseases Colon & Rectum*, **13**, 17–25.
80. Levy, A. G., Benson, J. W., Hewlett, E. L., Herdt, J. R., Doppman, J. L. and Gordon, R. S. (1976). Saline lavage: a rapid effective and acceptable method for cleansing the gastrointestinal tract. *Gastroenterology*, **70**, 157–161.
81. Gayet, B., Breil, P. and Fekete, F. (1991). Mechanical sutures in perforation of the thoracic esophagus as a safe procedure in patients seen late. *Surgery Gynecology & Obstetrics*, **172**, 125–128.
82. Gelder, P. A. van and Klopper, P. J. (1979). Healing of microvascular arterial anastomoses, as seen on corrosion casts by scanning electron microscopy. *Plastic and Reconstructive Surgery*, **64**, 59–64.
83. Smith, C. R., Cokelet, G. R., Adams, J. T. *et al.* (1981). Vascularity of gastrointestinal staple lines demonstrated with silicone rubber injection. *American Journal Surgery*, **142**, 563–566.
84. Gulik, T. M. van, Lygidakis, N. J., van Marle, J., James, J., Klopper, P. J. and van der Heyde, M. N. (1988). A scanning electron microscopic study of the common bile duct epithelium in the rabbit during experimental biliary tract obstruction. *Hepatogastroenterology*, **35**, 215–218.
85. Twisk, R. van, Claessens, R., Koeijers, V., Koenders, E. and Klopper, P. J. (1989). The lymphatic transport of colloid in omentum and skintransplants as a pedicled and free flap. In: *Microvascular tissue transfer*, pp. 233–238. Wesel: Quintisens Publ.
86. Halsted, W. S. (1891). Circular suture of the intestine, an experimental study. *American Journal Medical Sciences*, **94**, 436–461.
87. Lord, M. G., Broughton, A. C. and Williams, H. T. G. (1978). A morphologic study on the effect of suturing the submucosa of the large intestine. *Surgery Gynecology & Obstetrics*, **146**, 211–216.
88. Hunt, T. K. (1980). *Wound healing and wound infection: Theory and Surgical practice*. New York: Appleton .
89. Irvin, T. T. (1981). *Wound healing. Principles and practice* London: Chapman and Hill.

90. Jansen, A., Becker, A. E., Brummelkamp, W. H., Keeman, J. N. and Klopper, P. J. (1981). The importance of the apposition of the submucosal intestinal layers for primary woundhealing of intestinal anastomosis. *Surgery Gynecology & Obstetrics*, **152**, 51–58.
91. Peacock, Jr. E. E. (1984). *Wound repair*, 3d edn. Philadelphia: Saunders.
92. Klein, L. (1976). Collagen structure and metabolism.In *Symposium on basic science in plastic surgery*, edited by T. J. Krizek and J. E. Hoopes, pp. 80–86. Saint Louis: Mosby Comp.
93. Lord, M. G., Valies, P. and Broughton, A. C. (1978). A morphologic study of submucosa of large intestine. *Surgery Gynecology & Obstetrics*, **145**, 55–60.
94. Weil, P. H. and Scherz, H. (1981). Comparison of stapled and hand-sutured gastrectomies. *Archives of Surgery*, **116**, 14–20.
95. Taylor, D. E. M., Whamond, J. S. and Penhallow, J. E. (1987). Effects of haemorrhage on wound strength and fibroblast function *British Journal Surgery*, **74**, 316.
96. Post, S., Betzler, M., von Ditfurth, B., Schuermann, G., Kuepers, P. and Herfarth, C. (1991). Risks of intestinal anastomoses in Crohn's disease. *Annals of Surgery*, **213**, 37–42.
97. Chowcat, N. L., Savage, F. J., Hembry, R. M. and Boulos, P. M. (1988). Role of collagenase in colonic anastomoses: a reappraisal. *British Journal Surgery*, **75**, 330–334.
98. Hawley, P. R., Page, F. W., Hunt, T. K. *et al.* (1970). Collagenase activity in the gastrointestinal tract. *British Journal Surgery*, **57**, 896–900.
99. Stappen, J. W. J. van der, Hendriks, T. and de Boer, H. H. M. (1989). Collagenolytic activity extracted from intestinal anastomoses of the rat. *Matrix*, **9**, 238–243.
100. Thompson, J. S., Bragg, L. E. and West, W. W. (1990). Serum enzyme levels during intestinal ischemia. *Annals of Surgery*, **211**, 369–373.
101. Jiborn, H., Ahonen, J. and Zederfeldt, B. (1980). Healing of experimental colonic anastomoses. III. Collagen metabolism in the colon after left colonic resection. *American Journal Surgery*, **139**, 398–405.
102. Haukipuro, K., Melkko, J., Ristelli, L., Kairaluoma, M. I. and Ristelli, J. (1991). Synthesis of type 1 collagen in healing wounds in humans. *Annals of Surgery*, **213**, 75–80.
103. Rojkind, M. and Gonzalez, E. (1974). An improved method for determining specific radioactivities of proline 14C and hydroxyproline 14C in collagen and non collagenous proteins. *Analytical Biochemistry*, **57**, 1–7.
104. Irvin, T. T. (1976). Collagen metabolism in infected colonic anastomoses. *Surgery Gynecology & Obstetrics*, **143**, 220–224.
105. Khoury, G. A. and Waxman, B. P. (1983). Large bowel anastomoses. I. The healing process and sutured anastomoses. A review. *British Journal Surgery*, **70**, 61–63.
106. Farrar, J. J. and Ingelfinger, F. J. (1955). Gastrointestinal motility as revealed by study of abdominal sounds. *Gastroenterology*, **29**, 789–799.
107. Duthie, H. L. (1974). Electrical activity of the gastrointestinal smooth muscle. *Gut*, **15**, 669–681.
108. Cowles, V. E., Condon, R. E., Schulte, W. J., Woods, J. H. and Sillin, L. F. (1978). A quarter wheatstone bridge strain gauge force transducer for recording gut motility. *American Journal Digestive Disease*, **23**, 936–939.
109. Cullen, P. T., Storey, B. E., Cuschieri, A. and Campbell, F. C. (1989). Detection of clustered gastrointestinal contractions in partial intestinal obstruction by surface vibration analysis. *Annals of Surgery*, **210**, 234–238.
110. Coelho, J. C. U., Gouma, F. J., Moody, F. G. and Schlegel, J. F. (1986). Effect of feeding on myoelectric activity of the sphincter of Oddi and the gastrointestinal tract in the opossum. *Digestive Diseases Sciences*, **31**, 202–207.
111. Alvarez, W. C. and Mahoney, L. J. (1922). Action currents in stomach and intestine. *American Journal Physiology*, **58**, 476–493.
112. Code, C. F., Szurszowski, J. H. and Kelly, K. A. (1968). A concept of control of gastrointestinal motility. In: *Handbook of Physiology*, edited by C. G. Code and W. Heidel, vol. 5, sect. 6. Washington: Am. Physiol. Soc.
113. Calabuig, R., Seggerman, R. E., Weems, W. A., Weisbrodt, N. W. and Moody, F. G. (1990). Gallbladder and gastrointestinal motility after hemorrhagic shock. *Surgery*, **107**, 568–573.
114. Esser, M. J., Cowles, V. E., Robinson, J. C., Schulte, W. J., Gleysteen, J. J. and Condon, R. E. (1989). Effects of vagal cryo-interruption on colon contractions in monkeys. *Surgery*, **106**, 139–146.
115. Toouli, J., Dodds, W. J., Honda, R., *et al.* (1983). Motor function of the opossum sphincter of Oddi. *Journal Clinical Investigation*, **71**, 208–220.
116. Soper, N. J., Sarr, M. G. and Kelly, K. A. (1990). Human duodenal myoelectric activity after operation and with pacing. *Surgery*, **107**, 63–68.
117. O'Malley, V. P., Keyes, D. M., Cannon, J. P. and Postier, R. G. (1985). Longitudinal myotomy: a new reservoir for use with ileoanal anastomosis. *Current Surgery*, **42**, 113–117.
118. Janssens, J., de Wever, J., van Trappen, G. and Hellemans, J. (1976). Peristalsis in smooth muscle oesophagus after transection and bolus deviation. *Gastroenterology*, **71**, 100–1009.
119. Christensen, J. (1988). The enteric nervous system pp. 9–32. In *An illustrated guide to gastrointestinal motility*, edited by D. Kumar and S. Gustavsson. Chichester: Wiley and Sons.
120. Bucknall, T. E. and Ellis, H. (1984). *Wound healing for surgeons*. London: Bailliere Tindall.
121. Gambee, L. P., Garnjobst, W. and Hardwick, C. E. (1956). Ten years experience with a single layer anastomosis on colon surgery. *American Journal Surgery*, **92**, 222–225.
122. Hepp, J. (1957). Statistiques des sutures digestives en un plan. *Memoires Academie de Chirurgie*, **83**, 399–404.
123. Canalis, F. and Ravitch, M. M. (1968). Study of healing of inverting and everting intestinal anastomosis. *Surgery Gynecology & Obstetrics*, **126**, 109–114.

124. Rusca, J. A., Bornside G. H. and Cohn Jr, I. (1969). Everting versus inverting gastrointestinal anastomoses: Bacterial leakage and anastomotic disruption. *Annals of Surgery*, **169**, 727–735.
125. Trueblood, H. W., Nelsen, T. S., Kohatsu, S., Oberhelman, H. A. (1969). Investigative surgery. Wound healing in the colon. Comparison of inverted and everted closures. *Surgery*, **65**, 919–930.
126. Dunphy, J. E. (1970). The cut gut. *American Journal Surgery*, **119**, 1–8.
127. Wise, L., McAlister, W., Stein, T. and Schock, P. (1975). Studies on the healing of anastomoses of small and large intestines. *Surgery Gynecology & Obstetrics*, **141**, 190–194.
128. Goligher, J. C. (1976). Visceral and parietal suture in abdominal surgery. *American Journal Surgery*, **131**, 130–140.
129. Fazio, V. W., Sgalandiuk, H., Jagelman, D. G. and Lavery, I. C. (1989). Strictureplasty in Crohn's disease. *Annals of Surgery*, **210**, 621–625.
130. Takasaki, T., Kobayashi, S., Muto, H., Suzuki, S., Harada, M. and Nakayama, K. (1977). Transabdominal esophageal transection by using a suture device in cases of esophageal varices. *International Surgery*, **62**, 426–428.
131. Cooperman, M., Fabri, P. J., Martin, Jr. E. W., *et al.* (1980). EEA stapling for control of bleeding esophageal varices. *American Journal Surgery*, **140**, 821–824.
132. Penninckx, F. M., Kerremans, R. P. and Geboes, K. J. (1984). The healing of single and double-row stapled circular anastomoses. *Diseases Colon & Rectum*, **27**, 714–719.
133. McGinn, F. P., Gartell, P. C., Glifford, P. C. and Brunton, F. J. (1985). Staples or sutures for low colorectal anastomoses: a prospective randomized trial. *British Journal Surgery*, **72**, 603–605.
134. Kataoka, M., Masaoka, A., Hayashi, S., Honda, H., Hotta, T., Niwa, T. and Honda, K. (1989). Problems associated with the EEA stapling technique for esophagojejunostomy after total gastrectomy. *Annals of Surgery*, **209**, 99–104.
135. Griffen, F. D., Knight, C. D., Whitaker J. M. and Knight, Jr. C. D. (1990). The double stapling technique for anterior resection: results, modifications and observations. *Annals of Surgery*, **211**, 745–752.
136. Klopper, P. J. (1967). Experimental anastomoses of the ureter. *Archivum Chirurgicum Neerlandicum*, **18**, 319–331.
137. Polglase, A. L., Hughes E. S. R., McDermott, F. T. *et al.* (1981). A comparison of end to end s tapled and suture colorectal anastomosis in the dog. *Surgery Gynecology & Obstetrics*, **152**, 792–796.
138. Ravitch, M. M., Rivarola, A. and van Grov, J. (1967). Studies of intestinal healing 1. Preliminary studies of the mechanism of healing of everting intestinal anastomosis. *John Hopkins Medical Journal*, **121**, 343–348.
139. Ravitch, M. M., Canalis, F., Weinshelbaum, A. and McCormick, J. (1967). Studies on intestinal healing: III Observations on everting intestinal anastomoses. *Annals of Surgery*, **166**, 670–680.
140. Ravitch, M. M. and Steichen, F. M. (1984). Surgical stapling techniques. *Surgical Clinics North America*, **64**, no. 3.
141. Goligher, J. C., Lee, P. W. R., McFie, J., Simpkins, K. C. and Lintott, D. J. (1977). Experience with the Russian model 249 suture gun for rectal anastomoses. *Surgery Gynecology & Obstetrics*, **148**, 517–522.
142. Tagart, R. E. B. (1981). Colorectal anastomoses: factors influencing success. *Journal Royal Society of Medicine*, **74**, 111–118.
143. Wong, J., Cheung, H., Lui, R., Fan, Y., Smith, W. and Siu, K. F. (1987). Esophagogastric anastomosis performed with a stapler. The occurrence of leakage and stricture. *Surgery*, **101**, 408–412.
144. Didolkar, M. S., Reed, W. P., Elias, E. G., *et al.* (1986). A prospective randomized study of sutured versus stapled bowel anastomoses in patients with cancer. *Cancer*, **57**, 456–460.
145. Fischer, M. G. (1976). Bleeding from stapler anastomosis. *American Journal of Surgery*, **131**, 745–747.
146. Hemreck, A. S. and Crawford, D. G. (1976). The esophageal anastomotic leak. *American Journal Surgery*, **132**, 794–798.
147. Chassin, J. L., Rifkind, K. M., Sussman, B., Kassel, B., Fingaret, A., Drager, A. and Chassin, P. S. (1978). The stapled gastrointestinal anastomosis: Incidence of postoperative complications compared with the sutured anastomosis. *Annals of Surgery*, **188**, 689–696.
148. Karlstrom, L. H., Soper N. J., Kelly, K. A. and Phillips, S. F. (1989). Ectopic jejunal pacemakers and enterogastric reflux after Roux gastrectomy: Effect of intestinal pacing. *Surgery*, **106**, 486–495.
149. Livaditis, A. and Ivemark, B. (1969). Esophageal anastomosis in piglets. *Scandinavian Journal Thoracic Cardiovascular Surgery*, **3**, 174–180.
150. Irvin, T. T. and Goligher, J. C. (1973). Aetiology of disruption of intestinal anastomoses. *British Journal Surgery*, **60**, 461–464.
151. Wassner, J. D., Yohai, E. and Heimlich, H. J. (1977). Complications associated with the use of gastrointestinal stapling devices. *Surgery*, **82**, 3950–399.
152. Waxman, B. P. (1983). Large bowel anastomoses II. The circular staplers: a review. *British Journal Surgery*, **70**, 64–67.
153. Gordon, P. H. and Vasilevsky, C. A. (1984). Experience with stapling in rectal surgery. *Surgical Clinics North America*, **64**, 555–566.
154. Fazio, V. W. (1988). Cancer of the rectum-sphincter-saving operation. Stapling techniques. *Surgical Clinics North America*, **68**, 1367–1382.
155. Greven, A. T. (1988). Mechanische Darmanastomosen. *Thesis*, University of Amsterdam, Amsterdam.
156. Hoskins, I. A., Ordorica S. A., Frieden, F. J. and Young, B. K. (1991). Performance of caesarian section using absorbable staples. *Surgery Gynecology & Obstetrics*, **172**, 108–112.
157. Letwin, E. R. (1975). Evaluation of polyglycolic acid sutures in colonic anastomoses. *Canadian Journal Surgery*, **18**, 30–33.
158. Boerema, I. (1954). The technique of our method of transabdominal total gastrectomy in cases of gastric cancer. *Archivum Chirurgicum Neerlandicum*, **6**, 95–111.

159. Boerema, I., Klopper, P. J., Holscher, A. A. (1970). Transabdominal ligation-resection of the esophagus in case of bleeding esophageal varices. *Surgery*, **76**, 409–413.
160. Johnston, G. W. (1982). Six years experience of oesophageal transection for oesophageal varices, using a circular stapling gun. *Gut*, **23**, 770–773.
161. Wexler, M. J. and Stein, B. L. (1990). Nonshunting operations for variceal hemorrhage. *Surgical Clinics North America*, **70**, 425–448.
162. Jansen, A., Keeman, J. N., Davies, G. A. and Klopper, P. J. (1980). Early experiences with magnetic rings in resection of the distal colon. *The Netherlands Journal Surgery*, **32**, 20–27.
163. Jansen, A., Brummelkamp, W. H., Davies, G. A. G. and Klopper, P. J. (1981). Clinical applications of magnetic rings in colorectal anastomosis. *Surgery*, **153**, 537–545.
164. Koomen, A. and Klopper, P. J. (1977). Healing of the oesophagus after total ligation with the Boerema button and a similar apparatus acting with magnetic force: an experimental study. *Archivum Chirurgicum Neerlandicum*, **29**, 217–228.
165. Maney, J. W., Katz, A. R., Li, L. K., Pace, W. G. and Hardy, TH. G. (1988). Biofragmentable bowel anastomosis ring: Comparative efficacy studies in dogs. *Surgery*, **103**, 56–62.
166. Hopcroft, S. C. (1972). The use of an absorbable prosthesis in intestinal colonic anastomosis. *The Medical Journal of Australia*, **1**, 118–121.
167. Hardy Jr., T. G., Pace, W. G., Maney, J. W., Katz, A. R. and Kaganov, A. L. (1985). A biofragmentable ring for sutureless bowel anastomosis. An experimental study. *Diseases Colon & Rectum*, **28**, 484–490.
168. Verschuyl, M. A. (1965). Non suture anastomosis of the small intestine, an experimental study. *Archivum Chirurgicum Neerlandicum*, **7**, 95–109.
169. McCue, J. L. and Philips, R. K. S. (1991). Sutureless intestinal anastomoses Review. *Brit. J. Surg.*, **78**, 1291–1296.
170. Corman M. L. Prager, E. D. Hardy, T. G. Bubrick, M. F. (1989). Valtrac study group. Comparison of the Valtrac Biofragmentable anastomis ring with conventional suture and stapled anastomosis in colon surgery. *Dis. Col. Rect.*, **32**, 183–187.
171. Gullischen, R., Ovaska, J., Rantala, A., and Havia, T. (1992). Small bowel anastomis with the biofragmentable anastomosis ring and manual suture *W. J. Surg.*, **16**, 1006–1009.
172. Gullischen, R., Ovaska, J., and Rantala, A. (1992). Colonic anastomosis using the biofragmentable anastomotic ring and manual suture:a prospective, randomized study. *Brit. J. Surg.*, **79**, 578–580.
173. Smits, A. B. (1994). Compression anastomosis in the colon. *Ph.D. Thesis* University of Amsterdam.
174. Hendriks, T. and Mastboom, W. J. B. (1990). Healing of experimental anastomoses:parameters for repair. *Dis. Colon Rectum.*, **33**, 891–901.
175. Farkas, D. L. *et al* (1993). Multimode light microscopy and the dynamics of molecules,cells, and tissues. *Ann. Rev. Physiol.*, **55**, 785–817.
176. Hodde, K. C., Steeber, D. A. and Albrecht, R. M. (1990). Advances in corrosion casting methods *Scanning Microscopy*, **4**, 693–704.
177. Redmond, H. P., Hofman, K., Shou, J., Leon, P., Kelly, C. J. and Daly, J. M. (1992). Effect of laparotomy on systemic macrophage function. *Surgery*, **111**, 647–655.
178. Nieuwenhuizen, G. A. P., Deitch, E. A. and Goris, R. J. A. (1996). Infection of the gut and the development of the multiple organ dysfunction syndrome. *Eur. J. Surg.*, **162**, 259–273.
179. Martens, M. F. W. C, Man, B. M. de, Hendriks, T. and Goris, R. J. A. (1992). Collagen synthetic capacity throughout the uninjured and anastomosed intestinal wall. *Am. J. Surg.*, **164**, 354–60.
180. Alexander, J. W, Gianotti, L., Pyles, T., Carey, M. A. and G. F. Babcock, G. F. (1991). Distribution and survival of escherichia coli translocating from the intestine after thermal injury. *Ann. Surg.*, **213**, 558–567.
181. Lichtenberger, L. M. (1995). The hydrophobic barrier properties of gastrointestinal mucus. *Ann. Rev. Physiol.*, **57**, 565–583.
182. Graham, M. F. and Blomquist, P. The alimentary canal (1992). In *Woundhealing Biochemical and clinical aspects*, edited by Cohen, I. K. Diegelman, R. F. and Lindblad, W. J. pp. 433–449 Philadelphia Saunders Cy.
183. Playford, R. J. and Wright, N. A. (1996). Why is epidermal growth factor present in the gut lumen. *Gut*, **38**, 303–305.
184. Kingsnorth, A.N. and Slavin, J. (1991). Peptide growth factors and wound healing. *Brit. J. Surg.*, **78**, 1286–89.
185. Folkman, J. (1992). Is there a field of wound pharmacology. *Ann.Surg.*, **215**, 1–2.
186. Brown, G. L., Curtsingere, L. Jurkiewicz, M. J. *et al.* (1991). Stimulation of chronic wounds by epidermal growth factor *Plast. Reconstr. Surg.*, **88**, 189–196.
187. Beck, L. S., De Guzman L., Lee, W. P., XuY, Siegel, M. W., Amento, E. P. (1993). One systemic administration of tranforming growth factor 1 reverses age- or glucocorticoid-impaired wound healing. *J. Clin. Invest.*, **92**, 2841–2849.
188. Robson, M. C., Phillips, L. G., Lawrence, W. T., Bishop, J. B., Youngerman, J. S., Hayward, P. G., Broemeling, L. D. and Heggers, J. P. (1992). The safety and effect of topically applied recombinant basic fibroblast growth factor on the healing of chronic pressure sores. *Ann. Surg*, **216**, 401–408.
189. Cromack, D. T., Porraz-Reyes, B., Purdy, J. A., Pierce, G. F. and Mustoe, T. A. (1993). Acceleration of tissue repair by transforming growth factor $\beta 1$: Identification of in vivo mechanism of action with radiotherapy-induced specific healing deficits. *Surgery*, **113**, 36–42 .
190. Adzick, N., Scott and Longaker, M. T. (1992). Scarless fetal healing. Therapeutic implications. *Ann. Surg.*, **215**, 3–7.
191. Longaker, M.T., Chiu E.S., Adzick, N. S., Stern, M., Harrison, M. R. and Stern, R. (1991). Studies in fetal wound healing V.A prolonged presence of hy-

aluronic acid chracterizes fetal wound fluid. *Ann. Surg.*, **213**, 292–296.
192. Huang-Lee, L. L. H., Wu, J. H. and Nimni, M. E. (1994). Effects of hyaluronase on collagen fibrillar matrix contraction by fibroblasts. *J. Biomed. Mat. Res.*, **28**, 123–132.
193. Lowry, K. M., Ellington, M. B. (1994). Resistance of hyaluronate coatings to hyaluronidase. *J. Biomed. Mat. Res.*, **28**, 861–864.
194. King, R. S., Hickerson, W. L. and Proctor, K. G. (1991). Beneficial actions of exogenous hyaluronic acid on wound healing. *Surgery*, **109**, 76–84.
195. Sweeney, H. L. and Holzbaur, E. L. (1996). Mutational analysis of motor proteins. *Ann. Rev. Physiol.*, **58**, 751–792.
196. Evans, G. S., Flint, N. and Potten, C. S. (1994). Primary cultures for studies on cell regulation and physiology in intestinal epithelium. *Ann. Rev. Physiol.*, **56**, 399–443.
197. Horgan, P. G. and Gorey, T. F. (1992). Operative assessment of intestinal viability. *Surg. Clin. N. America*, **72**, 143–155.
198. Pierie, J. P. E. N., Graaf, P. W. de, Poen, H., Tweel, I. van der and Obertop (1994). Impaired healing of cervical oesophagogastrotomies can be predicted by estimation of gastric serosal blood perfusion by Laser Doppler flowmetry *Eur. J. Surg.*, **160**, 599–603.
199. Klagsbrun, M. and D'Amore, P. A. (1991). Regulators of angiogenesis. *Ann. Rev. Physiol.*, **53**, 217–239.
200. Gross, S. S. and Wolin, M. S. (1995); Nitric Oxide: Pathophysiological Mechanisms. *Ann. Rev. Physiol.*, **57**, 737–769.
201. Adams, W., Ctercteko, G. and Bilous, M. (1992). Effect of an omental wrap on the healing and vascularity of compromised intestinal anastomoses. *Dis. Col. Rect.*, **35**, 731–738.
202. Bemelman, W. A., Meier, D. W., Bellot, S. M. and Klopper, P. J. (1991). Is anastomotic protection by processed dermal sheep collagen possible ? *Surg. Res Comm.*, **10**, 95–105.
203. DeJongh, G. J. (1995). Protection of colonic anastomoses with collagen meshes. *PhD Thesis* University of Amsterdam.
204. Stead, R. H., Perdue, M. H. Cooke, H. Powell, D. W. and Barrett, K. E. (1992). Neuro-immuno-physiology of the gastrointestinal mucosa. Implications for inflammatory diseases. *Ann. N. Y. Ac. Sciences*, vol. 664 .
205. Birk, D. E., Silver, F. H. and Trelstad, R. L. (1991). Matrix assembly pp. 221–254 In: Hay E. D. Ed. Cell biology of extracellular matrix. Plenum Press, New York London 1991.
206. Pierie, J. P. E. N., Graaf, P. W. de, Poen, H., Tweel. I. van der and Obertop (1993). Incidence and management of benign anastomotic strctures after cervical oesophago-gastrotomy. *Brit. J. Surg.*, **30**, 471–474.
207. Hiolt, D. R., Kirk, S. J., Regan, M. C., Hurson, M., Lindblad, W. J. and Barbul, A. (1992). Effect of age on wound healing in healthy human beings. *Surgery*, **112**, 293–298.
208. Stoop, M.-J., Dirksen, R. and Hendriks, T. (1996). Advanced age alone does not surpress anastomotic healing in the intestine. *Surgery*, **119**, 15–19.
209. Traverso, L. W. (1996). Technology and surgery: Dilemma of the gimmick, the true advances, and cost effectiveness. *Surg. Clin. N. America*, **76**, 129–138.

PART III

6. Sphincters of the Gastrointestinal and Urinary Tract

P. J. Klopper

Professor of Experimental Surgery (emeritus), Academic Medical Centre, University of Amsterdam, Amsterdam, The Netherlands

Introduction

Sphincters are part of transport mechanisms in the body and are situated at the border of different parts of the gastrointestinal or urinary tract. Transport is accompanied by temporary storage and sphincters play an important role in this process. The main pathological features of disturbed function are: reflux of contents, incontinence or disturbance of transport.[1–3]

Efficient transport only takes place if the closing mechanism and peristalsis are coordinated,[156] otherwise dysphagia, distension or hydronephrosis follow. Insufficiency of the closing mechanism results in regurgitation, the mucosa is damaged and life is endangered by aspiration or changes in the balance between bacteria.

Severe complications can be the result of disturbance of the reservoir function, like dumping and mucosal damage in the case of the stomach and loss of quality of life if the bladder or anal mechanism is involved. Social behavior is severely influenced if incontinence is present,[157] not limited to practical problems but also making changes in the psyche of the individual. Defecation and micturition are still in the sphere of the taboo.

Disturbance of emptying is accompanied by the classical surgical symptoms of stasis, followed by distension, infection and loss of function of the proximal part. It can lead in a short time to perforation of the wall, accompanied by a decline in the general condition.[158]

The term "sphincter", derived from the greek suggests a circular structure that is capable of binding off certain parts of the bowel. Before functional studies gained the upper hand, anatomists had investigated all structures identified as sphincters and an impressive amount of names of investigators can be found related to border zones in the gut. Names of men like Gubarov, Bauhin and His indicate that extensive studies were made, but the correlation between physiological and anatomical findings is still subject to study.[4]

In the human body the main mechanisms are: the upper and lower esophageal sphincter (UES/UOS, LES/LOS), the pyloric region, the complex involved in the regulation of transport of bile and pancreatic juice by the sphincter of Oddi, the ileocecal junction with Bauhin's valve and the mechanism in the rectum.[5] In urology the pelvic-ureteral, the uretero-vesical junction and the urethral sphincter are of interest.

Attention is drawn in the pelvis to embryological development.[6–9] The bladder and the rectum are part of the cloaca and only in a later stage separation does take place as innervation and blood vessels are closely related.[10]

Surgical research is an applied science based on surgical practice in which patients need treatment for complications and repair of the original function. Research in this field is directed to knowledge of the normal function and morphology of the sphincter itself and to its related structures. Investigation of the disturbed function is part of diagnostics in a certain individual, but a standardized procedure in laboratory animals can often give more insight in pathology.

Inspired by the loss of function seen after surgical resection, techniques were developed to restore function. In particular in surgery of incontinence new methods were introduced in the last twenty years.[11–17]

Simple incision can be the treatment of choice in hyperactivity of sphincters as is done in pylorospasm, but in other sphincters the obtained result is only temporary. Recently myotomy became a subject of research to evoke the formation of a reservoir.[18]

The classic work of Pawlow stimulated physiologists at the beginning of this era to extensive work on the manner of transport in the digestive system. Physiology, however, is at present more directed to the cellular level. The investigations of Carrell and Dragstedt and his pupils are good examples of surgical physiology. The resection of the vagal nerve developed to lower acidity was accompanied by disturbance of the gastric closing mechanism and transport of bile.[19–24]

Radical resections in oncological surgery necessitated the removal of organs with a reservoir function like stomach, rectum and bladder and stimulated research in the repair of closing mechanism either artificial or natural and the construction of a new reservoir.[10,25,26,159,160]

At the same time new research in morphology of the mucosa with modern means like scanning and transmission electron microscopy developed.[161]

Planning of surgical research in this field depends on the local situation. Independent research by the surgeon is still attractive and the construction of new valves and reservoirs to ameliorate surgical technique and prevent complications with the subsequent methods for evaluation does not always evoke the interest of other disciplines. Besides the clinical results interesting work can still be done in view of new developments in technology, like mechanical suture instruments and the application of biomaterials. In some centers a joint effort of gastroenterologists and surgeons opened new fields.

Cooperation with morphologists is again actual. Three-dimensional presentation of myogenic components and bloodvessels contribute to practical results.[162,163] Ultrasonography and nuclear magnetic resonance are new methods with possibilities for the study of transit, structure and possibly treatment.[164–168]

In some countries surgical pathology is well developed but it is still amazing that the results of sphincter reconstruction like the Nissen fundoplication or related operations were seldom judged. The investigation of tissue reaction following reconstruction or implantation is also an open field, in particular after long term follow-up.[27]

Surgical history indicates that stomas of the bowel were already made very early on, deviation of contents was the only vital intervention possible up to the fifties, when the classic ileostoma was replaced by Brooke's technique, permitting a more convenient outflow, but the patient still remained incontinent.[28] In the sixties Kock introduced a new surgical technique including the construction of a reservoir and a nipple valve and got two remarkable advantages; the patient was continent and the reservoir could be emptied at regular times.[29–35]

From that time on a distinct line in development can be found. Many variations were introduced in the construction of the reservoir and the valve. However, complications of the valve are frequent and the question arose as to whether it was advisable to construct a valve of the nipple type, to use an artificial device or to develop techniques preserving the natural closing mechanism, as is done in the ileoanal procedure.[36,37] To obtain optimal functioning, techniques to construct a reservoir are still in progress. A number of variations is present, all intended to limit the frequency of defecation or, in the urostoma, the frequency of catheterization, while at the same time stimulating research on innervation.[38]

Parallel with this research, techniques for the surgical reconstruction of sphincters can be found, varying from folding techniques like the Nissen intervention,[39] sling techniques in the pelvic region to replacement of muscles performed with microsurgical methods. Knowledge of normal function can enhance the development of surgical methods preserving innervation.

In all sphincters muscular elements, mucosa and connective tissue are present. Muscular elements in different directions, mostly circular and longitudinal can be found. At present, sometimes a spiral pattern or local thickening are described. In some sphincters small separated muscular slings form part of the closing system.[21,40,41]

Most sphincters have smooth muscles, with exception of the proximal esophagus and the anus. In some animals, like the dog, striated muscles are also present in the lower esophagus. It shows that in planning and interpreting experiments related to the clinical situation the choice of the species has to be considered.[2,42,43]

Mucosa. Including the arterial and venous network in the submucosal layer, a specific vascular pattern can be identified in some sphincters, playing a role in the function of the closing mechanism; the regulation of this circulation is not well known.[44,45] The mucosa itself can show a pattern that is typical for that area like the Z-line in the esophagus or the linea dentata in the anal region.[46,47]

The epithelium in most sphincteric regions is marked by a sharp border of two different types of cell. Ectopic clusters are sometimes present. Of interest is the regeneration of epithelium after anastomosing.[23] Repair is mostly marked by a new well demarcated border of the two types of epithelium involved, sometimes separated by a single mucous cell. Malignant degeneration cannot be excluded at long term.[22]

In the sphincters cells can be found of the neural and neurohumoral system, playing a role in the

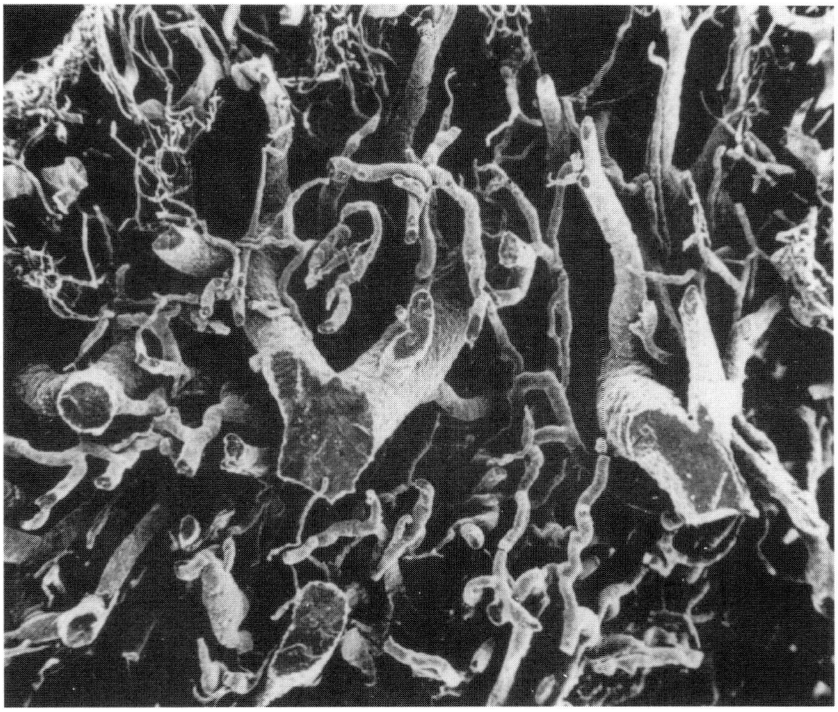

Figure 6.1. Corrosion-cast of the mucosal vessels of the canine ileocecal junction.

extrinsic and intrinsic regulation of sphincter action.[48,49] Glands are present in the epithelium as well as in the deeper layers. The Brunner glands are well known and they change considerably if a gastroduodenal connection is made. The function is not completely known, a contribution to the mucosal barrier, part of the defensive mechanism, is probable. The biophysical properties of mucin are better known and special peptides have been isolated.[169,170]

The connective tissue present in, or in the region of,the sphincters also plays a role in function.[50–52] In the esophago-gastric junction the outer structure was studied extensively[7,53,54] and an extrinsic and intrinsic mechanism was distinguished, the same is true for the ileocecal junction (see Fig. 6.1). The supporting connective tissue affects the choice of technique to treat regurgitation. Some authors state that the function of the sphincter is lost, if the normal connective tissue system is disturbed.[55]

Also in other sphincters special connective structures can be demonstrated, like the Waldeyer sheet in the uretero-vesical junction and the urogenital diaphragm. The presence of the three-dimensional constructed nets of collagen indicates that in sphincters active (myogenic) as well as passive (mechanic valve) factors determine its function.[163]

Reflectory motor activity of the sphincters is achieved in central or local pathways by signals evoked by stretch receptors.[56] The intramural nerve cells are located in ganglia in two areas of the wall, the submucous and the intermuscular myenteric plexus of Meisner and Auerbach.

Methodology

A. Functional Methods

1. Manometry

Pressure studies in the bowel with balloons , have been performed since the beginning of this century.[57–61] The observations concerning visceral pain can still form the basis for new research with more advanced techniques.

In the thirties Thomas and Quigley introduced balloons in the pyloric region to study gastric emptying and stated the hypothesis that the gastric closing mechanism was in reality a pump.[62,63]

In the fifties the free tip method with and without perfusion was introduced and compared with the registration with balloons. Electronic sensors became available later on. In urodynamics this type of sensors became standard. In studies of gastrointestinal motility free tip perfused catheters are used.[64–66] Microtips have a higher frequency response than most perfusion systems, but the high costs and vulnerability are a disadvantage, in particular when multiple registration is needed. For long-term pressure measurements in combination with other parameters such as pH in the lower esophagus, microtips are ideal.[67] Open tip catheters without perfusion are also used, but are often blocked by intestinal content, particulary in the rectum. If organs with a limited lumen are measured, the flow rate has to be low.

The compliance of a system with perfused catheters, consisting of perfusion pump, catheter, connection tubes and pressure-sensor is rather critical. At present most studies are made with an pneumohydraulic system. The perfusion fluid is kept under high pressure by gas in a reservoir and pressed into capillaries. The system introduced by Arndorffer *et al.*[65] guarantees an equal flow and low compliance; care has to be taken for the unfavorable influence of air bubbles in the system. Experience has shown that this can be ameliorated by using a bag filled with fluid thus preventing direct contact between fluid and gas. The small diameter of the capillaries does not permit fluids other than water.

Some influence of the perfusion fluid on the measured values cannot be excluded as the amount of fluid and its temperature may influence sphincter action. In experiments in dogs fluid with an elevated osmolarity increased analpressure, indicating the presence of osmosensors in the mucosa. The open tip measurement can be divided into stationary and dynamic methods.

a. Stationary methods:

Catheters. Measurements can be made with home-made catheters of PVC, with the advantage that variations can be made by the investigator as the material is easy to glue.[54,68] For pressure measurement tubes with an outer diameter of 1.45 to 2 mm and an internal diameter of 0.75 to 0.78 mm are commonly used[69] (see Fig. 6.2).

Combinations of tubes with a small and a larger internal diameter permit registration of pressure and aspiration. Several types of multi-lumen tube are now on the market, if wanted supplied with a radio-opaque marker.[66,70]

In most investigations free tip catheters are used with holes at the end or at the side and perfused with a flow of 0.5 to 3 ml/min, depending on the inner diameter of the tube. Loops can be attached to prevent contact with the mucosa. Multi-lumen tubes are used to register the propulsion of the peristaltic wave, as the distance between the holes is known (see Fig. 6.3).

A non-invasive catheter method can be used in the investigation of the deglutition mechanism. One nose-hole is closed and against the other a canula is held connected with the pressure sensor via a water-filled tube.

Sleeve Catheter. Specialized sensors have been developed to measure the tonic contractions of sphincters, the Dent sleeve catheter with variations (see Fig. 6.4). The device measures the absolute magnitude of sphincteric tone, variations in the

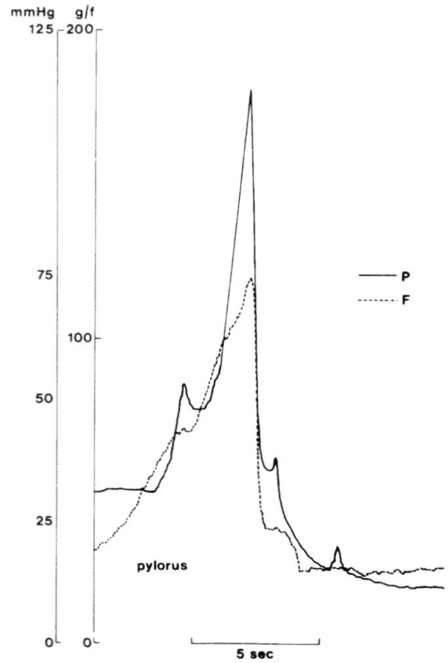

Figure 6.2. One-channel pressure and force profilometry of the canine pylorus.

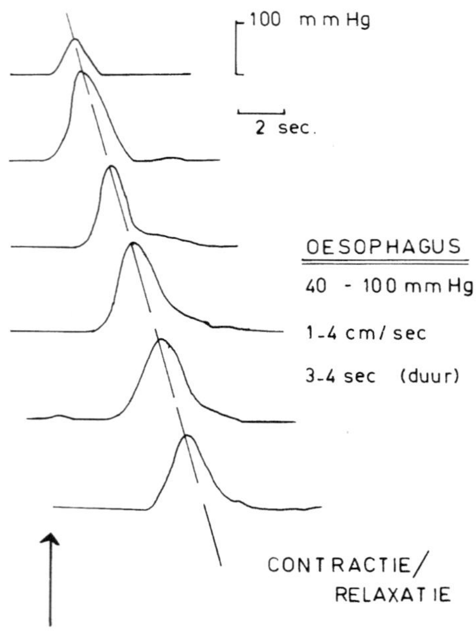

Figure 6.3. Propulsion of peristaltic wave in the esophagus. Arrow: Swallowing.

thickness of the membrane and the length (varying from 1.5 to 8 cm) influences the registration.[71,72]

For the same purpose microballoons are used, in deed, manometry started with balloonmeasurement.[73] The disadvantage is, that the measured pressure has been stimulated by distension of the balloon.

Combinations of perfused channels and microtips are also described.[74,75] Some authors showed that not only the tone of the sphincter is of importance, but that phasic burst can also appear and cannot be measured with this type of device. Perfused catheters or microsensors are then preferable but have the disadvantage of uncertain localization.

New methods (step-tests) were developed to measure the compliance of the system, also during clinical investigation.

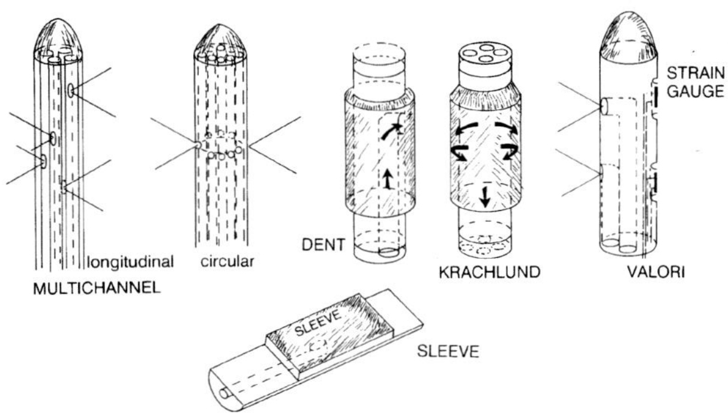

Figure 6.4. Various types of pressure sensors.

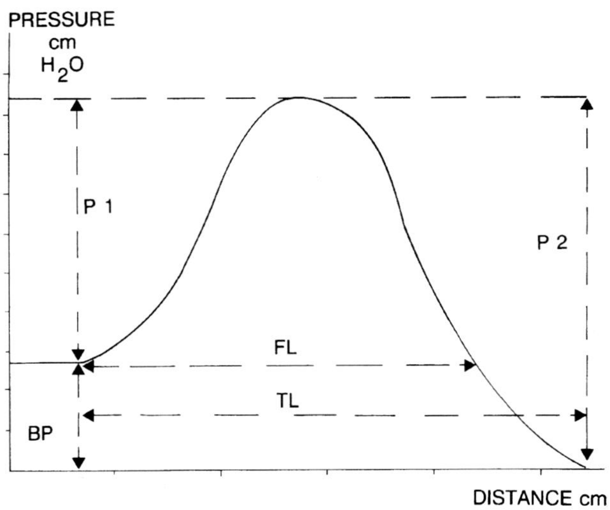

Figure 6.5. Standardized pressure profile of the urethral sphincter. TL and FL - Total and functional profile length. P1 - Maximum urethral closure pressure. P2 - Maximum urethral pressure.

b. Dynamic Measurements

In urology Brown and Wickham[76] introduced the rapid pull through method and this method became standardized in urodynamic investigations.[7,77] A one-channel perfused catheter with a diameter of 2 mm and with a side-hole is withdrawn through the sphincter with a continuous speed of 0.7 cm/sec. Experimentally the same method was used in the ureter. Later on the perfusion catheter was replaced by microtip measurements (see Fig. 6.5).

With one-channel perfused catheters or microtips withdrawn with at a certain speed a pressure-profile of the sphincter can be obtained.[78] The shape of the urethral profile is defined by the International Continence Society.[79]

When one-channel catheters are withdrawn repeatedly in different directions of the sphincter varying pressures can be registered suggesting an asymmetrical structure. Recently multi-channel perfused catheters were used to study the profile of the upper and lower esophageal sphincter, the anal region and the urethra simultaneously in four to eight directions, thus permitting three-dimensional pressure profiles (see Fig. 6.6).

Two types of measurement are mentioned, the continuous and the stationary pull-through or slow stepwise pull-through. With the latter pressure is measured for a certain time at different distances in the sphincter. This method of profilometry has also been performed in sphincters with a small diameter like the urethra, showing a good relation with the anatomical structure. A catheter with a small diameter also permits profilometry in pediatric surgery.[166] In the near future standardization of this profiles will be necessary to make a distinct comparison between the normal and pathologic profile and nomograms can contribute to further automatization of the recording.

Analyses up until now have been made by calculating volume-pressure index, an asymmetry index

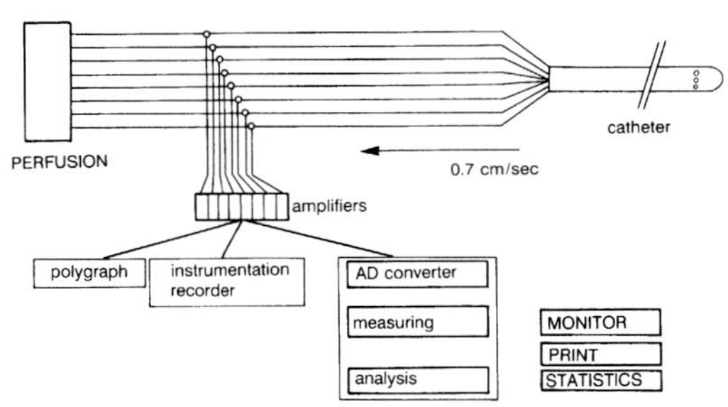

Figure 6.6. Technical scheme of multichannel pressure registration (rapid pull-through method).

and by standardizing the measured length in relation to pressure. The first shows a rather high spread in the esophagus, while the second is of value in diagnosing the damaged sphincter complex of the anus. The third can contribute to the localization of the different tops in the profile.[104,154,175,182]

Connection tubes, connecting the catheter with the sensor device, have to be short to main tain compliance. If possible, these tubes should be avoided.

Perfusion-Apparatus

The pneumohydraulic system was mentioned, systems acting with multiple syringes can be used if the force applied to move the piston is overcapacitated to guarantee a continuous flow.[70,80]

Recording

The recording of the pressures can take place with a galvanometric polygraph or with an instrumentation recorder.[81] If potentiometric recorders are used, attention has to be paid to frequency response.

The widespread use of small computers made it possible to digitalize the analogous signal and to store data on disks. The method also permits the analysis of peristalsis as well as of profiles. The use of software makes three-dimensional imaging possible by different methods: as wire frame, a presentation in planes or as a soft-shadowed image in which the color represents the pressures measured.

Storage of data in a computer has also contributed to analysis of the relation between electrical signals of the intestine and pressures. The ever complicated analysis of motor phenomena in the intestine has been facilitated. In programs in this development can be expected in future years can be expected, that are easy to use, provide statistical analysis and differentiate between propagated and non-propagated contractions.[68,82]

Radiology

X-ray investigation has played an important role in judging the function of sphincters, in particular in diagnosis of regurgitation. Particularly when an image intensifier (fluorography) is available, valuable information can be obtained on the function of the sphincter in transit and peristalsis and retrograde motor activity, for the image can be recorded by video or by a motor-driven camera. The image formation, in particular of the mucosa, can be enhanced by combination of X-contrast media with air-insufflation (see Fig. 6.7). Videodensitometry was seldom used in gastroenterological work.

New developments can be expected by application of nuclear magnetic resonance.

Defecography

Rectal filling with radio-opaque material can give insight into the dynamic process during defecation

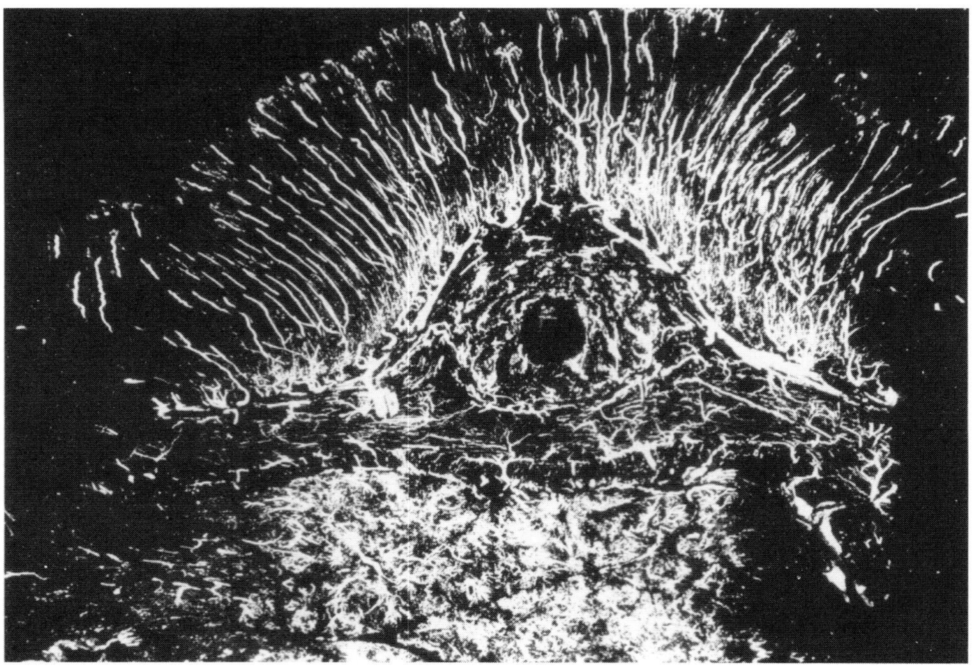

Figure 6.7. Vascular pattern at the distal part of the canine common bile duct.

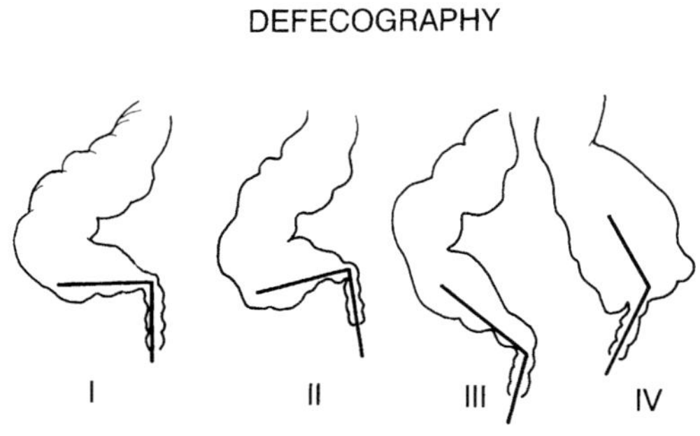

Figure 6.8. Schematic view of defecographic X-ray image.

and the relation of the shape to the anatomical structure in the pelvis, in particular to measure the anorectal angle[83–85] (see Fig. 6.8). The exposure to X-rays limits clinical use, but in a few years investigations with nuclear magnetic resonance can be expected in the exploration of sphincter function.

For animal research small apparatus have been already used in bone research. The costs of advanced instruments are too high for research institutes and use of clinical apparatus is limited for practical reasons. New information can be expected in clinical research.

Balloon proctography has been applied in the ampulla recti and deformation of the balloon informs about the expelling forces present. In man the balloon is filled with 100 ml of barium. In further development of this technique it seems there is a need to apply material with properties more comparable to the normal consistency of feces.

Electromyography

In experimental work electrodes can be attached to the muscular components of the bowel.[56] Chronic experiments have been performed in several species and multiple registration can be performed with the aid of a modified Pawlow cannula in which the contacts are fixed.[86]

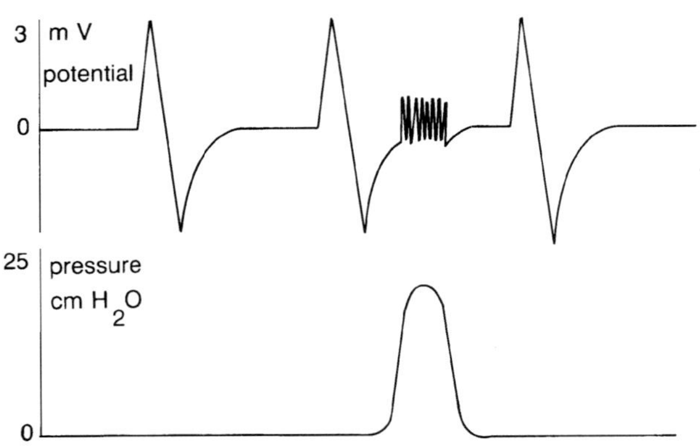

Figure 6.9. Electromyographic and pressure signal of intestinal peristalsis.

Experience was obtained in stomach, bowel, uterus and choledocho-duodenal junction.[172, 173] In humans electromyography of the esophagus was performed with suction-electrodes; recently measurements in a Roux- and Y-loop during one week postoperatively. It can be expected that measurements, without risk for the patient and in-tended to obtain more insight in human pathology, shall substitute experimental work in animals (see Fig. 6.9).

To diagnose dysfunction of urethra and anus electromyography has been used before and after surgical intervention. The development of fine needle electrodes stimulates this research and analysis of the action of the different myogenic components in combination with changes in pressure has been facilitated by the microcomputer.

Registration of electrical potentials can be combined with stimulation of nerves involved in innervation of the area. This has been done in the phrenic nerve and the nerves of the anal région. Registration of evoked potentials has been applied in orthopedic and abdominal vascular surgery. The application in organs with an autonomic innervation can be expected.

Potential Difference

The difference in type of epithelium in the border zones can be used to localize catheters by measuring the potential difference between the tip of a catheter placed in the border zone and a calomel electrode in contact with the skin or the buccal mucosa. The potential is measured by a millivoltmeter. The fluid within the catheter consists of saturated KCl-agar. The differences in potential vary from −2 mV in the duodenal mucosa to −30 mV in the gastric mucosa, and in the esophagus this difference can increase up to −40 mV.

Measurement of pH

In clinical gastroenterology 24-hours pH measurement is a standard method for detecting gastric reflux.[67, 174, 175] The application is enhanced by the use of portable systems with a microprocessor, permitting sampling of a large quantity of data.[87–89]

Glasselectrodes with a built-in reference electrode has been used to detect the border between esophagus and fundus, gastric fundus and antrum and the pyloric region. It is advisable to use a combined electrode for interference of the potential difference on the observed values is thus prevented. Prolonged measurement of intragastric pH was recently used to register duodenogastric reflux (see Figs. 6.10, 6.11).

In combination with pressure registration one group of investigators obtained a nomogram of length and pressure of the lower esophageal sphincter in man. Similar experiments were performed in the duodenal region.[52]

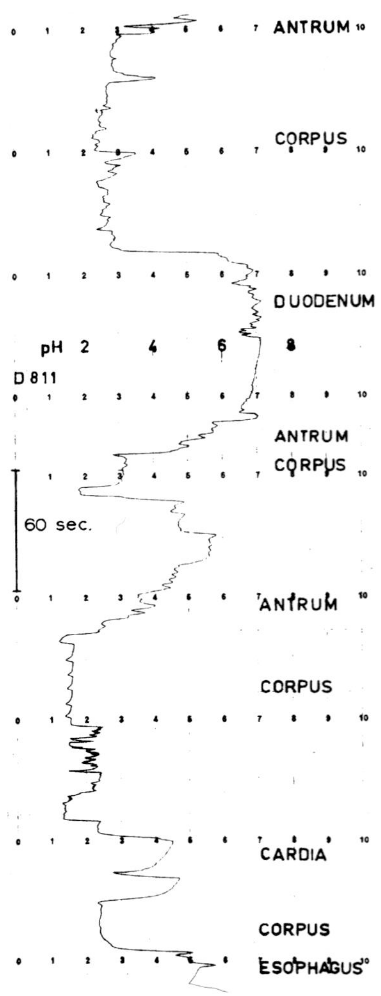

Figure 6.10. pH measurement of canine esophageal, gastric (corpus and antrum) and duodenal mucosa (continuous measurement).

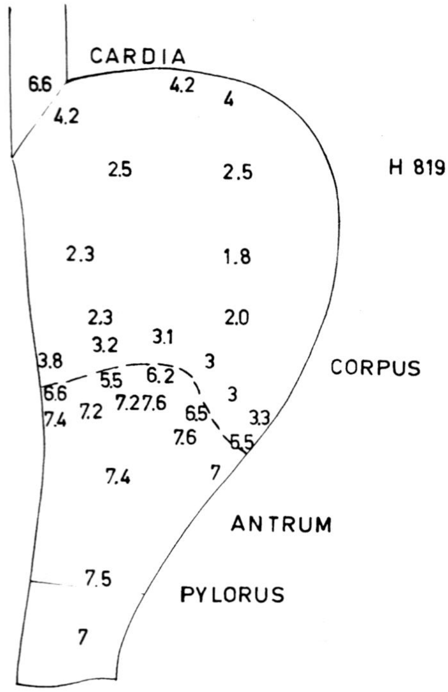

Figure 6.11. pH measurement of canine esophageal, gastric and duodenal mucosa (topical measurement).

Radiotelemetry has been done in the same areas but there is a disadvantage in that the sensor has to be fixed.

Scintigraphy

Radionucleoid tracers that can be detected by a gamma-camera are used in measurements of intestinal transit.[90] The method has the disadvantage that in several countries the use of radioactive tracers for diagnostic purposes is limited. In addition, in experimental research, the facilities to perform this technique are strictly regulated, including isolation of the animals for twenty four hours, the aid of authorized personnel and facilities.

For the detection of the transit of fluids 111 Indium DTPA, for meals 99 m Technetium are currently used.[91,92]

Ultrasonography

Extended information has been obtained in gynecological practice and since the seventies the method has been used to explore the gastrointestinal tract. Technological development in the imaging of the vessel wall is the subject of recent studies[91] and in near future three-dimensional imaging of morphological structures including the sphincters can be expected from this almost harmless technique.[164,165,166,167] in particular because miniaturization of the sondes is in full progress. Meals rich in fats permit ultrasonographic analysis of the pyloric region. The development of meals with particles detectable with ultrasound and related to normal food will enhance this field of research.[93] Combination of this method with measurement of impedance can be used in judging of the function of a reservoir, replacing scintigraphy. In view of recent developments in intravascular ultrasound, a similar development in intraintestinal techniques can be expected.

B. Morphology

Routine investigations with the light microscope are mostly part of the interpretation of changes in the sphincter evoked by surgical techniques.[176] Scanning electron microscopy (SEM) is a valuable tool for

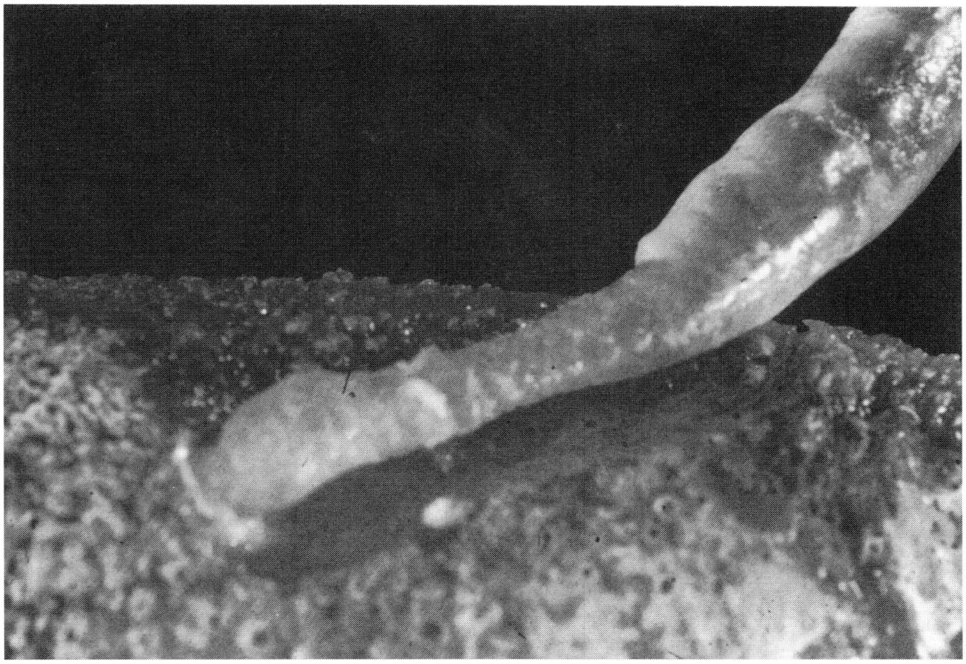

Figure 6.12. Corrosion cast of the choledocho-duodenal junction.

judging the changes caused by loss of sphincter function.[94]

Imaging intravascular casts (corrosion technique) with SEM increases the insight into the complex vascular structure present in some valves (see Fig. 6.12). The method can be compared with microradioangiography, for the arteries filled with a barium-suspension can be studied in slices and the obtained X-ray image can be combined to networks in space with suitable software, also giving information on the relation with other structures in the sphincter. The same slices are suitable for conventional histology. A disadvantage is that detailed information can only be obtained if a crystallographic X-ray tube is used.

Insight into the collagen structure of sphincters is provided by the use of polarization- or interference microscopy.

Special staining methods have been used to explore the neural innervation of sphincters. Silver impregnation has been applied for years and is supplemented by methods identifying acetylcholine and catecholamine in the myenteric plexus. Vital dyes were used to stain tissue and also to identify vagal innervation in the cardiac region. The investigation of neuropeptid cells in the sphincter regions with immuno-histochemical methods contributes to a better insight in the regulation of the smooth muscle elements and large differences between species have been observed. Up until now forty peptides have been described; well-known are gastrin, somatostatin, secretin, cholecystokinin, VIP (vasoactive internal polypeptides) and the pancreatic polypeptides.

The presence of stretch receptors in the organs separated by sphincters is related to the function. Distension of the gastric fundus increases the tone of the LES. In other regions the influence of the dimension of particles has been studied.[95]

Osmoreceptors are probably present in the anal region, which means that the form and amplitude of the measured profile can be influenced by temperature and osmolarity of the perfusate.

In animals standardization of anesthesia is necessary for the muscular activity of sphincters is distinctly influenced by a number of drugs.

In view of the complications of regurgitation during anesthesia, the tone of the LES has been particularly investigated.[96]

Upper and Lower Esophageal Sphincter

Upper esophageal sphincter (UES-UOS)

The upper esophageal sphincter has been studied by several disciplines, for the diagnostic deglutition mechanism is involved in neurological and laryngological disorders. The whole complex is composed of striated muscles and can be investigated with electromyographical, invasive and non-invasive pressure methods. Combinations with X-radiography are mentioned.[97]

The length of the sphincter in man varies from two to four cm and a distinct asymmetrical profile could be obtained with a slow step-wise pull-through method with a multichannel probe. The pressures are higher in the anterior posterior direction.

Lower Esophageal Sphincter (LES-LOS)

The distal esophageal closing mechanism is one of the most investigated areas in gastroenterology.[98–100,174,175] The main parts are the intrinsic muscular components, showing an asymmetrical aspect, mucosal rosette and the extrinsic components -ligaments from the diaphragm and stomach and esophagus- the right crus diaphragm and the angle of His. The closure mechanism is influenced by all components, whether active or passive.[72,101–103] The effectiveness of the sphincter is determined by the difference in pressure between the positive gastric and the negative pressure in the esophagus. The profile shows asymmetrical pressure and length in eight sectors and both are involved in pathology[104,105] (see Figs. 6.13, 6.14).

Regulation is influenced by extrinsic and intrinsic factors, as the parasympathetic and sympathetic system, local neural reflexes and humoral regulation, also by several neuropeptides like gastrin, VIP and CCK.[106–108] Three functional events can be distinguished, the basal tone, active relaxation following swallowing and active contraction.[109,110] To prevent regurgitation the most important factor is the resting tone, the difference in the positive gastric and negative intra-esophageal pressure is called the barrier pressure, reaching values higher than in other parts of the GI-tract.

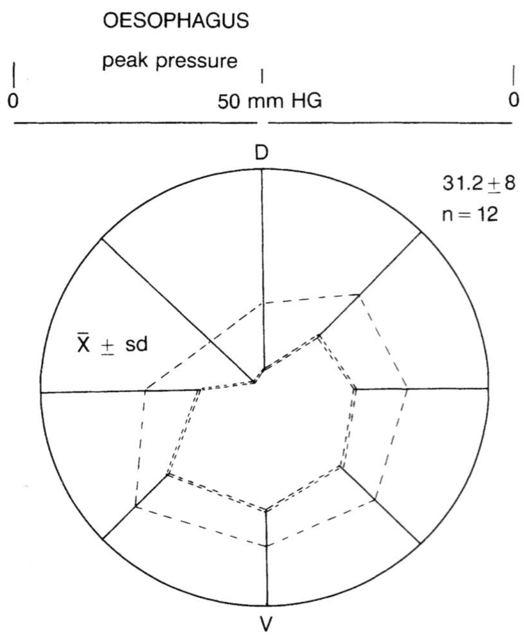

Figure 6.13. Asymmetric pattern of peak pressures of LES in 12 male volunteers.

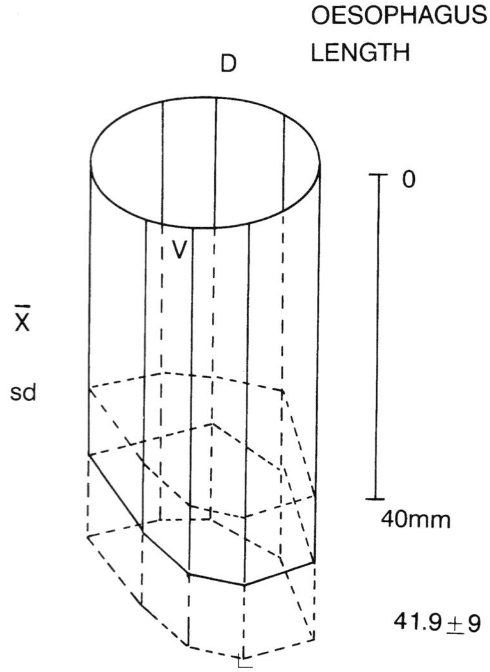

Figure 6.14. Asymmetric length of LES in 12 male volunteers.

Extensive reviews are present on the influence of drugs and the depth of anesthesia on LES-pressure.[96,111,177] Deep anesthesia decreases the tone of all sphincters, a factor of importance in experimental studies and difficult to standardize.

The pressure in both sphincters is influenced by respiration and the passage of peristalsis. In view of the increasing types of disorder determined by manometry, further research in this field can be expected[103] (see Figs. 6.15, 6.16).

In experiments several models were developed to simulate regurgitation and complications. In clinical research the correlation between pressure, the type of intervention and the effect on prevention of reflux is still under discussion. The number of surgical interventions decreased in favor of conservative treatment.

Pylorus

The closing mechanism of the gastroduodenal junction is still under discussion. The morphological and

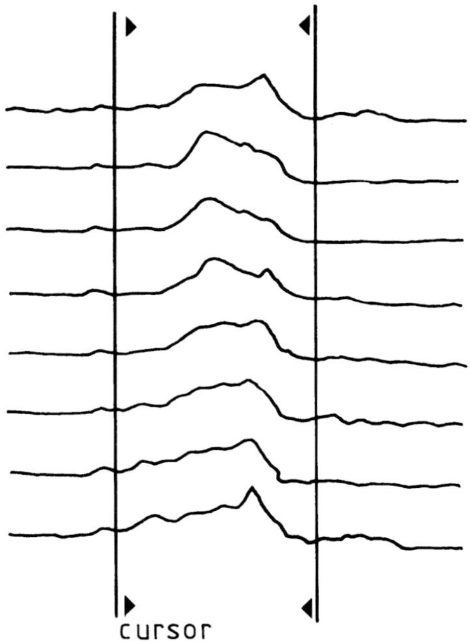

Figure 6.15. Eight pressure measurements of the LES as seen on the monitor. By replacing both cursors, the investigator can determine the side of analysis.

roentgenological aspect has been extensively studied. Besides the muscular parts, the structure of the connective tissue shows specific aspects.[63, 112, 113]

Functionally a difference for the transit of solid and fluid has been observed.[114] The peristalsis of antrum and duodenum appears to play an important role and distinct retropulsion has been observed.[115, 116] Already in 1935 different pressures between gastric antrum and duodenum were demonstrated,[117, 118] and coordination between both parts could be demonstrated with electromyography.[92, 119]

In man measurements with different types of catheters did not reveal a distinct high pressure zone as was previously found in dogs. A few years later an intermittent high pressure zone was demonstrated in man and in view of the contradictory results, the term fickle pylorus remains in use.[120] It shows that the junction has to be seen as an entity, the pylorus itself acting as a region of varying resistance against the propulsion of the gastric contents[24, 121–126] (see Figs. 6.16, 6.17).

Presence of fat in the upper intestine delays gastric emptying in dogs and it appears that the increased resistance of the pylorus against flow is partly responsible. The diameter of the pylorus was lower than after administration of saline, carbohydrate or protein.

Surgical intervention influences the coordination between stomach and duodenum considerably.

Biliary System

Cystic Duct

Pressure studies of this system were frequently made during operation by French authors.[127] Its value for diagnosis remained under discussion for a long time, even when electronic devices replaced the water-manometer.

Most authors are of the opinion that the well studied reservoir function of the gall-bladder is supported by a sphincter like mechanism in the cystic duct, influenced by neurohumoral and neural stimuli. The structure of the bile duct is dependent on the species. If present, the action of muscular elements to transit appears to be limited and the importance of the junction with the duodenum as a motor mechanism was one of the hypotheses to explain the transport of bile. In dogs some changes in tonus could be registered.

Sphincter Oddi

The anatomy of this area in different species has been described by Boyden.[128–130] He showed that the number of muscular slings varied from two to three.

In animals and in man a distinct high pressure zone can be found, in dogs with a biphasic pattern. In man a similar pattern can be found.[131–133]

In the bile system the parasympathetic innervation certainly plays a role as was found after vagotomy; as in other organs such as stomach and colon the main effect was a decrease in tone. In the integrated motor mechanism between bile duct and duodenum neurohumoral stimuli remain in the foreground. The investigation of this area on a large scale became possible after the introduction of endoscopic techniques and manometry can now be performed in a relatively non-evasive way.[134, 135]

In the non-anesthetized opossum, electromyographical registration showed differences in action potentials between sphincter and duodenum,[136] but the frequency was equal, while at the same time pressure in bile and pancreatic duct increased. Feeding was followed by an activity-pattern. Alcohol influences Oddi's sphincter.[132, 137]

Ileocecal Junction

This mechanism prevents reflux from the colon into the terminal ileum and regulates the transit from this loop into the colon.[32, 52, 75, 138–141]

Some experiments indicate, that the antireflux function is dependent on intact connective tissue, dissection was followed by reflux.[52, 142] Resection of the system in puppies caused high mortality in contrast to older animals. Distension of the terminal ileum decreased transit time in the entire intestine.[143, 144] Pressure studies in humans showed that

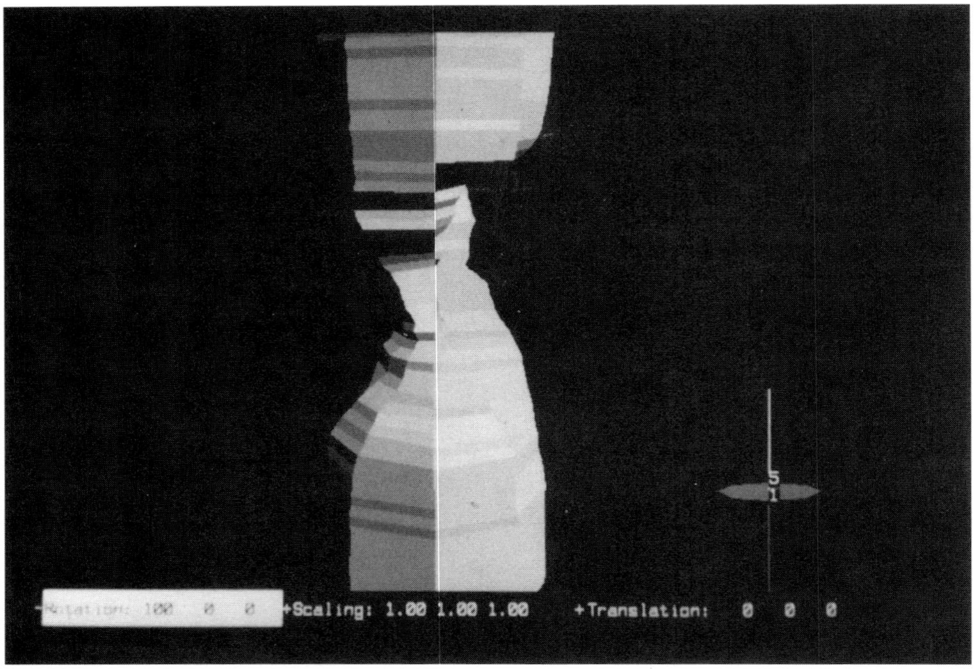

Figure 6.16. Three-dimensional colored shape of the human LES. The indicator at the right shows the projection.

the region had to be seen as a complex and that the small high pressure zone present is only part of this complex.[145,146] The length varied in dogs between 0.5 and 3 cm. Measurements with perfused catheters and a sleeve sensor revealed a distinct tone in this zone lasting for hours.

Casts of the submucosal area showed distended vessels and the filling of the submucosal plexus contributes to the function.

In view of the well-developed lymphatic system in this zone, the terminal ileum loop is part of the defense system.

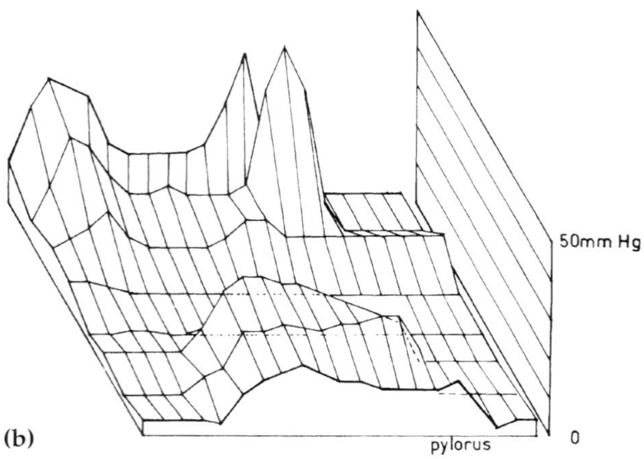

(b)

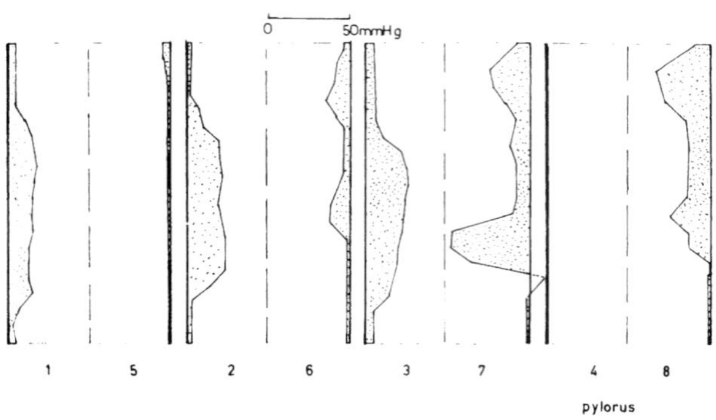

(a)

Figure 6.17. Three dimensional presentation of eight-channel profilometry of the canine pylorus. a) Cross-section b) Projection in horizontal plane.

Junction Between Sigmoid and Rectum

In earlier physiological literature this junction was seen as an area evoking reflexes that initiate filling of the ampulla. In more recent literature it was suggested that the junction can contribute to obstipation by changes of the muscular mass.[147] The area needs further investigation. The studies have to be performed in man since anatomy differs in most animals.

Anal Sphincter Complex

Publications concerning the function of the anal sphincter complex are numerous and still on the increase. The reappraisal of anatomy[178, 179, 180, 181] has contributed to operative methods for repair of specific parts of the complex; the introduction of mechanical suture instruments and methods to preserve the sphincter complex contributed to a better quality of life.[7, 8, 68, 85, 148–154,175]

In contrast to the standardization of methods to evaluate urethral function, the objectification of anal function shows a large variation. One of the reasons is that functional disturbance is not caused by failure of one part of the mechanism.[155] Intrinsic and extrinsic muscular components, the internal (IAS) or external (EAS) sphincter as innervation of muscles and mucosa are involved.

Manometry gives an impression of the pressure at rest and after squeezing. A distinct correlation of these values with continence and frequency of defecation is not clear. Manometry can be combined with dilatation of the ampulla recti by a balloon to control the reflex mechanism. Activity of specific parts can be evaluated with electromyography, also in combination with pressure studies. Information on the activity of muscles after surgical repair is scarce. Some of these methods were applied with success in the training of patients with disturbed function.

Recently multichannel manometry was introduced[153, 154, 169, 182] and an important finding is, that damage of the sphincter at partum can be followed by distinct asymmetrical profile. Further evaluation of this new technique and comparison between normal and pathological data is necessary.

In conclusion gastrointestinal research, being one of the oldest fields of investigation, will also remain a subject of high interest for surgeons in the future.

Reference

1. Didio, L. J. A. and Anderson, M. C. (1968). *Anatomical, functional and surgical considerations of the "Sphincters" of the digestive system*. Baltimore: Williams and Wilkins Cy.
2. Johnson, L. W. and Harmony, J. W. (1986). Experimental esophagitis in a rabbit model. Clinical relevance, *Journal of Clinical Gastroenterology*, **8**, suppl. 1, 26–44.
3. Raper, S. E., Barker, M. E., Jones, A. L. and Way, L. W. (1988). Anatomic correlates of bacterial cholangiovenous reflux, *Surgery*, **105**, 352–359.
4. Conte, M. and Bonfils, S. (1975). *Gastroenterologie* 3e edn. Paris: Flammarion.
5. Schulze-Delrieu, K. (1988). Sphincteric regions. In *An illustrated guide to gastrointestinal motility*, edited by D. Kumar and Gustavsson, pp. 243–251. Chichester: Wiley and Sons .
6. Krane, R. J. and Siroky, M. B. (1979). *Clinical neurourology* Boston: Little Brown and Cie.
7. Dalley, A. F. (1987). The riddle of the sphincters. The morphology of the anorectal mechanism reviewed. *American Surgeon*, **41**, 298–306.
8. Cherry, D. A. and Rothenberger, D. A. (1988). Pelvic floor physiology. *Surgical Clinics North America*, **68**, 1217–1230.
9. Ger, R. (1988). Surgical anatomy of the pelvis. *Surgical Clinics North America*, **68**, 1201–1216.
10. Aargaard, J., Gerstenberg, T. and Knudsen, J. J. T. (1986). Urodynamic investigation predicts bladder dysfunction at an early stage after abdomino perineal resection of the rectum for cancer. *Surgery*, **99**, 564–568.
11. Ballinger, W. F. (1964). *Research methods in surgery*. Boston: Little Brown Co.
12. Bass, P. and Wiley, J. N. (1967). Contractile force transducers for recording muscle activity in the unanesthetized animal. *Journal of applied Physiology*, **32**, 567–570.
13. Balfour, T., Giannakos, V. and Smith, A. N. (1972). Colomyotomy and intraluminal pressures studied in the rabbit. *British Journal Surgery*, **59**, 315–319.
14. Boer, J. de, Archibald, J. and Downie, H. G. (1975) *An introduction to experimental surgery*. Amsterdam: Excerpta Medica .
15. Wenderoth, U. K., Bachor, R., Egghart, G., Frohneberg D., Miller, K. and Hautmann, R. (1989). Drei Jahre Ileum-Neoblase - die ersten 108 Patienten. *Urologe* (A), **28**, 204–211.
16. Shoenut, J. P., Wieler J. A., Micflikier, A. B. and Teskey J. M. (1990). Esophageal reflux before and after isolated myotomy for achalasia. *Surgery*, **108**, 876–879.
17. Turnage, R. H., Coran, A. G. and Drongowski, R. A. (1990). The value of intestinal myotomy and myectomy in improving the reservoir capacity of the endorectal pull-through. *Annals of Surgery*, **211**, 463–469.
18. Montagne, G. J., Bekkers J. J. M., Hoitsma, H. F. and Klopper, P. J. (1991). Longitudinal myotomy of ileum for bladder replacements in dogs. *The Journal of Urology*, **145**, 384–388.
19. Allen, J. G. (1959). The influence of physiologic observations and treatment of peptic ulcer. In *The physiology and treatment of peptic ulcer*, edited by J. G. Allen, pp. 220–232. Chicago: University of Chicago Press .
20. Markowitz, J., Archibald, J. and Downie, H. G. (1964). *Experimental surgery including Surgical Physiology*, 5th edn. Baltimore: Williams and Wilkins Cy.
21. Harkins, H. N. and Nyhus, L. M. (1969). *Surgery of the Stomach and Duodenum*, 2nd edn. Boston: Little Brown and Cy.
22. Tansy, M. F., Mackowisk, R. C. and Chaffer, R. (1971). A vagosympathetic pathway capable of influencing common bile duct motility in the dog. *Surgery Gynecology & Obstetrics*, **133**, 225–236.
23. Attwood, S. E., DeMeester, T. R., Bremner C. G., Barlow, A. P. and Hinder, R. A. (1989). Alkaline gastro-

esophageal reflux. Implications in the development of complications in Barrett's columnar-lined lower esophagus. *Surgery*, **106**, 764–770.
24. Fuchs, K. H., DeMeester, T. R., Hinder, R. A., Stein H. J., Barlow, A. P. and Gupta, N. C. (1991). Computerized identification of pathologic duodenic reflux using 24-hour gastric pH monitoring. *Annals of Surgery*, **213**, 13–20.
25. Rothschuh, K. E. (1978) *History of physiology*, edited by G.R. Risse. New York: R. E. Krieger Publ. Cy.
26. Fazio, V. W. (1988). Cancer of the rectum-sphincter-saving operation. Stapling techniques. *Surgical Clinics North America*, **68**, 1367–1382.
27. Gadacz, T. R., McFadden, D. W., Gabrielson E. W., Uilah, A. and Berman, J. J. (1990). Adenocarcinoma of the ileostomy, the late risk of cancer after colectomy for ulcerative colitis and familial polyposis. *Surgery*, 107, 698–703..
28. Turnbull, R. B. and Whealey, F. L. (1969). Atlas of intestinal stoma's. St. Louis: Mosby Co.
29. Schiller, W. R., Didio, L. J. A. and Anderson, M. C. (1967). Production of artificial sphincters. Ablation of the longitudinal layer of the intestine. *Archives of Surgery*, **95**, 436–442.
30. Kock, N. G. (1969). Intraabdominal "reservoir" in patients with permanent ileostomy. *Archives of Surgery*, **99**, 223–231.
31. Kock, N. G. (1973). Continent ileostomy. *Progress Surgery*, **12**, 180–201.
32. Myrvold, H., Tindel, M. S., Isenberg, H. D., Steins, T. A., Scherer, J. and Wise, L. (1984). The nipple valve as a sphincter substitute for the ileocecal valve. Prevention of bacterial overgrowth in the small bowel. *Surgery*, **96**, 42–47.
33. Hulten, L. (1985). The continent ileostomy (Kock's pouch). versus the restorative proctocolectomy (Pelvic pouch). *World Journal of Surgery*, **9**, 952–959.
34. Martin, L. W., Sayers, H. J., Alexander, F., Fischer, J. E. and Torres, M. A. (1986). Anal incontinence follo-wing Soave procedure. Analysis of results in 100 patients. *Annals of Surgery*, **203**, 525–530.
35. Goheim, M. A., Kock, N. G., Lycke, G. and Shehab el-Din, A. B. (1987). An appliance-free, sphincter-controlled bladder substitute, the urethral Kock pouch. *Journal of Urology*, **138**, 1150–1154.
36. Chandler, J. G., Adams, R. B., Friedman, C. J., Marcella, K. L. and Guerrant, R. L. (1985). Assessment of an implantable ileostomy sphincter. *Surgery*, **98**, 72–80.
37. Meijer, D. W. and Klopper, P. J. (1990). Construction of a stable nipple valve with processed sheep dermal collagen for continent ileostomy and urostomy. *Urological Research*, **18**, 353–355.
38. Beahrs, O. H., Kelly, K. A. and Adson, M. A. (1974). Ileosto-my with ileal reservoir rather than ileostomy alone. *Annals of Surgery*, **179**, 5–9.
39. Thor, K. B. A. and Silander, T. (1989). A long-term rando-mized prospective trial of the Nissen procedure versus a modified toupet technique. *Annals of Surgery*, **210**, 719–724.
40. Gabella, G. (1987). Structure of muscles and nerves in the gastrointestinal tract. In *Physiology of the gastrointestinal tract*, edited by L. R. Johnson, pp. 335–381. New York: Raven Press.
41. Griffith, D. J. (1980). *Urodynamics*. The mechanism and hydrodynamics of the lower urinary tract. Medical Physics Handbooks 4. Bristol: Hilger Ltd.
42. Behar, J., Biancani, P. and Zabinski, M. P. (1977). Charac-terization of the feline pyloric sphincter by neural and hormonal stimulation. In *Gastrointestinal motility in health and disease*, edited by H. L. Duthie, pp. 481-484. Lancaster: MTP.
43. Valori, R. M., Collins, S. M., Daniel, E. E., Reddy, S. N., Shan-non, S. and Jury, J. (1986). Comparison of methodo-logies for the measurement of antroduodenal motor activity in the dog. *Gastroenterology*, **91**, 546–553.
44. Gannon, B., Browning, J., O'Brien, P. and Rogers, P. (1984). Mucosal microvascular architecture of the fundus and body of human stomach. *Gastroenterology*, **86**, 866–875.
45. Fisher, Jr., D. F. and Fry, W. J. (1987). Collateral mesen-teric circulation. *Surgery Gynecology & Obstetrics*, **164**, 487–492.
46. Lieberman-Meffert, D., Allgower, M., Schmid, P., *et al.* (1979). Muscular equivalent of the lower esophageal sphincter. *Gastroenterology*, **76**, 31–38.
47. Feurle, G. E., Tischbirek, K. and Baca, I. (1988). G. Cell hyperplasia in rats with pyloric stenosis. *Digestive Disease Sciences*, **33**, 795–800.
48. Modlin, I. M., Sank, A. and Albert, D. (1981). Current as-pects of gut hormones. *Journal of Surgical Research*, **30**, 602–618.
49. Deloof, S., Croix, D. and Tramu, G. (1988). The role of vasoactive intestinal polypeptide in the inhibition of antral and pyloric electrical activity in rab-bits. *Journal of Autonomous Nervous System*, **22**, 167–173.
50. Boyarsky, S. and Labay, P. (1972). *Ureteral dynamics. Pa-thophysiology, drugs and surgical implications*. Baltimore: Williams and Wilkins Cy.
51. Boyle, J. T., Altzschuler, S. M., Nixon, T. E., *et al.* (1985). Role of the diaphragm in the genesis of the lower esophageal sphincter pressure in the cat. *Gastroenterology*, **88**, 723–730.
52. Kumar, D. and Philips, S. E. (1986). The contribution of external ligamentous attachments to function of the ileocecal junction. *Diseases Colon & Rectum*, **30**, 410–416.
53. Cohen, S. and Harris, L. D. (1971). Does hiatus hernia affect competence of the gastroesophageal sphincter. *Gastroenterology*, **58**, 157–162.
54. Dodds, W. J., Steff, J. J. and Hogan, W. H. J. (1973). Factors determining pressure measurement accuracy by intra-luminal esophageal manometry. *Gastroenterology*, **14**, 983–987.
55. Waegemaekers, C. T. B. J. (1980). Profielmeting van de distale oesofagus sphincter. *PhD. Thesis*, University of Amsterdam.
56. Kelly, K. A. and Code, C. F. (1971). Canine gastric pacemaker *American Journal Physiology*, **220**, 112–118.
57. Cannon, W. B. (1933). A method of stimulating autonomic nerves in the unanaesthetized cat with observations on the motor and sensory effects. *American Journal Physiology*, **105**, 336–372.

58. Garry, R. C. (1934). The movements of the large intestine. *Physiology Review*, **14**, 103–132.
59. Hightower, N. C. (1963). Contributions of a gastrointestinal laboratory to the practice of gastroenterology In *Current Gastroenterology*, edited by G. McHardy, pp. 56–70. New York: Hoeber Harper.
60. Connell, A. M. (1978). Tests of gastrointestinal motility. *Clinical Gastroenterology*, **7**, 317–328.
61. Read, N. W. (1989). *Gastrointestinal motility, Which test?* Petersfield: Wrightson Ltd.
62. Thomas, J. E., Crider J. O. and Mogan C. J. (1934). A study of reflexes involving the pyloric sphincter and antrum and their role in gastric evacuation. *American Journal Physiology*, **126**, 28–38.
63. Thomas, J. E. (1935). Some physical and physiological factors involved in the regulation of gastric emptying. *Review Gastroenterology*, **2**, 32–38.
64. Pope, C. E. II (1967). A dynamic test of sphincter strength. Its application to the lower esophageal sphincter. *Gastroenterology*, **52**, 779–786.
65. Arndorffer, R. C., Steff, J. J., Dodds, W. J., *et al.* (1977). Improved infusion system for intraluminal esophageal manometry. *Gastroenterology*, **73**, 23–27.
66. Malagelada, J.R., Camilleri, M. and Stangehellini, V. (1986). *Manometric diagnosis of gastrointestinal disorders*. New York: Thieme Inc.
67. Vantrappen, G., Servaes. J., Janssens. J. and Peeters. T. (1982). Twenty-four-hour esophageal pH- and pressure recording in outpatients. In *Motility of the Digestive tract*, edited by M. Wienbeck, pp. 293–297. New York: Raven Press.
68. Dijkhuis, Th., Bemelman, W. A., Hulst, V. P. M. van der and P. J. Klopper (1990). Graphical representation of eight channel sphincter pressure profiles on a personal computer. *Medical biological engineering & computers*, **26**, 502–504.
69. Misiewicz, J. J. (1968). Measurements of intraluminal pressures. *American Journal Digestive Diseases*, **13**, 389–396.
70. Haas, D. de, Steinberg, P. J. and Klopper, P. J. (1989). Qualitative analysis of the four-channel, high-resolution pressure profile of the female dog. *Journal of Urology*, **142**, 855–859.
71. Dent, J. (1976). A new technique for continuous sphincter pressure measurement. *Gastroenterology*, **71**, 263–267.
72. Kahrilas, P. J., Dodds, W. J., Dent, J., Wyman, J. B., Hogan, W. J. and Arndorffer, R. C. (1987). A method for continuous monitoring of upper oesophageal sphincter pressure (UESP). *Digestive Disease Science*, **32**, 121–128.
73. Womack, N. R., Williams, N. S., Holmfield, J. H. M., Morrison, J. F. B. and Simpkins, K. C. (1985). New method for the dynamic assessment of anorectal function in constipation. *British Journal Surgery*, **72**, 994–998.
74. Mathias, J. R., Sninsky, C. A., Millar, H. D., Clench, M. H. and Davis, R. H. (1985). Development of an improved multip-ressor-sensor probe for recording muscle contraction in human intestines. *Digestive Disease Science*, **30**, 119–123.
75. Quigley, E. M. M., Dent, J. and Phillips, S. F. (1987). Manome-try of canine ileocolic sphincter, comparison of sleeve method to point sensors. *American Journal Physiology*, **253**, G585–G591.
76. Brown, M. and Wickham, J. E. A. (1969). The urethral pressure profile. *British Journal Urology*, **41**, 211–217.
77. Abrams, P. H. (1979). Perfusion urethral profilometry. *Urologic Clinics North America*, **6**, 103–109.
78. Bruijnes, E. (1978). The ureteral pressure profile, An experimental study in canine ureters. *Urologia internationalis*, **33**, 381–392.
79. Abrams, P., Blaivas, J. G. and Stanton, S. L. (1989). The standardization of terminology of lower urinary tract function. *World Journal Urology*, **6**, 233–245.
80. Klopper, P. J., Haas, D. de and Dijkhuis, T. (1990). Computerized experimental multi-channel urethral pressure profilometry. *World Journal Urology*, **8**, 159–162.
81. Hill, D. W. (1973). *Electronic techniques in anaesthesia and surgery*. 2d. edn. London: Butterworths.
82. Summers, R. W. (1988). The role of computers in the analysis of gut motility. In *An illustrated guide to gastrointestinal motility*, edited by D. Kumar and S. Gustavsson, pp. 131–141. Chichester: Wiley and Sons.
83. Mahieu, P., Pringot, J. and Bodart, P. (1984). Defecography. I. Description of a new procedure and results in normal patients. *Gastroenterological Radiology*, **9**, 247–251.
84. Mahieu, P., Pringot, J. and Bodart, P. (1984). Defecography. II. Contribution to the diagnosis of defecation disorders. *Gastroenterological Radiology*, **9**, 253–261.
85. Pescatori, M. and Ravo, B. (1988). Diagnostic anorectal functional studies. Manometry, sphincter electromyo-graphy and defecography. *Surgical Clinics North America*, **68**, 1231–1248.
86. Asoh, R. and Goyal, R. K. (1978). Manometry and electromyography of the upper esophageal sphincter in the opossum. *Gastroenterology*, **74**, 515–520.
87. Moe, R. E. and Klopper, P. J. (1965). Demonstration of the functional anatomy of the canine gastric antrum. II. Operative technics not requiring gastrotomy. *American Journal Surgery*, **111**, 80–88.
88. Klopper, P. J. and Moe, R. E. (1968). Peroperative determination of the gastric antral border. In *The physiology of gastric secretion*, edited by M. E. Semb and J. Myren, pp. 445–447. Oslo: Universitets forlaget.
89. Klopper, P. J. and Moe, R. E. (1968). Simultaneous dual tracings of pH variations in the canine stomach. In *The physiology of gastric secretion*, edited by M. E. Semb and J. Myren, pp. 448–452. Oslo: Universitets forlaget.
90. Sutton, J. A. and McClelland, G. R. (1985). Measurement of gastric emptying rates by radioactive isotope scanning and epigastric impedance *L ancet*, 898–900.
91. Bom, N., Lancee, C. T., Slager, C. J. and de Jong, N. (1987). Ein Weg zur intraluminaeren Echoarteriographie. *Ultraschall*, **8**, 233–236.
92. Ehrlein, H. J. (1988). Motility of the pyloric sphincter studied by the inductograph method in conscious dogs. *American Journal Physiology*, **254**, g650–g657.
93. Bolondi, L., Bortolotti, M., Santi, V., Calletti, T., Giani, S. and Labo, G. (1985). Measurement of gastric em-

ptying time by real time ultrasonography. *Gastroenterology*, **89**, 752–759.

94. Hodde, K. C., Steeber, D. A. and Albrecht, R. M. (1990). Advances in corrosion casting methods. *Scanning Microscopy*, **4**, 693–704.
95. Holstege, G., Griffiths, D., Wall, H.de and Dalm, E. (1986). Anatomical and physiological observations on supraspinal control of bladder and urethral sphincter muscles in the cat. *The Journal of Comparative Neurology*, **250**, 449–461.
96. Cotton, B. R. and Smith, G. (1984). The lower oesophageal sphincter and anesthesia. *British Journal of Anaesthesiology*, **56**, 37–46.
97. Monges, H., Salducci, J. and Naudy, B. (1978). The upper oesophageal sphincter during vomiting, eructation and distension of the cardia, an electromyographic study in the unanaesthetized dog. In *Gastrointestinal motility in health and disease*, edited by H. L. Duthie, pp. 575–582. Lancaster: MTP.
98. Fyke, F. E., Code, C. F. and Schlegel, J. F. (1956). The gas-trooesophageal sphincter in healthy human beings. *Gastroenterologia*, **86**, 135–139.
99. Welch, R. W., Luckman, K., Ricks, P. M., Drake, S. T. and Gates, G. A. (1979). Manometry of the normal upper esophageal sphincter and its alterations in laryngectomy. *Journal of Clinical Investigation*, **63**, 1036–1041.
100. Zaninotto, G., DeMeester, T. R., Schwizer, W., Johansson, K. E. and Cheng, S. C. (1988). The lower esophageal sphincter in health and disease. *American Journal of Surgery*, **155**, 104–111.
101. Itoh, Z., Honda, R., Aizawa, I., Takeuchi, S., Hiwatashi, K. and Couch, E. F. (1982). Interdigestive motor activity of the lower esophageal sphincter in the conscious dog. *Digestive Disease Science*, **23**, 239–247.
102. Jones, M. J., Mitchell, R. W. D. and Hindocha, N. (1989). Effects of the lower esophageal sphincter of cisapride given before the combined administration of atropine and neostigmine. *British Journal Anesthesiology*, **62**, 124–128.
103. Chakkaphak, S., Chakkaphak, K. and Ferguson, M. K. (1991). Disorders of esophageal motility. Collective Review. In *Surgery Gynecology & Obstetrics*, **172**, 325–334.
104. Bombeck, C. T., Vaz, O., DeSalvo, J., Donahue, P. E. and Nyhus, L. M. (1987). Computerized axial manometry of the esophagus. *Annals of Surgery*, **206**, 465–472.
105. Bemelman, W. A., Hulst, V. P. M. van der, Dijkhuis, Th, Hoeven, C. W. P. van der and Klopper, P. J. (1990). The lower oesophageal sphincter shown by a computerized representation. *Scandinavian Journal Gastroenterology*, **25**, 601–608.
106. Code, C. F., Craemer, B. and Schlegel, J. F. (1958). *An Atlas of esophageal motility in health and disease*. Springfield Illinois: C. Thomas.
107. Castell, J.A. and Castell, D.O. (1986). Computer analysis of human esophageal peristalsis and lower esophageal sphincter pressure. *Digestive Diseases & Sciences*, **31**, 1211–1216.
108. Ceccatelli, P., Janssens, J., Trappen, G. van and Cucchiiara, S. (1988). Cisapride restores the decreased lower oesophageal sphincter pressure in reflux patients. *Gut*, **29**, 631–635.
109. Welch, R. W. and Drake, S. T. (1980). Normal lower esophageal sphincter, a comparison of rapid vs slow pull through techniques. *Gastroenterology*, **78**, 1446–1451.
110. Yeatman, T. J. and Bland, K. I. (1989). Sphincter saving procedures for distal carcinoma of the rectum. *Annals of Surgery*, **209**, 1–18.
111. Horowitz, M., Maddox, A., Harding, P. E., Maddern, G. J., Chatterton, B. E., Wishart, J. and Shearman, D. J. C. (1987). Effect of Cisapride on gastric and esophageal emptying in insulin-dependent diabetes mellitus. *Gastroenterology*, **92**, 1899–1907.
112. Kelly, K. A. (1986). Effect of gastric surgery on gastric motility and emptying. In *Gastric and gas-trodudenal motility*, edited by L. M. Akkermans, A. G. Johnson and N. W. Read. Surgical Science Series vol. 4 pp. 241–262. New York: Praeger.
113. Kumar, D., Ritman, E. L. and Malagelada, J. R. (1987). Three-dimensional imaging of the stomach, role of pylorus in the emptying of liquids. *American Journal Physiology*, **252**, g79–g85.
114. Houghton, L. A., Read, N. W., Heddle, R., Maddern, G. J., Dowton, J., Toouli, J. and Dent, J. (1988). Motor activity of the gastric antrum, pylorus and duodenum under fasted conditions and after a liquid meal. *Gastroenterology*, **94**, 1276–84.
115. Blum, A. L., Sonneberg, A. and Mueller-Lissner, S. (1981). Der duodenogastrale reflux, ein Grenzphaenemenon zwischen Physiologie und Pathophysiologie. In *Magen und Magenkrankheiten*, edited by W. Domschke und K. G. Wormsley, pp. 58–69. Stuttgart: Thieme.
116. Bemelman, W. A., Brummelkamp, W. H., Lubbers, M. J., Royen, E. van and Klopper, P. J. (1990). Post operative intermittent positive pressure ventilation decreases duodenogastric reflux after intrathoracic esophagogastrostomy. *Surgical Research Communications*, **8**, 29–38.
117. Meschan, I. and Quigley, J. P. (1938). Spontaneous motility of the pyloric sphincter and adjacent regions of the gut in the unanaesthetized dog. *American Journal Physiology*, **121**, 350–357.
118. DeMeester, T. R., Fuchs, K. H., Ball, C. S., Albertucci, M., Smyrk, T. C. and Marcus, J. N. (1987). Experimental and clinical results with proximal end-to-end duodenojejunostomy for pathologic duodenogastric reflux. *Annals of Surgery*, **206**, 414–426.
119. Wienbeck, M., Jahornssen, H. and Kreuzpaitner, G. (1978). Nycthemeral variation of ileocolic myoelectrical activity in the cat. In *Gastrointestinal motility in health and disease*, edited by H. L. Duthie, pp. 399–405. Lancaster: MTP.
120. Winans, C. S. (1976). The fickle pylorus. *Gastroenterology*, **70**, 622–623.
121. Kaye, M. D., Mehta, S. J. and Showalter, J. P. (1976). Manometric studies of the human pylorus. *Gastroenterology*, **70**, 477–480.
122. Heintzen, P. H. and Buersch, J. H. (1978). *Roentgen-video-techniques*. Stuttgart: Thieme.

123. Schuurkes, J. A. J. and Nueten, J. M. van (1984). Gastroduodenal co-ordination. In *Gastric and gastrodudenal motility*, edited by L. M. Akkermans, A. G. Johnson and N. W. Read. Surgical Science Series vol. 4, pp. 103–130. New York: Praeger.
124. Gaffney, P. R., Gleeson, D. J., Hall, J. W. and Brady, M. P. (1987). The manometric findings at the human pylorus. The evidence against the presence of a tonic sphincter *Scandinavian Journal Gastroenterology*, **22**, 525–532.
125. King, P. M., Pryde, A. and Heading, R. C. (1987). Transpylo-ric fluid movement and antroduodenal motility in patients with gastro-oesophageal reflux. *Gut*, **28**, 545–548.
126. Heddle, R., Dent, J., Read, N. W., Houghton, L. A., Toouli, J., Horowitz, M., Maddern, G. J. and Downton, J. (1988). Antropyloroduodenal motor responses to intraduodenal lipid infusion in healthy volunteers. *American Journal Physiology*, **254**, g671–g679.
127. Mallet-Guy, P., Jeanjean, R. and Marion, P. (1947). *L achirurgie biliaire sous controle manometrique et radiologique per-operatoire*. Paris: Masson et Cie.
128. Boyden, E. A. and Schwegler, R. A. (1936). The sphincter of Oddi in the dog. *American Journal Physiology*, **116**, 14–21.
129. Boyden, E. A. (1957). The anatomy of the choledochoduodenal junction in man. *Surgery Gynaecology & Obstetrics*, **104**, 641–653.
130. Torsoli, A. and Corazziari, E.(1988). Biliary tract. In *An illustrated guide to gastrointestinal motilility*, edited by D. Kumar en S. Gustavsson, pp. 175–185. Chichester: Wiley and Sons.
131. Bruijnes, E. and Klopper, P. J. (1978). The pressure profile of the common bile duct, an experimental study in 15 mongrel dogs. *Acta Hepato-gastroenterologica*, **25**, 257–332.
132. Coelho, J. C. U., Moody, F. G. and Senninger, N. (1985). A new method for correlating pancreatic and biliary duct pressures and sphincter of Oddi electromyography. *Surgery*, **97**, 342–349.
133. Cameron, J. L. and Gadacz, T. R. (1990). Laparoscopic cholecystectomy. *Annals of Surgery*, **213**, 1–2.
134. Moody, F. G., Calabuig R., Vecchio R. and Runkel N. (1990). Stenosis of the sphincter of Oddi. *Surgical Clinics North America*, **70**, 6, 1341–1354.
135. Peters, J. E., Ellison, E. C., Innes, J. T., Liss, J. L., Nichols, K. E., Lomano, J. M., Roby, S. R., Front, M. E. and Carrey, L. C. (1990). Safety and efficacy of laparoscopic cholecystectomy. *Annals of Surgery*, **213**, 3–12.
136. Calbuig, R., Saccone, G. T. P. and Toouli, J. (1989). Inhibition of sphincter of Oddi motility in the Australian opossum by cholic acid. *Surgery*, **106**, 872–878.
137. Coelho, J. C. U., Gouma, D. J., Moody, F. G. and Schlegel, J. F. (1986). Effect of feeding on myoelectric activity of the sphincter of Oddi and the gastrointestinal tract of the opossum. *Digestive Disease Sciences*, **31**, 202–207.
138. Kellog, J. H. (1918). Repair of the ileocecal valve. *Annals of Surgery*, **67**, 83–93.
139. Kelley, M. L., Gordon, E. A. and de Weese, J. A. (1966). Pressure responses of canine ileocolonic junctional zone to intestinal distension. *American Journal Physiology*, **211**, 614–618.
140. Quigley, E. M. M., Philips, S. F., Dent, J. and Taylor, B. M. (1983). Myoelectric activity and intraluminal pressure of the canine ileocolonic sphincter. *Gastroenterology*, **85**, 1054–1062.
141. Quigley, E. M. M., Phillips, S. F. and Dent, J. (1984). Distinct patterns of interdigestive motility at the canine ileo colonic junction. *Gastroenteroly*, **87**, 836–844.
142. Nasmyth, D. G. and Williams, N. S. (1985). Pressure characteristics of the human ileocecal region. A key to its function. *Gastroenterology*, **89**, 345–351.
143. Richardson, J. D. and Griffen, W. O. (1972). Ileocaecal valve substitutes as bacteriological barriers. *American Journal Surgery*, **123**, 149–153.
144. Sanada, Y., Fonkalsrud, E. W. and Kojima, Y. (1982). Intermittent ileostomy occlusion for fecal storage using a balloon catheter. *Surgery*, **91**, 459–466.
145. Singleton, Jr., A. O., Redmond, D. C. and McMurray, J. E. (1964). Ileo-caecal resection and small bowel transit and absorption. *Annals of Surgery*, **159**, 690–694.
146. Spiller, R. C. (1989). Colonic manometry In *Gastrointestinal motility, Which test?*, edited by N. W. Read, pp. 191–202. Petersfield: Wrightson Ltd.
147. Baker, W. N. W. and Mann, C. V. (1981). The rectosigmoid junction zone, another sphincter? In *Alimentary sphincters and their disorders*, edited by P. A. Thomas and C. V. Mann, pp. 201–211. London: MacMillan Ltd.
148. Kerremans, R. (1969). *Morphological and physiological aspects of anal continence and defecation*. Brussels: Editions Arscia.
149. Oh, C. and Kark, A. E. (1972). Anatomy of the external anal sphincter. *British Journal Surgery*, **59**, 717–723.
150. Shavik, A. (1976). A new concept of the anatomy of the anal sphincter mechanism and the physiology of defecation *Investigative Urology*, **13**, 271–277.
151. Parks, A. G. (1977). Postanal pelvic floor repair and the treatment of anorectal incontinence. In *Operative Surgery Colon, rectum and anus*, edited by C. Robb and R. Smith, 3d edn, pp. 249–254. London: Butterworth.
152. Taylor, B. M, Bert, R. W. and Phillips, S. F. (1984). Longitudinal and radial variations of pressure in the human anal sphincter, *Gastroenterology*, **86**, 693–697.
153. Coller, J. A. (1987). Clinical application of anorectal manometry. *Gastroenterologic Clinics North America*, **16**, 17–31.
154. Perry, R. E., Blatchford, G. J., Christensen, M. A., Thorson, A. G. and Attwood S. E. A. (1990). Manometric diagnosis of anal sphincter injury - a new analytical technique. *American Journal Surgery*, **159**, 112–117.
155. MacNamara, M. J., Percy, J. P. and Fielding, I. R. (1990). A manometric study of anal fissure treated by subcutaneous lateral internal sphincterotomy, *Annals of Surgery*, **211**, 235–243.
156. Lundell, L., Meyers, J. C. and Jamieson, G. G. (1996). Is motility impaired in the entire upper gastrointestinal tract with gastro-oesophageal reflux disease *Scand. J. Gastroenterol.*, **31**, 131–135.
157. Sangwan, Y. P. and Coller, J. A. (1994). Fecal incontinence..*Surg. Clin. North America*, **74**, 1377–1398.

158. Nieuwenhuizen, G. A. P., Deitch, E. A. and Goris, R. J. A. (1996). Infection of the gut and the development of the multiple organ dysfunction syndrome. *Eur. J. Surg.*, **162**, 259-273.
159. Meijer, D. W. (1992). The continent ileal reservoir. *Eur. J. Surg.*, suppl. 568.
160. Salemans, J. M. J. I. and Nagengast, F. M. (1995). Clinical and physiological aspects of ileal pouch-anal anastomosis. *Scand. J. Gastroenterol*, **30**, suppl. 212, 3–12.
161. Johnston, B. T., Nunn, S, Sloan, J. W. Collins, J. S. A., McGarland s, R. J. and Parkin, S. (1996). The application of microridge analysis in the diagnosis of gastro-oesophageal reflux disease *Scand. J. Gastroenterol*, **31**, 97–102.
162. Verbeek, F. J., Huismans, D. P., Baeten, R. J. A. M., Schouten, N. J. C. and Lamers, W. H. (1995). Design and implementation of a database and program for 3D reconstruction from serial sections: A data-driven approach. *Microscopy research and technique*, **30**, 496–512.
163. Thomson, A. S. Dabhoiwala, N. F. Verbeek, F. J. Lamers, W. H. (1994). The functional anatomy of the uretrovesical junction *Br. J. Urol.*, **73**, 284–291.
164. Goldberg, B. B., Liu, J-B., Merton, D. A and Kurtz, A. B. (1990). Endoluminal US: Experiments with non vascular uses in animals *Radiology*, **175**, 39–43.
165. Cuesta, M. A. Meijer, S. Derksen, E. J. Boutkan, H. and Meuwissen, S. G. M. (1992). Anal sphincter imaging in fecal incontinence using endosonography. *Dis. Col. Rectum*, **35**, 59–63.
166. Senagore A. J. (1994). Intrarectal and intraanal Ultrasonography in the evaluation of colorectal pathology. *Surg. Clin. North America*, **74**, 1465–1473.
167. Ebert, T. *et al.* (1995). Hochenergetischer fokussierter ultraschall (HIFU)in der behandlung der benignen prostathyperplasie, *Urologe*, **34**, 404–408.
168. Haas, D.de (1993). Multichannel urethral pressure profilometry *Ph.D.Thesis*. University of Amsterdam.
169. Benninga, M. A. Wijers, O. B. Hoeven van der C. W. P., Taminiau, J. A. J. M. Klopper, P. J. Tijtgat, G. N. J. and Akkermans, M. A. (1994). Manometry, profilometry and endosonography Normal physiology of the anal canal in healthy children. *J. Paediatr. Gastroenterol. Nutr.*, **18**, 68–77.
170. Bansil, R., Stanley, E. and Lamont, J. T. (1995). Mucin biophysics *Ann. Rev. Physiol.*, **57**, 635–657.
171. Sands, B. E. and Podolsky, D. E. (1996). The trefoil peptide family *Ann. Rev. Physiol*, **58**, 253–273.
172. Soper, N. J. and Sarr, M. G. (1988). Electromyography. In *An illustrated guide to gastrointestinal motility*, edited by D. Kumar and S. Gustavsson, pp. 101–111 Chichester John Wiley and Sons.
173. Read, N. W. and Sun, W. M. (1989). Anorectal Manometry, anal myography and rectal sensory testing In: *Gastrointestinal motility: which test ?* Edited by N. W. Read pp. 227–241, Petersfield Wrightson Biomedical Publishing Ltd.
174. Stein H. J., Barlow, A. P., DeMeester, T. R. and Hinder, R. A. (1992). Complications of gastroesophageal reflux disease. Role of the lower esophageal sphincter,esophageal acid and acid / alkaline exposure, and duodenogastric reflux. *Ann. Surg*, **216**, 35–43.
175. Peters, J. H. and DeMeester, T. R. (1993). Gastroesophageal reflux *Surg. Clin. N. America.*, **73**, 1119–1144.
176. Farkas, D. L. *et al* (1993). Multimode light microscopy and the dynamics of molecules, cells, and tissues *Ann. Rev. Physiol*, **55**, 785–817.
177. Hoeven, C. W. P. van der, Attia, A., Deen, L. and Klopper, P. J.The influence of anaesthetic drugs on the lower oesophageal sphincter in propofol / nitrous oxide anaesthetized dogs.Pressure profilometry in an animal model. *Acta Anaesthesiol. Scand.*, **39**, 822–826.
178. VanDamme, J. P. J. (1993). Behavorial anatomy of the abdominal arteries. *Surg. Clin. North America*, **73**, 699–725.
179. Hull, T. L. and Milsom, J. W. (1994). Pelvic floor disorders. *Surg. Clin. North.America*, **74**, 1399–1413.
180. Hajivassiliou, C. A., Carter. K. B. and Tinlay, S. G. (1996). Anorectal angle enhances faecal continence. *Brit. J. Surg.*, **83**, 53–56.
181. Sorensen, M., Tetschner, T., Rasmussen, O. O., Bjarnesen, J. and Christiansen, J. (1993). Sphincter rupture in childbirth *Br. J. Surg.*, **80**, 392–394.
182. Hulst, V. P. M. van der (1995). 3-Dimensionale druk profilometrie van de anus sfincter:ontwikkeling van een nieuwe methode *Ph. D. Thesis*. University of Amsterdam.

PART III

7. Experimental Digestive Fistulae

José Ramon Morandeira

Service of Biomedicine and Biomaterials, University of Zaragoza, Spain

The Interest of Experimental Digestive Fistulae

The preparation of experimental fistulae in order to obtain digestive secretions has been a well-known process for a long time.[36,46]

This type of operation was first put into practice on the experimental animal with the intention of obtaining samples of gastric or intestinal content, without having to resort to oral catheterization,[36,37] and, at the same time, of studying intestinal motility.[4,7] Later, its use was extended to the obtention of biliary secretion and, at present, different models of fistulae are performed which permit the obtention of gastric, intestinal, biliary and pancreatic secretions[11]. These fistulous tracts, which communicate a secreting digestive organ with the exterior, not only permit the collection of the secretion of the said organ, but also make it possible to introduce diverse substances into the cavity through the fistula so as to study the systemic or localized repercussion which their stimulus and/or stasis produces in blood as well as in the digestive juices. Finally, they can also be used for the repeated practice of barytated radiographs, endoscopies, the taking of biopsic samples and others.

With this aim in mind, numerous methods have been proposed, such as the use of intestinal fragments exteriorized to the skin,[21] the implantion of various types of catheters[53] and the use of prostheses of metal or of other materials.[37]

All these procedures enjoy different prestige according to the different authors. In many cases it is the results, rather than their use, which are contradictory.

In any case, the ideal of all these techniques should be an optimum survival of the animal for an indefinite length of time and a minimum operatory morbi-mortality, all of this together with the simplicity of the use of the set up without having to resort to anaesthesia or any other act which alters the physiology of the animal, and with the minimum attention necessary of maintenance and conservation.

In the following pages we will go into the following:

a. A brief list of the best-known types of fistulae proposed by the different authors.
b. Which of these fistulae are most commonly used, either on their own or in association with diverse complementary experimental set ups, for the isolated obtention of digestive secretions. Such is the case of isolated gastric pouches, biliary and pancreatic derivations, etc.
c. What morbility and mortality is attributed by the authors to each type of digestive fistula.

At the same time, together with the experience reported by other authors, we will give an account of that accumulated in our laboratory over the last 20 years, during which we have worked with different models of these fistulae, comparing the efficiency and complexity of care necessary for their use,[11] and will try to explain which come closest to the ideal set out above.

Classification of the Different Experimental Models for the Preparation of Digestive Fistulae

The distinct procedures described for the individualized obtention of digestive secretions may be classified as:

1. According to the material used in their preparation 2. According to the type of digestive secretion obtained

1. According to the Material Used in their Preparation

Here we can distinguish between:

1.1. Those which are based on the preparation of isolated intestinal loops.
1.2. Those which use intestinal fistulae without cannulae.
1.3. Those which obtain digestive secretions by means of the preparation of intestinal fistulae with cannulae or other prosthetic means.

But many of these procedures can be used indistinctly to obtain one or various of the digestive secretions. This occurs with certain types of fistulae or catheters which can be implanted at different levels of the digestive tube and even in the gallbladder or in the pancreas. Because of this, we prefer to follow the classification which considers not only the material used for the preparation of the fistula, but also the type of digestive secretion to be obtained.

This classification may be subdivided according to:

(i) whether metal prosthesis or other materials are used.
(ii) the preparation of fistulae in which there is continuity with the digestive tract.
(iii) whether they are pouches in which a portion is separated definitively or the complete viscera of the digestive tract.

Thus, all the possibilities are admitted and it is especially useful for explanatory purposes.

This classification, which we will follow from now on studies the different types of fistulae.

2. According to the Type of Digestive Juice Obtained

2.1. Intestinal fistulae
  2.1.1. Without cannulae
  2.1.2. With cannulae

2.2. Gastric fistulae
  2.2.1. Simple gastric fistulae
  2.2.2. Total gastric fistulae
    2.2.2.1 The Walters and Bollman fistula
  2.2.3. Total gastric pouches
  2.2.4. Partial gastric pouches

2.3. Biliary Fistulae

The Most Commonly used Procedures for the Experimental Obtention of Digestive Secretions

I. Methods for the Obtention of Intestinal Secretion: Intestinal Fistulae

The preparation of experimental intestinal fistulae has three fundamental objectives:

(i) The obtention of the intestinal content at different levels.
(ii) The introduction of diverse substances as stimulants.
(iii) The isolation of intestinal loops in order to study their physiology and the reprecussion of their stimulus on other viscera or systems.

Firstly, we shall study the isolated intestinal loops, since, in many cases, they served as inspiration in the preparation to the different types of fistulae described later. In continuation we shall look at those models of intestinal fistulae performed in continuity with the digestive tract, using exclusively intestinal fragments of greater or lesser width. Finally, we shall analyse the different methods proposed based on metallic cannulae and catheters of rubber and other materials.

1. Intestinal Fistulae without Cannulae.

1.1. Isolated Intestinal Loops. The essential aim of the different techniques is to obtain an intestinal loop isolated from the rest of the intestine, placing it in a position which is easily accessible but conserving its vascular, lymphatic and nervous connexions in perfect condition.
The best-known procedures are as follows:

(i) Thiry's Procedure.[46]

(ii) Vella's Procedure.[50]

(iii) The Beibl and Douglas-Mann Procedure[4,7]**.** This technique is mainly used for the study of intestinal movements.

(iv) Puestow's Procedure.[40] The maintenance of this type of preparation over a long period requires great care and the use of a protection corset.

(v) The Barcroft and Robinson Procedure.[3]

(vi) The Pavlov[42] **and Omi**[35] **Procedures.** These procedures are used exclusively for experiments on isolated loops.

(vii) An intestinal fragment isolated and innervated beneath the skin: These procedures permit diverse experimental techniques to be performed easily such as:

- The perfusion of liquids of variable compositions so as to study the possible variations in intestinal absorption.
- The extraction of a pure intestinal juice without mixing it with other digestive secretions or with food.
- The temporary placing of electrodes in the intestinal wall or lumen.

(viii) The model of intermittent intestinal isolation proposed by Morandeira starting from two Mann and Bollman antiperistaltic fistulae[27,31] (Fig. 7.1).

This model, initially based on two antiperistaltic intestinal fistulae in the style of Mann and Bollman,[21] anastomosed to the first and fourth duodenal portions, may be applied to any section of the intestine.

The set up permits a simple structure capable of providing intermittent duodenal isolation, according to the needs of the experimental process. For this, two Foley catheters such as those normally used for humans for the catheterization of the urinary bladder are sufficient. The first of these catheters is introduced in the upper fistula as far as the pylorus and then the distal globe of the catheter is inflated (with 6 or 8 c.c of air or water). It is fixed to the skin in such way that it will not be shifted by the peristaltic movements; in this way the duodenal lumen is blocked at the level of the pylorus which prevents the gastric and duodenal secretions from mixing. The tip of the second Foley catheter is closed previously by strangling it between the terminal orifice of aspiration and the inflatable globe. A new orifice is opened near the globe, taking great care not to damage the fine tube which is used to inflate it. The inflation of the globe achieves the isolation of the duodenal segment between the two globes. In this way it is possible to collect the doudenal secretion in isolation without mixing it with the gastric or jejunal juices and, at the same time, it is possible to instil substances in the temporarily isolated duodenal segment and analyse the repercussions the stasis of these substances at the doudenal level provokes in the gastric juices and in blood.

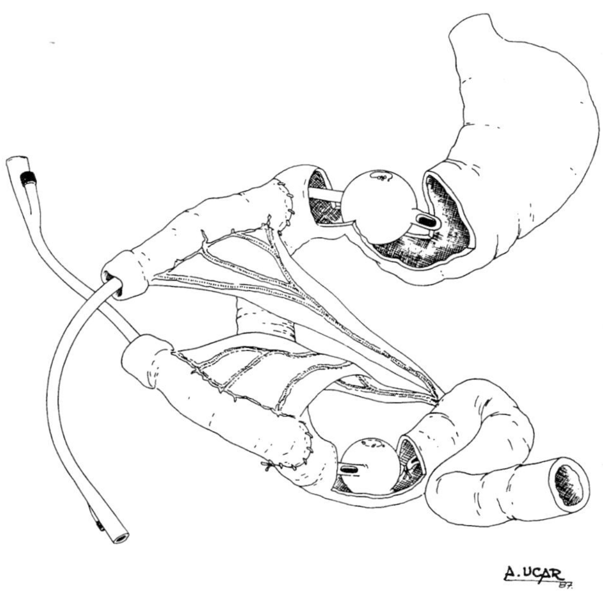

Figure 7.1 The model of intermittent intestinal isolation proposed by Morandeira starting from two Man and Bollman antiperistaltic fistulae.

Throughout the time of the experiments the animal is barely affected by any discomfort, and only in a few cases is it necessary to administer a transquillizer.

Once the experiment has finished the globes are deflated and removed, thus reestablishing the digestive tract to its normal conditions. The animal can be conserved for an indefinite time, with minimum care.

The authors have used other setups which permit the isolation of an intestinal fragment in a similar way to the one described. Such is the case of an intestinal loop prepared in the style of Vella and anastomosed latero-laterally over its whole length to the duodenal arc[31]. But the problem lies in the fact that one of the two exteriorized extrems of the loop is isoperistaltic which means that the reflux passing through it is very intense, leading invariably to the undernourishment and cachexia of the animal.

Similar problems are presented by the use of fistula such as that of Maydl. The same occurs with procedures which use metallic cannulae, as we shall describe in the corresponding section.

(ix) The Anderson and Unvas duodenal fistula.[1]
(x) The Hill, Neigus and Wilhelmj Procedure.[17]
(xi) The Walters and Bollman Procedure.[52]

1.2. Intestinal Fistulae in Continuity with the Digestive Tract

In certain studies of intestinal secretions, it may not be of interest of isolate an intestinal section of the general digestive tract definitively. For this reason fistulae are devised which directly communicate the contents of the intestine with the exterior.

Among these systems for the preparation of fistulae without cannulae, the following stand out:

(i) Maydl's Procedure.[22]
(ii) The Mann and Bollman fistula.[21] This fistula obtained from an intestinal fragment isolated anastomosed antiperistaltically from the skin to the intestine, may be implanted at any point of the gastrointestinal tract.

So as to achieve good results in the construction of this type of fistulae, it is fundamental to observe the following requirements:

- Intestinal fragments should be found which present good vascular pedicles which irrigate them (preferably terminal ileus).
- The fistulous intestinal fragment should be at least 10–12 cms. long. A shorter length means the risk of reflux and cutaneous ulceration.
- It is fundamental that the anastomosis of the fistula should be carried out in the antiperistaltic direction.
- The exterior orifice of the fistula through the abdominal wall should be wide enough to avoid the later strangulation of the stump.
- The fistula should be fixed both to the skin and the peritoneum with various silk stitches in order to avoid its retraction and invagination.

This type of fistula permits the aspiration of the intestinal content at any point of the digestive tract. Furthermore, it is the base of the models of intermittent duodenal isolation and of the isolated obtention of the gastric, biliary and pancreatic secretions described by Morandeira. Finally, the construction of one of these fistulae in the stomach permits the obtention of gastric juices without their being mixed with other secretions, by means of the insertion of a Foley catheter through the fistula; the inflation of the catheter globe at the level of the pylorus avoids the reflux of secretions towards the stomach and ensures that the juices obtained by aspiration by means of a second catheter introduced through the fistula will be purely gastric. Once the two catheters are removed, the normal continuity of the gastric tract is reestablished.

These models allow the instillation of diverse substances through the fistulae for the later analysis of the repercussions said substances produce in the digestive secretions and in blood, as well as the study of the modifications produced by the perfusion in blood of different drugs.

In our experience, the Mann Bollman fistula, with 20% morbility and the identical percentage of mortality, is the model which offers the best results of all those we have used. We believe that these good results must be attributed to various causes. In the first place, we are dealing with an antiperistaltic fistula in which, if the practical limits of the fistulous fragment are respected, there should be no reflux. Moreover, it is a fistula which requires little care or maintenance and the complications and infection, dehiscence, etc. are similar to those presented by laparotomies. The main complication, the thrombosis of the mesenteric vascular pedicle, is the principal cause of the morbomortality of this fistula, and is easily avoidable if care is taken during the intervention not to traumatize the vascular pedicule excessively, and to avoid tension. Finally, its versatility makes it useful in the majority of experimental procedures and setups for the isolated obtention of digestive juices.[27,31,41] for these reasons, we believe that it should be an obvious choice when constructing digestive fistulae.

2. Intestinal Fistulae with Cannulae

(i) Pavlov's Fistula.[22,36] The use of this cannulae, the simplest, oldest and best-known of all those that have been described, permits the obtention of samples of the gastrointestinal content at any level.

The use of Pavlov's cannula requires precise care, keeping it closed with a plug when intestinal content is not being extracted, so as to avoid the continual loss of said content and the consequent maceration of the skin and hydroelectrolitic disorders. Furthermore, it is necessary to perform daily aseptic cures around the cannula to prevent ascending infection, and to protect the zone intervened with corsets or similar methods to prevent the cannula from being pulled out.

Despite this care, and as our experience shows such complications appear frequently.

(ii) Thomas's Cannula[47] **– The Crocker-Markowitz Procedure.**[22] Our experience with this model presenting a morbility of 100% and a mortality of 75%, which is easily understandable if the implantation of two cannulae and a polyvinyl tube is considered, has led us to discard its use.

(iii) Murat's Fistula.[32] This procedure is intended to install an intermittent total intestinal fistula by means of an ingenious cannula which is made up of three elements:

(i) A 15 cms. long teflon tube with a variable diameter, with an oval lateral orifice 1 cm. from the end: at the other end there is an external disc to fix it to the skin.
(ii) A hollow chrome-plated metal tube of the same length as the cannula. It is used only during the experimental periods.
(iii) A solid teflon cylinder which is placed inside the cannula and which is of a shorter length.

The combination of two Murat cannulae, implanted at the level of the first and fourth duodenal sections, permits, after the insertion of two FOLEY catheters modified according to the procedure described by Morandeira, the temporary and intermittent isolation of the duodenum, the use of which has already been described.

In our experience, this is the procedure which is most efficient and gives the best results among those which use metallic prostheses. Its principal inconvenience is that of infections, which is why, once implanted, the care of these cannulae must be most meticulous.

(iv) Witzel's Enterostomy.[22,42] Witzel's classical ostomy performed on a catheter tunnelized in the wall of either the stomach (gastrostomy) or of the small intestine (jejunostomy, ileostomy) is a useful and relatively simple procedure for the instillation of substances in the digestive tract, or for taking samples of its content.

Our experience shows that the maintenance of this type of ostomies presents one important problem: the pulling out of the catheter, for which reason it is advisable to use a convenient corset. Should the catheter be pulled out, it must be replaced immediately before the fistulized tract stenoses.

(v) Coffey's Modification to Witzel's Method[22] **– a Catheter with Three Independent Aspiration Orifices.** Its use, with a zero morbi-mortality in our experience, makes it a good procedure which could be optimum if we had a radio-opaque and marked catheter, with small inflatable globes between the scaled orifices at our disposal, in order to carry out the intermittent isolation of different intestinal sections. Nevertheless, it should be understood that its use would require, apart from radioscopic control, the construction of a fistula for the introduction of the catheter or the anaesthesis of the animal each time the peroral catheterization is performed.

II. Methods for the Obtention of the Gastric Juice: Gastric Fistulae

Introduction

Experimental studies into the physiology of gastric secretion, the histological alterations of its mucus or its motility demand the construction of models which allow the repeated sampling of the gastric juice, endoscopies and seriated biopsies and radiological or other studies.

In man, these studies are usually carried out by approaching the gastric cavity orally. However, in the experimental animal, this kind of approach would imply repeated anaesthesia with its logical drawbacks. Various procedures have been proposed to avoid this problem. The most straightforward of these is the creation of simple gastric fistulae which permit intragastric access without troubling the animal with an esophagic catheter.

These simple gastric fistulae are probably the most commonly used preparation due to their easy construction, which maintains the integrity of the stomach without altering the phenomena of gastric digestion and which are perfectly tolerated. Because of this, they constitute an ideal method for the study of the secretory, nervous or humoral mechanisms in general, whether they are triggered by exogenous chemical agents (insulin, pentagastrin, etc.) or endogenous (ficticious food or "sham feeding").

However, the simple gastric fistula presents serious drawbacks which limit its use. In this sense, it is worth pointing out that this type of fistulae:

- Does not avoid the salivary and biliopancreatic contamination of the gastric juice
- Does not collect the totality of the gastric juice segregated by the stomach. An important part escapes through the pylorus.
- Does not permit the study of the gastric response to real feeding.

Therefore, total gastric fistulae or pouches were devised which completely separate the stomach from the digestive tract which is reconstructed by means of an esophagojejunostomy.

But these total gastric fistulae, apart from the fact that they do not permit the study of the gastric response to real feeding mean the consequences of a total gastrectomy for the animal, the loss of an important digestive secretion which incides on the hydroelectric balance, and, finally, the derivation of the acidic secretion of the duodenal circuit, this supressing an important gastric regulator of intestinal origin.

Because of this, partial gastric pouches or "little stomachs" have been proposed, both innervated and denervated, which are prepared from a small gastric portion (antral or fundal), completely separated from the rest of the stomach and connected to the exterior either directly or by means of a fistula. In this way, maintaining the majority of the gastric functions in the rest of the stomach, it is possible to study the physiology of gastric secretion under ideal conditions in these "little stomachs".

In this section we shall describe the simple gastric fistulae first, bearing in mind that these fistulae can be constructed on the whole stomach, on a total gastric pouch or on a "little stomach". In continuation we shall look at the total gastric fistulae and finally we shall study the isolated gastric pouches.

1. Simple Gastric Fistulae

1.1. Simple Gastric Fistulae without Cannulae

(i) The Mann and Bollman fistula.[21]

(ii) The Sbanajew-Franck procedure.[22]

(iii) Gavriliu's procedure.[6] This procedure consists in performing a tubulization of part of the main curvature so as to constitute a fistulous tract which is exteriorized to the skin. For this the region of the antrum is used, without peptic activity, to avoid the erosion of the skin.

The tube constructed in this way has the advantage that it is antral portion without peptic activity which is exteriorized to the skin. On the other hand, it has the drawback of being isoperistaltic and, therefore, favours the outward flow of the gastric content, apart from the fact that the base of implantation is at the level of the fundus and makes the collection of the gastric juice more difficult. To remove these problems there are two possible solutions:

(i) Constructing the tube inversely; that is to say, with the base of the implantation in the antrum.
(ii) Implanting a cannula of continence at the outlet of the fistulous tube.

1.2. Simple Gastric Fistulae with Cannulae

(i) Total Gastric Fistulae
(ii) The Walters and Bollman Fistula.[51]

2. Total Gastric Pouches These are the least used of all the gastric pouches due, firstly, to the fact that their indications are not at all frequent. Moreover, they mean the loss of all the gastric juice, which leads to grave homeostatic consequences unless a very strict vigilance and reanimation are set up. Finally, the removal of the stomach from the digestive tract causes considerable nutritional difficulties.

The essential indications of these total gastric pouches, whether they are innervated or not, lie in the study of the stimulation or inhibition mechanisms of intestinal origin. They are also excellent preparation for the obtention of large quantities of gastric juice in order to study the effect of the said secretion on the absorption of certain elements or for biochemical studies[53].

(i) The Innervated Total Gastric Pouch of Dragstedt *et al.*[9,10]
(ii) Fremont's Denervated Total Gastric Pouch.[18]

3. Partial Gastric Pouches Certain studies into gastric secretion makes it necessary to obtain samples of pure gastric juice, in exactly the same state as when it is segregated during digestion but without being mixed with food. Furthermore, it is necessary to maintain the stomach in the digestive circuit so as to avoid the inconveniences we mentioned when speaking of total gastric pouches.

To achieve this, it is possible to isolate a kind of tube from the stomach, exteriorizing it in a way that allows us to obtain its content. This operation does not interfere with the functioning of the rest of the stomach and can be carried out conserving the vagal innervation or interrupting it.

Experimentally, these isolated pouches allow us to obtain a considerable quantity of gastric juice; they reflect the chemical and humoral changes, it being possible to prepare antral or fundal isolated pouches according to whether the secretory control mechanism to be studied is vagal or gastrinic.

This type of perparation is highly favoured by researchers and they have been used to study the physiology of gastric juice since the time of Pavlov. Nevertheless, it is necessary to point out that no animal preparation, however perfect it may seem, can replace studies on a conscious man in his environment and with his normal physiological conditions. Without doubt, the use of the experimental animal, in particular the dog, has made it possible to further our knowledge of gastric physiology, but these methods must not make us forget that there are plenty of limitations, the most important of which we shall set out in continuation[16]:

(i) A fasting dog, and without any type of stimulus on its conditioned reflexes, has no basal secretion. This rarely occurs in man.
(ii) The interrelations between the digestive tract and the higher cerebral structure are far more complicated in man than in the dog.
(iii) It is undeniable that the "little stomach" are quite representative of what occurs in the complete stomach. However, it cannot be denied that their construction causes a reduction of the secretory mass of the principle stomach which influences its response and which, moreover, may affect the functioning of the pouch by the supression of the inhibition mechanisms of antral origin. Finally, the secretion of the pouch drains outside the duodenum, a fact which also alters the secretory mechanisms.
(iv) The act of operation implies anatomical alterations both on the pouch as well as on the rest of the stomach, due to the presence of cicatricial tissue, artificial ligatures, partial denervation, etc.
(v) Finally, the "little stomach" does not come into contact with the alimentory bolus, which removes the effect of mechanical distension and the possible local influences due to the composition of the food.

All these facts should not dampen the researcher's enthusiasm, but should be born in mind when extrapolating to man the results obtained from the animal.

3.1. Antral Pouches Antral pouches are to the exploration of the gastrinic mechanisms what ficticious feeding, after an esophagectomy, is to the study of the vagal phase. They are, therefore, of supreme interest as they allow us to study the liberation of endogenous gastrin in isolation.

Normally, these antral pouches produce little alkaline pH mucoprotein secretion, a fact of great importance since the appearance of an acid pH indicates the inclusion of parietal cells on the antral pouch which may be easily demonstrated by means of the elimination of stains (for example, neutral red).[26]

(i) Denervated antral pouches.[32] This type of pouch is the ideal preparation to achieve states of endogenous hypergastrinemia.

(ii) Innervated antral pouches.[32] These differ from the preceding ones in as much as a complete transversal section is not carried out at the level of the antrofundal union; the separation is performed by means of the creation of a double mucous septum. In this way, the innervation of the antral pouch remains integral. The rest of the innervation is idential to the previous description.

3.2. Fundal Pouches

The fundamental objective of these pouches is to collect the acidic secretion of the stomach, using both innervated and denervated pouches.

3.2.1. Denervated fundal pouches

(i) Heidenhain's pouch.[24] Is the most popular of the "little stomachs", and is the principle experimental preparation used for the study of the hormonal phase of gastric secretion. Its fundamental indication is the study of the gastrinic mechanism, whether its stimulation is endogenous (food, antral transplant of the colon, stimulation of an isolated antrum) or exogenous (gastrin or tissular extracts, gastrinic derivatives of synthesis such as pentagastrin).

We must insist on the importance of constructing a Heidenhain's pouch of the correct dimensions as its sensitivity will be greater.[14] This is why it is necessary to be especially meticulous when suturing so as to avoid the retraction of the rest of the stomach, with the possibility of mediogastric stenosis if a very large pouch is constructed, or of the Heidenhain's pouch if it is too small.

The Goldberg and Mann procedure and other methods of drainage of Heidenhain's pouch[12].

As we have already said, the main drawback from the surgical point of view of Heidenhain's pouch lies in the outward flow of irritating liquid which macerates the abdomen wall. In order to avoid this tendency of excoriation, Goldberg and Mann decided to leave the isolated pouch in the abdominal cavity and exteriorize it by means of a Mann and Bollman fistula. In this way a kind of valve is created which, as it functions antisperistaltically, drives the juice towards the gastric pouch.

However, in the proximities of the anastomosis of the fistulous intestinal fragment with the gastric fundal pouch there is an almost 80% incidence of a typical peptic mucous ulcer in the intestinal mucuous membrane, which, with time, has converted this method in an excellent model for the study of the development of peptic ulcers.

This is why, at present, the drainage of Heidenhain's pouches is usually carried out by means of cannulae implanted in its anterior face.

(ii) The Bickel and Gregory procedure for the total denervation of Heidenhain's pouch[24]**.**

(iii) Heidenhain's pouch associated with a gastric fistula on the remaining stomach.[32]

This particularly interesting mixed preparation is easy to construct in one session and permits the simultaneous study of the responses of the innervated and denervated stomach.

3.2.2. Innervated fundal pouches The numerous indications of innervated fundal pouches are, in principle, the same as those of simple gastric fistulae but with the advantage that the secretion obtained is not mixed with other digestive juices. Furthermore, they offer the possibility of studying the physiological response to various types of alimentation.

Innervated fundal pouches respond to vagal stimulation and their sensitivity to gastrin is greater than denervated pouches.

The same as with gastric fistulae, the main drawback of the innervated pouches lies in the difficulty of achieving authentic basal conditions, it being often necessary to wait various hours to obtain a zero secretion from the pouch.

(i) Pavlov's pouch[48] **– Neuwelt's procedure.**[34] Trying to achieve a better innervation of the fundal pouch initially described by Pavlov, in 1940, Neuwelt *et al.* proposed an interesting modification which, by conserving all the innervation of the posterior face of the stomach, manages to preserve 50% of the vagal fibres.

(ii) Perry's pouch.[39] This technical modification of Pavlov's pouch, introudced by Perry *et al.* in 1959, is based on a careful and atraumatic dissection of the anterior vagal trunk with the aim of maintaining all the vagal innervation of the pouch intact.[45]

(iii) Thomas' pouch[13,48]**–The Rudick-Hoexter procedure.**[44] This interesting variant of Pavlov's pouch is directly derived from the innervated total gastric pouches described by Dragstedt *et al.*[10]

The advantage of this pouch over those of Perry[39] and Neuwelt[34] is that it conserves innervation completely, and over that of Thomas[13,48] its advantage is that there is no risk of fistulization. Moreover, by conserving a gastric reservoir, it avoids the alimentary problems caused by total gastric pouches.

III. Methods for the Obtention of the Biliary Secretion: Biliary Fistulae

In humans, the exteriorization of the bile is usually performed because of an obstructive jaundice.

In experimental surgery, the bile is normally exteriorized with the aim of obtaining repeated samples so as to study its quantitive and qualitative variations with regards to the experiment being carried out.

The experimental models described for the draining of the bile to the exterior vary according to the aim of the experiment, but, in general, they may be classified in the following groups:

- (i) Those which use catheters and drainage tubes for the exteriorization.
- (ii) Those which transplant the biliary tracts to the abdominal wall.
- (iii) Those which drain the bile by means of duodenal cloacas or intestinal fistulae.
- (iv) Those which practise internal derivations through extradigestive organs such as the kidney.
- (v) Those which construct accessible internal fistulae.

It is well known that partial losses of bile are usually perfectly tolerated. However, in the total experimental derivations, because of the complete abscence of bile in the intestine and the frequent ascending infection, maintenance is occasionally difficult. In these cases a hyperglucidic and hyperlactate diet, the administration of capsules of dessicated bile (10 gr./day) or of bile through an endodigestive catheter (50 ml./day) is indispensable to avoid the important metabolic and hemorrhagic disorders (Vit.K) which make the correct evaluation of the results illusory.[32]

Finally, the majority of these experimental setups require the placing of receptacles to collect the segregated bile, which makes the continuous use of a corset and meticulous local care obligatory.

In continuation we shall describe the more significant methods proposed for the obtention of bile.

1. Permanent Drainage Procedures

(i) The McMaster, Elman and Rous Method.[43]

In 1993, these authors described a procedure consisting in the drainage of the choledochus by means of tubules and cannulae reinforced with a metallic spiral. The drainage tubes are exteriorized through the abdominal wall some distance from the laparotomy and connected to a rubber globe of some 200 cc. by means of a *T* cannula with an aseptic cotton filter in its free branch which allows the entry of air and avoids infection. This construction tries to reproduce the physiological conditions of drainage of the bilis to the intestine, in an open circuit with a distal valve which permits the drainage of bilis and the entry of air in order to avoid a negative pressure. It is indispensable to avoid any bending of the cannulae or drainage tubes, which is why it is convenient to use reinforced tubes.

The use of protective cover is of great importance, as well as the repeated washing of the tubes with benzalcone to avoid any precipition on them.

The reason for installing this complicated setup lies in the impossibility of draining the choledochus of the dog to the exterior with methods used normally in the treatment of humans. The implantation of a cannula in the choledochus which drains to the exterior by means of a rubber tube usually fails because the tube and the cannula are invariably expelled after a few days. If a *T* tube (Kher type) is used, better results are obtained but the technique of this operation does not lend itself to construction in small dogs whose choledochus is too narrow. On the other hand, with the McMaster, Elman and Rous setup, by placing the cannula well into the choledochus with a long tube attached to it, forming a large loop within the abdominal cavity, it is possible for the tube to be fixed by means of adherences and epiploic synechias which will prevent traction on the choledochus, thus avoiding its luxation.

This setup, with care, may last indefinitely, although it is true that even if the maximum precautions are taken, ascending infection usually manifests itself in half the animals after a few weeks. However, the other half continue in perfect conditions, which indicates that their bile is sterile and that they can live with no great problems if they are administered 50 cc. of bile daily.

(ii) Cholecystostomy with Pezzer's Catheter.[42]

This classical procedure consists in the drainage of the gallbladder by means of a number 20 Pezzer catheter which is introduced in the cholecyst through a small incision.

This model allows the collection of only a partial sample of vesicular bilis since it maintains the flow of said secretion to the duodenum through the principle tract. Its maintenance is simple and the main risk is that of ascending infection.

2. Intermittent Drainage Procedures

(i) The Elman and McMaster Method.[23]

Following the model of external draining of the bile proposed by McMaster, Elman and Rous previously described, these authors have described a two-way system of intubation in which the bile collected by the cannulation of the proximal protion of the choledochus reaches the external "T" cannula where it is either collected in the drainage globe or reintegrated into the intestine by means of a new tube connected to the distal portion of the choledochus.

(ii) Cholecystostomy with Pezzer's Catheter Associated with Bollman's Fistula.

In our laboratory, we have associated the two previous procedures which use catheters, the construction of an antiperistaltic intestinal fistula anastomosed to the second duodenal section. Through this fistula it is possible to reintegrate the bile into the intestine during the periods between experiments in a similar way to that described with regards to the intermittent pancreatic fistulae.

Our experience with this setup (Table no 10) demonstrates that, although it does not reduce the risk of ascending infection through the catheter, the maintenance of the animal is made simpler by returning the bile to the duodenum.

3. Other Procedures

(i) The Exteriorization of Vater's Ampulla – the Methods of Pavlov and Puestow.[36,41]

In his classical experiments on digestion, Pavlov[36] managed to transplant the choledochus to the skin in a similar way to that he had used to exteriorize Stenon's duct. In order to do so he transplanted to the exterior a small portion of the duodenum which surrounds Vater's ampulla.

This is a complex and difficult preparation whose fundamental purpose is to allow the study of the duodenal excretion of the bile through the papilla, although it may also be used for the obtention of the bilis itself.

(ii) Cholecystostomy.

This model consists in exteriorizing the gallbladder directly to the skin with the purpose of obtaining sample of bilis through the ostomy created.

It is a rarely used technique in which ascending infection and the continuous flow of bilis through the ostomy are the rule of the day.

(iii) Biliary Fistula by Means of a Duodenal Cloaca – Dragstedt's Method.[8]

The procedure was devised by Dragstedt with the aim of avoiding the appearance of the hepatic infection and osteoporosis which is produced in other models of biliary fistulae; however, it was later questioned when it was discovered that the presence of this fistula contributes to the elimination by the bilis and urine of two thirds more calcium than normal.

The result of this technique is that the excreted bile is collected in the duodenal cloaca, preventing it from flowing continously to the skin as happens in the cases where the gallbladder is exteriorized directly. However, the technical difficulty of the transplant of the principal pancreatic duct means that it is a rarely used procedure.

(iv) Biliary Derivation Outside the Intestine: Cholecystonephrostomy.

The direct derivation of the choledochus to the exterior or the exteriorization of the gallbladder (cholecystostomy) are used to study certain physiological aspects in experimental surgery.

However, when quantitative determinations of the bile are not necessary, and what is intended is a qualitative study, the choice may be to direct this secretion towards the urinary apparatus to collect it with the urine. Studies into total pigments and the global quantity of biliary salts may be performed very effectively by means of this operation. Furthermore, in this way the presence of a permanent external fistula, catheters, drainage bags and fixing dressings is avoided.

The first technique described with these aims in mind was that devised by Pearce and Eisenbrey which later provided Kapsinow, Engle and Harvey the base for the construction of a more satisfactory model.

– **The Pearce and Eisenbrey Method.**[38]
– **The Kapsinow, Engle and Harvey Method.**[19]

(v) Grossman's Method.[15] This procedure is similar to the transduodenal fistula described by this author for the obtention of activated pancreatic juice. The difference is that, in this case, the doudenal pouch is constructed by sectioning the duodenum 2 cms. from the pylorus and at an equal distance above the opening of the principal pancreatic tract. The extremes of this duodenal fragment are closed by means of an invaginating suture in two layers, after the previous ligature of the accessory pancreatic duct, and a drainage cannula is placed to collect the bilis which drains from the duodenal pouch. The continuity of the intestinal tract is reestablished by means of a termino-terminal anastomosis.

This method, which creates a similar situation to that of the duodenal cloaca described above, is simpler from a technical point of view and presents fewer complications.

(vi) Thomas's Method. This method is similar to that described by the same author for pancreatic fistulae, but here, the cannula is implanted on the duodenal wall opposite the papilla of the choledochus in order to carry out its catheterization when required and thus obtain the bile.

(vii) Colecystojejunostomy Associated with a Cannula Implanted in the Jejunum.[49]

This method, which we use in our laboratory, is based on a Roux's cholecystojenunstomy, after the ligature of the choledochus, and the implantation of a metallic cannula in the jejunal branch which decends from the cholecyst.

This setup is used by means of a modified Foley catheter, in which the distal orifice has been obturated and a new one has been made near the inflatable globe. The modified catheter is introduced through the cannula in the direction of the peristaltism and the globe is inflated to prevent the passing of the bilis from the colecystojejunal anastomosis. This will be aspirated through the catheter.

(viii) Cholecystojejunostomy Associated with a Mann and Bollman Triple Fistula – the Morandeira and Urtiaga Model.[49]

This procedure consists in associating the model of duodenal isolation proposed by Morandeira, previously described with regards to the intestinal fistulae, a Roux's cholecystojejunostomy to which a third Mann and Bollman antiperistaltic fistula is anastomosed in its descending branch.

In this way a setup is constructed which allows the isolated and independent obtention of the gastric, biliary and pancreatic secretions.

This procedure is of great simplicity and efficiency, permitting the collection in isolation of the gastric, biliary and pancreatic secretions. Although its mortality rate is somewhat superior to that of the simple model of duodenal isolation, it is much lower than that of other models described. The animals can be maintained for an indefinite time and the results, care and use may be superposed to those of the model of duodenal isolation.

Our Experience in the Management of Experimental Digestive Fistulae

1. Nursing and Care

As we have already stated, the ideal premises for any experimental model of a digestive fistula consist in maintaining the animal in conditions as near as possible to normal for an indefinite period of time, without the setup for the extraction of the secretions interfering in its normal physiologism, and requiring minimum husbandry during the interphase periods of the experiment under way.

Apart from the general care common to Experimental surgery and those related to the loss of organic liquids, the attention the animals require depends on the type of experimental model being performed.

In the case of the fistulae which do not use a prosthesis, the maintenance mainly depends on whether they anti or isoperistaltic to the skin. In the first case, (Mann-Bollman, May etc.) practically no special care is required from the periodic vigilace of the cutaneous stoma of the fistula, which will warn us to the possible appearance of retraction or strangulation. The isoperistaltic fistulae imply a continuous loss of digestive secretions which we attempt to avoid by plugging the cutaneous exteriorization of the fistula, or by placing collection bags and impermeable dressings on it. Despite this, the cutaneous complications caused by the constant loss of digestive secretions are usually important.

In the case of the fistulae which use metallic cannulae, the maintenance and care are greater. The cannulae require special conditions of cleanliness and asepsis so as to avoid pericannular infection and rejection. Furthermore, their closure must be perfect and they must be protected from any attempt on the part of the animal to pull them out (dressings, bandages, corsets, etc.).

Finally, the procedures which use catheters present problems similar to those of the cannulae. Apart from maintaining the exterior area of the catheter clean, we must check that the plug closes the lumen of the prosthesis hermetically during the interexperimental periods. These models require special vigilance since, in many cases, the drainage leaks either spontaneously or because the animal pulls it out in spite of the fact that a protective bandage or corset has been used in the same way as the previous case.

In short, all models of fistulae require definite care and maintenance in order to ensure that the animal will continue in optimum conditions throughout the experiment, but, in spite of all our care, this is not always possible.

2. Efficiency of Experimental Digestive Fistulae

We have already set out some particular results referring to different types of fistulae. To study the global results obtained by our group with this kind of experimental setups, we shall classify the different models proposed for the construction of digestive fistulae in three groups according to whether they are fistuale without cannulae, fistuale with the metallic cannulae or fistulae with non-metallic prosthesis or tubes. We see that the fistulae with the greatest morbimortality correspond to the procedures using metallic cannulae (mortality 70%, morbidity 75%). In comparison to these figures, the other two groups of fistulous techniques present similar global percentage: fistulae without cannulae 39% morbidity and 33% mortality, while the fistulae with non-metallic prostheses present 46% and 30% respectively.

But the global analysis of this classification is, in principle, misleading, since, within each group, there are techniques with a high morbimortality which augment the total percentages. Therefore, although the discussion of these results may give rise to general considerations about the different models of fistulae according to the material (prosthesic or not) used for its construction, we believe that the results must be discussed one by one.

(i) Fistulae without Cannulae.

In this case the fistulae without cannulae[21,22,50] owe their low morbimortality percentage to the inclusion within the group of the Mann and Bollman antiperistaltic fistula,[21] which, with a 20% mortality, is the type of fistula which offers the best results of all those we have used.

The causes which explain the morbimortality of this type of fistula have been sufficiently commentated "in extensus" when describing each one of them individually, which is why we shall not repeat them over again.

(ii) Fistulae with Metallic Cannulae.

With regard to the group of fistulae based on the implantation of metallic cannulae[15,32,37,48] our experience with the different models is limited, but these techniques, apart from being complicated and requiring the implantation of metallic cannulae, need a supremely careful postoperatory vigilance of the animals, since they tend to pull them out, in spite of the fact that a corset or bandage has been employed to prevent their doing so. This favours maceration and ascending infection the cannula.

Our experience shows that, with the exception of the highly complicated Crocker-Markowitz procedure with a morbidity of 100% and a mortality of 75%, which is easily understandable when taking into account the implantation of two metallic cannulae and a polyvinyl tube for the reestablishment of the continuity of the digestive tract, the other procedures we have employed demostrate a morbidity of between 65–70% and mortality higher than 55%. Normally, the pulling out of the device, dehiscence and infection are the cause of the appearance of evisceration and peritonitis which prevent the long-term conservation of the animals. All this, together with the continuous care and vigilance required by this type of fistulae, makes these procedures non-recommendable for long-term experiments. It is preferable, in the majority of cases to use fistulae without cannulae or those which employ non-metallic prosthesis.

(iii) Non-Metallic Prosthesic.

With regard to the third group of consideration, which uses non-metallic prostheses, we should note that Witzel's enterostomy[22] and the cholecy stostomy with Pezzer's catheter[49] demonstrate a minimum mortality (15% and 28% respectively). But the frequent pulling out of the drainage tubes, in spite of the use of protection bandages and corsets, makes the use of these procedures rather unreliable, especially in the cases in which the animal must be maintained for a long time.

Finally, the example of a catheter with three independent orifices of aspiration[31] could be a good procedure which would allow the isolated aspiration of the different secretions and also the intermittent isolation of intestinal sections, if we could count on a graduated radio-opaque catheter with inflatable globes between the scaled orifices of aspiration; nevertheless, its use would require, apart from the radioscopic control, the construction of a fistula for the introduction of the catheter, or the anaesthesis of the animal every time the catheterization is performed.

3. Conclusions

To sum up we may say that:

(i) Of the three different procedures we have used for the obtention of digestive secretions, the one which has offered the best results and the lowest morbimortality (20% in both cases) has been the antiperistaltic fistula of MANN and BOLLMAN. This procedure is simple to construct, highly versatile in its application and requires very little long-term maintenance.

(ii) Of the other procedures used, those with the highest morbimortality are those based on the implantation of metallic cannulae (morbidity 75% and mortality 70%) basically due to the frequency of the animals pulling out the device and perifistulous infection leading to peritonitis and the evisceration of the animal. Moreover, these procedures require constant vigilance and maintenance care.

(iii) The procedures which use drainage catheters demonstrate a low mortality (15–28%) in WITZEL's enterostomy and in the cholecystostomy with PEZZER's catheter, and is high (50%) in the pancreatic models principally due to the trypsic peritonitis which appears after the removal of the drainages. In all these cases, the exploitation of the setups is low owing to their being pulled out. Husbandry care, although less than that of metallic cannualae is still considerable and they require protection corsets.

References

1. Anderson, S. and Uvnas, B. (1961). Inhibition of postprandial gastric secretion in Pavlov pouches by instillation of hydrochloric acid into the duodenal bulb. *Gastroenterolog*, **41**, 486–490.
2. Babkin, B. P. and Sivelnikov, E. (1923). Isolation of different parts of the digestive tract as a method of studying its movements. *J. Physiol.*, **58**, 15–17.
3. Barcroft, J. and Ribinson C. S. (1929). Study of some factors influencing intestinal movements. *J. Physiol.*, **67**, 211–240.
4. Biebl, M. (1930). Graphischs darstellung der darmbewegusen an der in einem hautschlauch exventerierten darmschling. *Kilm Wochsr*, **9**, 1674–1675.
5. Cooke, A. R., Narhwold, D. L., Preshaw, R. M. and Grossman, M. I. (1967). Comparison of endogenous and exogenous gastrin in the stimulation of acid and peptid secretion. *Am. J. Physiol.*, **213**, 432–436.
6. Debin, D. M., Lataste, J. and Maillet, P. (1972). En Tratado de técnica quirúrgica de J. Patel y L. Leger. Tomo 10. Ed. Toray-Masson, S. A. Barcelona.
7. Douglas, D. M. and Mann, F. C. (1939). An experimental study of the rhythmic contraction in the small intestine of the dog. *Am. J. Digest. Diseases*, **6**, 318–322.
8. Dragstedt, L. R., Montgomery, M. L. and Ellis, J. C. (1930). A new type of pancreatic fistula. *Proc. Soc. Exper. Biol. & Med*, **28**, 109–110.
9. Dragstedt, L. R., Montgomery, M. L., Ellis, J. C. and Mathews, W. B. (1931). Pathogenesis of acute dilatation of the stomach. *Surg. Gynec. & Obst*, **52**, 1075–1086.
10. Dragstedt, C. R., Woodward, E. R., Storer, E. H., Oberhelman, H. A. Jr. and Smith, C. A. (1950). Quantitative studies on the mechanism of gastric secretion in health and disease. *Ann. Surg.*, **132**, 626–640.
11. García Gonzalez, A. (1985). Cuidados y mantenimiento de las fistulas digestivas experimentales en el perro. *Tesina.*, Facultad de Medicina de Zaragoza.
12. Goldberg, S. L. and Mann, F. C. (1931). Preparing pouches of the fundus of the stomach. *Ann. Surg.*, **94**, 953–954.
13. Gregory, R. A., Hallenbeck, G. A. and Code C. F. (1942). A modification of the method of Hollander and Jemerin for making Pavlov pouches. *Proc. Soc. Exper. Biol. & Med.*, **49**, 400–403.
14. Gregory, R. A. (1962). Secretory mechanisms of the gastrointestinal tract. Ed. Arnold Publishers Ltd., London.
15. Grossman, M. I. (1955). Experimental pancreatitis. *Arch. Int. Med.*, **96**, 298.
16. Grossman, M. I. (1967). Neutral and hormonal stimulation of the gastric secretion of acid. Handbook of Physiology, section VI, **2**, 835–864. American Physiological Society, C. F. Code Edit., Williams and Wilkins, Baltimore.
17. Hill, S. C., Neigus, I. and Wilhelmj, C. M. (1934). Method of preparing duodenal pouches for use in experimental work. *J. Lab. and Clin Med.*, **19**, 1126–1129.
18. Ivi, A. C., Lim, R. K. S. and McCarty, J. E. (1925). The intestinal phase of gastric secretion. *J. Exp. Physiol.*, **15**, 55–68.
19. Kapsinow, R., Engle, L. P. and Harvey, S. C. (1924). Intraabdominal biliary exclusion from intestines; cholecystonephrostomy, a new method. *Sur. Gynec. & Obst.*, **39**, 62–65.
20. Klopper, P. J. and Moe, R. E. (1968). Pre-operative determination of the gastric antral border. In the physiology of gastric secretion: 445–447 L. S. Semb. and J. Myrem Ed.; Universitest Forlaget Oslo.
21. Mann, F. C. and Bollman, J. L. (1931). A method for making a satisfactory fistula at any level of the gastrointestinal tract. *Ann. Surg.*, **93**, 794–797.
22. Markowitz, J. (1943). Cirugía experimental. Ed. Labor. Buenos Aires.
23. McMaster, P. D. and Elman, R. (1925). Studies on urobilin physiology and pathology. Derivation of urobilin. Relation of the bile to the presence of urobilin in the body. *J. Exper. Med.*, **41**, 513–534.
24. Mignon, M. and Bonfils, S. (1967). Technique de construction d'une poche de Heidenhain chez le chien. *Presse Med.*, **75**, 2337–2339.
25. Mignon, M., Russel, M. C., Semb, L. S., Morgan, E. H., Finch, C. A. and Nyhus, L. M. (1965). Role of gastric juice in the absorption of iron. *Surg. Forum.*, **16**, 319–321.
26. Moe, R. E., Klopper, P. J. and Nyhus, L. M. (1965). Demonstration of the functional anatomy of the canine gastric antrum. Operative techniques requiring gastrostomy. *Am. J. Surg.*, **110**, 227–285.
27. Morandeira, J. R. (1977). Fisiopatologia de las gastrectomías tipo Billroth II: revisión de conjunto y contribución a su estudio experimental. Tesis Doctoral. Facultad de Medicina de Zaragoza.
28. Morandeira, J. R., García, A. and Ucar, A. (1988). Fístulas digestivas experimentales. Edt. Universidad de Zaragoza, 190 p. Zaragoza.
29. Morandeira, J. R., Lanas, A., Morandeira, A., Sainz, R., Alcalde, J., García Cabo, R. and Lozano, R. (1986). Control of gastric acid secretion by the action of beta-adrenoreceptors. *Eur. Sur. Res.*, **18**(S1)., 47.
30. Morandeira, J. R., Lozano, R., Hernández, J., Ruiz Marcuello, F., Sádaba, A. and Revilla, J. (1977). A new model of double doudenal external fistula for the study of excluded duodenum on the long term living dog. *Eur. Sur. Res.*, **11**(S1)., 108.
31. Morandeira, J. R., Queralt, C., Urtiaga, J., Ucar, A. and Lozano, R. (1985). Nuevo método experimental de aislamiento temporal del duodeno: estudio comparado. *Rev. Esp. Enf. Ap. Digest.*, **67**, 3, 229–235.
32. Murat, J. (1972). Precis de chirurgie experimentale abdominale et thoracique chez le chien. Masson Edit.
33. Nasset, E. S. (1968). Secretions in the small intestine. Medical Physiology Vol. I, Chap. **25**, 427, 439. Saint Louis. Edit.
34. Neuwelt, F., Olson, W. H. and Necheles, H. (1940). A new and simple method for preparing large Pavlov pouches. *Proc. Soc. Explt. Biol. Med.*, 44–74.
35. Parsons, D. T. (1968). Methods for the investigation of intestinal absorption. In handbook of physiology. Code (C.L.). section 6 Alimentary canal Vol. III Chap., **64**, 1177–1216. *Am. Physiol. Soc. Edit.* Washington D. C.
36. Pavlov, I. P. (1902). Lectures on the work of the digestive glands. *Griffin*. Edit. London.

37. Pavlov, quoted by Babkin, B. P. (1928). In die anssers sekretion der verdavungsdrusen. Springer, V. Edt. Berlin.
38. Pearce, R. M. and Eisenbrey, A. B. (1913). A method of excluding bile from the intestine without external fistula. *Am. J. Physiol*, **32**, 417–436.
39. Perry, J. F., Salmon, P., Griffin, W. O., Root, H. D. and Wangensteen, O. H. (1959). A simple technique for preparing vagally innervated pouches in dogs. *Surgery*, **45**, 937–941.
40. Puestow, C. B. (1932). The activity of isolated intestinal segments. *Arch. Surg.*, **24**, 566–573.
41. Puestow, C. B. (1931). The discharge of bile into the duodenum. An experimental study. *Arch. Surg.*, **23**, 1013–1029.
42. Quenu, J. (1968). Intervenciones sobre la pared abdominal y el tubo digestivo. Edit. Toray-Masson. Barcelona.
43. Rous, P. and McMaster P. D. (1923). A method for the permanent sterile drainage of intra-abdominal ducts, as applied to the common duct. *J. Exper. Med.*, **37**, 11–19.
44. Rudick, J. and Hoexter, B. (1969). A vagally innervated pouch of the oxyntic gland area of the stomach. *J. Surg. Res.*, **9**, 336–338.
45. Rudick, J., Semb, L. S. and Nyhus, L. M. (1967). The preparation and maintenance of gastric pouches. Techniques of gastric pouch preparation in dogs. *J. Surg. Res.*, **7**, 383–398.
46. Thiry, L. (1864). Uber eine neve methods deu dündarm za isoliereu. Sitzugsberitchte der kaiserlischeu akad. Wissenschaften wien. Math-naturw, class, 50 (Part).; 77–96.
47. Thomas, J. E. (1941). An improved cannula for gastric and intestinal fistulas. *Proc. Soc. Exp. Biol. Med.*, **46**, 360.
48. Thomas, J. E. (1942). A simplified procedure for preparing an improved Pavlov pouch. *Proc. Soc. Exp. Biol. Med.*, **50**, 58–61.
49. Urtiaga, J. I. (1982). A propósito de un método original para la obtención aislada en el perro de las secreciones gástrica, biliary y pancreática. Tesis Doctoral. Facultad de Medicina de Zaragoza.
50. Vella, L. (1881). Nueva metode per avere succo enterico puro et stabilere la propista fisiologische. *Bull. Soc. Med. Bologna*, **7**, 441–443.
51. Walters, W. and Bollman, J. L. (1926). A study of experimental acute gastric fistula. *Arch. Surg.*, **18**, 578–587.
52. Walters, W. and Bollman, J. L. (1927). Toxemia of duodenal fistula: physiologic changes concerned in the production of characteristic chemical reactions of blood. *Jana*, **89**, 1847–1853.
53. White, I. and Magee, D. F. (1962). Recovery of pancreatic exocrine functions after controlled duct, obstruction. *Surg. Gyn. Obst.*, **114**, 463–469.

PART III

8. Experimental Models for Obtaining Pancreatic Juice from Rats

Eduardo M. Targarona[1], M. Teresa Ribera[2,*] and Laureano Fernández-Cruz[1]

[1]*Experimental Microsurgery Unit, Department of Surgery, Hospital Clínic, University of Barcelona, Barcelona, Spain*
[2]*Experimental Gastroenterology Laboratory, School of Digestive Pathology, Hospital de la Santa Cruz y San Pablo, Autonomous University of Barcelona, Barcelona, Spain*

The pancreas is the most productive of all the digestive glands having the greatest ability for synthesis and excretion of enzymes. The volume of exocrine secretion from the human pancreas ranges from 1500 to 2500 cc daily. This secretion is composed of water, ions and various proteins (mainly enzymes), all of which are regulated by a complex hormonal and humoral mechanism.[1]

Pancreatic function and physiology may be studied in different forms and with diverse models. Examination of isolated cells or acini and isolated perfused glands *ex vivo*, analysis of pancreatic homogenate, or the study of the changes that different experimental situations induce in the composition of pancreatic juice (PJ) have been the methods most frequently used. While study of isolated cells or acini is interesting for learning about physiology and metabolism at the cellular level, the fact that the pancreas is an exocrine gland means that functional changes are best reflected by changes in the composition of PJ.

The aims of this review are to describe briefly the experimental models for obtaining PJ from rats, their advantages and disadvantages, and the results obtained in our laboratory in various experiments using the model of obtaining PJ from anesthetized animals.

Experimental Model in Rats

The advantages of using small animals in an experimental surgery laboratory are obvious. The animals are easy to handle and to anesthetize, either by intraperitoneal (ketamine, urethane) or inhalation (ether) anaesthetic administration. Moreover, biologically identical animals may be used and the number of animals may be increased in each experimental series.

PJ may be obtained in two ways,[2–5] either from conscious animals, which requires prior surgery to insert a cannula, or from anesthetized animals, thus obtaining the PJ peroperatively. Both models have been used for the study of changes in PJ in different experimental situations, and both present advantages and disadvantages. In our opinion, the conscious animal model may be more useful for studying changes in the composition of the PJ in acute studies (administration of drugs, hormones, infusion of dietary elements in the digestive tract, etc.), where it is desirable to compare the results obtained with previous basal values in the same animals.

The model for obtaining PJ from anesthetized animals may be more useful in situations in which a chronic biological change has been produced (hyperlipidemia, hypercalcemia, parenteral nutrition), in which the effects on the composition of the PJ can be evaluated by comparing the results with those of a control group in which no experimental changes have been induced. In the anesthetized model, it is important to remember that anesthetics agents, in particular ether, affect the composition of the PJ, reducing its basal secretion[6]. The conscious animal model has the disadvantage that it is technically more complex, with high morbidity and mortality as a result of the surgical intervention. Furthermore, the animals must be kept immobilized in cages and exocrine pancreatic secretion does not return to normal until 3 or 4 days after surgery when the animals

*Author for Correspondence.
Supported by Grants No. 82/302 and 83/811 of Fis.

have recovered and resumed their diets. All these facts must be considered when selecting the model to be used or when making comparisons between the results of different studies which use different methods for obtaining PJ.

Anatomy of the Rat Pancreas

In rats, the pancreas is an ill-defined structure which lies within the duodenal loop and extends to the splenic hilum above the transverse mesocolon. The bile duct is enclosed in the pancreatic parenchyma and enters the second part of the duodenum through a prominent papilla of Vater. The common bile duct is about 20 mm long and 1 mm thick, and is easily cannulated through the papilla or by a choledochotomy. From a surgical point of view, the most important aspect of rat pancreatic anatomy is that this organ does not have a single pancreatic duct, but multiple canaliculi that empty directly into the biliary tract. There are two larger pancreatic ducts, one of which empties a few millimeters above the papilla of Vater, which makes it necessary to take great care not to obliterate it when inserting a catheter. The common outlet of biliary-pancreatic secretions makes it necessary to perform the cannulation away from of the biliary tract at the level of the hepatic hilum to divert the bile and thus obtain pure PJ[7] (Fig. 8.1).

Surgical Model in Anesthetized Animals[8–25]

After inducing anesthesia with the method of choice (ether, ketamine, etc.), a midline laparotomy is performed, the duodenal loop is exteriorized and the papilla of Vater is located. The duodenal wall is punctured and a 0.9 mm diameter polyvinyl catheter is introduced through the papilla of Vater. This catheter should be inserted only 2–3 mm to avoid obliterating the outlet of the larger pancreatic duct. The bile duct should be ligated at the hepatic hilum in order to avoid the entry of bile. The bile duct may either be ligated, or cannalized to avoid bile stasis within the liver (Fig. 8.2 and 8.3). A central (jugular, femoral) or peripheral (dorsal vein of the penis) venous line may be inserted for the duration of the experiment in order to inject drugs, hormones to stimulate pancreatic secretions or administer physiological saline to maintain circulatory volume. Additionally, the pylorus is ligated in order to avoid the passage of gastric juice to the duodenum during the obtention of PJ. The temperature of the animal should be maintained by a lamp or electric blanket and be controlled by means of a rectal thermometer. The PJ is collected in containers surrounded by ice or dry ice to avoid inactivation of the pancreatic enzymes.

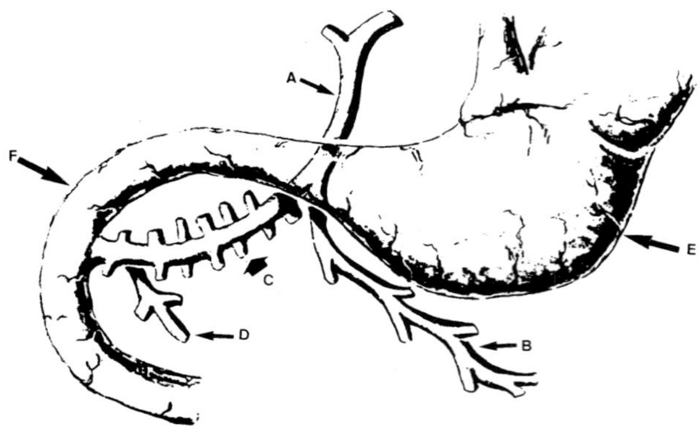

Figure 8.1. Morphology of the pancreatic ducts in the rat. A. Common bile duct. B. Main pancreatic duct. C. Small pancreatic ducts draining directly in the common bile duct. D. Large pancreatic duct drainig near de papilla of Vater. E. Stomach. F. Duodenum.

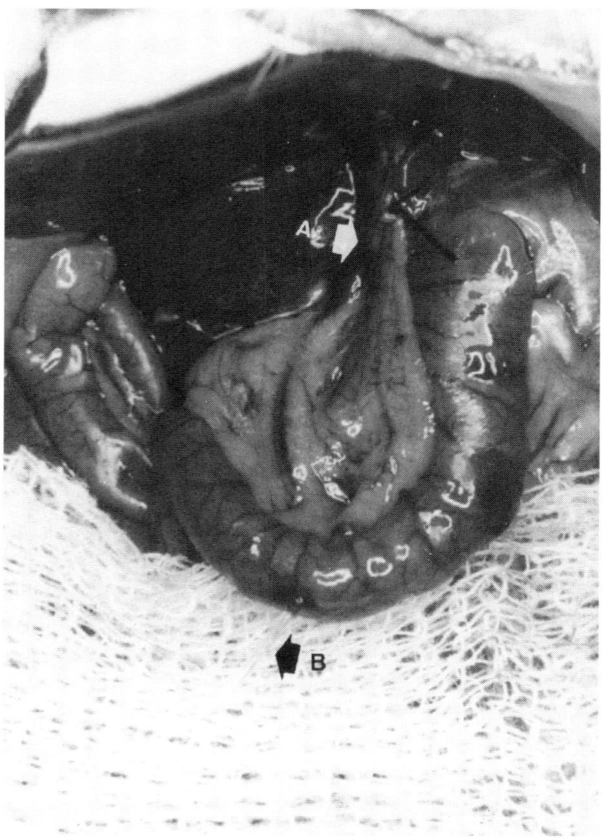

Figure 8.2. Model of obtention of pancreatic juice in the anesthetized rat. **A**. Main bile duct ligated near the hilum in order to obtain pure pancreatic juice. **B**. Cannula located at the end of the bile duct, through the papilla of Vater.

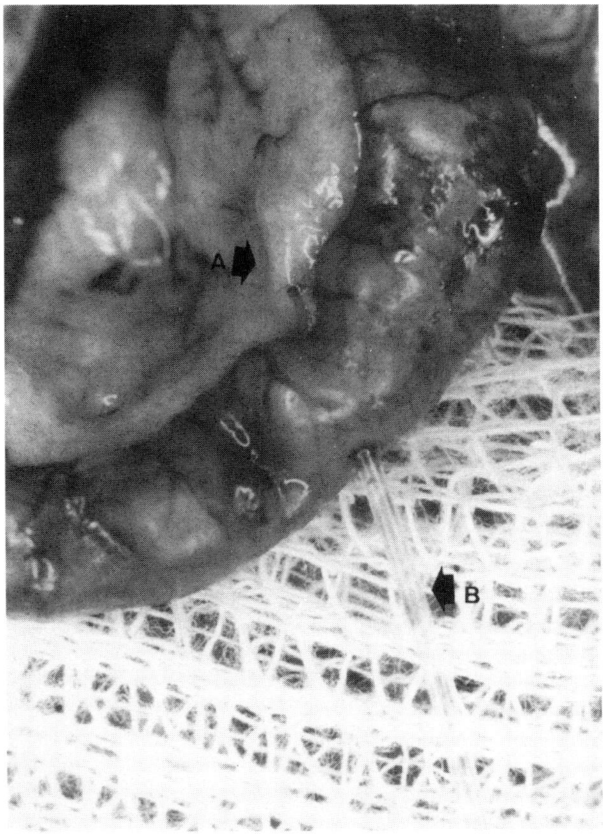

Figure 8.3. Model of obtention of pancreatic juice in the anesthetized rat. **A**. Detail of the cannula located in the papilla of Vater. **B**. Catheter used to cannulate the bile duct (polyvinile, 0.9 mm of diameter).

Surgical Model in Conscious Animals [26–40]

The surgical model for obtaining PJ from conscious animals is based on that described by Colwell[2] and Grossman,[3] which continues to be worth today. After performing the laparotomy, the duodenal loop is exteriorized. The biliary tract should be tied just above the upper edge of the pancreas, and later cannulized with a 0.89-mm catheter to divert the bile. The common bile duct is then cannulized through the papilla of Vater, in order to obtain pure PJ. At the level of the duodenum, 2 catheters which will be used to perfuse drugs or nutrients into the lumen of the intestine are placed 1 centimeter above and below the papilla of Vater. Finnally, a gastrostomy is fashioned in the gastric fundus, in the non-secreting zone of the stomach, in order to collect gastric secretions during the experiments. All catheters are exteriorized through the abdominal muscle wall and later tunneled under the skin to emerge through the back. Animals are placed in Bollman boxes to allow feeding but restrict movement. It is recommended that PJ be obtained 3–4 days after surgery since the animals do not resume normal feeding for several days.[35] At this time, the biliary and pancreatic cannulae are connected to the duodenal ones and the biliary-pancreatic secretions may be reinfused into the intestine by a pump. This model allows the animals to be kept for up to 9 days, although it is necessary to constantly monitor catheter patency.

Study of the Composition of Pancreatic Juice

The study and collection of PJ may be performed under basal conditions or after stimulating the gland (CCK, secretine, caerulein). The model for obtaining PJ from conscious animals allows the collection from 0.2 cc/100 g/hour under basal conditions to a maximum of 1 cc/100 g/hour. In the ether anesthetized model the volume ranges from 0.06 cc/100 g/hour under basal conditions to 0.8 cc/100 g/hour when stimulated with caerulein.

The PJ should be obtained while the animal is fasting (12–24 h) to ensure significant accumulation of zymogen granules and to allow evaluation of the exocrine response of the pancreas to hormonal stimulation.

It is possible to analyze the volume of PJ secretion as well as the output and concentration of proteins (which are general markers of protein synthesis and the secretory ability of the pancreas), as well as the output or concentration of secretory pancreatic enzymes or ions (bicarbonate, calcium, zinc, etc.).

Experience of the Microsurgery Unit of the University of Barcelona

The aim of the work performed in our center has been to study the changes in the composition of PJ in different experimental models that attempt to simulate different clinical situations, the correlation of which with pancreatic physiology remains controversial (hypercalcemia, hyperlipidemia, prolonged parenteral nutrition or chronic administration of immunosuppressive agents). For this we used the anaesthetized rat model, in groups of 10 animals, and we compared the results obtained with those from a healthy control group. Ether was the anesthesia of choice, due to ease of control of depth of anesthesia throughout the experiment. The animals fasted 24 hours prior to surgery. Samples of PJ were obtained over a 90 minute period under basal conditions and after stimulation with caerulein. The caerulein was perfused through a perfusion pump at a concentration of 33.75 ng/min/kg, which is the concentration considered to be optimum for maximum stimulation of the gland. We measured the total concentration and output of proteins, which are a global parameter of the pancreas' synthesizing ability, and also determined the levels of chymotrypsin and amylase in the PJ.

Pancreatic Secretion and Calcium Metabolism[41–43]

The relationship between hyperparathyroidism and pancreatitis is generally well accepted, but the pathogenic mechanism has never been satisfactorily explained and some authors have questioned this association. Postoperative pancreatitis has also been observed after exploration of the neck for hyperparathyroidism. The aim of this work was to study the influence of parathyroidectomy and thyroparathyroidectomy on the composition of the exocrine pancreatic secretion. We used 50 Sprague-Dawley rats, divided into 3 groups: Group I, control (C); Group II, parathyroidectomy (PT); and Group III, thyroparathyroidectomy (TPT). All groups were studied for 6 weeks. Serum calcium was measured every 2 weeks. Pancreatic juice was obtained using the anesthetized model previously described, under basal and stimulated (caerulein) conditions. Amylase, chymotripsin and total proteins were measured in pancreatic juice and parathormone and thyrocalcitonine in blood serum.

The concentrations of the three protein compounds measured in pancreatic juice were significantly depressed after PT. When thyroidectomy was added, the amylase concentration was more pronounced. We concluded that thyroid and parathyroid glands exert an influence on the composition of the exocrine pancreatic secretion, and the surgical excision of these glands is followed by a decrease in their enzymatic components. This study underlines the relationship between the pancreas and the thyroid and parathyroid glands.

Hyperlipidemia and Exocrine Pancreatic Function[44–45]

Although the relationship between hyperlipidemia and pancreatitis is well known, the pathogenic mechanism is not clear. In this work we studied the exocrine pancreatic secretion (EPS) in animals after short and longer periods of hyperlipidemia. We used 100 Sprague-Dawley rats. Group I, control; Groups II and III, hyperlipidemia induced by diet during 4 and 15 days of hyperlipidemic diet administration respectively. Animals in Group IV were treated intraperitoneally with a single 900 mg/kg dose of Triton WR 1339 (Tyloxapol). In Group V,Triton WR 1339 at dose of 900 mg/kg was injected intraperitoneally every 48 hours over a 15 days period. In Groups III and V, the exocrine pancreatic secretion was studied after the normalization of blood lipids 2 weeks after resotring the normal diet and 2 weeks after the last doses of Triton. In all groups blood samples were centrifuged and total serum lipids, total serum cholesterol and triglycerides levels determined. Pancreatic juice was obtained with using the anesthetized model, under basal and stimulated (caerulein) conditions. Amylase, chymotrypsin and total proteins were measured in pancreatic juice. The pancreatic glands were then removed for histological study.

Hyperlipidemia induced either by diet or by Triton WR 1339 injection resulted in a significant decrease in the concentration of enzymes in the caerulein stimulated animals. These changes were more pronounced in chronic states of hyperlipidemia induced by diet. When blood levels returned normal in these animals, the pancreatic juice enzyme levels were similar to those in control animals. Histological study shows a fatty infiltration and vacuolization of the acinar cells in both forms of hyperlipidemia, but focal necrosis was only seen in hyperlipidemia induced by Triton.

Our results suggest a constant modification of EPS after short and long periods of hyperlipidemia. The exocrine pancreatic secretion was restored to normal values when the serum lipids returned to normal. Also, the microscopic pattern of acinar cells alterations were constant.

Immunosuppressive Drugs and Exocrine Pancreatic Secretion[46–49]

It has been shown that cyclosporine produces a reversible impairment of the endocrine pancreas with hyperglycemia and low levels of insulin in plasma. It has also been demonstrated that cyclosporine at high doses (50 mg/kg, p.o.) modifies the exocrine pancreatic secretion. However, this dose was 4 times higher than that needed to achieve an immunosuppressive effect. Other immunosuppressive drugs such as azathioprine and corticosteroids have been suggested as inducers of acute pancreatitis. The aim of this work was to study the influence of cyclosporine, azathioprine, corticosteroids and cyclosporine plus corticoids at immunosuppressive doses on the exocrine pancreatic secretion. We used 80 Sprague-Dawley rats: Group I, control; Group II, cyclosporine, 15 mg/kg/day/P.O. for 15 days and 6 weeks; Group III, azathioprine, 1.5 mg/kg/day/I.M. for 15 days and 6 weeks; Group IV, corticosteroids, (methylprednisolone) 1 mg/kg/day/I.M. for 15 days and 6 weeks; and Group V, cyclosporine, 10 mg/kg/day/V.O. plus corticosteroids, 0.75 mg/kg/day/I.M. for 15 days and 6 weeks.

Pancreatic juice was obtained using the anesthetized model, in basal and stimulated (caerulein) conditions. Amylase, chymotrypsin, and total proteins were measured in pancreatic juice. The pancreatic glands were removed for histological study. Glucose and cyclosporine levels were measured in blood serum.

Animals under cyclosporine or corticosteroids treatment had elevated blood glucose levels. Cyclosporine levels at the end of the study ranged between 437 and 1474 ng/ml. The composition of the exocrine pancreatic secretion was not modified by cyclosporine alone or azathioprine. Corticosteroids alone or in association with cyclosporine significantly depressed the concentration of enzyme and proteins in pancreatic juice (Tab. 8.1). There was no histological change in the exocrine pancreas in the groups studied.

We could conclude that corticosteroid therapy has a direct effect on pancreatic secretion in the rat producing a decrease in pancreatic enzymes. At low doses of cyclosporine the endocrine pancreatic function is impaired but the normal composition of the exocrine pancreatic secretion is preserved. At these low immunosuppressive doses, the immunosuppressive drugs do not change the morphological pattern of the exocrine pancreas.

Exocrine Pancreas and Parenteral Nutrition[50–52]

Total parenteral nutrition (TPN) is a well-established treatment in the management of patients with acute pancreatitis. The beneficial role of TPN in patients with acute pancreatitis is to facilitate bowel and pancreatic rest. There are two satisfactory sources of non-protein calories, carbohydrate as glucose solution and fat as lipid emulsion, either by long chain-triglycerides (LCT) or medium-chain triglycerides (MCT). The question continues as to which solution achieves the minimum stimulation of pancreatic structure and function. The aim of this work was to study the exocrine pancreatic secretion in rats with prolonged TPN using 3 different solutions: Group I, control (standard diet); Group II, glucose (66 mg/g) and aminoacids; Group III, glucose (38.8 mg/g)] + LCT (11 mg/g) and aminoacids; Group IV, glucose (38.8 mg/g)] + LCT and MCT (5.5 mg/g each) and aminoacids. These solutions provided 1.46 mg/N_2/g resulting in improved nitrogen balance. In each group, exocrine pancreatic secretion (amylase, chymotrypsin, total proteins) were studied 3, 7 and 14 days after administration of these solutions. The pancreas glands were removed for histological study. Animals with TPN showed a clear-cut decrease in the pattern of pancreatic secretion throughout the duration of TPN. Groups III and IV had a significant reduction in the EPS compared with Group I (Control), in periods ranging between 3 and 14 days of TPN (Tab. 8.2). An interesting finding was the observation of acinar cell death by apoptosis after 3 days of TPN, that significantly increased in Groups III and IV compared with group II.

In conclusion, the provision of calories with either glucose or lipid emulsion resulted in nitrogen retention in animals with prolonged TPN. Exocrine pancreatic secretion was significantly reduced in animals in which the source of calories was mainly in lipid emulsions, associated with apoptosis as an index of acinar cell deletion.

Table 8.1. Exocrine pancreatic secretion and immunosuppressive drugs administered for 45 days.

Group	*n*	*Amylase* (UI/ml)	*Chymotrypsin* (mUI/ml)	*Total Proteins* (mg/100 ml)
I Control	9	23.756 ± 5.625	1.429 ± 156	9.180 ± 2.144
II Cyclosporine (15mg/kg/day)	9	21.516 ± 562	1.177 ± 716	7.536 ± 1.130
III Azathioprine (1 mg/kg/day)	9	19.150 ± 2.710	1.308 ± 101	10.281 ± 3.453
IV Corticoids (10mg/kg/day)	9	14.350 ± 6.650*	1.140 ± 170**	6.100 ± 1.270*
V Csa + Corticoids (10mg/0.75mg)	9	13.144 ± 3.409**	622 ± 272*	4.894 ± 735**

Mean ± SD
*$p < .001$
**$p < .005$

Table 8.2. Exocrine pancreatic secretion (caerulein stimulated) after 14 days of total parenteral nutrition.

Group	*n*	*Amylase* (UI/ml)	*Chymotrypsin* (mUI/ml)	*Total Proteins* (mg/100 ml)
I Control	10	14.350 ± 6.067	1.234 ± 454	11.359 ± 3.025
II Glucose + AA[1]	10	10.205 ± 4.235	370 ± 41*	7.799 ± 3.243*
III Glucose + LCT + AA[2]	10	5.030 ± 3.783**	670 ± 192*	5.821 ± 1.766*
IV Glucose + LCT + MCT + AA[3]	10	3.032 ± 1.531**	687 ± 314*	5.764 ± 2.329*

Mean ± DS, [1]Glucose 66 mg/g NPT + Aminoacides,
*p < .05 vs Group I
[2]Glucose 38 mg/g NPT + LCT (Long chain triglycerides) 11 mg/g NPT
**p < .005 vs Group I and II
[3]Glucose 38 mg/g NPT + LCT (5.5 mg/g NPT) + MCT (5.5 mg/g NPT), (Medium chain triglycerides).

References

1. Rinderknecht, H. (1986). Pancreatic secretory enzymes. In "The exocrine pancreas: Biology, pathobiology and diseases. "Edited by V.L.W. Go *et al.*, pp. 163–183. Raven Press, New York.
2. Colwell, A. R. (1951). Collection of pancreatic juice from rats and consequences of its continued loss. *Am. J. Physiol.*, **164**, 812–21.
3. Grossman, M. I. (1958). Pancreatic secretion in the rat. *Am. J. Physiol.*, **194**, 535–9.
4. Thomas, J. E. (1959). Methods for collecting pancreatic juice, *Gastroenterology*, **36**, 362–7.
5. Debray, Ch., De La Tour, Ch., Roze, Cl. and Souchard, M. (1961). Etudes sur la sécrétion pancréatico-biliaire du rat. *La. Presse Med.*, **56**, 2061–3.
6. Petersen, H. and Grossman, M. I. (1977). Pancreatic exocrine secretion in anesthetized and conscious rats. *Am. J. Physiol.*, **233**, E530–6.
7. Nyhus, L. M. (1965). Surgery of the digestive system in the rat. C. H. Thomas, Publisher. Springfield, Illinois.
8. Linari, G. and Balderi, M. (1978). The action of caerulein on the pancreatic secretion of the rat. Il *Farmaco*, **32**, 611–6.
9. Otsuki, M., Sakamoto, Ch., Maeda, M., Yuu, H., Morita, S. and Baba, S. (1979). Effect of caerulein on exocrine and endocrine pancreas in the rat. Endocrinology, **105**, 1396–9.
10. Lu, L., Louie, D. and Owyang, C. (1989). A cholecystokinin releasing peptide mediates feedback regulation of pancreatic secretion. *Am. J. Physiol.*, **256**, G430–5.
11. Renner, I. G., Wisner, J. R. and Lavigne, B. C. (1986). Partial restoration of pancreatic function by exogenous secretin in rats with ceruletide- induced acute pancreatitis. *Dig. Dis. Sci.*, **31**, 305–13.
12. Rosch, Th., Schudziarra, V., Wolf, A. and Classen, M. (1989). Effects of CCK-8 in combination with natural or synthetic secretin on amylase, lipase, trypsin and chymotrypsin secretion in rats. *Int. J. Pancreatol.*, **5**, 249–61.
13. Petersen, H., Solomon, T. E. and Grossman, M. I. (1979). Pancreatic secretion in rats after chronic treatment with secretin plus caerulein. *Gastroenterology*, **76**, 790–4.
14. Shiratori, K., Watanabe, S. and Takeuchi, T. (1989). Effect of fatty acid on secretin release and cholinergic dependence of pancreatic secretion in rats. *Pancreas*, **4**, 452–8.
15. Imamura, M., Yamauchi, H. and Chey, W. Y. (1994). Effect of intraduodenal infusion of tocamphyl on pancreatic exocrine secretion and gastrointestinal hormone release in rats. *Int. J. Pancreatol.*, **15**, 187–93.
16. Jo, Y. H., Yoon, S. H., Hahn, S. J., Rhie, D. J., Sim, S. S. and Kim, M. S. (1994). Effect of medial amygdaloid stimulation on pancreatic exocrine secretion in anesthetized rats. *Pancreas*, **9**, 117–22.
17. Dios, I. D. and Manso, M. A. (1994). Effect of L-364,718 (CCK receptor antagonist) on exocrine pancreatic secretion of hydrocortisone treated rats. *Pancreas*, **9**, 212–18.
18. Nakano, S., Tachibana, I. and Otsuki, M. (1995). Effect of the CCK receptor antagonist loxiglumide on pancreatic exocrine function in rats after acute pancreatitis. *Pancreas*, **10**, 287–94.
19. Park, H. J., Lee, Y. L., Kwon, H. Y., Suh, S. W. and Yon, J. H. (1995). Pancreatic exocrine secretion in response to median raphe stimulation in anesthetized rats. *Pancreas*, **10**, 407–12.
20. Okumura, T., Taylor, I. L. and Pappas, T. N. (1995). Microinjection of TRH analogue into the dorsal vagal complex stimulates pancreatic secretion in rats. *Am. J. Physiol.*, **269**, G328–G334.
21. Tachibana, I, Watanabe, N, Shirohara, Akiyama, T, Nanano, S. and Otsuki, M. (1995). Effects of tetraprenylacetone on pancreatic exocrine secretion and acute pancreatitis in two experimental models in rats. *Int. J. Pancreatol.*, **17**, 147–54.
22. Balaghi, M. and Wagner, C.(1995). Folate deficiency inhibits pancreatic amylase secretion in rats. *Am. J. Clin. N*, **61**, 90–6.
23. Manso, M. A., Gallego, H. A. and Dios, I. (1992). Effect of acute and chronic administration of ethanol on the pancreatic exocrine response to CCK in rats fed different diets. *Clin Sci*, **82**, 433–438.

24. Moriyoshi, Y., Shiratori, K., Takeuchi, T. and Watanabe, S. (1995). Effect of a novel CCK receptor antagonist, FK 480, administered intraduodenally on pancreatic secretion in rats. *Pancreas*, **10**, 295–300.
25. Masuda, M., Miyasaka, K. and Funakoshi, A. (1995). Cholinergic stimulatory effect of intragastric administration of a Prostaglandin E2 analogue on pancreatic exocrine secretion in conscious rats. *Pancreas*, **10**, 395–400.
26. Takcás, T., Nagy, I., Pap, A. and Varro, V. (1990). The effect of CR 1409, a potent CCK receptor antagonist on basal and stimulated pancreatic secretion in rat. *Pancreas*, **5**, 60–4.
27. Kelly, T. R. (1968). Relationship of hyperparathyroidism to pancreatitis. *Arch. Surg.*, **97**, 267–74.
28. Green, G. M. and Nasset, E. S. (1980). Importance of bile in regulation of intraluminal proteolitic enzyme activities in the rat. *Gastroenterology*, **79**, 695–702.
29. Rozé, C., Dubrasquet, M., Chariot, J. and Vaille, C. (1980). Central inhibition of basal pancreatic and gastric secretions by beta-endorphin in rats. *Gastroenterology*, **79**, 659–4.
30. Green, G. and Miyasaka, K. (1983). Rat pancreatic response to intestinal infusion of intact and hydrolized protein. *Am. J. Physiol.*, **245**, G394–8.
31. Pap, A. and Sarles, H. (1986). Effect of atropine on pancreatic secretion in conscious rats. *Digestion*, **33**, 17–25.
32. Miyasaka, K., Kitani, K. and Green, G. (1986). The sequential changes in pancreatic exocrine function after abdominal surgery in the rat. *Pancreas*, **4**, 347–53.
33. Murayama, K. M., Drew, J. B., Nahrwold, D. L. and Joehl, R. J. (1990). Acute edematous pancreatitis impairs pancreatic secretion in rats. *Surgery*, **107**, 302–10.
34. Miyasaka, K. and Green, G. M. (1984). Effect of partial exclusion of pancreatic juice on rat basal pancreatic secretion. *Gastroenterology*, 1984, **86**, 114–9.
35. Miyasaka, K., Kitani, K. and Green, G. M. (1986). The sequential changes in pancreatic exocrine function after abdominal surgery in the rat. *Pancreas*, **1**, 347–353.
36. Miyasaka, K., Funakoshi, A., Kitani, K., Tamamura, H., Funakoshi, S. and Fujii, N. (1990). Inhibitory effect of pancreastatin on pancreatic exocrine secretions. *Gastroenterology*, **99**, 1751–6.
37. Hara, H., Narakino, H. and Kiriyama, S. (1994). Enhancement of pancreatic secretion by dietary protein in rats with chronic diversion of bile-pancreatic juice from the proximal small intestine. *Pancreas*, **9**, 275–9.
38. Okubo, K., Miyasaka, K., Matsumoto, M. and Funakoshi, A. (1994). Mechanism of stimulatory effect of Neuromedin C on pancreatic exocrine secrection in conscious rats. *Pancreas*, **9**, 585–90.
39. Hara, H., Narakino, H. and Kiriyama, S. (1995). Gastric acid-independent enhancement of pancreatic secretion by dietary protein in chronic bile-pancreatic juice diverted rats. *Pancreas*, **11**, 173–8.
40. Fischer, H., Konturek, J. W., Szlachcic, Konturek, S. J. and Domschke, W. (1995). Plasma amino acid consumption and pancreatic secretion during and after cerulein-induced pancreatitis in rats. *Int. J. Pancreatol.*, **18**, 127–34.
41. Ribera, M. T., Targarona, L., Fernández-Cruz, L., Fuster, J., Hotter, G. and Codina, J. (1983). Study of exocrine pancreatic secretion in response to calcium homeostasis changes. *Digestion*, **28**, 56.
42. Targarona, E. M., Ribera, M. T., Hotter, G. and Fernández-Cruz, L. (1985). The influence of the thyroid and parathyroid glands on the exocrine pancreatic secretion. *Eur. Surg. Res.*, **17**, 9.
43. Targarona, E. M. (1988). The influence of calcium changes on the exocrine pancreatic secretion. ScD Thesis. *University of Barcelona*.
44. Targarona, E. M., Ribera, M. T., Fernández-Cruz, L., Codina, J., Canovas, M. and Hotter, G. (1984). Influence of hyperlipidemia on the composition of the exocrine pancreatic secretion. *Digestion*, **30**, 80.
45. Targarona, E. M., Ribera, M. T., Fernández-Cruz, L., Casas, A. and Hotter, G. (1986). Morphological and pancreatic function after hyperlipidemia. *Eur. Surg. Res.*, **18**, 2.
46. Fernández-Cruz, L., Targarona, E. M., Ribera, M. T., Casas, A. and Hotter, G. (1986). The effect of immunosuppressive drugs on rat exocrine pancreatic secretion. *Eur. Surg. Res.*, **18**, 53.
47. Ribera, M. T., Targarona, E. M., Casas, A., Hotter, G., Mart, M., Ribalta, M. T. and Fernández-Cruz, L. (1986). The influence of cyclosporine and cyclosporine plus corticoids on the exocrine pancreas in the rat. *Digestion*, **35**, 49.
48. Fernández-Cruz, L., Targarona, E. M., Ribera, M. T., Colomer, J., Pi, F., Saenz, A., Casas, A. and Hotter, G. (1988). The effect of cyclosporine and the standard immunossupressive drugs on the exocrine rat pancreas. *Transplantation Proc*, **20**, 453–5.
49. Amorin, P. (1989). The effect of immunossupressive drugs on rat exocrine pancreas. ScD Thesis, *University of Barcelona*.
50. Fernández-Cruz, L., Benarroch, G., Amorin, P., Astudillo, E., Ribera, M. T. Bombi, J. A. and Condon, E. (1988). The influence of total parenteral nutrition on the pancreatic morphology and exocrine pancreatic secretion changes. *Br. J. Surg.*, **75**, 1255.
51. Ribera, M. T., Ozalla, N. D., Benarroch, G., Amorin, P. and Fernández-Cruz, L. (1988). The influence of prolonged total parenteral nutrition with different sources of calories on exocrine pancreatic secretion. *Int. J. Pancreatol.*, **3**, S276.
52. Benarroch, G. (1990). Influence of parenteral nutrition on the structure and exocrine secretion of the pancreas. ScD Thesis. University of Barcelona.

PART III

9. Animal Experiments in Inflammatory Bowel Disease

Eckhard Nagel u. Rudolf Pichlmayr,

Klinik für Abdominal- und Transplantationschirurgie der Medizinischen Hochschule in Hannover, Germany

Under the synonym "ideopathic inflammatory bowel disease" (IBD) two collected entities, Crohn's disease and Ulcerative Colitis are summarized. Both are characterized by the phaenotype of a nonspecific inflammation of the gut – differing mainly in extension and some morphological features. The histological findings of epithelial-cell destruction, microulceration, granulocyte and monoclear cell infiltration, and crypt abcesses are relevant in ulcerative colitis. The presence of lymphocytes, macrophages and granulomas through all layers of the intestine resembles more the changes seen in Crohn's disease. The lack of specific biological markers, e.g. within the scope of morphological pathology complicates the attempt of answering questions concerning the pathogenesis and the etiology of the disease process. In this context it is also unknown if both diseases reflect totally different gastrointestinal illnesses which simply appear with a similar clinic.

Clarification of Pathomechanisms in Experimental Animal Models

"Ideal" animal experiments were introduced, e.g. by Alter *et al.* in 1978 for non-A, non-B-hepatitis in monkeys[1] or by Storrs (1971) in his model for mycobacterium leprae[2] and did fulfill the following demands:

1. The same diseases were produced in animals as in man by a suspicious etiological agent.
2. The overall feature of the diseases was congenerous.
3. Parameters of illness were similar to those seen in man.
4. Similar therapeutic possibilities and reactions as in man were tested.

It has been considered that such "ideal" experimental animal studies were classified as models for diseases with a monocausal etiology and a known pathogenetic process.

The question arises what constitutes an "ideal" animal experiment of IBD. The requirement has been stated that the animal disease should have the same cause and maintain on the basis of the same pathophysiology as in man. Trying to identify a spectrum of specific clinical and pathological manifestations in IBD it becomes clear that one cannot define a model of an entity that is poorly defined itself. This leads to the statement that a useful animal experiment on IBD does not exist.

However, one has to object that many diseases are due to a large variety of different factors which act independently but combined to produce a pathological state. In such cases other principles in planning and evaluating of an animal experiment are required than in a disease with a monocausal pathomechanism. Considering that a multifactorial event happens in an array of single steps–and one may hypothesize those steps occur more or less in sequence–the question of an experiment should focus on the primary event or any other specific stage of the process.

Since the "ideal" animal model does not exist and the pathogenesis is most likely based upon a multifactorial array, models known today should be properly called "models of gastrointestinal inflammation in general". This does not mean under any circumstances that animal experience for IBD should be abandoned. In contrast knowledge of gastrointestinal inflammation in animals is doubtless greatly useful for the investigation of different etiological agents, immune mechanisms, the physiology and biochemistry of the mucosal and transmural inflammation and at the same point possibly various forms of treatment in the human disease process.

In the following we tried to classify some existing concepts of experimental models of gastrointestinal inflammation. A classification is very often difficult and somewhat contraproductive, because most of the examples will show overlapping physiological, biochemical, microbiological or immunological aspects of gastrointestinal inflammation. This again makes it difficult to differentiate between models of Crohn's disease and those for Ulcerative colitis – having in mind that both diseases may have a totally different background and resemble different pathogenetic concepts of human gastrointestinal disorders.

Naturally Occuring Gastrointestinal Inflammation

Animal Models Of Infectious And Spontaneous Ileitis

Naturally occuring models of spontaneous ileitis and ileocolitis have been observed in a variety of different species as listed in Table 9.1. When the Scottish surgeon Dalziel in 1913[3] described for the first time 6 cases of terminal ileitis comparable to those seen in 1932 by Crohn, Ginsburg and Oppenheimer[4] he suggested that the disorder might be a result of M. paratuberculosis, an organism known to cause granulomatose ileitis. Johne and Frothingham had published in 1985 a case of a cattle with chronic inflammatory reactions, principially of the

Table 9.1. Animal Models of Spontaneous Regional ileitis

Animal species	*Pathogenesis (Hypothesis)*	*Clinic/Histology*	*References*
Cattle	Mycobacterium johnei (paratuberculosis) was isolated from affected animal. No transmision possible	- diarrhoea - wasting disease process - thickening of bowel wall predominantly terminal ileum, cecum - Corrugation of mucosal surface Histol.: - epitheloid cell accumulation - macrophages major cell-type of infiltration	Johnes a. Frothingham (1985), 5
Horse	Origin unknown, animals with altered immune status, no transmissability	- chronic weight loss - intermittend diarrhoea - plagues, nodules predominate ileum, often segmental - mucosa cobblestoned, thickened, edematous with fissures Histol.: - clubbering and fusion of villi, villi atrophy - transumural inflammatory cell infiltration - lymphoctyes, epitheloid cells, scattered plasma cells infiltrated in lamina propria	Cimprich (1974), 95
Dog	Only in one case positive culture of Camphylobacter supp.	- diarrhoea (4 weeks before death) - ileal mucosa thickened, corrugated Histol.: - hyperplastic epithelium - reduction of goblet cells	Collins *et al.* (1983), 8

Table 9.1. (*Continued*)

		- extended crypts - inflammatory cells predominantly lamina propria	
Pig	Isolation of Camphylobacter sputorum subsp. mucosalis from intestinal lesion (regional ileitis → intestinal adenomatosis)	- hypertrophic ileal wall - ulceration of hyperplastic mucosa Histol.: - Immature hyperplastic epithelium - granulation-tissue proliferation in the lamina propria	Lawson a. Rowland (1974), 9
Lambs	Unknown. No signs for bacteriological or parasitological agents.	- diarrhoea - clinical signs of peritonitis - thickened and rugose ileal wall, destruction of epithelium Histol.: - many mitotic figures and infiltration of lymphocytes, plasmacytes, histiocytes, and neutrophils in submucosa and lamina propria	Wensvoort (1962), 96
Hamster	Positive culture of Camphylobacter fetus subspecies jejuni. No transmissability → questionable predisposing factor? (proliferative ileitis!)	- sudden onset of illness, diarrhoea → death - hyperplastic ileal lesions, adherent masses Histol.: - enlarged peyer's patches - multiple intramural inflammatory foci	Lentsch *et al.* (1982), 6
Rat	Similarites to M. Hirschsprung (no histopathologic evidence). No specific bacteriological agent	- diarrhoea, abdominal distension - clinical sign of sepsis - dilated ileal segments - enlarged mesenteric lymphodes Histol.: - leucocytic inflammatory infiltration, lymphocytic infiltrate in lamina propria - enlarged myenteric plexus	Geil *et al.* (1961), 97

terminal ileum but also of the cecum, colon and the surrounding mesenterium.[5] Macrophages were the major cell type involved in these lesions and epitheloid-cell granulomas caused gross thickening of the bowel wall and distorted corrugation of the mucosal surface. Even though the test for M. Tuberculosis was positive, Johne and Frothingham did not succeed in transmitting the disease experimentally. The variety of animal transmission studies on Crohn's disease will be discussed later on in detail. So far this case shows accordingly that the pathological picture of IBD so far described in most instances does not reflect a unique aspect of the disease, but rather an inflammatory response of the intestine that can be induced by a variety of infectious agents.

Other infectious agents which seem to cause ileitis in animals are different types of Campylobacter. Campylobacter species could be isolated from hamsters with proliferative ileitis,[6] from lambs with regional enteritis[7] as well as from pups.[8] In addition regional or terminal ileitis occuring in swine has been associated with a vibrio-like bacterium, Camphylobacter sputorum subsp. mucosalis.[9] Granulomatous inflammation was not a histological feature but the porcine ileitis has always drawn special attention since Crohn himself compared similarities with human ileitis.[10]

The typical anatomical distribution, the inflammatory thickening of the terminal ileum and the acute or subacute necrotizing process primarily limited to the lamina muscularis mucosae and the serosal tissue was also seen in pigs with ideopathic ileitis (Tab. 9.2). The list of further examples of naturally occuring animal models of intestinal inflammation could be continued. They are of indirect relevance to Crohn's disease and parallels between the diseases described and the human pathological and clinical feature are limited.[11]

Animal Models Of Infectious And Spontaneous Colitis

Similar problems in the differentiation between infectious and ideopathic origin of inflammation are seen in animal models for colitis (Tab. 9.3). While e.g. one variant of canine colitis–a granulomatous colitis observed in boxer dogs[12]–appears to be caused by an intracellular organism that may be related to Chlamydia organism, there exists also a report about spontaneous canine ileitis, colitis and proctitis of unknown origin.[13] The feature was macroscopically accompanied with perianal fistulas, hepatic granulomas, and ulcerative dermatitis observed in affected dogs and comparable to the extraintestinal lesions of regional enteritis in man. However, the

Table 9.2. Examples of spontaneous terminal ileitis in pigs

Number of examined animals	*Pathogenesis/ Hypothesis*	*Clinical details*	*Pathology*		*References*
			Macroscopic	*Microscopic*	
58		- diarrhoea, constipation - peritonitis	- rigid ileal wall - hypertrophy of the muscularis - proliferation of the mucosa - mesenterial edema		Emsbo (1951), 98
12 + 5		*3–4 month old pigs* - severe short illness of sudden onset characterized by loss of appetite, difficulties to stand up and to breath - death after 1 or 2 days due to acute peritonitis	- thickened ileal wall with stenosis and irregular folded mucosa - perforation, peritonitis *Clinical healthy animals*: 3 cases with macroscopic and microscopic, 2 cases with only microscopic lesions were seen, which were similar to the lesions of clinical ill animals	- chronic inflammation covered by acute peritonitis - infiltration by mononuclear cells, plasma cells, lymphcytes, epitheloid cells and eosinophils - ulcerations - granulation tissue - hyperplastic lymphatic follicles	Field et al (1953), 99
25	A functional lesion could cause the muscular hypertrophy: Hyperplastic lymphatic follicles narrow the opening. Increased pressure is necessary for the passage of feces → muscular hypertrophy	*3–6 month old pigs* - normal developed, - acute short illness - death after 1 or 2 days caused by acute peritonitis	- thickened ileal wall wall (especially proximal the ileocecal valve) - in some pigs cecum involved	- thickening of all layers of the ileal wall especially of lamina muscularis mucosa and tunica muscularis - infiltration with eosinophils in the lamina propria - fibrosis of the submucosa	Szabo Szeky (1964), 100

Table 9.2. Continued

7		Two groups were differentiated: *6–12 weeks old*: - loss of appetite - intermittend diarrhoea changing with constipation - pain, sometimes fever - feces contains mucus and fibrin, seldom blood - perforation leads to death due to peritonitis *6 months old:* incidental finding after slaughtery: chronic lesions without symptoms	- the wall of the terminal ileum (sometimes of cecum and colon) was thickened and rigid - hypertrophy especially of the tunica muscularis mucosa - ulcerations - enlarged ileal lymphnodes	- mucosa: hyperaemic and hyperplastic, irregular folded, "cobble stone" like mucosa - pseudomembranes - inflammatory infiltration contains: lymphoctyes, neutrophile granulocytes, eosinophils - tortous and dilated crypts - enlarged peyer's patches	Nielsen (1971), 101
39		5–10 *month old* clinical healthy pigs - incidental finding after slaughtery	- thickened ileal wall, rigid, non elastic - enlarged peyer's patches - cysts - regional lymphodes infiltrated	- acute non-specific inflammation - hyperaemic, edematous mucosa - proliferation of the mucosa - enlonged branched crypts - cysts, crypt abcesses all layers of the wall changed: muscular-hypertrophy, chronic transmural inflammation with epitheloid cell like macrophages, Langhans giantcells, round cells, deepened branched crypts, ulcerations, crypt abscesses, fissures, fibrosis of the submucosa	Rhakoa. Saloniemi (1972), 102
	Proliferative intestinal adenomatosis leads if repaired to RI, if not repaired to NE Relations to camp. sputorum		Regional ileitis: ileal wall thickened - ileum: rigid, stiff, opening narrowed, mucosa only irregular present Necrotic Enteritis: Ileum: - gray, necrotic, zones, fissured, often edematous	RI: muscular hypertrophy granulation tissue camp. sputorum mucosalis NE: non-specific inflammation granulation tissue hyperplastic epithelium	Rowlan d.a. Lawson (1995), 103

Table 9.3. Animals models in spontaneous colonic inflammation

Animal species	*Pathogenesis (Hypothesis)*	*Clinic/Histology*	*References*
Subhuman primates: - Siamang - Gibbons	Emotional stress	- loss of apetite, lethargy - small amount of blood and mucus in stool - extensive mucosal ulceration limited to colon and rectum Histol.:- destruction of glandular architecture - crypt abscesses - lamina propria necrotic and infiltrated with leucocytes, plasma cells, lymphoctyes	Stout a. Snyder (1969), 14
cotton-top-tamarins	- No typical pathological bacterium - 18 out of 68 animals carried corona viruses - psychological stress (questionable neuro-endocrine or neuro-immunologic factors?)	- diarrhoea, bloody stools - wasting disease process - after 4 years development of adenocarcinoma - extensive mucosal ulceration predominantly colon ascendens Histol: - crypt abscesses - monoclear and neutrophil cell infiltration	Chalifox a. Bronson (1981), 16
Dog	Familial (breed) predisposition is considered a significant etiological factor	- bloody diarrhoea, weight loss - depression, fever - mucosal ulceration differing in all cases: ileocolitis, colo-proctitis, proctitis Histol: - excessive mucus production - crypst abscesses, cystic crypts - excessive mononuclear cells in basal lamina propria	van Kruiningen (1972), 13
Ferret	Isolation of Camphylobacter jejuni from crypt-epithelium	- suden onset, diarrhoea - loss of weight, rectum prolaps - hyperplastic colonic wall mucosal ulcerations Histol: - proliferation of epithelial cells	Fox *et al.* (1982), 104

Table 9.3. *(Continued)*

		- goblet cell depletion - infiltration of leucocytes, plasma cells, and lymphocytes in lamina propria	
Rat	Origin unknown. Comparison to chronic ulcerative colitis in man	- sudden onset, diarrhoea - wasting disease process - mucosal ulcerations of the colon, fissures - lymphstasis Histol: - perivascular inflammation - lymphostasis	Stewart a. Jones (1941), 105

dogs had been examined as a group and there was a considerable variability between the cases represented, and any one animal had only a few characteristics of regional enteritis. The diversity of clinical presentation and pathological findings limits their potential for a consistent animal model.

Colitis in subhuman primates has been first described by Stout and colleagues (1969) in gibbons[14], and has also been reported as ileitis in gorillas as well as orangutans.[15] In 1981 Chalifoux and Bronson described ulcerative colitis in the cotton-top-tamarin.[16] In the following a lot of attention has been paid to this model since the disease also threatened the species seriously. 50 to 70% of cotton-top-tamarins have been demonstrated to have colitis. The histological characteristics were crypt abscesses, mononuclear and neutrophil cell infiltrations. In contrast to ulcerative colitis the inflammation can occur also in the small intestine but predominantly involves the ascending colon as in humans. After 4 years of colitis most animals develop adenocarcinoma of the colon. In terms of treatment sulfasalazin and forced tube feeding affect the disease positively, but it reoccurs after stopping the application. The disease could start early in life, be very severe and lead to death within days. It seemed that the disease observed was not likely to be caused by a bacterial organism since stool cultures were negative in most of the cases, but when moving the Yale primate colony to the New England Regional Primate Center in Southbaro, Mass., within 3 months new cases of ulcerative colitis and carcinoma were identified and the search for an infectious agent was undertaken again[17]. No typical pathological bacterium was identified (Campylobacter in 15 out of 71 animals) and instead in 18 out of 68 coronavirus was found. Though it is known that coronaviruses can cause enteritits in different animal species the role of the virus in cotton-top-tamarins remains unclear–leaving it open if the virus is a secondary invader that perpetuates an immune response initiated by another agent or is of no significance at all. A relation between periods of psychological stress and the onset of the colitis in primates has also been observed. Since similar psychological relationships have been reported for ulcerative colitis this gives an interesting impact to the possible role of neuroendocrine or neuroimmunologic factors on the inflammation of the gastrointestinal tract.

Experimentally Induced Intestinal Inflammation in Animals

The Search for a Transmittable Agent

It has already been mentioned that from the beginning research work in IBD has focused on the goal to fullfill Koch's postulates by isolating a specific agent from affected patients and transmit the disease experimentally (Tab. 9.4).

In the case of the Mycobacterium johnei (today renamed as M. paratuberculosis) numerous studies have been undertaken.[18] Even early efforts have been disappointing. In 1984 Chiodini and colleagues identified an unclassified mycobacterium from two patients with Crohn's disease which could be related to M. paratuberculosis.[19] The isolation seemed to be very difficult and was only successful through long incubation periods and complex media with mycobactin. The organism was thought to be a pathogen since inoculating it to goats produced thickening of the ileum, enlargement of mesenteric lymph nodes and showed microscopically multiple non caseating tuberculoid granulomas with giant cells in the ileum as well as in the regional lymph nodes.[20] A mycobacterium species with characteris-

Table 9.4. Transmission studies in Crohn's disease

Animal species	*Pathogenesis*	*Clinic/Histology*	*References*
Rabbit	Inoculation of virus into the ileum of a rabbit	Reisolation of the virus; no specific gross or histopathologic lesions	Schneierson (1962), 108 Aronson (1975), 109
Mouse	Injection of ileal or mesenteric lymphnode homogenates from patients with Crohn's disease into mouse footpads	Transmission of epitheloid granulomas into mouse footpads	Mitchell a . Rees (1970), 26 Mitchell a. Rees (1986), 106
Rabbit	Injection of affected tissue homogenates of two patients with Crohn's disease into ileum of a rabbit	Transmission of epitheloid granulomas into wall of the gut	Cave *et al.* (1973), 27 (1975), 29
Goat	Unclassified mycobacterium from patients with Crohn's disease given to goats orally	- lesions resembling those of Johne's disease - granulomatous changes in mesenteric lymphnode - thickening of the intestinal wall	Chiodini *et al.* (1984), 19 Morgan (1987), 18
Rat	Distributed group A or group D streptococcal cell wall fragments injected intramurally into the small intestine and cecum	- no clinical changes - thickening of the intestinal wall - contraction of adjacent mesentery - adhesions located between the tip of the cecum and distal ileum	Sartor *et al.* (1985), 33
Monkey	Inoculation of C. trachomatics, Inflammation associated with the presence of suppressor T- cells	- proctitis - mucosal erythema - mucosal triability and ulceration - hyperplastic lymphoid follicles - conjunctional follicles persisting 4 weeks	James *et al.* (1987), 42

tics identical to those of the inoculum was reisolated from the mesenteric lymph nodes, not from intestine. No lesions did occur when rabbits, chickens, rats or guinea pigs were used as experimental animals. Even the finding of mycobacterial antibodies in Crohn's disease did support the importance of the possible role of mycobacteria in IBD,[21] a treatment trial with antimycobacterial therapy did not show any specific improvement for the patients[22] and further studies suggest that the M. species is only an intestinal resident unrelated to Crohn's disease[23, 24] - giving support to the idea that the mycobacterium is a secondary invader of the IBD lesion which becomes manifest only because of the preexisting inflammation.

This idea is of course applicable in all studies where animals receive material derived from patients with IBD. It must be recognized that transmission of a granulomatous inflammation tissue reaction in not necessarily transmission of CD[25]. With this in mind one has to look at the studies first introduced by Mitchell and Rees in 1970 who had injected affected tissue homogenates into the footpads of mice.[26] They did find epitheloid-cell granulomas upon injection of mesenteric lymph node homogenates from a patient with CD in several animals. Further investigation proclaimed the reproduction of CD-like lesions in the ileum after injection of tissue homogenates into the ileal wall.[26,27] The findings could, however, not be reproduced by others[28] and became even more complicated to interpret when studies observed the possibility of granuloma transmission from ulcerative colitis colonic homgenates as well as from normal colon that had been interposed into the esophagus for two years.[29] That the experimental animal model which has been used successfully in the study of leprosy and sarcoidosis did not give conclusive result in IBD points towards the need of an adequate unique design as discussed in the beginning.

Several reports have implicated viruses as etiological agents in IBD, but also in this respect the results vary extensively.[30] Gitnick and his colleagues isolated a small RNA-virus like particle from several patients with CD and a different agent from a patient with UC.[31] But again Koch's postulates could not be fulfilled nor have electron microscopic studies demonstrated viral particles either in the bowel wall of patients with CD or UC or of animals to whom CD or UC filtrates have been passaged. Further extensive research did not provide any evidence for the presence of conventional viral agents in tissue filtrates from IBD patients.[32]

Another interesting model showed a granulomatous entercolitis induced in rats by purified bacterial cell wall fragments.[33] Granulomas of the intestine, mesentery, and mesenteric lymph nodes were seen in nearly 45% of rats injected intramurally group A or group D streptococci cell walls compared to 20% of the controls injected with albumin. Although none of the animals developed symptoms of bowel disease the microscopic appearence after necropsy showed a chronic inflammatroy process after 3 to 6 months. Even if it seems accepted today that IBD lesions cannot be passed into animals by injection of homogenates of filtered material derived from patients, these experiments point towards a better understanding of the inflammatory process of the intestine. It might be possible e.g. that luminal bacterial cell wall fragments leaking across the mucosa could induce acute or chronic granulomatous inflammation of the gut. Cell wall fragments entering the lymphatics and systemic circulation could produce extraintestinal inflammation. Similarly, any other uptake of macromolecules with toxic or antigenic potential could lead to an gastrointestinal, even general inflammatory process.

Immunological Alterations

The search for understanding the process of gastrointestinal inflammation more in detail is reflected by the numerous studies during the last 40 years which have concentrated on immunological manipulations of the intestine. The Arthus and the Shwartzman phenomenon were described in rabbits, first sensitizing the animals to an antigen by systemic antigen administration and then provoke an intestinal inflammation by local injections of the antigen into the intestinal tissue.[34,35] Ulcerations and hemorrhagic lesions were seen and after a short time period left residue of granulomata and site of fibrosis. The "Auercolitis" was based on the same concept where antibodies were the main disease producing factors.[36] The local immune complex reaction was characterized by a feature containing macroscopically signs of acute inflammation as well as microscopically infiltration of neutrophils, macrophages, eosinophils, lymphocytes, and plasma cells. This model has also been used to investigate the role of prostaglandins in experimental colitis showing that the changes of blood flow, mediated by the induced inflammation, could be stopped by e.g. indomethacin, but a thromboxane inhibitor was not effective.[37]

The concept of an increased pathogenic idea of "Auercolitis" is comparable to increased uptake of toxic or antigenic macromolecules. An area of altered vascular permeability is created where circulating antibodies easily can invade the gastrointestinal "barrier" which again can react with locally applied antigens. All experiments have shown that only in cases when the local irritant and the antigen is regularly reapplied the process of inflammation will continue–which does not resemble the idea of a self-sustaining inflammatory process in IBD. Nevertheless it gives again further information on particular aspects of the intestinal function related to details of the clinical feature as seen in IBD - e.g. vascular changes. Direct manipulations of vascular or lymphatic permeability have also been tried to induce experimental colitis[38,39] but did not produce comparable pathological features as seen in ideopathic IBD.[40]

The idea, that local immune complex deposition functions as a point in a inflammatory cascade[41] leads to the question of what is known about the first event in the disease process. Turning later to the current ideas of early morphological changes, the presence of circulating antigen-non-specific suppressor T-cells seems immunologically to indicate that Crohn's disease is in its early stage.[42] Such T-cells are also found in experimental inflammation caused by Chlamydia organisms in the gastrointestinal tract of subhuman primates.[43,44] Even if this model leads primarily to increased knowledge of sexually transmitted diseases (lymphgranuloma venerum) it suggests, that suppressor T-cells are markers of an underlying and persistent immune response to a yet unidentified antigen. Further investigation of the specific mucosal immune response could give better understanding of gastrointestinal inflammation.

Another immunological model of IBD looking at an exogenous antigen was contributed by Bicks and Rabin.[45,46] Rabbits and guinea pigs were initially sensitized to Dinitrochlorobenzene (DNCB) by skin painting followed by colonic installation (Tab. 9.5). The cell mediated lesions induced colonic lesions including crypt abscesses, depletion of mucus in the cells lining the crypts (a feature of ulcerative colitis), and infiltration of the lamina propria with plasma cells, lymphocytes, and polymorphonuclear leucocytes including eosinophils. After repeat application of DNCB, delayed hypersensitivity skin tests became positive when challenged by colonic antigen. The model shows that exogenous antigen may initiate a cellular immune response to endogenous antigen and that such an immune response can appear in gastrointestinal inflammation but the process in not sustained by self-continuing mechanisms.

In contrast, Mee and coworkers demonstrated an immune complex colitis which was not dependent on reexposure.[47] Rabbits were first immunized with a common E. coli antigen. By injection of soluble immune complexes following irritation of the rectum with diluted 1% formalin the colitis was established. The animals developed an acute colitis within the first week but in contrast with unsensitized rabbits the inflammation persisted longer than 6 weeks. Therefore it was concluded that hypersensivity to colonic bacterial antigens might be a mechanisms whereby an acute colitis becomes chronic.

Table 9.5. Transmission studies in Crohn's disease

Animal species	*Pathogenesis*	*Clinic/Histology*	*References*
Rabbit	Immune complexes damaged mucosa → increased absorption of bacterial antigens local antigens - antibody reactions	- no clinic Histol.: - chronic inflammatory process - gland and epithelial distortion - atrophy	Kraft *et al.* (1963), 36 Mee (1979), 47
Rabbit/ Guinea pig	Dinitrochlorobenzene (DNCB) sensitization → DNCB reapplication → cell-mediated immunity. Colon lesions compatible with IBD	- no clinic Histol.: - crypt abscesses - mucosa ulcerations - edema	Bicks (1964), 45 Rabin (1980), 46
Guinea pig	Carrageenan allows increased absorption of bacterial antigen into the mucosa → cell-mediated immunity → inflammatory process	- loss of weight within 2–3 weeks - blood and mucus in the stools Histol.:	
Rat	Colonic ulceration and inflammation by metabolic induction through an enzymatic inhibition of β-oxidation, especially Ketothiolase, by exposure to a bromiated fatty acid	- weight loss, bloody diarrhoea Histol.:- mucus cell depletion - vessel dilation - mucosa ulcerations	Roediger a Nance (1986), 87
Rat	Colonic ulceration and inflammation induced by intracolonic administration of TNB (PAF-acether mediated inflammation)	- diarrhoea - intraabdominal adhesions - epithelial inflammation ulceration	Wallace (1989), 85

In an animal model of HLA-B 27 associated human disorders Hammer demonstrated the occurrence of spontaneous gastrointestinal inflammation.[115] The development of inflammatory lesions in the bowel of rats transgenic for human HLA-B 27 and β2m-globulin pointed out the importance of this locus for intestinal inflammation.

The discoveries that patients suffering from inflammatory bowel disease have low levels of some cytokines.[116] lead to employment of genetic 'knock out' mice (KO) of study the influences of these deficits. Mouse with defective T-cell receptor or class II major histocompatibility complex (MHC)[117] and mouse with defective expression of IL2.[118] IL1O[119] or TGFβ have been designed to study the influence of these proteins involved in immune response. All of these "knock out mice" with dysfunction in the mucosal immune system developed a chronic colonic inflammation. These KO lines emphasize the importance of an uncontrolled or dysregulated local immune response of the mucosal immune system to normal antigens in the gut.

With the help of adoptive transfer subsets of CD4$^+$ T-cells to immunodeficient hosts various inflammatory diseases can be induced. Moreover those expriments show that IBD can be induced by transfer of a selective subpopulation of T-cells into healthy host animals and by the way demonstrate their pathogenic potential.[120]

In a profound review about immunological aspects of animal models in IBD Strober summarized that the existing experiments of gastrointestinal inflammation showed that T-cell mediated as well as B-cell mediated events can be the background of colitis-like lesions.[41] This leads to the view, which is primarily supported by the similarities of the morphological features in intestinal diseases, that the inflammatory response of the gut is a rather monotypical picture independent of the primary event.[48] Like it has been stated for the permeability of the mucosa one might speculate that the immunological manipulation is one part of the array of events which may serve as the initiator of a common immunopathological pathway.

Food Related Irritants

Epidemiological studies over the last few decades have suggested that the incidence rates of ulcerative colitis decline while in CD rates have increased sharply during the 1950's and 1960's and reached a plateau in the 1980's.[49] Primarily an improvement in diagnosis and recognition of CD as an entity that differs from UC has been made responsible for those changes. However, this did not explain by itself the described increase. In addition the prevalence of CD have risen more in urban settings while the growth has been much slower in rural areas. According to the observed differences in the incidence of the diseases it was suggested that for CD and UC the causal agent(s) or factor(s) may not be the same. Also, diseases that change incidence over a few decades are likely to be affected by environmental causes, especially if the distribution is very high in specific countries (e.g. United States, Sweden, Germany) while low in others (e.g. France, Italy, or countries in Africa).

Under these circumstances the search for possible etiological agents have concentrated on different environmental factors and here specifically on the daily food intake of patients with CD. Many different foods have been related to the origin of the disease[50–52] while only a few have been investigated in an animal model.

Carrageenan–a sulfated polygalactose widely used in the food industry–has been used for the development of large intestine ulceration.[53,54] A 5% aqueous solution of degraded carrageenan, derived from red seaweed, was fed in the drinking water to guinea pigs over a period of 20–45 days. After two weeks symptoms of ulcerative colitis occurred with occult blood in the feces and multiple ulcers in the cecum, colon and rectum. This process could also be induced in the mouse,[55], rhesus monkey[56] and rabbit.[57] On the other hand, the induction of symptoms and morphological changes was not successful in rats, hamsters, squirrels, monkeys and ferrets in dependency of the type and the degree of degradation of carrageenan.[55,58]

The intestinal ulceration induced in guinea pigs could be prevented through the elimination of anaerobic bacterica by metronidazol or clindamycin,[59,60] and does not develop in germ free rats.[61]

Interestingly rat fed levels of 1 to 10% carrageenan in their diets develop colorectal adenomas and adenocarinomas, thereby supporting theories connecting chronic inflammation and epithelial cell transformation.[62]

Another ubiquitous agent which can cause experimental inflammation of the intestine is lectin. The pathways of epithelial damage after lectin within diets is unknown, mainly irritation of small bowel is seen.[63,64]

Dietary Fats And Fatty Acid Metabolism

In recent years a variety of studies have been conducted to investigate in more detail the relevance of dietary fats and the fatty acid metabolism in gastrointestinal inflammation. Dietary fats are relevant for the quite often appearing fat-incompatibility syndrome,[65] but also accused to be linked with the genesis of colorectal cancer,[66] changes in the immunological response[67] as well as to be relevant for changes in the arachidonic-acid metabolism as seen also in IBD.[68] In this context epidemiological data, historical and biochemical investigations have been reported[69,70] which support the hypothesis of Guthy[71] that chemically processed fat could play a causal role in the pathogenesis of CD.

Mucosal fatty acids' profiles have been determined by dietary studies in pigs to examine the connection between chemically processed dietary fats, e.g. trans-isomeric fatty acids, and alterations of the mucosal architecture. Samples were collected by way of an ilecutaneous fistula and analyzed gaschromatically.[121] The effect of partially hydrogenated chemically processed fats (CPF) on the fatty acid profile of the ileal mucosa was then investigated. Diets rich on heated CPF caused higher mucosal incorportion of trans-isomeric fatty acids and extended intestinal transit time of cyme. The accumulation of these fatty acids could enhance the formation of epithelial lesions, characteristic after long term ingestion of CPF. Changes in the distribution of polyunsaturated fatty acids have also been demonstrated in inflamed mucosa of patients with active CD.[122]

In an animal model in swine it could be shown that after a diet high a chemically processed fats morphological changes in the terminal ileum occurred which resemble those seen as early lesions in CD[72]. After induction of bowel stasis by creating a 40 cm retroperistaltic ileal loop, 20 cm proximal the ileocaecal valve, all animals developed in a time period of 3 months edema formation, lympho-reticular hyperplasia, changes of villious structures as well as destruction of basal goblet cells after fat feeding. An increase of intraepithelial lymphocytes followed.[73] While the histological findings are comparable to the primary preulcerous stage in the morphological classification of CD[74] the ultrastructural changes seemed to be more relevant. The mucosal surface was characterized by goblet cell hyperplasia, villi degeneration and epithelial-bridge formation[75] – morphological alterations which have been called characteristic of CD by different experts of intestinal pathology[76–78]. Further experiments have suggested the importance of bowel stasis for the inflammatory process seen in this model and tried to specify in more detail substances in chemically processed fats which may be responsible for the feature seen.[79]

The animal model with the retroperistaltic ileal segment was designed to study effects of prolonged contact time between iteal chyme rich on hydrogenated fatty acids and epithelium regarding disorders of the intestinal barrier function.[123] Transmission electron microscopic widening of the intercellular space demonstrated in part total loss of the funcional structure of the epithelial cell layer in these animals. Examination of defined areas in the lamina mucosa showed dense infiltration and activation of lympho-plasmocytotic cells. Moreover focal infiltration of mast cells and degranulation of cytoplasmatic granula was observed. In the neighborhood of these focal areas dilatation of axons with depletion of axon organelles could be noted.[124] The ultrastructural findings of an inflammatory process of ileal mucosa had close parallels with CD.

Looking at early pathological changes of IBD one has to focus at the mucosal "barrier function", being the organ which obviously seems to be the place of the primary event in the disease process. Endoscopically aphthous ulcers are the first macroscopic manifestation, while histologically a focal accumulation of inflammatory cells occupying an area adjacent to an ephithelial crypt and changes in the intestinal gland structure are primary. Ultrastructural analyzing techniques have made it possible to define even early steps of structural alterations. But it is difficult to classify documented early lesions. For example, lysosomal inclusions could be a hint for a distorted macrophage function through an external or internal insult when induced in an experimental animal model,[48] but at the same time could resemble partially degraded bacteria as a secondary phenomenon in inflamed intestinal tissue from CD patients.[80]

In this context one should recognize animal models which analyze in detail physiological and biochemical reaction of the mucosa. For example intestinal damage can be induced by intravenous infusion of Platelet-activating-factor (PAF-acether) which is a low-molecular-weight endogenous phospholipid and seems to be a mediator of the gastrointestinal damage associated with endotoxin shock.[81,82] PAF-acether can be released from different celltypes such as neutrophils, macrophages, basophils and monocytes, and is suggested to have a potent proinflammatory action.[83] In this context interactions of PAF-acethers and e.g. eicosanoids and products of the lipooxygenase enzymes including LTB4, LTL4 or LTD4 could be carefully investigated and lead to a better understanding of the inflammatory cell metabolism which is related to the arachidonic acid metabolism in IBD as well.[84,85]

The study of fatty acid oxidation, primarily in the colon, turns to a somewhat similar direction. The impairment of fatty acid oxidation has been directly implicated in the genesis of experimental colitis. One model concerns diminution of free CoA which is essential for activation of short-chain fatty acids and subsequent oxidation to CO_2 and ketones.[86] Another experiment is based upon an enzymatic inhibition of β-oxidation by exposure to a brominated fatty acid.[87] Interestingly the degree of colitis of CoA-deficiency is species dependent, therefore suggesting that a genetic factor might be necessary for expression of the disease.[88] Roediger concludes in his review about colonic mucosal metabolism in the pathogenesis of

Table 9.6. The pig as a model for human gastrointestinal disease

Pig	*Human*	*References*
Physiology and morphology of the gastrointestinal system very much alike in pigs and humans		Miller a . Ullrey (1987), 114
Small intestinal maturation and intestinal enzyme distribution highly similar to humans	Intestinal enzyme development and regulation are excellent to investigate in pigs	Shulman *et al.* (1988), 113
Highly dependent on dietary quality since symbiotic microorganisms within the gut play minor role in modifying the nutrients	In healthy subjects microorganisms play a relatively minor role in modifying the nutrients and for the digestive efficiencies	Miller a. Ullrey (1987), 114
As an omnivorous animal a variety of dietary carbohydrates and fats can be fed and manipulations of dietary composition and intake will be tolerated	Large variety of dietary components characterize individual diets and do effect carbohydrate, and lipid metabolism	Mersmann (1985), 112
Breath hydrogen and volatile fatty acid excretion, as well as glucose, insulin and enteroglucagon responses, are related qualitatively in both species	Physiological interactions of digestion and fermentation are similar to those in pigs	Fleming a. Arce (1985), 111
Relatively large size of the pig permits surgical interventaions which are unreasonable for smaller animals. Samples of an adequate amount can be drawn and over longer time intervals without endangering the animal		Swindle (1985), 110

ulcerative colitis that the disease is based primarily on an impaired fatty acid metabolism in colonocytes resulting in a failure of these cells of the redox-function. This capability is supposed to be necessary to maintain a homeostasis between the intraluminal content of the colon - e.g. specific macromolecules - and the mucosal immune system.[86]

Prospect Of Animal Experiments In IBD

As it has been stated in the paragraph concerning the clarification of pathomechanisms in experimental animal models in IBD one has to keep in mind present concepts about a multifactorial pathogenesis of the disease process. In this context the intestinal "barrier function" as primary affected region takes a central position. One has to concentrate on what is known about gastrointestinal inflammation, making it obvious that the functional mechanisms of the gut are still in part widely unknown. Aspects to look at have to derive from the clinical observation of patients with IBD and on this basis experiments have to be designed.

In this context further investigations in CD should look at food and food metabolism as one main topic.

Especially since it became clear that elemental diets in the therapy of acute exacerbation of Crohn's disease are highly effective,[89] the hypothesis has been strongly supported that the antigen load of normal daily diets for those patients has particular importance for the pathogenesis of the disease, respectively for the persistance.[90]

One hypothesis which tried to combine the most possible factors of the etiology was published by Shorter and colleagues in 1972.[91] Although they centered on an intestinal cell-mediated immune reaction to bacterial antigen, cross reacting with antigens on colonic mucosal cells, the underlying theory would fit for all intraluminal distributed antigens. The local immune sensitization of the gut-associated lymphoid tissue (GALT) to antigens absorbed occurs early in life because of immaturity of the newborn mucosal barrier. Macromolecular uptake across the intestinal epithelium is an essential component of mucosal immunity[92] and is significantly higher, e.g. in neonatal animals.[93] A secondary, non specific intestinal injury results in renewed augmented antigen exposure to the GALT and a local hypersensitivity, cell mediated immune response may theoretically ensue, perpetuated by a continued absorption of the cross-reacting macromolecule/colonic-epithelial-antigen, resulting in chronic inflammation.

Paying special attention to food-related intestinal interactions the pig seems to be the animal species to recommend if looking at the small intestine.[94] Because the pig is a true omnivore, it has been described as the best model of the gastrointestinal physiology of man (Tab. 9.6). Also for all gastrointestinal surgical procedures the abdominal organs are very useful. This excludes studies of the colon, since a spiral colon characterize porcine antomy. Therefore studies on experimental colitis should be preferred on other animal species. The very tempting natural occuring model in the cotton-top-tamarin has advantages, but is mitigated by the fact that they are among the endangered species and therefore cannot be used as experimental animal species. In any case subhuman primates are expensive, very difficult to handle and should be reserved for institutions with special expertise.

In summary it must be stated that no reliable animal model of IBD has been established. Specific knowledge about the nature of the human illness can be gained from the study of naturally occuring and induced gastrointestinal inflammation seen in animals. Recent results should encourage and design have been encouraging and designing further experimental animal models for Crohn's disease and Ulcerative colitis. This seems to be necessary and substantially helpful to learn more about the complex disease processes for which no specific treatment does exist and which today still is a serious, long lasting burden for the mostly young patient.

Acknowledgement

We are indebted to Susanne Schattenfroh and Dr. Sabine Bühner for the intensive discussion of the subject, to Angelika Köhler for typing the manuscript, to Petra Schmidt and Markus Kondruweit for assistance in preparing the tables as well as to Heidi Frölich for the excellent techniqual management.

References

1. Alter, H. J., Holland, P. V., Pucell, R. H. and Popper, H. (1978). Transmissible agent in Non-A, Non-B-Hepatitis. *Lancet*, **I**, 459–463.
2. Storrs, E. (1971). The nine bande Armadillo: A model for leprosy and other biochemical research. *Int. J. Leprosy*, **39**, 703–714.
3. Dalziel, T. K. (1913). Chronic interstitial enteritis. *Brit. Med. J.*, **2**, 106–111.
4. Crohn, B. D., Ginzburg, L. and Oppenheimer, G. D. (1932). Regional ileitis: A pathological and clinical entity. *J. Americ. Med. Ass.*, **99**, 1323–1328.
5. Johne, and Frothingham (1895). Ein eigentümlicher Fall von Tuberkulose beim Rind. *Dtsch. Ztschr. f. Thiermed. u. vergl. Pathol.*, **25**, 438–454.
6. Lentsch, R. H., Mc Laughlin, R. M., Wagner, J. E. and Day, T. P. J. (1982). Campylobacter fetus subspecies jejune isolated from Syrian hamsters with proliferative ileitis. *Lab. Anim. Science*, **32**, 511–514.
7. Van der Berghe, J. and Hoorenss, J. (1980). Camphylobacter speices and regional enteritis in lambs. *Res. in Vet. Science*, **29**, 390–91.
8. Collins, J. E., Libal, M. C. and Brost, W. (1983). Proliferative enteritis in two pups. *JAVMA*, **183**, 886–889.
9. Lawson, G. H. and Rowland, A. C. (1974). Intestinal Adenomatosis in the pig: Bacteriological study. *Res. of Vet. Science*, **17**, 331–336.
10. Crohn, B. B. and Turner, D. A. (1952). Porcine ileitis. *Gastroenterology*, **20**, 350–351.
11. Mayberry, J. F., Rhodes, J. and Heatley, R. V. (1980). Infections which cause ileocolic disease in animals: Are they relevant to Crohn's disease? *Gastroenterology*, **78**, 1080–1084.
12. Van Kroiningen A. J. (1967). Granulomatous colitis of boxer dogs. Comperative aspects. *Gastroenterology*, **53**, 114–122.
13. Van Kruiningen A. J. (1972). Canine colitis comparable to regional enteritis and mucosal colitis of man. *Gastroenterology*, **62**, 1128–1142.
14. Stout, C. and Snyder, R. L. (1969). Ulcerative colitis like lesion in Siamang gibbons. *Gastroenterology*, **57**, 256–260.
15. Scott, G. B. D. and Keymer, I. F. (1975). Ulcerative colitis in apes: A comparison with the human disease. *J. Americ. Med. Ass.*, **115**, 241–244.
16. Chalifoux, L. V. and Bronson, R. D. (1981). Colonic adenocarcinoma associated with chronic colitis in cotton-top-marmoset (sanguinus oedipus).. *Gastroenterology*, **80**, 942–946.

17. Beeken W. L. (1988). Experimental inflammatory bowel disease. In: Kirsner J. B., Shorter R. G.: Inflammatory bowel disease, 3rd edition, Philadelphia: Lea & Febiger, pp. 37–49.
18. Morgan, K. L. (1987). John's and Crohn's: Chronic inflammatory bowel disease of infectious etiology? *Lancet*, **2**, 1017–1019.
19. Chiodini, R. J., Van Kruiningen H. J., Thayer, W. R., Merkal R. S. and Coutu J. A. (1984). Possible role of micobacteria in inflammatory bowel disease I: An unclassified micobacterium species isolated from patients with Crohn's diesease. *Dig. Dis. Sciences*, **29**, 1073–1079.
20. Van Kruiningen, A. J., Chiodini, R. J. and Thayer, W. R. (1986). Experimental disease in infant goats induced by a microbacterium insolated from a patient with Crohn's disease: A preliminary report. *Dig. Dis. Sciences*, **31**, 1351–60.
21. Thayer, W. R., Coutu, J. A., Chiodini, R. J., van Kruiningen, A. J. and Merkal, R. S. (1984). Possible role of microbacteria in inflammatory bowel disease II: Microbacterial antibodies in Crohn's disease. *Dig. Dis. Sciences*, **29**, 1080–1085.
22. Shaffer, J. L., Hytes, S., Limaker, B. D., R. D. and Tunberg, L. A. (1984). Control trail of rifampicin and ethambutol in Crohn's disease. *Gut*, **25**, 203–205.
23. Sang-Nae Cho, Brennan, P. J., Yoshimura, H. H., Korelitz, B. I. and Graham, D. Y. (1986). Mycobacterial aetiology of Crohn's disease: serologic study using commun mycobacterial antigens and a species-specific glycolipid antigen from Mycobacterium paratuberculosis. *Gut*, **27**, 1353–1356.
24. Graham, D. Y., Markesich, D. C. and Yoshimura, H. H. (1987). Mycobacteria and Inflammatory Bowel Disease. Results of Culture. *Gastroenterology*, **92**, 436–442.
25. Sachar, D. B., Taub, R. N. and Janowitz, A. W. (1975). A transmissable agent in Crohn's disease? A new pursuit of an old concept. *New Engl. J. Med.*, **293**, 354–55.
26. Mitchell, D. N. and Rees, R. J. W. (1970). Agent transmissable from Crohn's disease tissue. *Lancet*, **2**, 489–490.
27. Cave, D. R., Mitchell, D. N. and Cane, S. B. (1973). Further animal evidents of a transmissable agent in Crohn's disease. *Lancet*, **2**, 1120–1122.
28. Bolton, P. M., Owen, E. and Heatley, R. V. (1973). Negative findings in laboratory animals for transmissable agent in Crohn's disease. *Lancet*, **2**, 1122–1124.
29. Cave, D. R., Mitchell, D. N. and Brook, B. N. (1975). Observation on the transmissability of Crohn's disease in ulcerative colitis. *Gastroenterology*, **68**, 871.
30. Beeken, W. L. (1980). Transmissable agents in inflammatory bowel diseases. *Med. Clin. of North Amer.*, **64**, 1021–1035.
31. Gitnick, G. L., Rosen, V. J., Arthur, M. H. and Hertwick, S. A. (1979). Evidence for the isolation of a new virus from ulcerative colitis patients. *Dig. Dis. Sciences*, **24**, 609–619.
32. Yoshimura, A. H., Estes, K. and Graham, D. J. (1984). Search for evidence of a viral etiology for inflammatory bowel disease. *Gut.*, **25**, 347–355.
33. Sartor, R. B., Cromartie, W. J., Powel, D. W. and Schwab, H. J. (1985). Granulomatous enterocolitis induced in rats by purified bacterial cell-wall fragments. *Gastroenterology*, **89**, 587–595.
34. Goldgraber, N. B. and Kirsner, J. B. (1959). The Arthus phenomenon in the colon of rabbits: A serial histological study. *Arc. Pathol.*, **67**, 556–560.
35. Goldgraber, N. B. and Kirsner, J. B. (1959). The Shwartzman phenomenon in the colon of rabbits: A serial histological study. *Arc. Pathol.*, **68**, 539–543.
36. Kraft, S. C., Fitch, F. P. and Kirsner, J. B. (1963). Histologic and immunohistochemical features of "Auer-colitis" in rabbits. *Arc. Pathol.*, **43**, 913–922.
37. Brown, J. A. and Szipser, R. D. (1985). Regulation of colonic blood flow in experimental colitis: Role of prostaglandines. *Gastroenterology*, **88A**, 1336.
38. Marsten, A., Marcuson, R. W., Chapmann, M. and Arthur, J. F. (1969). Experimental study of devascularisation of the colon. *Gut*, **10**, 121–130.
39. Reichert, F. L. and Mathes, N. B. (1936). Experimental lymph edema of the intestinal tract and its relation to regional cicatrizing enteritis. *Ann. Surg.*, **104**, 601.
40. Wellmann, W. and Tsatchev, K. (1983). Lymphatic obstruction in Crohn's disease: In: Bartos, V., Davidson, J. W.: Advances in lympology. Prag: Avicenom press, pp. 567–572.
41. Strober, W. (1985). Animals models of inflammatory bowel disease: An overview. *Dis. Dis. Sciences*, **30 suppl**, (12). 3s–10s.
42. James, S. P., Strober, W., Quinn, T. C. and Danovitch, S. H. (1987). Crohn's disease: New concept of pathogenesis and current approaches to treatment. *Dig. Dis. Sciences*, **32**, 1297–1310.
43. James, S. P., Graeff, A. S., Quinn, T. C. and Strober, W. (1985). Natural killer cell activities decreased in lamina propria lymphocytes isolated from non-human primates with colitis due to lymphgranuloma venerum. *Gastroenterology*,, **A88**, 1430.
44. Quinn, T. C., Taylor, A. I. and Schacter, J. (1986). Experimental proctitis by rectal infection with Chlamydia trachomatis in non-human primates. *J. Infectious Disease*, **154**, 833–841.
45. Bicks, O. and Rosenberg, E. W. (1964). A chronic delayed hypersensitivity reaction in the guinea pig colon. *Gastroenterology*, **46**, 543–549.
46. Rabin, B. S. (1980). Immunological model of inflammatory bowel disease. *Arc. Pathol.*, **63**, 253–256.
47. Mee, A. S., Mc Lochlen, J. E., Hodginson, H. and Jewel, D. P. (1979). Chronic immune colitis in rabbits. *Gut*, **20**, 1–5.
48. Schattenfroh, S., Bartels, M. and Nagel, E. (1994). Early morphological changes in Crohn's disease. Transmission electron-microscopic findings and their interpretation: an overview. Acta Ant. Basel, **149**(4), 237–46.
49. Mendeloff, A. I. and Calkins, D. M. (1988). The epidemiology of ideopathic inflammatory bowel disease. In: Kirsner, J. B.: Inflammatory bowel disease. Shorter, R. G., 3rd edn., Philadelphia: Lea & Febiger, pp. 3–34.
50. Martini, G. H. and Brandes, J. E. (1976). Increased consumption of refined carbohydrates in patients with Crohn's disease. *Klin. Wschr.*, **54**, 367–371.

51. Kasper, H. and Sommer, H. (1979). Dietary fiber and nutrient intake in Crohn's disease. *Amer. J. Nutr.*, **32**, 1898–1901.
52. Thornton, J. R., Emmett, P. M. and Heaton, K. W. (1979). Diet in Crohn's disease. Characteristics of the pre-illness diet. *Brit. Med. J.*, **2**, 762–764.
53. Marcus, R. and Watt, J. (1969). Series in ulcerative colitis in laboratory animals. *Lancet*, **2**, 489–490.
54. Watt, J. and Marcus, R. (1971). Carrageenan induced ulceration of the large intestine in the guinea pig. *Gut*, **12**, 164–171.
55. Watt, J., Mc Lean, C. and Marcus, R. (1979). Degration of Carrgeenan for the experimental production of ulcers in the colon. *J. Pharmacology*, **31**, 645–646.
56. Benitz, K. F., Goldberg, L. and Coulston, F. (1973). Intestinal Defects of carrageenans in Rhesus monkey. *Food Cosmet. Toxicol.* **11**, 565–675.
57. Grasson, P., Sharrat, N., Carpanini, F. and Gangolis, D. (1973). Studies on carrageenan in large bowel ulcerations in mamels. *Food Cosmet. Toxicol.* **11**, 555–564.
58. Norris, A. A., Louis, A. J. and Zeitlin, I. J. (1981). Inability of degraded carageenan fractions to indue inflammatory bowel ulcerations in the guinea pig. *J. Pharmacology*, **33**, 1206–1213.
59. Van der Waaij, D., Cohen, B. J. and Anver, M. R. (1974). Mitigation of experimental inflammatory bowel disease in guinea pigs by selective elimination of the aerobic gram negative intestinal microflora. *Gastroenterology*, **67**, 460.
60. Onderdonk, A. B., Hermos, J. A., Zink, J. L. and Bartlett, J. G. (1978). Protective effect of metronidazol in experimental ulcerative colitis. *Gastroenterology*, **74**, 521–526.
61. Onderdonk, A. B. and Bartlett, J. G. (1979). Bacteriological studies of experimental ulcerative colitis. *Am. J. of Clinic Nutrition*, **32**, 258–265.
62. Oohashi, J., Ishioka, T., Wakabayaski, K. and Kowabara, N. (1981). A study on carcinogenesis induced by degraded carrageenan arising from squamous cells metaplasia of the rat colorectum. *Cancer letters*, **14**, 267.
63. King, T. P., Posztai, A. and Clerk, N. W. (1984). Kidney bean lectin-induced lesions in rats small intestine: 1. Light microscopic studies. *J. Comp. Pathology*, **90**, 585.
64. Wilson, A. B., King, T. P., Clerk, N. W. and Posztai, A. (1984). Kidney bean lectin-induced lesions in rats small intestine: 2. Micro biological studies. *J. Comp. Pathology*, **80**, 597.
65. Schmidt-Wilcke, H. A. (1987). Der Einfluss von Nahrungsfetten und Sekretion des Gastrointestinaltraktes. Ein Beitrag zum Fettunverträglichkeits- Syndrom. *Med. Klinik*, **82**, 356–359.
66. Mc Cay, P. B. (1983). Dietary fat and cancer: an overview. In : Diet, Nutrition and Cancer, edited by D. A. Roe, New York: A. R. Liss, pp. 8–16.
67. Erickson, K. L. (1986). Dietary-fat modulation of immuneresponse (review).. *Int. J. Immuno pharmac.*, **8**, 529–543.
68. Donowitz, M. (1985). Arachidonic acid metabolites and their role in inflammatory bowel disease. *Gastroenterology*, **88**, 580–587.
69. Guthy, E. (1983). Ätiologie des Morbus Crohn. Was spricht für Fette als mögliche Ursache? *Dtsch. med. Wschr.*, **108**, 1729–1733.
70. Heckers, H., Melcher, F. E., Kamenich, N. and Henneking, G. R. (1988). Chemisch aufbereitet tes Fett und Morbus Crohn. Z. *Gastroenterologie*, **26**, 259–264.
71. Guthy, E. (1982). Morbus Crohn und Nahrungsfette. *Dtsch. med. Waschr.*, **107**, 71–73.
72. Nagel, E., Guthy, E., Schattenfroh, S. and Pichlmayr, R. (1988). Ultrastructural abnormalities in swine ileum induced by high dietary fat intake resemble those seen in patients with Crohn's disease. *Gastroenterology*, **94**, 1526.
73. Nagel, E., Schattenfroh, S., Strobi, K. and Pichlmayr, R. (1988). Morphologische und morphometrische Kriterien zur Beurteilung des Einflusses von Fetten auf die Dünndarmschleimhaut. *Fat. Sci. Technol.*, **90**, 18.
74. Amman, R. and Cottier, H. (1960). Die primären histologischen Veränderungen der Enteritits regionalis. *Bibl. gastroenterol.*, **2**, 69–79.
75. Nagel, E. (1989). Zur Bedeutung von Nahrungsfetten in der Pathogenese des Morbus Crohn: Tierexperimentelle Studie an Sweinen. *Z. Gastroenterologie*, **24**, 7–9.
76. Dvorak, A. N., Connel, A. D. and Dickersen, G. R. (1979). Crohn's Disease: A scanning electron microscopic Study. *Human Pathology*, **10**, 165–177.
77. Myllaerniemi, A. and Nickels, J. (1985). Scanning Electron microscopy of Crohn's disease and ulcerative colitis of the colon. *Virchow's Arch.*, **385**, 343–350.
78. Rickert, R. R. and Carter, H. W. (1980). The "early" ulcerative lesion in Crohn's disease: Correlative light and scanning electron microscopic study., *J. of Clin. Gastroenterology*, **2**, 11–19.
79. Nagel, E., Bühner, S., Henke, C., Bartels, M. and Pichlmayr, R. (1989). Zur Bedeutung der intestinalen Stase in der Pathogenese entzündlicher Darmerkrankungen. *Z. Gastroenterologie*, **27**, 494.
80. Thyberg, J., Graf, W. and Klingenström, P. (1981). Intestinal fine structure in Crohn's disease. Lysomal inclusions in epithelial cells and macrophages. *Virchow's Archiv A*, **391**, 141–152.
81. Wallace, J. L. and Whittle, B. J. R. (1986). Profile of gastrointestinal damage induced by platelet-activating factor, *Prostaglandins*, **32**, 137–141.
82. Rosam, A.-C., Wallace, J. L. and Whittle, B. J. R. (1986). Potent ulcerogenic actions of platelet-activating factor on the stomach. *Nature*, **319**, 54–56.
83. Wallace, J. L., Steel, G., Whittle, B. J. R., Lagente, V. and Vargaftig, B. (1987). Evidence for platelet-activating factor as a mediator of endotoxin-induced gastrointestinal damage in the rat. Effects platelet-activating factor antagonists. *Gastroenterology*, **93**, 765–773.
84. Stenson, W. F. (1988). Leucotrien B4 in inflammatory bowel disease. In: Inflammatory bowel diseases. Basic reasearch and clinical implications, edited by Goebell, H., Peskar, D. M., Malchow, H., Lancester: M.T.P. Press, pp. 143–152.
85. Wallace, J. L., Braguet, P., Ibbotson, G. C., Mac Naughton, W. K. and Cirino, G. (1989). Assessment

of the role of platelet-activating factor in an animal model of inflammatory bowel disease. *J. Lipid Mediators*, **1**, 13–23.
86. Roediger, W. E. W. (1988). The role of colonic mucosal metabolism in the pathogenesis of ulcerative colitis. In: Inflammatory bowel diseases. Basic research and clinical implications, edited by Goebell, H., Peskar, D. M., Malchow, H., Lancester: M. T. P. Press, pp. 69–78.
87. Roediger, W. E. W. and Nance, S. H. (1986). Metabolic induction of experimental colitis by inhibition of fatty acid oxidation. *Brit. J. Exper. Pathol.*, **67**, 773.
88. Berg, D. N. (1959). Duodenitis and duodenal ulcer produced in rats by pantothenic acid deficiency. *Brit. J. Exper. Pathol.*, **40**, 371.
89. Clark, M. L. (1986). Role of nutrition in inflammatory bowel disease: An overview. *Gut*, **27**, 72–75.
90. Nagel, E., Canzler, H. and Pichlmayr, R. Welche Rolle spielt die Ernährung bei Morbus Crohn? Ein Beitrag zur Bedeutung der diätetischen Therapie der Enteritis Regionalis. (submitted).
91. Shorter, R. G., Huizenga, K. A. and Spencer, R. J. (1972). A working hyperthesis for the etiology and pathogenesis of non specific inflammatory bowel disease. *Amer. J. of Digestive Disease*, **17**, 1024–1032.
92. Dobbins, W. O. (1982). Gut Immunophysio- logy: A gastroenterologists's view. Emphasis on pathophysiology. *Am. J. of Physiology*, **242**, g. i..
93. Udall, J. N. and Walker, W. R. (1982). The physiologic and pathologic basis for transport of macromolecules across the intestinal tract. *J. of Pediatric Gastroenterology and Nutrition*, **1**, 295–301.
94. Graham, H. and Aman, P. (1987). The pig as a model in dietary fiber digestion studies. *Scand. J. Gastroenterology*, **22**, 55–61.
95. Cimprich, R. E. (1974). Equine granulomatous enteritis. *Vet. Pathology*, **11**, 135–145.
96. Wensfoort, B. A. (1962). Rekkers of Strekkers, een aandoening bij Pammerem. Tijdschrift foor Divergeneenkunde, **87**, 841–846.
97. Geil, R. G., Davis, C. L. and Thompson, S. W. (1961). Spontaneous ileitis in rats: A report of 64 cases. *Am. J. Veter. Research*, **22**, 932–936.
98. Emsbo, P. (1951). Terminal or regional ileitis in swine. *Nord. Vet. Med.*, **3**, 1–28.
99. Field, H. I., Buntain, D. and Jennings, A. R. (1953). Terminal or regional ileitis in pigs. *J. Comp. Pathology*, **63**, 153–158.
100. Szabo, I. and Szeky, A. (1964). The incidence of regional ileitis in pigs and in a rabbit. *Acad. Sci. Hong.*, **14**, 219–226.
101. Nielsen, K. (1971). Regional Enteritis in domestic animals. In: Regional enteritis (Crohn's disease)., edited by Engel, A. and Larson, T. Stockholm: Desselte Tryck, pp. 266–282.
102. Rahko, T. and Saloniemi, T. (1972). On the pathology of Regional Enteritis in the pig. *Nord, Vet. Med.*, **24**, 132–138.
103. Rowland, A. C. and Lawson, G. H. K. (1975). Porcine intestinal adenomatosis: A possible relationship with necrotic enteritis, regional ileitis and proliferative haemorrhagic enteropathy. *Vet. Res.*, **97**, 178–180.
104. Fox, J. G. and Murphy, J. C. (1982). Proliferative colitis in ferrets. *Am. J. of Vet. Res.*, **43/5**, 858–864.
105. Steward, H. L. and Jones, B. F. (1941). Pathologic anatomy of chronic ulcerative colitis. A spontaneous disease of the rat. *Arch. Pathology*, **31**, 37–42.
106. Mitchell, D. N. and Rees, R. J. W. (1986). Further observations on the transmissibility of Crohn's disease. *Ann. Ny. Acad. Sci.*, **278**, 546–559.
107. Cave, D., Kirsner, J., Mc Laren, L., Strickland, R. and Philips, A. et al. (1980). Infectious agents in inflammatory bowel disease (IBD).: a status report. *Gastroenterology*, **87/5**, 1185.
108. Schneierson, S. S., Garlock, J. H., Schore, B., Stuart, W. D., Steinglass, M. and Aronson, D. (1962). Studies on the viral etiology of Regional Enteritis and Ulcerative colitis: a negative report. *Am. J. Dig. Dis.*, **7**, 839–843.
109. Aronson, M. D., Philips, C. A., Beeken, W. L., and Forsyth, B. R. (1975). Isolation and charac- terization of a viral agent from intestinal tissue of patients with Crohn's disease and other intestinal tissue of patients with Crohn's disease and other intestinal disorders. *Progr. med. Virol.*, **21**, 165–176.
110. Swindle, M. M. (1985). Procine models in surgical research. An overview. In: ME Tumbleson (Hrsg.). **1**, New York: Plenum Press, pp. 235–243.
111. Fleming, S. E. and Arce, D. (1985). Using the pig to study digestion and fermentation in the gut. In: ME Tumbleson (Hrsg.)., **1**, New York: Plenum Press, pp. 123–132.
112. Mersmann, H. J. (1985). The pig as a model for abberations associated with carbohydrate and lipid metabolism. In: ME Tumbleson (Hrsg.)., **2**, New York: Plenum Press, pp. 981–991.
113. Shulman, R. J., Henning, S. J. and Nichols, B. L. (1988). The miniature pig as an animal model for the study of intestinal enzyme development. *Pediatric Research*, **23/3**, 311–315.
114. Miller, E. R. R. and Ullrey, D. E. (1987). The pig as a Model for Human Nutrition. *Ann. Rev. Nutr.*, **7**, 361–382.
115. Hammer, R. E., Maika, S. D., Richardson, J. A., Tang, Y. P. and Taurog, J. D. (1990). Spontaneous inflmmatory disease in transgenic rats expressing HLA-B 27 and human β2m. An animal model of HLA-B 27 associated human disordres. *Cell*, **63**, 1099–1112.
116. Flocchi, C. (1993). Cytokines and animal models: a combined path to inflammatory bowel disease pathogenesis. *Gastroenterology*, **104**(4), 1202–5.
117. Mombaerts, P. Mizogueli, E. Grusby, M. J., Glimeker, L. H., Bhan, A. K. and Tonegawa, S. (1993). Spontaneous development of inflammatory bowel disease in T cell receptor mutant mice. *Cell.* **75**, 275–282.
118. Sadlach, N., Merz, H., Schorte, H., Schimpl, A., Feller, A. C. and Horak, I. (1993). Ulcerative colitis-like disease in mice with disrupted interleukin 2 gene. *Cell*, **75**, 253–261.
119. Kühn, R., Löhler, J. Rennick, D., Rajewsky, K. and Müller, W. (1993). Interleukin 10-deficient mice develop chronic enterocolitis, *Cell*, **75**, 263–274.
120. Reimann, J., Rudolphi, A. and Claesson, M. H. (1995). Novel experimental approaches in the study of the

immunopathology in inflammatory bowel disease. *J. Mol. Med.*, **73**, 133–140

121. Bühner, S., Nagel, E., Stockhorst, H., Körber, J., Sagredos, A. N. and Pichlmayr, R. (1995). Influence of heated and nonheated partially hydrogenated dietary fats on ileal chyme fat and fatty acid composition of illeal mucosa in pigs. *Digestice Dis. and Sci.*, **40**(2), 338–345.
122. Bühner, S., Nagel, E., Krber, J., Vogelsang, H., Linn, T. and Pichlmayr, R. (1994). lleal and colonic fatty acid profiles in patients with active crohn's disease. *Gut*, **35**, 1424–1428.
123. Nagel, E., Schattenfroh, S., Bühner, S., Bartels, M., Guthy, E. and Pichlmayr, R. (1992). Experimentelle Schädigun der Epithelzellsicht des lleums durch Nahrungsfette: Transmissionselektronenmiroskopische Befunde und deren Vergleich zur Zellpathologie beim Morbus Crohn. *Z. Gastroent.*, **30**, 403–410.
124. Nagel, E., Schattenfroh, S., Bühner, S., Bartels, M., Guthy, E. and Pichlmayr, R. (1993). Tierexperimentelle Untersuchungen zu ultrastrukturellen Veränderungen der Lamina propria des lleums durch Nahrungsfette und deren Vergleich zur Zytopathologie beim Morbus Crohn. Z. Gastroent., **31**, 727–734.

PART III

10. Experimental Models in Acute Pancreatitis

Michael H. Schoenberg, Marcus Büchler and H. G. Beger

Department of General Surgery, University of Ulm

Introduction

Pancreatic diseases have increased continuously in the Western industrial countries. As a result during the last years major efforts have been made to elucidate the causes of the acute inflammation and their possible implications for therapy. The pathogenesis of acute pancreatitis, however, still remains unclear although some interesting but controversial concepts have been developed. In order to study the pathomechanisms involved in the acute inflammation of the pancreatic gland, various experimental models of acute pancreatitis in animals have been developed. In this chapter we summarize the major experimental models for the study of acute pancreatitis and indicate their clinical relevance.

Doudenal Reflux

Methods

The close duodenal loop method, first described by Seidel[1] and Pfeffer[2] is considered to be the classic experimental model inducing acute pancreatitis secondary to duodenal reflux. Thereby, firstly the common bile duct which in the dog enters the duodenum seperately from the pancreatic duct, is ligated and the bile diverted to the jejunum. The duodenum is then surgically closed at the pylorus and distal to the orifice of the pancreatic duct. Chetty *et al.*[3] described a modification of this surgical procedure in the rat. In this experimental model acute pancreatitis develops rather rapidly. Within one to two hours the close duodenal loop shows marked distension due to the accumulation of the duodenal and pancreatic secretion. Within 4 hours the head and the body of the pancreas are edematous and small amounts of hemorrhagic peritoneal fluid and fat necrosis can be found. By 9 to 11 hours a severe hemorrhagic pancreatitis is observed. Within 24 hours most experimental animals die.

The pathophysiological events leading to acute pancreatitis, however, are not completely clear. Isolation of the pancreas from the close loop either by pancreatic duct cannulation[4] or by ligation[5] prevents the development of the acute inflammation in the pancreatic gland. Moreover, further studies could show a substantial reflux of duodenal fluid including active hydrolases[5] and bacteria[6] into the pancreatic duct. It was hypothesized that bacteria were instrumental for the development of acute pancreatitis in this model. This hypothesis was supported by the study of Bryne and Joison[7] who could show that antibiotic treatment prevented the development of acute pancreatitis. In contrast to the above mentioned study, Nance and Cain[8] could induce acute pancreatitis by the close duodenal loop method even in *germ-free dogs*. Species differences occur since in monkeys a close duodenal loop leads to a duodenal reflux and pancreatitis only following stimulation with secretin or cholecystokinin (CCK)[9].

Histology

In ultrastructural studies it was shown that within 5 hours after closing the duodenal loop the majority of occluding junctions between the duct cells were destroyed thus leading to an enlargement of the intracellular space and free passage of pancreatic enzymes, bacteria and duodenal juice into the interstitial compartment. Shortly thereafter, PMN leukocytes and macrophages accumulate and migrate into the pancreatic tissue. At the same time histology revealed membrane damage and beginning cell necrosis.

Clinical Relevance

Duodenal reflux into the pancreatic duct possibly occurs in patients under clinical conditions where duodenal pressure is increased (e.g. during vomiting episodes after alcohol ingestion and in patients with gallstone, blind loop syndrome, duodenal diverticulum or gastrectomy).[8–10] Recently, it has been demonstrated that in 38% of the dogs the postprandial duodenal pressure increases transiently above the pressure of the pancreatic ducts leading to a reflux of duodenal markers into the duct of the pancreatic gland.[12] Since the amount of reflux was with 0.07% rather low an antireflux mechanisms possibly related to the presence of sphincter muscle tissue at the side where the pancreatic ducts empties into the duodenal lumen was postulated. Therefore, the experimental model of duodenal reflux seems of minor clinical relevance since most likely the amount of duodenal reflux is negligible in humans. It allows, however, to study the harmful effects of duodenal contents on the pancreas.

Bile-Induced Pancreatitis

Methods

It is a well-established fact that biliary calculi may induce pancreatic diseases. The mechanisms by which biliary tract diseases might lead to acute inflammation of the pancreas is not clear. Since Opi's description[13] of 2 cases of lethal hemorrhagic pancreatitis due to gallstone obstruction, subsequent studies have focused not only on the obstructive event but on the effect of bile brought in contact with the pancreatic parenchyma. Several studies could demonstrate that pancreatitis can be induced in experimental animals by a retrograde injection of bile or bilesalts into the pancreatic duct. Thereby, the solutions were injected with increased pressure into the duct system. In all studies, carried out in several animal species, this retrograde injection induced a lethal hemorrhagic pancreatitis with a fat necrosis within one to three days. The deleterious effect of bile and bilesalt injections depends on the number of factors. Not only the bile salt concentration is of importance but also the type of bile salt. Thereby non-conjugated bile salts are more toxic than conjugated bile salts. Additionally, the dihydroxycholanic acids were up to 8 times more toxic than the trihydroxcholanic acid.[14] Combinations of bile and trypsin or pancreatic juice as well as bacterial contamination increased the severity of the inflammatory process.[15] Recently, sodiumtaurocholate has been used for induction of the severe type of acute pancreatitis in rats.[16] The mortality of this model correlated with the concentration of bile salt. Using a 3% or 4.5% solution lead to a lethality of 23.5 and 71.4% of animals within 72 hours, respectively. Following injection of 5% sodiumtaurocholate (volume 0.2 ml) over a period of 60 seconds all animals died within 30 hours.[16]

Histology

Early structural changes consisted of an immediate interstitial edema and subsequent necrosis of duct cells followed by adjacent acinar cells.[16] Within 5 minutes after injection of the bile by manual pressure into the main pancreatic duct of the dog, ultrastructural findings indicated destruction of the plasma membrane of duct cells and disruption of the occluding junction. Due to the disruption of this membrane bile and pancreatic proteins could enter the interstitial space where detergent and enzyme effects led to extensive membrane damage and subsequent acute hemorrhagic pancreatitis.

Clinical Relevance

Clinically the bile and pancreatic duct is often drained into the duodenum by a common channel which would of course supports Opi's theory of bile-induced pancreatitis. However, careful anatomical and physiological studies revealed only little evidence that human pancreatitis is induced by an increase in choledochal pressure exceeding the secretory pressure of the pancreas and resulting in movement of bile into the pancreatic duct. Still at the clinical level 94% of patients with acute pancreatitis associated with biliary calculi had gallstones in their feces.[17] Possibly recurrent passages of biliary stones through the common bile and pancreatic duct sphincter could lead to repeated but transient obstructions of the pancreatic duct. Therefore, the role of such repetitive obstructive events in the pathogenesis of acute pancreatitis associated with biliary tract diseases deserves to be studied.

Enzyme Induced Pancreatitis

Methods

Another method to induce acute inflammation in the pancreatic gland is retrograde infusion of various pancreatic enzymes into the pancreatic duct.[18] This method was used in various species and various enzymes in its activated and inactivated form were injected. The enzymes used were trypsin, trypsinogen, chymotrypsin, amylase, carboxypeptidase, elastase, phospholipase A_2, and phospholipase A_2 together with bile. Trypsin and chymotrypsin led to a severe hemorrhagic pancreatitis and about 30% of animals died within 48 hours. Inactivated trypsin and trypsinogen, however, induced only minimal histological changes and the same holds true for amylase and carboxypeptidase. Retrograde infusion of elastase induced vascular damages similar to that observed in experimental hemorrhagic pancreatitis.[19] Moreover, the introductal instillation of phospholipase A_2 did not lead severe morphological changes in the pancreas.[20] The injection of a mixture of phospholipase A_2 and bile which can be expected to generate lysolecithin, however, resulted in hemorrhagic pancreatitits. Similarly, necrotic pancreatitis has also been induced by instillation of lipase together with triglycerides.[21]

Clinical Relevance

The clinical relevance of these studies is not completely clear, since, in most experimental models the injection pressure was rather high which does not correspond to any clinical situation. Nevertheless, observations made in these experimental models are of pathophysiological importance since it is clear that inappropriate translocation of activated enzymes into the interstitial space is a necessary prerequisit for the development of pancreatitis.

Experimental Obstruction of the Pancreatic Duct

Methods

In a variety of clinical conditions including biliary stones, spasm of the sphincter, sphincter edema, tumor, etc., the underlying disease leads to a partial or complete obstruction of the main pancreatic duct.[22] Under experimental conditions, surgical ligation of the pancreatic duct followed by post-operative secretin stimulation or feeding resulted in an increase in the pancreatic duct pressure,[23] appearance of interstitial edema,[24] and a subsequent increase in the pancreatic enzyme levels.[25]

Histology

Subsequent light microscopic studies after *complete* sphincter obstruction demonstrated a widening of intraacinar and interlobular spaces, indicating the presence of edema (see above). Shortly thereafter, a large number of macrophages and fibroblasts and a small number of PNM-leukocytes accumulated and migrated into the interstitial space indicating an inflammatory reaction. If the pancreatic duct obstruction was kept up for several weeks and months, the organ showed a dramatic loss of acinar tissue and the appearance of fibrosis. Islet of endocrine tissue, blood vessels and nerves, however, remained largely undamaged by the inflammatory fibrotic process. Electronmicroscopy studies indicated that 24 hours following duct obstruction the lumina of acini and intracalated ducts were dilated.[26] Acinar cells showed a marked stimulation in lysosomal degradation of intracellular organelles. Zymogen granules were rapidly degraded by crinophagy. The rough endoplasmic reticulum showed an initial disruption with segmentation, dilatation and subsequent degradation by autophagic vacuoles. Moreover, mitrochondria initially exhibited swelling and later extensive degradation of inner membranes leading to residual bodies positive for acid phosphatase activity. Concomitant with luminal dilatation and degradation of intracellular organelles, the height of acinar cells markedly decreased. Necrosis of acinar cells was observed but was considerably less evident than cellular atrophy.

Partial duct obstruction which would resemble more closely the situation encountered clinically, has been incompletely studied. In one study[27] the pancreatic duct of cats was obstructed to varying degrees and the effects of such manoeuvres were studied histologically and biochemically after two to three months. Similar to the findings of total obstruction, atrophy and fibrosis was the most consistent finding. Duct obstruction which exceeded 85% resulted in extensive atrophy and fibrosis as well as impairment of both protein and bicarbonate secretion in response to CCK or secretin, respectively. If the obstruction narrowed the lumen by less than 75% only the protein secretion was reduced. Follow-up studies examining the functional disturbances after severe obstruction (over 75%) showed no improvement of either bicarbonate or enyzme secretion three months after release of the duct stenosis. In some cases, histological examination showed progressive parenchymal changes. Additional studies, however, are needed to further characterize the effects of partial duct obstruction.

Clinical Relevance

As stated before several diseases and changes within the sphincter region may lead to complete or partial obstruction. Moreover, a valuable clinical observation[22] had been made in patients who received after the revision of the ductus choledochus a long-limbed T-tube. If the T-tube passed down the common bile duct through the ampullary region into the duodenum, it usually resulted in partial or complete obstruction of the pancreatic duct as it joins the bile duct to form a common ampulla. The consequence of this manoeuvre was a high incidence of clinically acute pancretatitis.

Diet Induced-Pancreatitis

Methods

An experimental model of acute hemorrhagic pancreatitis in mice was established first by Lombardi and Rao.[28, 29] Acute pancreatitis is induced by feeding yound female mice or males treated with estrogens[30] a choline-deficient diet which is supplemented with 0.5% DL/ethionine (CDE). On the 4th day of this diet the animals developed a hypovolemic shock leading to death in 100% of the experimental animals. The reason for this lethal outcome of acute pancreatitis is probably due to massive hemorrhagic necrosis of the pancreas with concomitant fat necrosis throughout the peritoneal cavity. Regurgitation of bile into the pancreas, activation of the complement cascade or hyperlipidemia were not observed and seemed not to be important ethiological factors. This generation of pancreas necrosis leads to increased amylase and lipase levels in the serum and to intraparenchymal activation of the protease zymogens.

Histology

After 24 hours of CDE diet ultrastructural analysis of the pancreas reveals that enzyme discharge at the luminal plasma membrane (exozytosis) is blocked which leads to an accumulation of zymogen granules within the cell.[34] These first alterations are followed by an aggressive autophagocytosis within the cells. Within 48 hours lightmicroscopy reveals interstitial edema, enlargement of the pancreatic gland and beginning hemorrhage which leads after 72 hours to acute hemorrhagic pancreatitis with extensive fat necrosis. The pathophysiology leading to the development of the acute inflammation remains unclear, however, intravenous administration of various proteinase inhibitors may attenuate considerably the damages normally see.[32]

Clinical Relevance

The diet induce acute inflammation resembles a hemorrhagic necrotizing pancreatitis other than of biliary origin. The value of this models lies in the fact that no surgical manipulation is necessary in order to initiate this acute inflammation which within 4 days leads to the severe and lethal hemorrhagic pancreatitis. It has been mainly used for pathophysiological and therapeutical studies. The pathomechanisms, however, are still unclear and certainly warrants future studies. This experimental model, however, is a *chronic* experimental model in which within 4 days the mice developed this severe and painful pancreatic disease. In the last years it has become increasingly difficult to use this particular model since the state authorities are rather reluctant to permit experimental procedures which submit the animal to such considerable stress and pain over a longer period of time.

Hormon Induced Pancreatitis

Methods

As early as 1895 Mouret reported that excessive neural stimulation led to vacuolisation of the acinar cells and damage to the exocrine pancrease.[33] Ever since then, the concept of excessive stimulation of the pancreas as a possible cause for acute pancreatitis has been studied. Recently, intravenous infusion of synthesis peptides such as Ceruletid into conscious rats has been used to establish a detailed dose-response relationship for these agonists on the structure and the function of the exocrine pancrease.[34] Thereby, rats received after implantation of an intravenous catheter Ceruletid in various dosages. Prolonged and maximal hormon stimulation can be obtained with 0.25 μg/kg/h Ceruletid and resulted in 80 to 90% depletion of enzymes stores within 1 to 3 hours. Prolonged stimulation (1–3 days) with optimal concentrations of ceruletic did not result in acinar or ductal cell necrosis. Further increases in the dose of Ceruletid by a factor of two resulted in a significant increase of pancreatic enzymes in the blood and signs of acinar cell destruction.[35] These findings have been analyzed in detail with higher dosages until 5 μg/kg/h Ceruletid. Infusion of these supraoptimal concentrations of Ceruletid led within 15 to 30 minutes to the appearance of large vacuoles within the apical portion of the acinar cells. Such vacuoles appeared to be derived both from the Golgi complex possibly as a result of an arrest and condensing vacuole maturation.[36] This model of secretagogue-induced pancreatitis originally described in the rat has been extended to other species such as dog, mouse and hamster. In all these animals species course and severity of tissue damages are similar to those seen in the rat.

Histology

Lightmicroscopy reveals a progressive interstitial edema and inflammation as evidence by massive accumulation of PMN-leukocytes, lymphocytes and macrophages. Both primary events lead to destructive changes in the acinar cells including cells necrosis.[35] With supraoptimal stimulation with 5 μg/kg/h Ceruletid beginning cell necrosis occur 12 hours after onset of the infusion. Detailed ultrastructural studies employing freeze fraction techniques verified that an early event in this form of experimentally induced pancreatitis is a dramatic change in the site of granule exocytosis. Under physiological conditions exocytosis of zymogen granules occur at the luminal plasma membrane. Infusion of supraoptimal concentrations of Ceruletid immediately alter the route of secretion from exocytosis at the luminal membrane to exocytosis at the lateral membrane. Moreover, as mentioned above, large vacuoles are formed within the apical portion of the acinar cells possibly as a result of an arrest and condensing vacuole maturation and from the random intracellular fusion of zymogen granules. Vacuole formation within the cell leads to marked increases in autophagocytosis resulting in extensive digestion of secretory products collected within these vacuoles. Watanabe *et al.* in fact demonstrated the presence of both secretory enzymes and lysosomal hydrolases in these vacuoles.[37] The ultrastructural changes are reversible if the stimulation with Ceruletid is ended. Therefore, this model has been also used to study the regenerative capacity of individual cell types in the exocrine pancreas.

Clinical Relevance

In humans excessive neural stimulation resulting in severe interstitial pancreatitis has been observed in two clinical settings: Accidental ingestion of anticholinesterase insecticides[38] and the exposure to venum from certain scorpion species in Trinidad.[39] These forms of acute pancreatitis in humans are rather seldom. Nevertheless, the significance of this model lies in the fact that the toxic agent induces disruption of cellular function and integrity without resulting in severe pancreatic necrosis leading to death of the experimental animal. As such the model allows careful and timed studies concerning the onset and development of tissue disruption and its resolution following withdrawal of the toxic agent. Furthermore since the toxic effects of hormonal stimulation are irreversible, this model allows study of the pathological consequences of repeated injury. For example, if a 12 hour toxic dose of Ceruletid is infused 12 days after the first treatment the resulting degree of tissue disruption not only is increased but new changes occur and the disruption and alterations persist for a long period of time.

Summary (See Table 10.1)

Acute pancretitis in humans is still clinical problem, specific therapies are still not at hand. Therefore, the physiological and therapeutical studies employing various models of acute pancreatitis are necessary. All the above-mentioned models are "in vivo" models. There exist also other "ex vivo" models of acute pancreatitis where the pancreatic gland is explanted, isolated, and perfused. With this isolated pancreas preparation various forms of pancreatitis were initiated by free fatty acid infusion, partial duct obstruction, and secretion stimulation as well as ischemia.[40] The experimental model to

Table 10.1. Summarising the different animal models of use in the study of pancreatitis and their clinal relevance

Model	Species	Form of pancreatitis	lethality	Clinical relevance
Closed duodenal loop	dog, rat monkey	hemorrhagic (necrotizing)	~ 80% within 24h	low
retrogr. injection of bile, bile salts or sodium-taurocholate into the pancreatic duct	rat, dog, goats cat, pig, opossum	hemorrhagic-necrotizing	depending on substance and concentration	acute biliary pancreatitis
retrograde injection of enzymes	rat, dog	hemorrhagic	30% within 48h (activated trypsin)	unclear
complete / incomplete } obstruction pancreatic duct	dog, rat, cat	edematous, later: tissue fibrosis loss of acinar cells tissue fibrosis, atrophy	no lethality	partial obstruction of pancreatic duct due to calculi or sphincter alterations
choline-deficient ethionine-supplemented diet	mice(female)	hemorrhagic	100% within 96h	alcohol-induced pancreatitis
ceruletid infusion ceruletid intraperitoneally	rat, dog, mice hamster	edematous	no lethality	acete edematous pancreatitis

be chosen depends of course as always on the questions which are to be answered. In the last years, most frequently pathophysiological and therapeutical studies have been performed with the Ceruletide and diet-induced pancreatitis. Also sodium taurocholate pancreatitis was employed to study bile-induced acute inflammation of the pancreatic gland.

References

1. Seidel, H. (1910). Bemerkungen zu meiner Methode der experimentellen Erzeugung der akuten haemorrhagischen Pankreatitis. *Zentralblatt für Chirurgie*, **37**, 1601–1604.
2. Pfeffer, R. B., Stadior, O. and Hinton, J. W. (1957). The clinical picture of the sequential development of acute hemorrhagic pancreatitis in the dog. *Surgical Forum*, **8**, 248–251.
3. Chetty, U., Gilmour, H. M. and Taylor, T. V. (1980). Experimental acute pancreatitis in the rat - A new model. *Gut*, **21**, 115–117.
4. Paulino-Netto, A. and Dreiling, D. A. (1960). Chronic duodenal obstruction: A mechanovasucular etiology of pancreatitis. *American Journal of Digestive Diseases*, **5**, 1006–1018.
5. McCutcheon, A. D. (1964). Reflux of duodenal contents in the pathogenesis of pancreatitis. *Gut*, **5**, 260–265.
6. Byrne, J. J., Reilly, P. S. and Toutounghi, F. M. (1964). Regurgitation in experimental pancreatitis. *Annals of Surgery*, **159**, 27–31.
7. Byrne, J. J. and Joison, J. (1964). Bacterial regurgitation in experimental pancreatitis. *American Journal of Surgery*, **107**, 317–320.
8. Nance, F. C. and Cain, J. L. (1968). Studies of hemorrhagic pancreatitis in germ-free dogs. *Gastroenterology*, **55**, 368–374.
9. Johnson, R. H. and Doppman, J. (1967). Duodenal reflux and the etiology of pancreatitis. *Surgery*, **62**, 462–467.
10. Dreiling, D. A., Kirschner, P. A. and Nemser, H. (1960). Chronic duodenal obstruction: A mechanovascular etiology of pancreatitis: I. Report of 6 cases illustrating this clinical variety. *American Journal of Digestive Diseases*, **5**, 991–1005.
11. McCutcheon, A. D. (1968). A fresh approach to the pathogenesis of pancreatitis. *Gut*, **9**, 296–310.
12. Hendricks, J. C., DiMagno, E. P., Go, V. L. W. and Dozois, R. R. (1980). Reflux of duodenal contents into the pancreatic duct of dogs. *Journal of Laboratory & Clinical Medicine*, **96**, 912–921.
13. Opie, E. L. (1901). The etiology of acute hemorrhagic pancreatitis. *Bulletin of Johns Hopkins Hospital*, **12**, 182–190.
14. Hansson, K. (1967). Experimental and clinical studies in the aetiologic role of bile reflux in acute pancreatitis. *Acta Chirurgica Scandinavia* (*Supplement*), **375**, 1–102.
15. Sum, P. T., Bencomse, S. A. and Beck J. T. (170). Pathogenesis of bile-induced acute pancreatitis in the dog. Experiments with detergents. *American Journal of Digestive Diseases*, **15**, 637–646.
16. Aho, H. J. and Nevalainen, T. J. (1980). Experimental pancreatitis in the rat. Ultrastructure of sodium taurocholate-induced pancreatic lesions. *Scandinavian Journal of Gastroenterology*, **15**, 417–424.
17. Acosta, J. M. and Ledesma, C. L. (1974). Gallstone migration as cause of acute pancreatitis. *New England Journal of Medicine*, **290**, 484–487.
18. Wanke, M. (1970). Experimental acute pancreatitis. In *Current Topics in Pathology*, Vol. 52, edited by E. Grundmann and W. H. Kirsten, pp. 64–142. Berlin, Heidelberg, New York: Springer.
19. Geokas, M. C. (1968). The role of elastase in acute pancreatitis. III. The destructive capacity of elastase on pancreatic tissue in vivo and in vitro. *Archives of Pathology*, **86**, 135–141.
20. Schmidt. H. and Creutzfeldt, W. (1969). The possible role of phospholipase A_2 in the pathogenesis of acute pancreatitis. *Scandinavian Journal of Gastroenterology*, **4**, 39–48.
21. Kimura, T., Zuidema, G. D. and Cameron, J. L. (1980). Experimental pancreatitis: Influence of secretory state of the pancreas. *Surgery*, **88**, 661–666.
22. Schiller, W. R., Suriyapa, C. and Andersson, M. (1974). A review of experimental pancreatitis. *Journal of Surgical Research*, **16**, 69–70.
23. Elliot, D. W., Williams, R. D. and Zollinger, R. M. (1957). Alterations in the pancreatic resistance to bile in the pathogenesis of acute pancreatitis. *Annals of Surgery*, **146**, 669–682.
24. Zelander, T., Ekholm, R. and Edlund, Y. (1964). The ultrastructure of the rat exocrine pancreas after experimentally occuluded outflow. *Journal of Ultrastructural Research*, **10**, 89–102.
25. Papp, M. (1976). Pathogenesis of acute pancreatitis: Pancreatic ductal-interstitial-vascular and lymphatic pathways. *Acta Medical Academic Science of Hungary*, **33**, 191–206.
26. Churg, A. and Richter W. R. (1971). Early changes in the exocrine pancreas of the dog and rat after ligation of the pancreatic duct. *American Journal of Pathology*, **63**, 521–546.
27. Austin, J. C., Roberts, C., Rosenholtz, M. J. and Reber, H. A. (1980). Effects of partial duct obstruction and drainage on pancreatic function. *Journal of Surgical Research*, **28**, 426–433.
28. Lombardi, B., Estes, L. W. and Longnecker, D. S. (1975). Acute hemorrhagic pancreatitis (massive necrosis) with fat necrosis induced in mice by DL-ethionine fed with a cholone-deficient diet. *American Journal of Pathology*, **79**, 465–480.
29. Rao, K. N., Tuma, J. and Lombardi, B. (1976). Acute hemorrhagic pancreatic necrosis in mice. Intraparenchymal activation of zymogens, and other changes in pancreas and serum. *Gastroenterology*, **70**, 720–726.
30. Rao, K. N., Eagon, P. K. and Okamura, K. (1982). Acute hemorrhagic pancreatic necrosis in mice. Induction in male mice treated with estradiol. *American Journal of Pathology*, **109**, 8–13.
31. Koike, H., Steer, M. and Meldolesi, J. (1982). Pancreatic effects of ethionine: Blockade of exocytosis and appearance of crinophagy precede cellular necrosis. *American Journal of Physiology*, **242**, G297–G307.

32. Lombardi, B. and Rao, K. N. (1982). Acute hemorrhagic pancreatic necrosis in mice. Effect of proteinase inhibitors on its induction. *Digestion*, **83**, 57–64.
33. Mouret, J. (1895). Contribution a l'étude des cellules glandulaires (pancreas). *Journal of Anatomy & Physiology*, **31**, 221–236.
34. Adler, G., Gerhards, G., Schick, J., Rohr. G. and Kern, H. F. (1983). Effect of in vivo cholinergic stimulation in the rat exocrine pancreas. *American Journal of Physiology*, **224**, G623–G629.
35. Adler, G., Rohr, G. and Kern, H. F. (1982). Alteration of membrane fusion as a cause of acute pancreatitis in the rat. *Digestive Disease & Science*, **27**, 993–1002.
36. Scheele, G. A., Adler, G. and Kern, H. F. (1984). Role of lyosomes in the development of acute pancreatitis. In *Pancreatitis: Concepts and Classification*, edited by K. Gyr, M. Singer and H. Sarles, pp. 17–23. Amsterdam: Elsevier.
37. Watanabe, O., Baccino, F. M., Steer, M. L. and Meldolesi, J. (1984). Supramaximal caerulein stimulation and ultrastructure of rat pancreatic acinar cell: Early morphological changes during development of experimental pancreatitis. *American Journal of Physiology*, **246**, G457–G467.
38. Dressel, T. D., Goodale, R. L., Arneson, M. A. and Borner, J. W. (1978). Pancreatitis as a complication of anticholinesterase insecticide intoxication. *Annals of Surgery*, **189**, 199–204.
39. Bartholomew, C. (1970). Acute scorpion pancreatitis in Trinidad. *British Medical Journal*, **1**, 666–668.
40. Sanfey, H., Bulkley, G. B. and Cameron, J. L. (1985). The pathogenesis of acute pancreatitis: The source and role of oxygen-derived free radicals in three different experimental models. *Annals of Surgery*, **201**, 633–638.

PART III

11. Animal Models of Intra-Abdominal Sepsis

Karl-Erik Giercksky* and Runar Lundblad

The Norwegian Radium Hospital and Institute of Cancer Research, 0310 Oslo, Norway

Introduction

The advent of antibiotics led many to believe sepsis would soon become a part of medical history instead of an ubiquitous problem. Only a decade later severe problems with resistance to antibiotics demonstrated the unique adaptability of bacteria. The stringent and delicate balance of "microcosmos" became clearly evident. Again, nature had taught man an important lesson of ecology that should never be forgotten. More than fifty years after the discovery of antibiotics sepsis is still an important problem in all clinical disciplines. From an abundance of clinical experience, newer and better antibiotics are no longer regarded as the sole solution to treatment for septic shock. Development within the field of molecular biology has revealed an intriguing complexity of cellular responses to a septic challenge and experimental models have become decisive tools in carrying this knowledge into clinical progress.

Cell Culture Versus Animal Experiments

Before a decision of using animal experiments to study a certain problem is taken, the possibility of using cell culture experiments instead should always be evaluated (Table 11.1). A number of early drug screenings, where inhibition or stimulation of cell types or intracellular signals are to be considered, can almost certainly be elucidated as more appropriate in cell culture systems. The militant attitude of some anti-animal experiment groups is definitely inappropriate, but if serious animal research is going to get future official approval and funding, every experimental series reported should give the scientific reasons for choosing animals instead of cell culture.

Table 11.1. Considerations that should be done during the planning of animal experiments

1. Cell culture alternatives
2. Available animals
3. Relevance of difference species to the reactions of interest
4. "Small" or "large" animal models
5. Pain treatment
6. Number of animals needed to give a scientifically relevant answer

In spite of all its superiority in reducing the number of uncontrolled factors, cell culture remains an *"in vitro"* situation. Interactions between different cell systems are possible to study, but do not accurately reflect *in vivo* conditions. Organ to organ interactions and animal survival are, as a rule, almost impossible to study with cell culture models. In experimental surgery, our experience is that cell cultures can be used instead of animal experiments whenever clinical observations or previously reported animal research points to an unknown stimulation/inhibition of an organ or cell type by a certain substance. The more or less specific reactions can then be elucidated in culture systems and putative second or third target cells studied in multiple-cell incubating systems. The most promising of these reactions can then be tested in animal experiments. Such an interaction between cell cultures and animal experiments based on sound scientific reasoning and "hard research data" instead of inconvenience (on behalf of the researcher) is bound to give a

*Author for correspondence.

much needed sobering impact on regulatory authorities and public opinion.

Choosing the Appropriate Experimental Animal

General Consideration

A disturbing amount of important clinical questions have remained obscure following extensive animal experiments because an inappropriate species was studied. The animals are too often chosen due to availability instead of suitability. It is obvious to most researchers that transplantation and transfusion studies carried out in species where little is known about major histocompatibility genes can be misleading, but it is not common knowledge that reactions to endogenously introduced micro-organisms may vary considerably between species.

When the decision to embark upon animal experiments has been taken either as an appropriate extension to cellular experiments or due to the multiorgan nature of the problem, the relevance of available animal species to the reaction(s) of interest has to be established. The practical question of using a "small" or "large" animal model remains. Small animals (e.g. mouse, rat, hamster, guinea-pig) are inexpensive, easily available and larger groups of animals can be handled simultaneously. Studies on small and "traditional experimental animals" bred in captivity for this sole purpose are more likely to avoid criticisms or overt attack by activists. Using such animals should not prevent the researcher from conducting the experiments as painlessly as possible.

We mainly use small animal studies when screening for reactions observed in cell culture experiments or clinical practice to see if this is a relevant and reproducible *in vivo* problem. When the reaction is established *in vivo* and the variabilities narrowed as far as possible, we prefer to leave the small animal and home in on the specific problem in a larger species. Today it is possible to give complete parenteral nutrition to rats and even to germ free rats.[1,2] An extensive degree of cardiorespiratory monitoring can now be done with appropriate instruments. Extensive monitoring, long term intravenous nutrition and repetitive blood sampling are the hallmarks of large animal models, and their availability in rats has reduced the need for larger animals in many studies. There are still some problems with small animals that severely limit their usefulness. Repetitive blood sampling is an essential part of many experimental protocols. The amount of blood that can be taken from such animals is often not sufficient to allow time studies and even more inconvenient is the number of blood cells that can be repeatedly studied from the same animal after exposure to an agent or micro-organism. Even if cardiorespiratory monitoring in the rat has been repeatedly reported we find it cumbersome and not well suited to elucidate the many subtle organ reactions to septic shock. It should be remembered that careful monitoring of shock states in rats can be almost as expensive and time-consuming but seldom of the same quality as in larger animals.

Considerations Related to Intra-Abdominal Sepsis

When mortality is the endpoint and dosage and combination of different agents are the main problems, small animal models should be the first choice. Even if dosage is not a simple problem to extrapolate from small to larger animals and man, we regularly prefer rat models to study dosage and intervals of any combination therapy and then refine the approach in a pig model before clinical suggestions are made. When advanced genetic or immunological problems are studied, the choice between experimental animals is narrowed considerably. Probes, monoclonal antibodies and genetic structures are tools almost exclusively available for mice and man. The degree of cross-reaction and unspecific reactivity must be interpreted carefully when other species are studied with these often very species-specific tools and drugs.

Experimental models of sepsis usually include exposure of the animals to either exogenous or endogenous microorganisms on inappropriate sites. The animals reactions to such manipulation are heavily dependent on their host defense status (HDS). HDS is a poorly understood dynamic equilibrium of a multitude of reactions and can be severely disarranged by many experimental conditions. Nutritional status and cellular reactions to anesthesia are powerful modulators of host defense often not taken into consideration when animal experiments are planned.

The animal's individual, not species-related, reactions to a micro-organism due to previous exposure to the organism or antigens shared by related organisms are often overlooked. Escherichia coli related endotoxin colitis (oedema-illness of the pig) is a common disorder in the domestic pig. The resistance to lethal doses of endotoxin following exposure to previous sublethal doses is known for many species including man and illustrates the problem of individual HDS. When a large series of rat experiments are done under controlled conditions the problems of previous individual exposure are less important than whenever smaller series of large animal experiments are carried out. Our best advice to circumvent this really intriguing problem is a steady state test during anesthesia but before experimental exposure to the organism of both humoral (antibody titer to the actual bacteria) and cellular (phagocytosis or procoagulant induction) reactivity to the organism.[3]

When exogenous micro-organisms are introduced it should not be forgotten that some characteristics of the organisms can be altered during multiple passages in the laboratory. Each experimental series should include randomized parallel controls to evaluate the organism's disease inducing factors. Comparison with groups previously exposed to the organisms even if only days before, can be insufficient to rule out such changes. This can be impossible to achieve in a large animal model, but with the double animal technique described later this problem is significantly reduced. Even in expensive and elaborate large-animal studies we prefer to use double studies with two animals exposed in parallel to the micro-organisms simultaneously.[4] The installed dosage of micro-organisms is split into two halves immediately before introduction into the animals. When using live bacteria one should remember that under appropriate conditions they rapidly increase their number; doubling-time for E.coli is twenty minutes. A sample for detection of colony forming units (CFU) should be taken immediately before installation into the animals.

The ultimate goals of medical research are to understand the mechanisms of disease and find a cure with acceptable side-effects. Given unlimited resources and denying the accelerating problem of endangered species, primates studied in a combination of a surgical theatre and intensive care unit would seem to be the ultimate animal research facility. Both ethical and practical reasons make this an impossibility for almost all medical researchers today, but there is no reason for despair. The major part of important animal experiments have been done using traditional experimental animals and equipment. Experiments on primates, if at all necessary, should be reserved for an extremely small number of problems where the close genetic relationship to man is of decisive importance. In our experience most problems of intra-abdominal sepsis can be conveniently studied in different models using traditional experimental animals; mice, rats and pigs.

Mice are cheap and easily handled, but their organ reactions to a septic stimulus are sometimes impossible to monitor. They have one important advantage that in certain problems make them "the animal of choice"; during the last forty years an impressive amount of genetic and immunological experimental mice data have been accumulated. The mice hybridoma technique opened up new strategies in diagnosis and treatment. Due to their limited size, models of abdominal sepsis are usually limited to coecal perforation or fecal or bacterial installation. The mice models of intra-abdominal sepsis are of special importance when the main objects of study are immunocompetent cells and intercellular reactions.

Rats and pigs are rapidly becoming the most widely used animals in intra-abdominal sepsis research. They clearly satisfy the aims of having a "small" and "large" animal model choice. An increasing number of important scientific problems are now being addressed using combinations of these models and in the future we are likely to see a rapidly accumulating amount of basic data on the genetics of these animals and the commercial production of species-specific biological response modifiers (BRM) and monoclonal assay systems. The handling, experimental set-up and monitoring of these two species cover a wide general area of experimental design easily exported to other animal-models, making them in many ways prototypes of animal models of intra-abdominal sepsis. A more detailed description of models of intra-abdominal sepsis in these two animals are therefore given.

In all experimental models of intra-abdominal sepsis it should be remembered that introduction of foreign material into the abdominal cavity will increase the response to the stimuli. Regular surgical gloves contain starch powder that, together with, a surgical trauma, stimulates macrophages to a granulomatous reaction.[5] Powder free gloves should always be used in such experiments.

Adequate pain treatment is an essential but sometimes difficult aspect to achieve. Instead of more or less occasional injections we try to include a continuous intravenous line in all experimental animals using buprenorfin (Temgesic®) to relieve pain. Some experience (especially in small animals) and swivel equipment are necessary but once established it does not add much to the experimental workload. The equipment is essentially the same as for parenteral nutrition.[1]

Experimental Intra-Abdominal Sepsis in the Rat

General Considerations

The rat is well suited as a basic "screening" laboratory animal in the study of abdominal sepsis (Table 11.2). It is inexpensive, and age and size differences within an experimental population are easily avoided. They can be ordered as genetically identical if necessary and are available in large numbers as specific pathogen free animals. Its lack of aggressiveness makes it easy to house and to handle. Most importantly, its anatomy and reactions to sep-

Table 11.2. Some basic physiological data for the rat

Rectal temperature (C)	38–39
Heart rate (beats/min)	320–480
Blood pressure (mmHg)	
systolic	75–120
diastolic	60–90
Respiratory frequency (breaths/min)	85–110
Respiratory minute volume (ml/min)	160–240
Respiratory tidal volume (ml)	1.5–1.8
Urine volume (ml/day)	10–15

sis resemble many of the reactions observed in man. The gastrointestinal, vascular and lymphatic systems are large enough for operative procedures and indwelling catheters. We regularly use males to avoid interference from sexual cycle and hormonal changes.

Any animal model should fullfil some basic demands to ensure that it is a true sepsis model (Table 11.3).[6]

Clinical Signs of Sepsis in the Rat

Early signs are slow motor activity, raised body temperature, piloerection, tachypnoe and tachycardia. Subsequently the animal becomes dehydrated, best visualized on the back and neck. It develops a crusting exudate in the medial eye corners and sometimes epistaxis. Loss of the "righting" reflex appears a few hours before death; the animal does not try to turn around when put on its back. Reaction to pain stimuli usually disappears less than one hour before death.

Hemodynamic and Metabolic Changes in Sepsis

Patients in the mild or early phase of gram-negative sepsis present increased cardiac output (CO) and oxygen consumption, decreased systemic vascular resistance (SVR) and maintained or slightly decreased mean arterial blood pressure (MAP). This hyperdynamic or high-flow state has been shown difficult to provoke in subhuman primates. The hypodynamic, low-flow or shock phase is commonly observed in sepsis models. The effect of a substantial endotoxin injection is rapid myocardial depression. This is not only a direct effect of lipopolysaccaride (LPS) on the myocardial cells, but some unknown "cardiac depressant factors" seem to be implicated.[7] The lack of high-output models in the rat may partly be due to species difference, but the administration form of the inoculum and depth of anesthesia are important.

Bolus injections of endotoxin or live micro-organisms do not result in a high-flow state in the rat, while continuous, slow infusion of endotoxin[8] or intraperitoneal fecal inoculation[9] has been reported to do so. Normovolemia with optimal preload seems to be necessary for the increase in cardiac output .[10] Sepsis involves a variety of metabolic alterations. These are dependent both on the sepsis model and the time course of illness. In early sepsis the rat shows hyperglycemia and hyperinsulinemia, mainly due to hepatic glycogenolysis. Worsening of the septic state leads to hypoglycemia and decreased insulin level due to impaired gluconeogenesis. Decreased peripheral glucose metabolism is followed by increased blood levels of pyruvate, lactate and alanine. Increased muscle catabolism results in utilization of amino acids as an energy source and substrate for acute phase proteins.

Despite decreased glucose oxidation increased oxygen consumption is regularly seen in the hyperdynamic septic state, possibly due to an uncoupling in the intracellular oxidation system, the so called "mitochondrial failure".

Table 11.3. The diagnostic criteria for experimental sepsis (all three should be fulfilled)

1	Microorganisms should be isolated from blood or local sites of infection
2	One or more objective signs of infection should be present: fever, tachycardia, tachypnoe, hyperfusion, oliguria, lethargy, or death
3	Alterations in cellular or biochemical parameters

Monitoring the Septic Changes

Sepsis studies beyond pure mortality or bacteriological observations demand some kind of intravasal access. Intravascular catheters are mandatory for uncontaminated and repeated blood sampling, hemodynamic monitoring and easy manipulation of the animal by fluids or drugs. Modern pain treatment is also greatly facilitated by continuous intravasal access.

For sampling or infusion, a central venous catheter is sufficient. A reliable technique is the upper caval catheterization through a venotomy of the internal jugular vein.[11] Aseptic technique is important to avoid interference with the controlled septic challenge by additional bacteria from catheter infections.[12]

Hemodynamic studies usually make one or more intravasal catheters neccessary. A number of methods for advanced cardiovascular monitoring of rats have been reported, but we are reluctant to recommend the rat as a suitable animal for the purpose of studying cardiovascular and respiratory mechanism in sepsis in detail. The small blood volume and the interference of multiple catheters with flow make it a cumbersome and often unprecise model. We prefer a "global" monitoring approach in the rat followed by a more detailed problem oriented porcine model. But when judged appropriately some of these problems can be studied in the rat. With a caval and an ascending aortic catheter, cardiac output can be measured with the dye-dilution technique.[13,14] When MAP and CO are measured, SVR can be calculated.

Blood flow values can be measured with wash-out techniques.[13] Radioactive labelled molecules are injected into the artery supplying the organ (or the portal vein when the liver is to be studied). The wash-out of isotopes from the organ is dependent on organ flow. This method demands vascular calibers suitable for cannulation, so flow in small viscera is not easily studied by this technique.

Perhaps more attractive is the microsphere technique for determination of cardiac output and flow distribution.[15,16] Microspheres are molecules large enough to be trapped in the capillary bed. Flow distribution can be measured when the microspheres are labelled radioactive. When cardiac output is known, the flow to each organ can be calculated. Usually three arterial catheters are used. One is for injection of microspheres into the ascending aorta or left ventricle, one for sampling of microspheres in the abdominal aorta and finally one in a femoral artery for blood pressure monitoring. With this method all the important hemodynamic parameters can be measured or calculated: CO, MAP, SVR and regional organ blood flow. A substantial drawback of this method is that longitudinal monitoring in the same animal is impossible.

Peritoneal lymph can be studied by catheterization of ductus thoracicus.[17] This demands thoracotomi (sternum split), tracheostomi and peroperative assisted ventilation. However, thoracic duct content is mainly nutrition lymph from the GI tractus. Peritoneal lymph is mostly drained through diaphragmatic lymphatic channels into other mediastinal lymphatics than the thoracic duct.

Oxygen consumption can be measured by the Kipp and Zonen diaferometer.[18]

Anesthesia and Analgesia

We have been quite satisfied with midazolam/ fentanyl/fluanisone (MFF) anesthesia in abdominal rat surgery. Short induction time and a beneficial physiological response with good tissue blood flow[20] makes it a safe alternative. Hypnorm® (fentanyl 0.2 mg/ml + fluanisone 10 mg/ml) and Dormicum® (midazolam 5 mg/ml) are diluted separately 1:1 with sterile water. A user solution is made by mixing equal parts of diluted Hypnorm and Dormicum. The final concentrations in the solution are 1.25 mg/ml midazolam, 0.05 mg/ml fentanyl and 2.50 mg/ml fluanisone. The solution should be stored at room temperature and can be used for at least a month from preparation. Two to three ml/kg as a subcutaneous or intraperitoneal injection is appropriate for surgery lasting up to thirty to forty minutes. For prolonged surgery, continuous infusion in the lateral tail vein (1.5 ml/kg/hour) is convenient. We use Temgesic® (buprenorphine 0.3 mg/ml) for postoperative analgesia. 0.2 – 0.3 mg/kg as subcutaneous injections every eight to twelve hours give continuous analgesia. In situations with reduced subcutaneous microcirculation (sepsis, shock etc), intravenous injections should be preferred. Depending on pain severity, 0.1 – 0.2 mg/kg every six to eight hours is suitable.

Induction of Sepsis

Three major types of model will be discussed. Although not a true sepsis model, the endotoxin models have been widely used since Borden and Hall[19] forty years ago suggested that gram-negative septic shock was caused by endotoxin. Subsequently it became obvious that endotoxinemia and sepsis were not identical entities, and models more resembling human sepsis were developed. Intravasal or intraperitoneal injection of live micro-organisms and the numerous methods of intraperitoneal inoculation are regarded as more appropriate sepsis models. However, it should not be forgotten that the reaction to intravenous and intraperitoneal installation can differ considerably.

Unfortunately, few, if any, of these models have fulfilled the demands of an ideal sepsis model. Such a model would have to demonstrate the full scale of the sepsis syndrome as seen in the human which is very unlikely to be achieved in an animal model, but certain very important aspects of the problem can be adequately studied.

Endotoxin Models

The ultimate role of LPS in gram-negative sepsis is still controversial. Substantial evidence has proven LPS to be an important mediator in at least some of the pathophysiological reactions. Injections of purified LPS in the rat, however, need doses in the mg/kg range to cause shock and death. Plasma endotoxin levels in such models are in the range of μg/ml,[8] while in true rat sepsis models, the values are in the g/ml range.[23] This indicates other links between gram-negative bacteria and sepsis than introduction of LPS in the general circulation alone. Models using LPS in bolus injections do not produce hyperdynamic sepsis. The usual result is a typical low-flow state. Knowledge about the hyperdynamic phase is important, because advanced sepsis treatment can shift the patient population from a severe hypodynamic to a less dangerous hyperdynamic situation. The lack of high-flow models is partly due to species difference. However, continuous LPS infusion with a more chronic endotoxemia can bring about at least some of the high-flow characteristics[8]. It seems that the substantial LPS load used in bolus injection models overwhelms the animal, and host defence mechanisms are not expressed to their full scale.

The endotoxin model is an appropriate basis for studies of LPS kinetics and hemodynamic changes in low-flow states. Nonetheless, one should take great care in using experimental results from rat endotoxin models alone as the basis for changing treatment modalities in human sepsis.

Intravenous Injection of Live Bacteria

Injection of a defined amount of micro-organisms should accomplish a simple and optimal standardization of the inoculum and consequently a highly reproducible model.

Nevertheless, there are at least two disadvantages to this approach. The virulence can vary from one bacterial strain to another. Maximal standardization implicates an identical bacterial strain to be used if different laboratories should be able to compare their results. Intravenous bolus injection of live micro-organisms does not resemble any situation in clinical medicine. The problem is in many ways the same as with endotoxin models. The bacterial challenge is too massive for development of the complete range of all the host defense mechanisms seen in cases of human sepsis. A hyperdynamic phase is nearly impossible to provoke with this technique. A more continuous, low-graded bacterial infusion may in part overcome this host exhaustion. However, even this leaves out the hosts natural local and systemic reaction of isolating the septic focus by inflammation.

Intraperitoneal Inoculation

These methods have two principal advantages:

1. They produce peritonitis and abdominal sepsis with a course resembling clinical sepsis.
2. Intraperitoneal fecal contamination is the dominating cause of gram-negative sepsis next to urosepsis. The clinical relevance is evident.

Before concentrating on the different peritonitis models, some facts regarding human and rodent fecal bacterial flora must be emphasized.

In human feces the ratio of anerobic to aerobic micro-organisms is about 100:1 and more than 500 species have been identified. Total bacterial number is about 10^{12} colony forming units (CFU)/g feces. The different Bacteroides species are major constituents of the normal fecal flora (10^{8-10} CFU/g) B. fragilis is a dominating anerobe in clinical cases of intra-abdominal sepsis. It accounts for only 0.5 % of the colonic bacterial flora,[24] but is present in a disproportionately large number of clinical isolates. Kasper has demonstrated the importance of the bacterial capsule as a virulence factor in B. fragilis.[21] Bacteroides lack biological active LPS. The usual pattern after intraperitoneal inoculation is abscess formation, not endotoxemic shock or death. Next to Bacteroides, Clostridium perfringens predominantes in abdominal anerobic infections. The concentration in feces is about 10^8 CFU/g.[25] Other main fecal anerobic constituents are gram-positive rods (Eubacteriae and Bifidobacteriae) or the gram-positive cocci (Peptococci and Peptostreptococci). These are of less clinical importance during acute intra-abdominal sepsis, but their role as pathogens can be significantly increased during inadequate antibiotic treatment.

Among aerobes, E. coli is the major pathogen in endogenous peritonitis. The fecal concentration is 10^7 CFU/g. Other frequent pathogens are Enterobacter, Proteus, Serratia and Str. fecalis. Klebsiella pneumoniae and Pseudomonas aeroginosa are not frequently found in normal human feces. Gastrointestinal colonization due to antibiotic therapy or immunological impairment seems to be necessary for these bacteria to produce severe abdominal sepsis.[26]

Normal fecal flora in meat-fed rats resembles human stools having a similar anerobe to aerobe ratio, with anerobic predominance. The concentrations of the main pathogen are: B. fragilis 10^6 CFU/ml, Cl. perfringens 10^6 CFU/ml and E. coli 10^5 CFU/ml.[24–26] In grain-fed rats, the aerobes predominate in feces. It contains B. fragilis 10^{3-5} CFU/ml, Cl. perfringens 10^7 CFU/ml and E.coli 10^{4-5} CFU/ml. This decreased anaerobe to aerobe ratio, especially the B. fragilis to E. coli ratio, in grain-fed versus meat-fed rats increases mortality in fecal peritonitis.[24]

Independent of the rat food, the concentrations of the important pathogens B. fragilis, Cl. perfringens and E. coli are a hundred to a thousand times lower in rat than human feces.[25]

From the original 400–500 bacterial species in the feces, after introduction into the abdominal cavity, an important selection is regularly found resulting in only a few species in a polymicrobial peritonitis or abscess.[25] This selection reflects an interesting aspect of virulence and selection among the fittest of the involved bacteria. In the bacteremic phase of the disease, the aerobes predominate in blood culture isolates.[25] The blood pattern is usually less polymicrobial than the peritoneal exudate.

Adjuvants are various non-bacterial substances with the ability to enhance mortality in bacterial peritonitis. Hemoglobin, fibrin clots, bile, bran, barium sulphate powder and sterile feces have all been used as adjuvants. Adjuvants are installed intraperitoneally in combination with the infecting micro-organisms. The mechanisms involved are not yet fully understood. It may be that phagocytes saturate on the adjuvant, leaving more bacteria for growth and multiplication.

Intraperitoneal Fecal Implantation

In 1974, Weinstein *et al.* published their two-phased peritonitis-abscess rat model.[24,25] Pooled rat feces mixed with barium sulphate in a gelatin capsule was introduced into the pelvic part of the peritoneal cavity. Bacteriological tests of the inoculum revealed differences dependent on grain or meat diet in the preoperative period, as earlier described. Mortality was 80 % in grain-fed males, and 100 % when blood was added as an adjuvant to the inoculum. Meat-fed rats showed a biphasic disease. The initial phase with peritonitis and positive blood cultures was associated with 43 % mortality. All survivors developed intra-abdominal abscesses.

Bacteriological tests of the infected animals showed a predominance of the aerobes E. coli and Enterococci in the acute peritonitis exudate.[25] The anerobes B. fragilis and Fusobacterium dominated in the abscess phase. E. coli, and to a lesser degree, Proteus morganii, were predominants in blood culture isolates, usually being positive the first three days after induction of peritonitis.

The same authors demonstrated that antibiotics against aerobic coliforms reduced mortality in the peritonitis phase, while antibiotics against anerobes reduced abscess formation.[26] This strongly suggests that aerobic coliforms cause death in acute peritonitis and the anerobes promote the more indolent abscess formation. In more recent experiments the animals were infected with bacterial monocultures.[27] E. coli was required to produce mortality. Intra-abdominal abscess occurred only when aerobes and anerobes were combined. Bacteroides alone did not result in mortality or abscess formation. Subsequent studies, however, revealed that this B. fragilis strain was unencapsulated. Repeating the experiment with encapsulated B. fragilis showed that this subtype alone can induce abscess formation. Lang and colleagues[9] used the same model as Weinstein *et al.* for hemodynamic and metabolic studies. Grain-fed rats were used, and in contrast to Weinstein, the inoculum was prepared under aerobic conditions. The inoculum was predominantely aerobic, with aerobes to anerobes 100:1. The mortality was 76 % in five days. This correlates with the 80 % in Weinstein's work. Lang concludes that this is one of the few hyperdynamic sepsis models in the rat.

Nichols *et al.*[22] modified the technique of Weinstein. Their aim was a more human relevant peritonitis and thus human feces with barium sulphate was used instead of pooled rat feces. The reason for this was the observed difference between normal human and rat fecal flora. Mortality rates varied as expected with the size of the inoculum. All survivors developed intra-abdominal abscess. With the same inoculum size as Weinstein, 0.5 ml, the cumulative mortality was 100 % after twelve hours, corresponding to Weinstein's 80 % after five days. Thus, the mortality was higher and the time course faster. An inoculum size of 0.1 ml produced nearly identical mortality curves to Weinstein's work. The predominants in both peritoneal exudate and blood were E. coli, Str. fecalis, B. fragilis and Cl. perfringens.

Rink *et al.*[28] induced peritonitis by installation of fresh feces (1.25 gram/kg wet weight suspended in 2.0 ml 0.9 % saline) through an abdominal incision. Feces was obtained from the coecum of a rat on the same pellet diet. No gelatin capsule was used and no adjuvant added. Mortality was 100 % after twenty four hours. Blood culture isolates showed the presence of both aerobes and anerobes.

Induction of peritonitis by needle injection of feces or pure bacterial cultures into the peritoneal cavity have not been regarded as reliable methods. The location of the needle tip is uncertain and the inoculum may end in the abdominal wall, bowel wall, bowel lumen or other intraperitoneal organs. Consequently this may not lead to free peritonitis.

Intraperitoneal Inoculation by Bacterial Monoculture

Many workers have induced peritonitis with this method in the rat.[29–32] The main problem has been their inability to produce significant mortality. This demands a mixture of bacterial cultures or an adjuvant added to a single culture.[32,34,35]. Rotstein and Kao demonstrated the important synergy between E. coli and B. fragilis in their model with such bacterial cultures in fibrin clots.[31]

Coecal Ligation Models

Ryan *et al.*[33] first proposed the model with peritonitis due to necrosis of ligated coecum distal to the ileocoecal junction. Chaudry *et al.*[34] combined this ligation with puncture of the coecum, coecal ligature puncture (CLP). They found that ligation alone produced no mortality, only abscess formation. Combined with coecal puncture it gave significant mortality, dependent on the puncture size or number.

Using a eighteen gauge needle, two punctures gave 94 % mortality after two days. One puncture resulted in 90 % mortality after five days.

This is a commonly used model with sufficient reproducibility. The size of the inoculum is not standardized in its true meaning. However, trapping of the feces below the ligature and standardized puncture may result in a controlled escape of feces from the coecum into the peritoneal cavity. Important details include milking of colonic content down into the coecum and avoiding bleeding from the puncture on the antimesenteric side.

In his original work, Chaudry did not perform full hemodynamic monitoring. However, liver blood flow measurements indicated that this model implicates both hyper-and hypodynamic sepsis phases.

Bowel Ischemia Models

Interrupted blood supply to a bowel segment results in necrosis of the affected part. The bowel wall permeability increases, resulting in bacterial and endotoxin escape from the bowel. The ileocoecal part is commonly used. As already stated, devascularization of the coecum alone does not seem to be sufficient for producing significant mortality.

The reproducibility of such models depends on both a standardized ischemic bowel length and the bowel content of bacteria and endotoxin.

Olofssen *et al*[35] performed extensive endotoxin studies in an ileocoecal ischemia model. They concluded that endotoxinemia was due to leakage through the bowel wall into the peritoneal cavity and lymphatic transport into the systemic circulation. The portal vein transported ten times as much endotoxin as the thoracic duct. The portal transport, however, never exceeded the filtration and inactivation capacity of the liver.

Free Bowel Segment Models

Lahnborg and colleagues[36] reported a model of peritonitis originating from an isolated ischemic ileum segment.

Five centimetres from the ileocoecal junction, a one cm ileum segment was isolated on its vascular pedicle. This was interrupted with a ligature and intestinal continuity reestablished by end-to-end anastomosis.

This resulted in acute peritonitis with 50 % mortality after one day, exceeding to the final 77 % after four days. The predominants in peritoneal exudate were E. coli, Proteus mirabilis, Str. fecalis, B. fragilis and Cl. perfringens. Blood cultures or endotoxin measures were not performed. The authors did not report whether the ileum segment was empty or not. The rats, however, fasted twenty four hours before surgery, and this probably means that the ileum was empty.

We have tried to reproduce the findings in this model, but with quite different mortality rates.[23] With a one cm empty segment our mortality was only 17 %, increasing to 25 % when the segment was filled with a 3*8 mm fresh fecal cylinder. Surprisingly, mortality reached 58 % when the vascular pedicle of the segment was not interrupted.

With a two cm segment, the corresponding numbers were 25 %, 33 % (cylinder 5*15 mm) and 92 %. Nearly all the survivors developed an abscess around the segment.

Bowel Perforation Models

Gastrointestinal perforation is a common cause of peritonitis and abdominal sepsis. Therefore, the clinical relevance of such models is clearly evident. The uncontrolled inoculum amount is the main argument against them.

The most widely used technique in our laboratory is coecal perforation.[23] In our hands this is a simple and reproducible method but some technical details must be followed:

1. The coecum should be filled with bowel content from ascending colon.
2. A standardized, antimesenteric hole should be excised with a perforator. Puncture alone is not sufficient.
3. There must be minimal fecal soiling when the coecum is put back into the peritoneal cavity after operative manipulation. This is achieved by temporarily occluding the hole.
4. Bleeding from the coecotomy should be avoided. This can interfere with the results in two ways: Hypovolemia or postoperative anemia and sealing of the hole by a blood clot, resulting in no free peritonitis.

The mortality rate depends mainly on the diameter of the hole. Thus, the method can be designed to the requested mortality level.

Experimental Abdominal Sepsis in the Pig

General Considerations

Due to practical problems in handling the animals, young pigs weighing between 15 – 30 kg are preferred. The "mini-pig" is a genetic variation bred exclusively for research purposes. We have found the price too high and the size too small for our experimental protocols, but others have been quite satisfied with these animals. If you need multiple blood samples for processing of cells or plasma factors, the small blood volume of mini pigs can be a severe problem. The choice is usually one of monitoring equipment. If you have the equipment and the experience needed to monitor premature babies you are likely to be doing well. Otherwise the size will be limiting and the reasons for choosing a large animal model will be challenged. The handling of a pig before the experiment should not be underestimated. A severe stress reaction and a lethal hyperthermic reaction are well known to experimental surgeons. The hyperthermic response usually following halothan anesthesia can be difficult to distinguish from an acute reaction to endotoxin or micro-organisms and is also sometimes unjustified related to an antibody or biological response modifier. The gene responsible for this reaction has now been cloned and the preponderance to a hyperthermic reaction can be tested in very young piglets using a small test dose of halothan, making stocks, free of this troublesome problem, easily available.[37] A gentle handling in a quiet environment immediately before and when handling the anesthesia are important steps in reaching a steady state before the experimental intervention.

Anesthesia

A number of anesthetic procedures are available for porcine experiments and they all interfere more or less with the animals "normal" function.[38]

Cardiovascular and respiratory functions are impaired during septic shock but are also dependent on the depth of the anesthesia. These parameters are typical shock parameters often studied in experiments using injections of barbiturate intravenously "when necessary" or a steady infusion of barbiturate. The well known interference of barbiturates with cardiac and respiratory functions are seldom discussed and the less well known kinetics of barbiturates by an organism in shock are usually avoided. It is impossible to avoid this problem totally but the interpretation of the experimental results is made easier if you can monitor the level of the anesthetic drug and keep it constant, or use a drug (e.g. ketamin) known to give minimal interference with these functions. We try to reduce these problems using either halothan on a special pig-face mask shortly for induction followed by a quick intubation and continuous halothan anesthesia kept at a steady level measured in the expired gas mixture[37] or induction and surgery with ketamin continued with i.v. diazepam during the steady state and observation period.[37] The first method depends on some expensive hardware (gas monitor) and an elaborate ventilation system to avoid environmental pollution. The other method is simpler and less expensive as long as the dosage of ketamin is kept low. One point often forgotten is that not only is the animal immobilized by anesthesia, but some of the experimental animal cell systems are also attenuated. Mobility and migration of circulating cells have been demonstrated to be reduced during halothan anesthesia.[38]

Induction of Intra-abdominal Sepsis

The most widely reported model of intra-abdominal sepsis has been injection of E. coli endotoxin LPS. In spite of the interesting reaction to LPS this is not an appropriate model of true intra-abdominal sepsis in pigs. Situations where LPS is injected into a vein or abdomen are never encountered in clinical or veterinary medicine. Its closest clinical reaction may be a case of fulminant overwhelming meningococcal endotoxinemia.

Intra-abdominal sepsis is caused either by micro-organisms carried to the abdomen by the blood or lymphatics (seldom and usually in young children, the organisms being gram-positive streptococci), infected material introduced into the peritoneal cavity (surgery, trauma or dialysis) or perforation of an intra-abdominal organ containing micro-organisms. Variations of these main themes are spontaneously occurring peritonitis in portal hypertension with ascites and abscesses entering the abdomen from infected neighboring organs.

Translocation of endotoxin and bacteria from the bowel to the general circulation has been found in some animal models but so far not been demonstrated in man. Injection of LPS results in a number of interesting reactions but should not be called intra-abdominal models of sepsis.

In clinical situations of intra-abdominal sepsis, infectious material is located within the abdomen giving a certain reactive pattern.[4,39,40] The most widely used porcine models include either installation of bowel micro-organisms, fecal matter or a combination of these, sometimes combined with blood or foreign material. The amounts of bacteria and feces are dependent on the purpose and end-points of the study. Bowel perforation, so widely used in small animals, is seldom reported in porcine models.

One problem with large-animal models is the difficulties contained with making a model of protracted severe intra-abdominal sepsis. Authority regulations in most countries have rightfully imposed a ban on painful experiments without continuous anesthesia and supervision. A model of prolonged bacterial peritonitis in pigs would have to include anesthesia and mechanical respiratory support for days with continuous medical support. In fact this would be and should be an animal intensive care unit. Thus the majority of models are designed to give the expected end points (e.g. mortality, cardiorespiratory failure, kidney failure) during two to twelve hours. Installation of $2.5{*}10^9$ CFU of E. coli intraperitoneally is a typical bacterial load for such an experiment. The experiments can easily be fine tuned dependent on special end points by regulating this dosage or combining the bacteria with foreign matter, sterile feces or other bowel micro-organisms. The bacteria and LPS will be detectable in the general circulation and handled in the same way as a severe intra-abdominal intra-operative contamination is handled.

As a basic model we prefer to use two pigs simultaneously (one control receiving placebo and one receiving active substance) and split the dose of bacteria between them (Figure 11.1). This makes the interpretation of intervention after induction of intra-abdominal sepsis easier.[41] The exact number and condition of micro-organisms in laboratory specimens of feces can be very difficult to repeat for different animals on different occasions and can be of decisive importance when different treatment modalities are studied.

Monitoration

The monitoration of a large-animal model has two different aspects; the basic monitoration of the animal to define a steady state before intervention and a meaningful description of the changes followed by intervention, and subsequent monitoration of a special organ or cellular function. The young pig (15–30 kg) lends itself easily to multiorgan monitoration using regular slightly modified clinical equipment or original pediatric equipment.

Basic Monitoration

In our experiments basic monitoration includes controlled mechanical support (Servo 900B, Elema-Siemens) and regular blood gas sampling using standard clinical equipment (Radiometer, Copenhagen). Swan-Ganz catheterization via the right external jugular vein (5 French with thermodilution, Edwards Lab. S. Ana, California) are used for cardiac output (Model 9520A, Edwards Lab.) and pulmonary artery pressure (Pap). Via the left femoral artery an aortic pressure catheter is inserted (Pao).

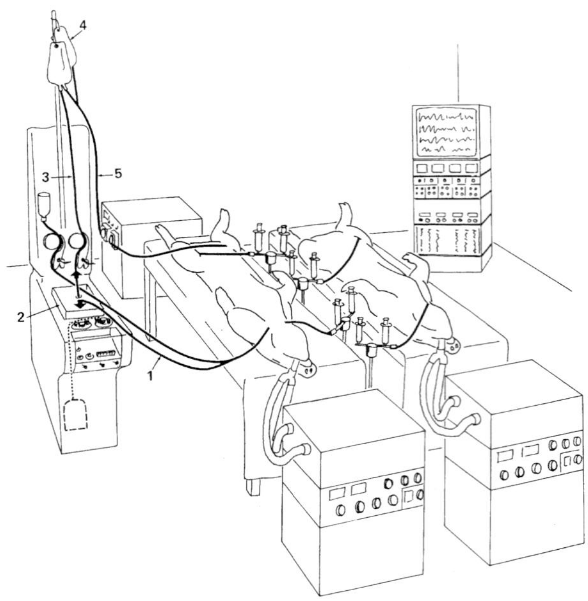

Figure 11.1

Intravenous cannulas are always placed in both ears for infusion and a central venous catheter in the internal caval vein for easy blood sampling. Saline of 0.9 % is continuously infused at a rate of 20 ml/kg*h during surgery and 10 ml/kg*h under the observation period.

Urine output is registered with a catheter, ademeure placed in the bladder by a small cystotomy. Female pigs can be successfully catheterized but it is extremely difficult to insert a catheter into the male pig.

We prefer intubation to tracheostomy, but there is little or no significant difference during an observation period of six hours that can be related to this in our experience. For regular success in intubation of pigs some training and a long straight laryngoscope are useful.

This "basic" monitoration set-up makes it relatively easy to reach an acceptable steady state before any intervention is done and is usually adequate for describing acute reactions to intra-abdominal sepsis and different treatment modalities .[39, 42–45] A definition of steady state before embarking upon complicated experimental protocols is of considerable importance. Sometimes a few pilot experiments have to be done before steady state criteria can be established, but this is an extremely valuable investment. After all monitoring devices are in place we demand a one or two hour period (depending upon the type of experiment) of regular heart rate not altering by more than 10 beats/min, a steady pulmonary pressure not altering by more than 2 mmHg, urinary output > 75 ml/h and a cardiac output not altering by more than 0.25 l/min.[39]

Special Monitoration

This is of course dependent upon the end points of the study. From a technical point of view almost any vital sign measured in man can be used in a porcine model. The advent of sophisticated ultrasonic flow and volume measuring systems has now given us the possibility of monitoring both portal and arterial perfusion of the liver (and virtually all other intra-abdominal organs) and volume changes of the liver and spleen during intra-abdominal sepsis.[37] A method of long term total parenteral nutrition has been constructed for both young and older pigs.[1] Only imaginative and monetary limitations are restrictive when it comes to advanced monitoration in the pig.

Intravasal catheters and monitoring devices certified for clinical use are usually disposable. This is a necessary precaution against blood borne contaminants, but can impose severe economical problems upon an experimental laboratory. Disposable equipment (often obtained after previous use in patients) are used until they no longer function in many laboratories. We have experienced bacteriemia developing due to such contaminated catheters during experiments and now use either ethylenegas sterilization or a ten minute wash in cidex (as used with endoscopes).

The blood volume in pigs is usually sufficient for repeated blood cell and plasma studies. A large number of reactions occuring in man during sepsis have been demonstrated in pigs using the same or slightly modified procedures. The extensive menu of RIA and ELISA methods offered today covers many of the classical problems investigated in porcine models. In spite of their attractive features, many of these tests are based on monoclonal antibodies so specific that the minor grade of cross reactions to porcine tissue will not give meaningful answers (e.g. tumor necrosis factor). Before buying quantities of expensive RIA or ELISA kits, always ask for documentation of porcine sensitivity or a free sample and do your own tests. All the serious manufacturers of such kits are aware of this problem and are usually willing to help you out. Functional assays for enzymes, cytokines and cells are not usually a problem as they are not dependent upon immunologic reactions. One minor but troubling problem has been the visual detection of monocytes. Porcine monocytes are not sufficiently stained by the non-esterase method used for human cells. We have developed a simple functional assay for porcine monocytes that has served us well for years.[39]

RIAs for gastrointestinal hormones are very often originally based upon porcine antigen and seem to work excellently.[44] It is very likely that the rapidly increasing use of porcine models in experimental medicine will lead to a development of advanced special assay systems making the study of porcine sepsis even more likely to result in progress in clinical medicine.

References

1. Nordstrand, K. (1988). Organ changes and mortality following parenteral nutrition. An experimental study in rats and pigs. Thesis, Department of Surgery, University of Tromsø, Norway.
2. Nordstrand, K., Midtvedt, T. and Giercksky, K.-E. Parenteral nutrition in germ-free rats. (in preparation)
3. Andersen, O. K. (1986). The influence of E. Coli endotoxin and trauma on some aspects of host defence. Thesis, Dpt. of Surgery, University of Tromsø., Norway.
4. Røkke, O. (1990). Gram-negative septicemia. The influence of a preceeding trauma and different treatment modalities. Thesis, Dpt of Surgery, University of Tromsø., Norway.
5. Nordstrand, K., Melhus, O., Eide, T. J., Larsen, T. and Giercksky, K.-E. (1987). Intraabdominal granuloma reaction in rats after introduction of maize-starch powder. *Acta Path. Microbiol. Immunol. Scand. sect A*, **95**, 93–98.
6. Poole, T. (1989). UFAW Handbook on the Care and Management of Laboratory Animals, 6th edition. Longman Scientific & Technical, England, p 313.
7. Abel, F. C. (1990). Does the heart fail in endotoxin shock? *Circ Shock*, **30**, 5–15.
8. Fish, R. E. and Spitzer, J. A. (1984). Continuous infusion of endotoxin from an osmotic pump in the conscious, unrestrained rat: a unique model of chronic endotoxemia. *Circ Shock*, **12**, 135–149.
9. Lang, C. H., Bagby, G. J., Bornside, G. H., Vial, L. J. and Spitzer, J. J. (1983). Sustained hypermetabolic sepsis in rats: characterization of the model. *J. of Surg. Res.*, **35**, 201–210.
10. Wichterman, K. A., Baue, A. E. and Chaudry, I. H. (1980). Sepsis and septic shock-a review of laboratory models and a proposal. *J. of Surg. Res.*, **29**, 189–201.
11. Nordstrand, K. *et al.* (1989). Mortality in rats on long-term parenteral nutrition. *Eur. Surg. Res.*, **19**, 40–52.
12. Bopp, M. B. and Brennan, M. F. (1981). Long-term vascular access in the rat: importance of asepsis. *Am. J. of Physiol.*, **241**, 606–612.
13. Pedersen, P., Biber, B., Martinell, S., Seeman, T. and Hasselgren, P. O. (1984). Hemodynamic and hematologic changes in a standardized trauma-sepsis model in rats. *Circ Shock*, **14**, 13–23.
14. Albrecht, I. *et al.* (1975). Arterial pressure, cardiac output and systemic resistance before and after pithing in normotensive and spontaneously hypertensive rats. *Acta Physiol. Scand.*, **94**, 378–385.
15. Idvall, J., Aronsen, K. F., Nilsson, L. and Nosslin, B. (1979). Evaluation of the microsphere method for determination of cardiac output and flow distribution in the rat. *Eur. Surg. Res.*, **11**, 423–433.
16. Olofsson, P. (1988). Evaluation of the effects of lymph drainage by a thoracic duct fistula in experimental peritonitis. *Acta Chir. Scand.*, **154**, 453–459.
17. Saldéen, T. and Linder, E. (1960). A method for long-term collection of lymph from the thoracic duct in rats. *Acta Pathol.*, **49**, 433–437.
18. Robinson, K. M. and Miller, H. I. (1981). Free fatty acid turnover and oxidation after burn shock in guinea-pigs. *Circ Shock*, **8**, 283–290.
19. Borden, C. W. and Hall, W. H. (1951). Fatal transfusion reactions from massive bacterial contamination of blood. *N. Eng. J. of Med.*, **245**, 760–765.
20. Almdahl, S. M., Nordstrand, K., Melby, K., Østerud, B. and Giercksky, K. E. (1985). Fecal peritonitis in the rat. *Acta Chir. Scand.*, **151**, 213–216.
21. Kasper, D. L. and Onderdonk, A. B. (1982). Infection with Bacteroides fragilis: pathogenesis and immunoprophylaxis in an animal model. *Scand. J. Infect. Dis. suppl.*, **31**, 28–33.
22. Nichols, R. L., Smith, J. W. and Balthazar, E. R. (1978). Peritonitis and intraabdominal abscess: an experimental model for the evaluation of human disease. *J. of Surg. Res.*, **25**, 129–134.
23. Lundblad, R., Sandven, P. and Giercksky, K.-E. (1995). The Physical nature of a large bowel perforation pre-

dicts severity of the subsequent inflammatory response. *Shock*, **3**, 455–461.
24. Weinstein, W. M., Onderdonk, A. B., Bartlett, J. G. and Gorbach, S. L. (1974). Experimental intra-abdominal abscesses in rats: development of an experimental model. *Infect. Immun.*, **10**, 1250–1255.
25. Onderdonk, A. B., Weinstein, W. M., Sullivan, N. M., Bartlett, J. G. and Gorbach, S. L. (1974). Experimental intra-abdominal abscesses in rats: quantitative bacteriology of infected animals. *Infect. Immun.*, **10**, 1256–1259.
26. Weinstein, W. M., Onderdonk, A. B., Bartlett, J. G., Louie, T. J. and Gorbach, S. L. (1975). Antimicrobial therapy of experimental intraabdominal sepsis. *J. of Infect. Dis.*, **3**, 282–286.
27. Bartlett, J. G., Onderdonk, A. B., Louie, T. J., Kasper, D. L. and Gorbach, S. L. (1978). Lessons from an animal model of intra-abdominal sepsis. *Surgery*, **113**, 853–857.
28. Rink, R. D., Short, B. L., Van, N. and Fry, D. E. (1977). Role of colonic bacteria in the pathophysiology of fecal peritonitis. *Circ Shock*, **4**, 259–270.
29. Hau, T. and Simmons, R. L. (1977). Surgical pros and cons. *Surg. Gynaecol. & Obstet.*, **144**, 755–756.
30. Dunn, D. L., Barke, R. A., Ewald, D. C. and Simmons, R. L. (1987). Macrophages and translymphatic absorption represent the first line of host defence of the peritoneal cavity. *Arch. of Surg.*, **122**, 105–110.
31. Rotstein, O. D. and Kao, J. (1988). The spectrum of Escherichia coli-Bacteroides fragilis pathogenic synergy in an intraabdominal infection model. *Can. J. Microbiol.*, **34**, 352–357.
32. Andersson, R., Tranberg, K. G., Alwmark, A. and Bengmark, S. (1989). Factors influencing the outcome of E. coli peritonitis in rats. *Acta. Chir. Scand.*, **155**, 155–157.
33. Ryan, N. T., Blackburn, G. L. and Glowe, G. H. (1974). Differential tissue sensitivity to elevated endogenous insulin levels during experimental peritonitis in rats. *Metabolism*, **23**, 1081–1089.
34. Chaudry, I. H., Wichterman, K. A. and Baue, A. E. (1979). Effect of sepsis on tissue adenine nucleotide levels. *Surgery*, **85**, 205–211.
35. Olofsson, P., Nylander, G. and Olsson, P. (1986). Endotoxin-transport routes and kinetics in intestinal ischemia. *Acta Chir. Scand.*, **151**, 635–640.
36. Lahnborg, G., Hedstrøm, K. G. and Nord, C. E. (1982). Efficacy of different antibiotics in the treatment ofexperimentally induced intra-abdominal sepsis. *J. Antimicrobial Chemotherapy*, **10**, 497–504.
37. Myhre, U., Rise, C., Pettersen, J. and Giercksky, K.-E. (1991). Liver perfusion and volume during endotoxinaemia. Proceedings of Norwegian Surgical Association (International edition in preparation).
38. Røkke, O., Revhaug, A. and Giercksky, K. E. (1990). Effects of Halothane and nitrous oxide on chemiluminescence of polymorphonuclear granulocytes. Acta Anaesthesiologica Scandinavica.
39. Andersen, O. K., Lundgren, T. I., Revhaug, A., sterud, B. and Giercksky, K. E. (1984). Controlled endotoxinemia in pigs. *Acta Chir. Scand.*, **150**, 599–606.
40. Skau, T. Intraabdominal infections. *Linköping University medical dissertations*. **184**, 186: 20.
41. Røkke, O., Lundgren, T. I., Revhaug, A., Giercksky, K. E., Osterud, B. and Rekvig, O. P. (1989). Plasma exchange performed during severe gram negative septicemia: The influence on circulatory performance, plasma levels of endotoxin and bacterial counts in blood. *Surg. Res. Comm.*, **189**, Vol. 5, 59–73.
42. Røkke, O., Giercksky, K. E. and Revhaug, A. (1989). Depression of plasma endotoxin levels during gram-negative septicemia subsequent to moderate trauma. *Acta Chir. Scand.*, **155**, 145–149.
43. Røkke, O., Revhaug, A., sterud, B. and Giercksky, K. E. (1988). Increased plasma levels of endotoxin and corresponding changes in circulatory performance in a porcine sepsis model: The effect of antibiotic administration. Bacterial Endotoxins: Pathophysiological Effects, Clinical Significance, and Pharmacological Control, 247–262.
44. Revhaug, A., Lygren, I., Lundgren, T. I., Jorde, R., Andersen, O., Burhol, P. G. and Giercksky, K. E. (1985). Release of gastrointestinal hormones in Cardiodepressive Shock. *Acta Anesthesiol. Scand.*, **29**, 371–374.
45. Revhaug, A., Lygren, I., Lundgren, T. I., Andersen, O. K., Burhol, P. G. and Giercksky, K. E. (1985). Increased plasma levels of vasoactive polypeptide in pigs during endotoxinaemia. *Eur. Surg. Res.*, **17**, 75–82.

PART III

12. Animal Models of Experimental Liver Failure

Rosemary Hickman

University of Cape Town Medical School, Cape Town, South Africa

Introduction

The liver has interested physicians for centuries probably because of its very large size. During the last 100 years understanding the liver function has progressed from almost nothing to fine details of molecular structure and activity. The early history of this understanding includes names such as van Eck,[1] Hahn and Pawlow [2] and Mann, Bollman and Magath,[3,4] who performed pioneering studies of various types–diversion of portal blood, total or partial extirpation of the liver in various animals including frogs, geese and dogs, or induction of acute necrosis by administration of carbon tetrachloride[5] or yellow phosphorus.[6] These early studies stimulated a persistent interest in the effects of total or partial liver dysfunction and methods of study have advanced as biochemical techniques for measurement of liver function have evolved.

Need for a Model

What began as studies based largely on curiosity, soon became clinically relevant as the mortality rate from fulminant hepatic failure became apparent. From the 1950's to the 1970's there was intense effort both at finding means of artificial liver support together with a very active interest in the causes of hepatic coma, the most confounding sign of liver failure. At that time, the principal cause of fulminant hepatic necrosis was acute viral hepatitis;[7] gradually the effects of overdosage of paracetamol[8] and of sensitivity to halothane (9) have also become important causes. While the mortality from this complication was quoted as 1% of the 10% of patients who develop fulminant hepatic necrosis after viral hepatitis, this figure has remained almost unchanged despite almost forty years of intense study. Such data present an immediate challenge and it soon becomes apparent that clinical material is too diverse and variable to form the basis for evaluation of methods of management and thus the search begins for animal models to represent the human disease.

The subject of animal models of liver failure can be investigated as follows:

Animal Models of What?

The principal feature of liver failure which has been investigated by researchers is hepatic encephalopathy. Hahn and Pawlow first showed that dogs with portacaval shunts were particularly sensitive to protein intoxication,[2] the effects of nitrogenous substances and more specifically ammonia became apparent.[10,11] Ammonia has remained the substance most often suggested to be the cause of hepatic encephalopathy but proof that it is the cause is not absolute.[12] Other substances such as short chain fatty acids and mercaptans have been proposed[13,14] alone and in concert with ammonia. Fischer and Baldessarini presented the elements of their theory of increased levels of false neurotransmitters in 1971 and studies began of changes in branched chain amino acids[15] and in the permeability of the blood brain barrier.[16] James et al. drew together the hypothesis of the involvement of ammonia and false neurotransmitters in hepatic encephalopathy in 1979.[17] From the concept of accumulation of false neurotransmitters evolved the theories to which most attention is currently being given - the use of antagonists of gamma amino isobuytric acid (GABA)–the studies particularly being made by Jones and various co-workers.[18,19]

Numerous workers have investigated the problem of hepatic encephalopathy from a different view point, that of treatment, the reversal of encephalopathy and provision of methods of support while the liver undergoes regeneration. There was particular effort between 1950 and 1970, as extraordinary and desperate methods of liver support were applied–these included exchange transfusion of at least the total blood volume,[20,21] cross circulation with humans with terminal disease[22,23] or with non human primates which had already been exchange transfused with compatible human blood,[24] the so called "laundering of the blood" by isolated liver perfusion, pioneered by Eiseman and co-workers[25] and using livers from various farmyard animals,[26–28] non human primates[27] and even the occasional human.[29] More recently attempts at treatment have concentrated upon the reversal of hepatic coma, especially as it has been shown that short chain fatty acids and mercaptons,[30] incriminated in the pathogenesis of encephalopathy may themselves interfere with liver regeneration. Thus methods of charcoal haemoperfusion have been studied especially by Williams *et al.*[31–33] and various forms of dialysis by Opolon *et al.*,[34] culminating in the development of an "artificial liver machine" by Brunner *et al.*[35] which detoxifies certain substances.

This review of the uses to which animal models have been and will be used, now allows definition of some of the requirements of such models. These requirements were summarised by Terblanche *et al.* in 1975 (36) and more recently updated, although not changed, in 1989.[37]

Requirements for Animal Models

(i) use of a *large animal* – to allow repeated and often considerable blood sampling and also for application of various forms of assist devices. For the study of the biochemistry of hepatic encephalopathy however, smaller models including rats and rabbits have proved suitable.

(ii) *reversibility* – there must be the potential for complete recovery of liver function otherwise the efficacy of any forms of management becomes impossible to evaluate.

(iii) *reproducibility* – for the same reasons as above, the model needs to have a high rate of reproducibility in order that the effects of treatment may be evaluated against a standard background.

(iv) *death* in liver failure, to stimulate the clinical syndrome.

(v) *an established period of coma* - to allow implementation and evaluation of any form of treatment.

(vi) *safety for personnel* - this requirement is also absolute because some of the agents which have been proposed for the experimental induction of liver necrosis are hazardous.

When these requirements are presented in this tabular form they would seem simple and easy to conform to. However, it needs to be stressed that only a very limited number of models which have been proposed is found to conform completely.

What Models are Available?

1. Total Hepatectomy

Removal of the liver is one of the oldest surgical manoeuvres to be practised in hepatic research, the technique being developed almost eighty years ago to investigate liver function.[38] There were significant limitations to the operation at that stage – no vascular grafts, arterial suture materials, anti-coagulants, and mechanical bypass/assist apparatus. Biochemical methodology was limited – measurements were available only for glucose, urea, uric acid and some components of bile. Nevertheless a technique was evolved – this included the reverse portacaval shunt with ligation of the infrahepatic inferior vena cava distal to the shunt to promote the development of portosystemic collaterals. After a period of weeks the portal vein and hepatic artery were divided and the animal thus rendered "anhepatic" at least in terms of any circulation through the liver. Initially these models were used to discover and then demonstrate their total dependence upon administration of glucose.[39] Once this was realised survival became longer and hepatectomised animals could be used. The techniques were modified and the livers were definitely removed;[40] later, vascular grafts were used to restore the circulation;[41] in dogs, livers could be resected off the vena cava.[42,43] Finally in the dog,[44] the pig[45,46] and the rat[47,48] one stage hepatectomy has been perfected. The models were principally studied for their resemblance[47] to hepatic encephalopathy in terms of intermediary metabolism[49] and changes in coagulation.[50] Very recently, hepatectomy has come into sharp focus as surgeons need to know more about the anhepatic phase of liver transplantations.[51] It has also recently been suggested by the group from Hannover[52] that the damaged liver of the recipient with fulminant hepatic necrosis should be removed early, before the donor replacement liver is fully available, in order that the anaesthetised recipient may be stabilised as far as possible. There has been new understanding of the early changes in acid-base balance, glucose homeostasis, levels of glucoregulatory hormones and coagulation function.[45] Very recent studies have shown rapid and sustained increases in plasma insulin and catecholamine levels in hepatectomised pigs; the fact that these animals are peculiarly sensitive to the administration of 5 or 10% dextrose intravenously may have relevance in the management of patients with fulminant hepatic failure or during the anhepatic phase of liver transplantation.[45]

2. Partial Hepatectomy

The total hepatectomy technique as a model for hepatic failure has deficiencies - it is not reversible and only recently has it become reproducible. In addition, it represents only absence of liver function and not the effects of venous return from a damaged liver which spills a variety of toxic substances into the circulation. Thus, modifications of the technique of total hepatectomy were evolved as a 90% and the lesser forms of partial hepatectomy. The 90% hepatectomy which was first described sixty years ago has recently been performed again for studies of nucleosides and oxypurines by Berlinger.[53,54] This model is potentially reversible because of the described ability of even the 90% partially hepatectomised liver to regenerate.[55] So far this appears to be the only description of similar use of the technique - perhaps further studies will follow.

Lesser partial hepatectomy, conventionally of 50–66% has long been used for studies of liver regeneration. Since the original description by Higgins and Anderson[56] numerous studies have been conducted of liver regeneration using the partially hepatectomised model in the rat,[57] rabbit[58] pig,[59] and dog.[60] All animals recover well and do not manifest evidence of encephalopathy or liver insufficiency and thus partial hepatectomy alone cannot be used for studies of acute hepatic failure. However, there are recently described methods in which partial hepatectomy is combined with portacaval shunt[60] in order to achieve a hepatic failure mode. This combination will be described later.

3. Portacaval Shunt

Portacaval shunt alone results in encephalopathy in dogs and is apparently directly related to protein intoxication.[61] Rats and pigs show only limited variable features of encephalopathy perhaps because the laboratory diet of both species contains comparatively little protein. Another possible reason is an excellent response of the hepatic arterial response in both species to interruption of portal inflow. The model of the portacaval shunted rat has been used for a variety of studies related to the integrity of the blood-brain barrier in encephalopathy[62] and also for further understanding of the role of portal blood components rather than reduced volume of flow in the changes which follow portacaval shunt. This has been accomplished by comparing portacaval transposition with portacaval shunt and observing that many of the changes which occur are common to both procedures.[63] Thus it is diversion of constituents of portal blood rather than hepatic ischaemia which to a great extent contribute to encephalopathy. The topic of portal hypertension is dealt with elsewhere by Dr Jensen and only a very limited discussion of experimental liver cirrhosis and chronic portosystemic encephalopathy will be given.

4. Ischaemia Permanent

The original methods of performing total hepatectomy were indeed no more than total devascularisation of the liver. When it became possible to remove the liver totally with reconstitution or circulation, comparisons were made of the advantages and suitabilities of each mode. Total devascularisation was accomplished by establishing an end to side portacaval shunt and by ligation of the hepatic artery[64] with the liver remaining *in situ*. A little doubt has been expressed about this model as it might be thought that sufficient blood might reflux from the inferior vena cava into the liver to support some function.

Temporary Ischaemia

Early work suggested that temporary occlusion of the inflow to the liver might result in enough damage to simulate clinical hepatic necrosis. These studies were conducted in dogs and showed initially that thirty minutes total ischaemia was lethal.[64] However, it was soon found that death was due to congestion of the splanchnic circulation which was not decompressed during ischaemia in early procedures. As soon as this was realised and decompression was provided by means of an external or an internal portosystemic shunt, it became apparent that the liver was remarkably tolerant of warm ischaemia. Early studies during the isolated liver perfusion era which showed that such livers could remain ischaemic on the bench for three hours and could then, during perfusion, yield a normal clearance of bromsulphthalein[65] were confirmed in intact animals. Gradually the concept that the liver was extraordinarily sensitive to ischaemia was replaced by observations that hepatocytes would tolerate up to three hours' ischaemia with almost 100% subsequent survival of animals.[66] More recent studies show that a period of 5 hours is necessary to result in a fatal outcome.[67] This procedure is acceptably reproducible and seems to result in uniform liver damage. Similar results have been obtained in studies in pigs[68,69] and recently there has been clinical application of the observations in the use of hepatic inflow occlusion during liver resections.[70] The use of ischaemia for production of models of liver failure has also been combined with administration of toxins.

5. Drug or Toxin Induced Hepatic Failure

In many cases, models of drug induced hepatic necrosis are tested for biochemical changes especially in terms of causes of encephalopathy, but such studies as published may not necessarily give details of survival or evidence of encephalopathy.

Carbon tetrachloride has long been recognised as a hepatotoxin and has been studied intermittently as a means of inducing liver failure. The studies of McLean et al.[71] have shown that prior administration of phenobarbitone results in more uniform damage by induction of detoxifying microsomal enzymes but carbon tetrachloride is now more frequently used for induction of experimental cirrhosis.[72] This model is a reproducible and simple means of inducing chronic liver damage and animals survive for variable long periods. There is, however, only limited evidence of encephalopathy and the model is not representative of human fulminant hepatic necrosis. Nevertheless some attempts have been made to use larger, more acute doses given intravenously with prior phenobarbitone induction. These studies in pigs[73] were conducted in the early 1970's and were abandoned because of an unacceptably wide range of survival times. More recently it has become apparent that such survival times are not uncommon in studies using large animals and carbon tetrachloride may yet have a role to play, especially in combination models (described later in the Chapter).

Galactosamine was studied almost twenty years ago for its property of causing hepatic necrosis – extensive studies of the biochemical defects were conducted before application was made of the toxin to models of liver failure. The initial studies were made in inbred rabbits and were found to be acceptably reproducible.[74] Initial applications of the drug to rats were found to be more difficult to reproduce[75–77] but recently more studies have been published which suggest that results are now more acceptable and a number of researchers are now using the model.[75–78] In the original work, Blitzer et al. cited 74% mortality between 21 and 44 hours and a further 15% by 91 hours with a period of 2.6 hours encephalopathy.[74] Others have reported 80–100% mortality within 3–8 days postoperatively[73] or 26% survival after 2 weeks.[77] At present this is one of the most widely used models but its one disadvantage is that, except in exceptional circumstances, large animal studies are precluded by the cost of the drug.

Acetaminophen came to the attention of those searching for drugs to induce hepatic necrosis when the clinical problem of paracetamol induced liver damage arose.[79] Initial studies in animals fifteen years ago did not define the model clearly and this has only recently been published (80). In the initial studies there were problems determining the correct dose to induce necrosis without other complications. For example, Miller et al. encountered severe intrahepatic haemorrhage as well as hepatocyte necrosis.[81] In the most recent study published in 1989, Francavilla et al.[80] aimed to determine blood levels and the correct levels of drug necessary to produce consistent necrosis as well as using the easier method of subcutaneous infection. They evolved a model of fulminant hepatic necrosis by using three subcutaneous injections of acetaminophen dissolved in dimethyl sulphoxide (600mg.ml) within twenty-four hours. This resulted in death of 80% of animals between forty-eight and seventy-two hours which coincided with the peak of plasma transaminase levels but which was not complicated by methemoglobinuria. Miller et al.[80] reported a degree of reversibility by administration of Ranitidine. This technique is proposed as the most reproducible for use in large animals fulfilling all the proposed "ideal characteristics" except, apparently, for a specific period of encephalopathy during which therapy for coma could be instituted.

Other Drugs

A number of other substances has been used over the years to provoke hepatic necrosis but all these have certain disadvantages. Dimethylnitrosamine[82] and yellow phosphorus (6) are hazardous to those using them; fluoroxene (83 or halothane (84) thoioacetamide[85] and the injection of heat inactivated anaerobic bacteria with lipolysaccharide (86) have only been applied to rats; and the induction of animal viral hepatitis has been applied only to mice which are too small for much use[87] and once to dogs.[87] Infusion of ammonia causes a specific type of coma which is related only to the infusion time (89, 90).

6. Combination Models

There is increasing evidençe to suggest that a combination of methods may be more successfully applied in large animals. The basis for many of these models is hypoxia caused by the original portacaval shunt or by hepatic arterial occlusion and to this are added a selection of drugs given by different routes. These agents include carbon tetrachloride[91] and mushroom toxins.[92] In a recent further attempt to cause diffuse hepatic damage in the pig we have combined temporary arterial occlusion with intragastric administration of carbon tetrachloride. This resulted in 90% mortality between 15 and 52 hours after administration of carbon tetrachloride; there was a variable period of 6 to 36 hours of encephalopathy before death and biochemical and histological evidence of severe hepatocyte damage. Reversibility was demonstrated by survival of 40% of animals given prostaglandin F2.[93]

The Future

There is as yet no ideal animal model of hepatic failure and particularly for induction in large animals. There have been few innovations and most of the current models have been derived from origins at least twenty years ago. The need for appropriate models continues unchanged, however, as patients continue to die of fulminant hepatic necrosis or more recently of primary graft failure after liver transplantation[94] It remains essential to test machines which may take over the manifold functions of the liver in order to be able to provide essential liver support. Thus far most attention has been paid to damage to hepatocytes but it seems clear that several new avenues merit exploration. It is becoming apparent that other hepatic cells such as Kupffer cells may be involved in the development of hepatic necrosis[95] as well as in the process of liver regeneration.[96] Attention should, therefore, probably become more focussed upon influencing Kupffer cell function and perhaps interfering with liver regeneration. Also, preliminary studies have been made of the effects of endotoxin in combination with galactosamine (95) – this effect has not yet been clearly defined but this is certainly an area which warrants further study.

References

1. Von Eck, N. (1964). Jahrgang SS cited in Experimental Surgery. *Militar Medicinisches J*, eds Markowitz J, Archibald J, Downie H. G., Baltimore Williams and Williams Company, **1877**, 130, 559.
2. Hahn, Pawlow., Hahn, M., Massen, O. and Nenchi, M. (1893). Die Ecksche Fistel zwischen der unteren Hohlvene und der Pfortader und inre Folgen fur den organismus. *Arch. Exp. Pathol. Pharmakol.*, **32**, 161–210
3. Mann, F.C. (1927). The effects of complete and partial removal of the liver. *Medicine*, **6**, 419–511.
4. Bollman, J. L., Mann, F. C. and Magath, T. B. (1924). Studies on the physiology of the liver VIII. *Amer. J. Phys.*, **69**, 371–392.
5. Lamson, P. D. and Wing, R. (1926). Early cirrhosis of the liver produced in dogs by carbon tetrachloride. *J. Pharm. Expt. Therap.*, **29**, 191–202.
6. Burnell, J. M., Dennis, M. B., Claysen, K, Smuckler, E. A. and Clift, R. A. (1976 Evaluation in dogs of cross circulation in the treatment of acute hepatic necrosis-induced by yellow phosphorus. *Gastroenterology*, **71**, 827–831.
7. Trey, C. G. and Davidson, C. S. (1970). The management of fulminant hepatic failure. In *Progress in Liver Disease*, Vol 3, Eds .Popper, H., Schaffner, F. p. 382, New York: Grune and Stratton.
8. Gazzard, B.G., Hughes, R.D. and Mellon, P.J. (1975). A dog model of fulminant hepatic failure produced by paracetamol administration. *Br. J. Expt. Path.*, **56**, 408–414.
9. Miller, D. J., Dwyer, J. and Klatskin, G. (1978). Halothane hepatitis: benign resolution of a severe lesion. *Ann. Int. Med.*, **89**, 212–215.
10. Walshe, J. M. (1951). Observations on the symptomatology and pathogenesis of hepatic coma. *Quart. J. Med.*, **20**, 421–438.
11. Denis, J., Delorme, M. L., Boschat, M., Nordlinger, B. and Opolon, P. (1983). Respective roles of ammonia, amino acids and medium sized molecules in the pathogenesis of experimentally induced acute hepatic encephalopathy. *J. Neurochem.*, **40**, 10–19.
12. Eriksson, L.S. and Conn, H.O. (1989). Branched chain amino acids in the management of hepatic encephalopathy. *Hepatol.*, **10**, 228-246.
13. Zieve, L. (1981). The mechanism of hepatic coma. *Hepatol*, **1**, 360–365.
14. Zieve, L., Doizaki, W. M. and Zieve, F.J. (1974). Synergism between mercaptans and – ammonia or fatty acids in the production of coma. *J. Lab. Clin. Med.*, **83**, 16–28
15. Fischer, J. E. and Baldessarini, R. J. (1971). False neurotransmitters and hepatic failure. *Lancet.*, ii,75–79.
16. Livingstone, A. S., Potvin, M. and Goresky, C. A. (1977). Changes in the blood brain - barrier in hepatic coma after hepatectomy in the rat. *Gastroenterology*, **73**, 697-704
17. James, J. H., Ziparo, V. and Jepsson, B. (1979). Hyperammonaemia, plasma amino imbalance and blood brain amino acid transport. *Lancet.*, **ii**, 72–775.
18. Bassett, M. L., Muller, K. D., Skolnik, P. and Jones, E.A. (1987). Amelioration of hepatic encephalopathy by pharmacologic antagonism of the GABA – Benzodiazepine receptor complex in a rabbit model of fulminant hepatic failure. *Gastroenterology*, **93**, 1069–1077
19. Rzepczynski, D., Zieve, L., Lindblad, S. and La Fontaine, D. (1986). In vivo studies of GABA - ergic effects in experimental hepatic encephalopathy. *Hepatol.*, **6**, 902–905
20. Saunders, S. J., Hickman, R., MacDonald, R. and Terblanche, J. (1972). The treatment of acute liver failure. In *Progress in Liver Disease*, Eds. Popper., H., Schaffner, F. pp. 333–334, New York, Grune and Stratton.
21. Jones, E. A., Clain, D. and Klink, H. M. (1967). Hepatic coma due to acute hepatic necrosis treated by exchange transfusion. *Lancet*, **11**, 169–172.
22. Burnell, J. M., Dawborn, J. K., Epstein, R. B., Gutman, R. A., Leribach, G. E. and Thomas, E.D., et al. (1967). Acute hepatic coma treated by cross circulation or exchange transfusion. *New. Eng. J. Med.*, **276**, 935–943.

23. Chamelou, R. A. F. M., Popken, R. J., Beyerbacht, E. C. and de Koning, W. M. (1983). Problems in treating experimentally induced acute liver failure by haemoperfusion or cross circulation. *Hepatol.*, **3**, 696–700.
24. Saunders, S.J., Terblanche, J., Bosman, S.C.W. (1968). Acute hepatic coma treated by cross circulation with a baboon and by repeated exchange transfusions. *Lancet*, **ii**, 585–588.
25. Eiseman, B., Liem, D. S. and Raffuci, F. (1965). Heterologous liver perfusion in treatment of hepatic failure. *Ann. Surg.*, **162**, 329–345.
26. Hickman, R., Parker, J. R. and Saunders, S. J. (1972). Heterologous liver perfusion: A comparison of the use of livers from 4 different species. *Brit. J. Surg.*, **59**, 881–884.
27. Abouna, G. M., Boehmig, H. G., Serrou, B., Amemiya, H. and Martineau, G. (1970). Long term hepatic support by intermittent multi species liver perfusion. *Lancet*, **ii**, 391–396.
28. Hickman, R., Saunders, S. J., King, J. B., Harrison, G. G. and Terblanche, J. (1971). Pig liver perfusion in the treatment of fulminant hepatic necrosis. *Scand. J. Gastroent.*, **6**, 563–568.
29. Abouna, G. M. (1971). Extracoporeal liver perfusion for hepatic coma. *Lancet*, I,1185–1189.
30. Zieve, L., Dozeman, R., La Fontaine, D. and Draves, K. (1985). Effect of hepatic failure toxins on liver thymidine kinase and ornithine decarboxylase activity after massive necrosis with acetaminophen in the rat. *J. Lab. Clin. Med.*, **106**, 583–588.
31. Gimson, A. E. S. (1979). Prostacyclin to prevent platelet activation during charcoal haemoperfusion during fulminant hepatic failure. *Lancet*, ii, 173–175.
32. Gimson, A. E. S., Braude, S. and Mellon, P. J. (1982). Earlier Charcoal haemoperfusion in fulminant hepatic failure. *Lancet*, ii, 681–683.
33. Hughes, R. and Williams, R. (1986). Clinical experience with charcoal and resin haemoperfusion. *Seminars in Liver Disease*, **6**, 164–173.
34. Denis, J., Opolon, P. and Nusinovici, V. (1978). Treatment of encephalopathy during fulminant hepatic failure by haemodialysis with higher permeability membrane. *Gut*, **19**, 787–793.
35. Brunner, G., Tsikas, D., Otto, V. and Matthieu, B. (1990). An Apparatus for the removal of lipophilic and hydrophilic toxins from blood of patients in fulminant hepatic failure. *Proceedings Internat. Symp. on Artificial Liver Support*, Celle.
36. Terblanche, J., Hickman, R., Miller, D. J. and Saunders, S. J. (1975). Animal Experience with support systems: Are there appropriate animal models of fulminant hepatic necrosis? In *Artificial Liver Support*, Eds. Williams, R., Murray Lyon, I., pp. 163–172. London: Pitman Medical.
37. Terblanche, J. and Hickman, R. (1990). Animals models of fulminant hepatic failure. *Dig. Dis. Sci.*, **36**, 770–774.
38. Mann, F. C. (1944). Studies on the dehepatized animal. *J. Mount. Sinai. Hosp.*, **11**, 1–22.
39. Pollack, H. and Millet, R. F. (1933). Glucose tolerance of dehepatized animals. *Mayo. Clin. Proc.*, **8**, 557.
40. Markowitz, J., Lotto, W., Archibald, J. and Downie, H. G. (1952). Simple one stage method of hepatectomy in the dog. *Archs Surg.*, **64**, 766–772.
41. Hickman, R., Dent, D. M. and Terblanche, J. (1974). The anhepatic model in a pig. *S. Afr. Med. J*, **48**, 263–264.
42. Starzl, T. E., Bernhard, V. M. and Benvenuto, R. (1959). A new method for one stage hepatectomy for dogs. *Surg.*, **46**, 880.
43. Stewart, J.D., Williams, J.S., Kluge, D.N., Drapanas, T. (1963). Effect of cross circulation on metabolism following hepatectomy. *Ann. Surg.*, **158**, 812–819.
44. Daloze, P., des Rosiers, C., Arnoux, R., Daloze, T., Smeesters, C. and Brunengraber, H. (1990). One stage hepatectomy in the dog. *J. Surg. Res.*, **48**, 33–37
45. Hickman, R., Bracher, M., Tyler, M., Lotz, Z. and Fourie, J. (1992). The effect of total hepatectomy upon coagulation and glucose homeostasis in the pig. *Dig. Dis. Sc.*, 37,328–334.
46. Engelbrecht, G. H. C., Hickman, R., Winters, Z. and Phillips, G. A one state total hepatectomy in the rat using microvascular anastomoses. *Europ. Soc. Surg. Res. Proc. Vol II Research Methodology*. (In press)
47. Yamaguchi, Y., Bollinger, R., de Faria, E., Landis, B. and Zuarfordt, S. (1989). A simplified single stage total hepatectomy in the rat with maintenance of gastrointestinal absorptive function. *Hepatol*, **9**, 69074.
48. James, M. F. M., Hickman, R., Janicki, P., Mets, B. and Fourie, J. (1996). The early effects of total hepatectomy (the anhepatic state) upon haemodynamics and organ uptake of catecholamines in the pig. *Brit. Jnl. Anaest.*, **76**, 713–720.
49. Drapanas, T., McMenamy, R. H., Adler, W. J. and Vang, J. O. (1965). Intermediary metabolism following hepatectomy in dogs. *Ann. Surg.*, **162**, 621–633
50. Tonnesen, K. (1977). Experimental liver failure. *Acta. Chir. Scand.*, **143**, 271–277.
51. Khoury, G. F., Kleindorff, J., Brems, J., Busuttil, R. W. (1989). Pancreatic hormonal responses to hypoglycaemia during orthotopic liver transplantation in man. *Transpl*, **47**, 892–893.
52. Ringe, B., Pichlmayr, R., Lubbe, N., Bornscheuer, A. and Kuse, G. (1988). Total hepatectomy as a temporary approach to acute hepatic or primary graft failure. *Transpl. Proc.*, xx, 1, Suppl. **1**, 552–557.
53. Berlinger, W. G., Stene, R. A., Spector, R. and Al-Jurf, A. J. (1987). Plasma and csf nucleosides and oxypurines in acute liver failure. *J. Lab. Clin. Med.*, **110**, 137–144.
54. Muller, K. D. (1987). Resurrection of the 90% hepatectomy rabbit model of acute liver failure. *J. Lab. Clin. Med.*, **110**, 128–129.
55. Weinbren, K. and Tarsh, E. (1964). Mitotic response in rat liver after different regenerative stimuli. *Brit. J. Expt. Path.*, **45**, 475–480.
56. Higgins, G.M. and Anderson, R.M. (1931). Experimental pathology of the liver. *Archs. Path.*, **12**, 186–198.
57. Nakatani, T., Yasuda, K. and Ozawa, K. (1988). Hyperglycaemic hyperosmolar non ketotic dehydration in relation to blood ketone body ratio in partially hepatectomized rabbits. *Amer. J. Surg.*, **155**, 559–563.
58. Kahn, D., Hickman, R., Terblanche, J. and von Sommoggy, S.V.S. (1988). Partial hepatectomy and liver regeneration in pigs - the response to different resection sizes. *J. Surg. Res.*, **45**, 176–180.

59. Starzl, T. E., Porter, K. A., Hayashida, N., Schechter, P. and Terblanche, J. (1980). Further studies on hepatic stimulator substance (SS) after partial hepatectomy. *J. Surg. Res.*, **29**, 471–474.
60. Roessle, M., Jones, D. B., Mullen, K. D., Grun, M. and Jones, E. A. (1986). Induction of unequivocal encephalopathy in the portacaval shunted rat. *Hepatol.*, **6**, 1222
61. Thompson, J.S., Schaefer, D.F., Haun, J., Schafer, G.J. (1986). Adequate diet prevents hepatic coma in dogs with Eck fistulas. *Surg. Gynae. & Obstets.*, **162**, 126–130
62. Mans, A. M., Saunders, S. J., Kirsch, R. E. and Biebuyck, J. F. (1979). Correlation of plasma and brian amino acid and putative neurotransmitter alterations during acute hepatic coma in the rat. *J. Neurochem.*, **32**, 285–292.
63. Benjamin, I. S., Engelbrecht, G. H. C., Saunders, S. J. and Hickman, R. (1988). Amino acid imbalance following portal diversion in the rat. *J. Hepatol.*, **7**, 208–214.
64. Raffucci, F. L. (1953). Effects of temporary occlusion of afferent hepatic circulation in dogs. *Surgery*, **33**, 342–351.
65. Van Wyk, J., Liem, D.S. and Eiseman, B. (1965). Function of cadaver liver. *Surgery*, **58**, 120–124.
66. Swenson, O., Grana, L., Inouye, T. and Donnellan, W. L. (1967). Immediate and longterm effects of acute hepatic ischaemia. *Archs. Surg.*, **95**, 451–463.
67. Reuvers, C. B., Terpstra, O., Boks, A. L., de Groot, G. H., Jeekel, J. and ten Kate, F. W. J., et al. (1987). A reproducible model of acute hepatic failure by transient hepatic ischaemia in the pig. *J. Surg. Res.*, **42**, 92–100.
69. Kahn, D., Hickman, R., Dent, D.M. and Terblanche, J. (1986). For how long can the liver tolerate ischaemie? *Europ. Surg. Res.*, **18**, 277–282.
70. Terblanche, J., Krige, J. E. and Bornman, P. C. (1991). Simplified hepatic resection with the use of prolonged vascular inflow occlusion. *Archs. Surg.*, **126**, 298–301.
71. Garner, R. C. and McLean, A. E. M. (1969). Increased susceptibility to carbon tetrachloride poisoning in the rat after pretreatment with oral phenobarbitone. *Biochem. Pharmacol.*, **18**, 645–650.
72. Tamajo, R. P. (1983). Is cirrhosis of the liver experimentally produced by carbon tetrachloride an adequate model of human cirrhosis? *Hepatol.*, **3**, 112–120.
73. Van Leenhoff, J. W., Hickman, R., Saunders, S. J. and Terblanche, J. (1974). Massive liver cell necrosis induced in the pig with carbon tetrachloride. *S Afr. Med. J*, **48**, 1201–1204.
74. Blitzer, B. L., Waggoner, J. G., Jones, E. A., Gralnick, H. R., Towne, B. and Butler, J., et al. (1978). A model of fulminant hepatic failure in the rabbit. *Gastroenterol*, **74**, 664–671.
75. Muller, K.D., Schafer, D.F., Maynard, T.F., De Knagt, R., Jones, D.B. and Roessle, M., et al. (1986). Galactosamine induced fulminant hepatic failure in the rat may be an unsuitable model for acute hepatic encephalopathy. *Gastroenterol.*, **90**, 1750.
76. Roy, S., Pomier Layrargues, G., Butterworth, R.F. and Huet, P.M. (1988). Hepatic encephalopathy in cirrhotic and portacaval shunted dogs. Lack of changes in brain GABA uptake, brain GABA levels, brain glutaminic acid - decarboxylase activity and brain post synaptic GABA receptors. *Hepatol*, **8**, 845–849.
77. Makowka, L., Falk, R. E. Rotstein, L. E., Falk, J. A., Nossal, N. and Langer, B., et al. (1980). Reversal of experimental acute hepatic failure in the rat. *J. Surg. Res.*, **29**, 479–487.
78. Zieve, L., Anderson, R. and Dozeman, R. (1988). Hepatic regenerative enzyme activity after diffuse injury with galactosamine. *J. Lab. Clin. Med.*, **112**, 575–581.
79. Clark, R., Thompson, R. P. H., Borirakchanyavat, V., Widdop, B., Davidson, A. R. and Goulding, R., *et al.* (1973). Hepatic damage and death from overdose of paracetamol. *Lancet*, I,66–69.
80. Francavilla, A., Makowka, L., Polimeno, L., Barone, M., Demetris, J. and Prelich, J., et al. (1989). A dog model for acetaminophen-induced fulminant hepatic failure. *Gastrolenterol.*, **96**, 470–478.
81. Miller, D. J., Hickman, R., Fratter, R., Terblanche, J. and Saunders, S. J. (1976). An animal model of fulminant hepatic failure. *Gastroenterol.*, **71**, 109–113.
82. Aguirre, A., Yoshimura, N., Westman, T. and Fisher, J. E. (1974). Plasma amino acids in dogs with 2 forms of liver damage. *J. Surg. Res.*, **16**, 339–345.
83. Harrison, G. G. and Smith, J. S. (1973). Massive lethal necrosis in rats anaesthetised with fluoroxene after microsomal enzyme induction. *Anaesthesiol.*, **39**, 619–624.
84. Eade, O. E., Millward Sadler, G. H., Mitchell, J. and Wright, R. (1980). Hepatic necrosis in rats following halothane administration: protective effect of diethyl dithiocarbonate. *Scand. J. Gastroent.*, **15**, 859–864.
85. Hilgier, W., Zitting, A. and Albrecht, J. (1985). The brain octopamine and phenyl ethanolamine contect in rats in thioacetamide induced hepatogenic encephalopathy. *Acta. Neurol. Scand.*, **71**, 195–198.
86. Yashi, K., Yamamoto, S. and Morisawa, S. (1987). The protective effects of prostaglandin E1 in an experimental massive hepatic cell necrosis model. *Hepatol.*, **7**, 1184–1188.
87. Farivar, M., Wands, J. R., Isselbacher, K. J. and Bucher,N. L. R. (1976). Effects of insulin and glucagon on fulminant murine hepatitis. *New Eng. J. Med.*, **295**, 1517–1519.
88. Morris, T. Q., Goche, D. J. and Sardi, G. F. (1957). Exchange transfusion treatment of fulminant viral hepatitis in the dog. *J. Clin. Investig.*, **46**, 1090.
89. Fick, T. E., Schalm. S.W. and de Vlieger, M. (1989). Continuous intravenous ammonia infusion as a model for the study of hepatic encephalopathy in rabbits. *J. Surg. Res.*, **46**, 221–225.
90. Joning, T., Rigotti, P., Jeppsson, B., James, J. H., Peters, J. C. and Fisher, J. E. (1984). Methionine sulfoxime prevents the accumulation of large neutral amino acids in brain of hyperammonaemic rats. *J. Surg. Res.*, **36**, 349–353.
91. Shiboyeme, Y. (1986). Potentiation of carbontetrachloride hepatotoxicity by hypoxia. *Br. J. Expt. Path.*, **67**, 907–914.
92. Stroetter, L., Erhardt, W., Fischer, M., Gartmaier, P., Kirscher, G. and Schwabe, M. (1974). Acute liver failure by ischaemia and intoxication with mushroom poisons. *Digestion*, **10**, 381.

93. Hickman, R. and Alp, M. H. (1986). A predictable pathophysiological model of porcine hepatic failure. *Europ. Surg. Res.*, **18**, 283–292.
94. Greig, P. D., Woolf, G. M., Sinclair, S. B., Abecassis, M., Strasberg, S. M. and Taylor, B. R., et al. (1989). Treatment of primary liver graft nonfunction with prostaglandin E1. *Transplantation*, **48**, 447–453.
95. Ktell, H., Schuster, R. and Schroer, W. (1989). Early increase in paracellular permeability in hepatitis induced by endotoxin/galactosamine in rats. *Hepatol.*, **10**, A592.
96. Yonomura, K., Miyahara, S., Takahashi, H. and Kimura, Y. (1983). Regeneration and the immune system (i) and (ii). *Europ. J. Immunol.*, **13**, 787–889.

PART III

13. Liver Regeneration

Antonio Lopez-Pena[1], Bernard de Hemptinne[1] and Bengt Jeppsson[2]

[1]*Bernard de Hemptinne, Dienst voor Levertransplantatie, Digestive Heelkunde, Universitair Ziekenhuis Gent, De Pintelaan 185, B-9000 Gent, Belgium*
[2]*Department of Surgery, Malmö University Hospital, Lund University, S-20502 Malmö, Sweden*

Introduction

The liver has a marked capacity to regenerate after damage, whether infective, toxic or traumatic. In rodents partial hepatectomy (70 %) is the operation most commonly used to stimulate liver regeneration and changes occur very soon afterwards. Within ten minutes of the operation there is a hyperpolarization of the hepatocyte plasma membrane. After 30–60 minutes there is an increased expression of several cell cycle associated genes and within 30 minutes enhanced expression of more than 50 new genes.[1,2] Increase in RNA synthesis can be measured about 6–9 hours after liver resection, while DNA synthesis starts after 12–16 hours and peaks after 22–24 hours. DNA synthesis in non-parenchymal cells, endothelial cells, macrophages (Kupffer cells), lipid storing cells (Ito) and bile duct cells begins on average 24 hours later than in hepatocytes.[3,4]

Cell Cycle–Kinetics of Liver Regeneration

The starting point of hepatoproliferation in the regeneration process is not a fully quiescent situation but a state (G0) in which a very small number of cells enter the cell cycle so as to balance the cell deaths.

When a hepatoproliferogenic stimulus such as partial hepatectomy acts, the G0 phase changes to an S phase, i.e., a DNA replication phase; there is, however, a period of 12–16 hours during which cells are exposed to growth stimulus, the St phase. Usually, the St phase is independent of stimulus strength.

Hepatocytes do not enter the cell cycle at the same time. One should have in mind the concept of a cellular fraction entering the S phase per unit time, which reflects the number of committed cells that begin synthesizing DNA.[5] These entry rates vary according to the intensity of the growth stimulus.

The transition to cell division is a temporal gap called G2, which is normally a 4–6 hour pre-mitotic phase. During regeneration, the inter-cycles temporal gap is known as G1, shorter than the prereplicative phase (6–8 h).

The mitotic (M) phase comes at the very end of the cell cycle, two new hepatocytes ultimately being formed within 30–60 minutes.

The quantitation of cells entering the S phase is expressed by the labelling index (LI), while the labelled mitotic index (LMI) indicates the M-phase fraction that duplicate its DNA. The whole cycle must be seen as proliferative transitions expressed by dynamic fluctuations of both index frequencies.

Proliferative events are also spatially different, for periportal hepatocytes enter the S phase before centrilobular cells. This is a consequence of the structural and functional differences between cells in the portal zone and in the centrilobular area. In fact, the structural differences reflect functional requirements, with a lobular intracellular concentration gradient.

Extracellular matrix plays an important role in determining phenotypic expression of cells. Each molecule has a different effect on cell function.

During regeneration, the matrix produced is that which is needed for the reconstruction of tissue. Under physiological conditions, there is an important structural / functional relationship.

The extracellular matrix is composed of three classes of macromolecules:

a) collagen – type I and II
b) non-collagenous glycoproteins – fibronectin, laminin
c) proteoglycans – heparan sulfate

The synthesis and deposition of matrix components is an active process which requires cell-cell and cell-matrix interactions. Any alteration in the amount, type and/or distribution of matrix components will disturb the interactions and transmit a signal to the cell. The latter responds by modifying its phenotypic expression.

Laboratory Animals

Despite the possibility of using different species in experimental laboratory work, most experiments use male Wistar rats, 6–9 weeks old, weighing 250–350 g. Clearly hepatic liver regeneration proceeds at higher speed in young animals than in older animals.[6] The problem of impaired liver regeneration in the clinical situation after liver surgery or other noxious insults must relate to an elderly patient group. Data on the impact of advancing age on recovery from liver injury are extremely limited. It is therefore important to bare in mind that most experimental data are retrieved from studies in young animals.

The use of male individuals is supported by several facts. First, females are weaker and present a higher mortality rate. Then, male rats are not affected by beta-antagonists in their increase of the activity of thymidine kinase during regeneration, while females are, with subsequent decrease of DNA content in the regenerating liver.[7]

Several functions could be influenced by sex;[8] there is a demasculinization of hepatic oxidative metabolism in regenerating rat liver and estrogens have a possible role in the liver regeneration process.

Laboratory conditions must be given a great importance: number of animals per cage (already regulated in Europe: Council Directive 86/609/EEC), feeding, etc. The room must be temperature and light controlled; a light cycle from 6 a.m. to 6 p.m. is recommended. Food and water supply *ad libitum*, either prior to or after hepatectomy, is advised. The role of feeding could be discussed, nevertheless, because liver energy metabolism is given priority in the regenerating liver, even in malnourished rats.[9] Previous reports, however, showed retarded liver regeneration during malnutrition.[10,11]

Surgical Techniques

It is important that the animals should not undergo stress either before or during surgical intervention. The animals must remain in their cages right up to the time of the operation and be kept in a separate room. Noise and abrupt handling must be avoided. Operations should be performed between 8 a.m. and 11 a.m. to minimize diurnal variation in metabolic activity.[12] The choice of anaesthesia is important since some drugs may influence the regenerative response.[13]

There are two ways of inducing a regenerative response surgically in the liver, e.g. liver resection and portal branch ligation.

Liver Resection

This operation is usually performed as originally described by Higgins & Andersson.[14] The surgical technique itself must be as gentle as possible and the intestines should not be displaced from the peritoneal cavity and must be kept irrigated with saline solution throughout the intervention. Care should be taken when placing the ligature around the lobes to be resected not to cause liver ischemia, portal stasis or outflow occlusion. There must be a sham operated group as a control, because of the impact of portal and bowel manipulation as well as dehydration. The operations must be developed under aseptic conditions.

This procedure is usually well standardized in most laboratories. After a 70 % resection no mortality should be expected. A 90 % liver resection will cause mortality. The size of the resection will influence the regenerative response.[15]

Portal Branch Ligation

Sometimes it may be desireable to induce liver regeneration without inflicting a major tissue damage at the same time. Under an operative microscope it is possible to ligate a portal branch supplying the two anterior lobes by a 7/0 ligature. In order to avoid denervation, special care must be taken to leave all the structures intimately adjacent to the vessels. This procedure will induce a rapid atrophy in the ligated lobes and a regenerative response in the non-ligated lobes with increased DNA synthesis at about the same time as after a corresponding liver resection.[16,17] Portal branch ligation to the two anterior liver lobes results in similar hemodynamic alterations as excision of the same liver mass. Similarly the regenerative response can be graded by ligation of smaller amount of hepatic tissue.

This model seems ideal for study of hepatotrophic factors and also to study liver regeneration in tumour models and cirrhotic models where one wants to avoid a major surgical trauma and tissue loss.

Study Techniques in Liver Regeneration

Macroscopic

Non-invasive methods are important for the sequential study of liver growth after partial hepatectomy in the rat. The use of CT-scans and calculation of liver volumes is reliable and efficient in the study of liver growth in operated animals.[18] The examination can be performed repeatedly in the same animal, which reduces the variants in the determinations.

Nuclear magnetic resonance is also a useful tool for studying hepatic regeneration. NMR is able to detect the pathophysiological changes occurring in liver parenchyma during the regeneration process and to monitor different stages in the hyperplastic process,[19,20] showing an increase in signal intensity, as well as to assess periportal changes and inhomogeneous appearance of the liver.[21]

Liver volume is well assessed by single-photon emission computed tomography (SPECT), which gives the limits of the organ (isocounter) and detects oedema.[22,23]

Injection of radioactive microspheres[24] into the left ventricle is a reference sample technique using a gamma scintillation counter which measures some hepatic circulation parameters like hepatic arterial blood flow and total hepatic blood flow. This method also assesses the porto-systemic shunt. Thus, it measures quantitatively splanchnic haemodynamics after partial hepatectomy (PH). Results indicate that porto-systemic shunt is negligible though portal pressure increases after hepatectomy.

Metabolic

The study of total bile acid concentrations in portal and peripheral blood, as well as calculating the clearance index, seems useful for evaluating hepatic regeneration after PH, because of the correlation of the clearance index with the rate of regeneration as well as with total bile acid (TBA) concentration in peripheral blood.[25]

Hepatocellular proliferation gives an obvious increase of serum TSA and biliary tauro-beta-muricholic acid (TbetaMCA); these substances can be measured by high performance liquid chromatography (HPLC), a useful indicator of hepatic regeneration.[26]

Another marker for hepatic regeneration can be the plasma fibronectin, which shows a response after partial hepatectomy.[27]

Circulation

Partial hepatectomy involves a decrease of hepatic blood flow depending on the extent of hepatectomy. Immediately after PH the following effects can be seen:

- decrease of hepatic arterial blood flow;
- increase of portal venous inflow, with concomitant increase of portal venous pressure;
- increase of total hepatic blood flow[24]

Post-sinusoidal resistance also increases proportionally to the percentage of PH.

Nevertheless, the hepatic blood flow per unit of weight increases after PH and there is an increase of hepatic oxygen extraction ratio as well.[28]

Liver blood flow changes are not constant, since the flow decreases below baseline the first day, showing an increase at day 4 and returning to baseline at day 7. The changes could be related to liver regeneration stimulus as the functional demand increases after hepatectomy.[29]

At all events, it must be remembered that portal blood flow is more important for liver regeneration than arterial flow.[30]

Another interesting point is that remaining lobes must have intact draining veins to share liver function and show regeneration after PH.[31]

Regeneration Rate

Regeneration rate can be measured with several methods. The most common is to study DNA synthesis be single injection of ^{3}H-labelled thymidine at different times after liver resection. DNA synthesis can then be observed during the short period between injection and sacrifice.[11] Usually a dose of 0.5 μCi per gram of body weight is used. A too high dose of radioactivity should not be used since that may by itself induce cellular changes.[32] A continuous infusion of tritiated thymidine can also be used.[33] The advantage of a continuous infusion is that cell replication can be measured after a longer time, giving a summation of regeneration during that time. Furthermore, as more cells are labelled, the results will be more reliable. The use of 5-bromo-2-deoxouridine (BUdR), another precursor can also be used and the two methods are probably equally efficient.

The radioactivity can be analyzed after homogenization of liver tissue, extraction of DNA and a measurement of radioactivity in a liquid scintillation counter. It can also be preformed by autoradiography. Autoradiography permits selective evaluation of regeneration in hepatocytes and other cells, while scintigraphic counting measures a replication rate in all liver cells. On histological examination one can also measure mitotic index or labelling index.

Growth Factors

Hepatocyte proliferation, which will eventually lead to organ regeneration, possesses its own kinetics influenced by growth factors and some other substances.

Over the last decade, the early scheme made of a few factors[1] has been enriched by much experience, the results of which cast new light on the subject.

The current state of the art (Fig. 13.1), far more complex, can be considered neither as something

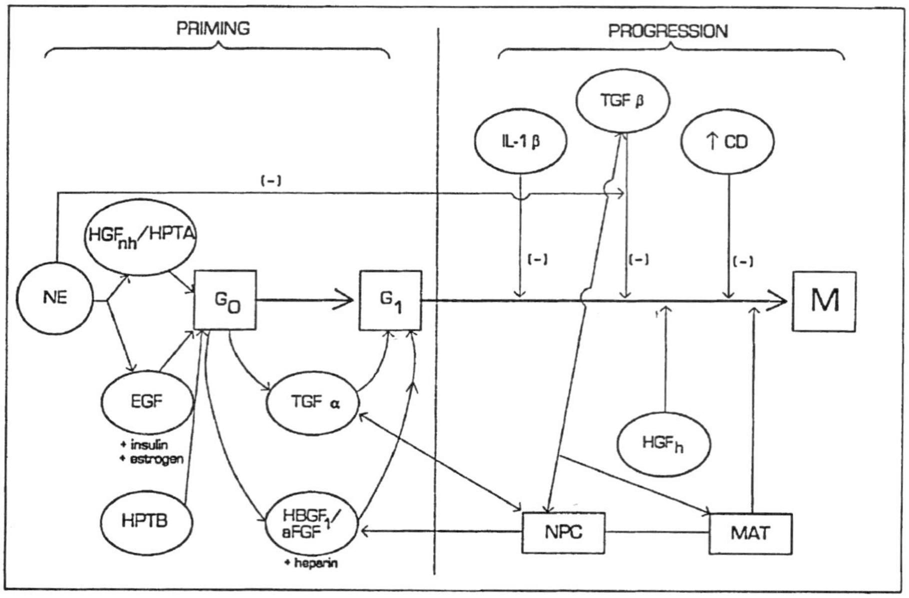

NE	norepinephrine	IL-1β	Interleukin 1β
HGF_{nh}/HPTA	(non-hepatocytic) Hepatocyte Growth Factor/Hepatopoietin A	TGFβ	Transforming Growth Factor β
		↑CD	(Increase) Cell Density
EGF	Epidermal Growth Factor	HPI	Hepatocyte Proliferation Inhibitor
HPTB	Hepatopoietin B	HGF_h	(hepatocytic) Hepatocyte Growth Factor
TGF_α	Transforming Growth Factor α	NPC	Non-Parenchymal Cells
$HBGF_1$/aFGF	Heparin Binding Growth Factor/acidic Fibroblast Growth Factor	MAT	Matrix
		(-)	Inhibitory mechanism

Figure 13.1.

fixed nor as a full explanation of the liver regeneration phenomenon. Nevertheless, some substances and interactions are now known which allow the hepatocyte to follow the cell cycle until proliferation. The same factors which control and maintain the hepatic mass under physiological conditions must synchronize the response of liver regeneration and its end point. During the process, cells respond to intra- and extrahepatic signals and may synthesize growth factors.[34]

Transcripts of several protooncogenes may be found during the process which can explain or highlight the regulatory mechanism of liver regeneration. This protooncogene expression is specific, sequential and transient.[35]

Several ways by which these factors work could be identified: trigger signal(s), complete mitogens, comitogens, inhibitors.

It is also evident that there is a substantial difference between the results of the *in vivo* and the *in vitro* studies. Hepatocytes *in vivo* are part of a complex system in which they interact with other parenchymal and nonparenchymal cells and, at the same time, are under the influence of different humoral factors needed for the activity of growth factors (heparin, insulin, estrogens) as well as that of the different components of the extracellular matrix.[36]

On the other hand, the study of these phenomena *in vitro* allows us to reach conclusions regarding an individual function, as they isolate the cell from the wide range of factors that exist *in vivo*. However, these conclusions must be interpreted very carefully when extrapolated to a functional system in animal or man.

In fact, there are several factors that act either *in vivo* or *in vitro*, or show a different action in each case.

Cell cultures allow us to assess the importance of cell density as a modulator of hepatic regeneration[36–38] and to study cell biology phenomena such as the function of the gap junctions, its opening and closing mechanism, its influence on normal and tumoral regenerative growth, the synchronization of endocrine functions, and so forth.

Hepatic growth factors are part of a system, so their production is influenced by homeostatically regulated autocrine, paracrine and endocrine circuits. Their action will depend, sometimes, on their mutual equilibrium, so that they can modulate the hepatic regenerative growth (TGFalpha & TGFbeta). There are organ-specific factors like HGF/HPTA, and general stimulatory growth factors like EGF.[39,40] According to their source of production, the factors can originate in the hepatocyte itself (hepatocytic HGF), in NPC (HSS), in the extracellular matrix or in other organs or cells (pancreas, platelets).

Depending on the mechanism of action, we currently know some enhancers like EGF, HGF (HPTA), HPTB and HBGFs, and some inhibitors like IL-1 beta and TGF-beta.[41] Besides, norepinephrine (NE) and other hormones can act as triggers, despite not being mitogenic.

In order of importance, HGF comes first, for HBGFs need heparin and have less activity; EGF enhanced by NE would be the earlier factor and alpha-adrenergic receptors act as a protooncogene. Hepatic HGF, i.e., from Kupffer cells, does not control liver regeneration after PH. HGF synthesized in the remnant liver could sustain the subsequent waves of hepatocyte proliferation.

NE and HGF[42] amply precede the initiation of DNA synthesis; they could constitute the natural signal for the first round of hepatocyte proliferation.

The action of all these factors could constitute the first step in the process of liver regeneration, as it can be artificially divided into two separate parts, competence (or priming) and progression (Fig. 13.1). The first part, competence, encompasses the action of several factors that take the hepatocytes out of the quiescent state and make them able to follow the cell cycle, while other mechanisms act in the further stage of progression, leading to DNA synthesis and hepatocyte proliferation. This stage seems to be controlled within the liver, through positive and negative loops.[34]

There is an acceptable distinction between agents that can induce replicative competence, progression or both. Liver cell replication can be initiated by a number of different agents, but a second set of agents is required to move the cells into DNA synthesis.[43]

In the regenerating liver, TGF alpha probably functions as an important physiological growth factor that contributes to DNA synthesis. It has the same receptor as EGF, but TGF alpha is made by hepatocytes, thus acting in an autocrine manner.[43]

Protooncogenes c-fos, c-jun and c-myc have an important expression soon after partial hepatectomy and, at the same time, high levels of TGF alpha mRNA can be found. Has TGF alpha any relationship with increased expression of these protooncogenes? What initiates TGF alpha synthesis?. Koch found that TGF alpha stimulates c-fos and c-jun expression.[35] Nevertheless, TGF alpha does not appear to initiate the regenerative response, but comes into play later, probably when hepatocytes are in the G1 phase.

Gene products can be considered as specialized regulatory proteins that stimulate (or attenuate) the mitogenic promotor sequence.

There could be two principal signalling events sequentially triggered by different sets of mitogens. Signal I would be ionic fluxes like Na^+ influx and signal II would be phosphorylations.[35].

Na^+
↓
Go.......................→G1.............................→M
↑ ↑
c-fos PKA-dependent phosphorylation
c-jun

Thus, the complete chain would be:
"priming" initiated by (e.g.) NE----- → transcripts of protooncogenes------- → specific action of

/TGF alpha as a positive effector
/TGF beta as an inhibitor

DNA synthesis in hepatocytes is also stimulated and modulated by cell-cell interaction between parenchymal and nonparenchymal liver cells. The stimulation is due to a proteinaceous material, acting even in the absence of growth factors, the so-called Hepatic Stimulatory Substance.[44] There might be a difference in action from different cells, epithelial or endothelial.[45,46]

TGF beta, produced by Kupffer cells, is an inhibitor, a negative regulator[3,43] of liver regeneration, via a paracrine mechanism. Its production could be balanced with that of the TGF alpha by changing proportions.[47] It starts early (4 h) after partial hepatectomy and peaks at 72 h. But how do hepatocytes manage to proliferate in the presence of an elevated TGF beta production? Is there, as Michalopoulos postulates,[3] a resistance to TGF beta, or are the hepatocytes programmed to stop after two rounds of cell proliferation? Another possibility is that production of TGF beta is due to the needs for NPC replication. The exposure of hepatocytes to such a factor could lead to a two-stage resistance to TGF beta. It must not be forgotten that TGF beta has also an inactive or "latent" form.[36] By which mechanism this form is activated is not known. Moreover, DNA synthesis inhibition is also a consequence of TGF beta effects on other functions like composition and abundance of extracellular matrix and hepatocyte attachment to the matrix.[3] TGF beta might also be a source of questions as it plays a dual regulatory role during liver regeneration. It inhibits the reinitiation of hepatocyte DNA synthesis stimulated by TGF alpha but, by itself, stimulates DNA synthesis in the St (prereplicative) phase *in vitro*.[35] At least, that is what is indicated by the fact that TGF beta mRNA levels increase within 4 hours after partial hepatectomy.

Interleukin 1 beta (IL-1beta) from Kupffer cells also plays an important role as a growth inhibitor. It may act at two different stages in the G1 phase.[47]

Although the preceding pages may be a fairly close description of what actually happens and the ideas expressed may be not very far from the real situation, it should be remembered that the whole process of liver regeneration is complicated by other still poorly understood factors, especially in vivo.

The importance of the microcirculation must not be forgotten; the immunological processes as well play a role and so does the integrity of the hepatic innervation.[48–50] Last but not least, we would underline as a matter for further study the role of the hepatic ischemia accompanying surgical procedures, which is without doubt another factor in the process.

References

1. Michalopoulos, G. K. (1992). Liver regeneration and growth factors: old puzzles and new perspectives. *Lab Invest*, **67**, 413–415.
2. DuBois, R. D. (1990). Early changes in gene expression during liver regeneration: what do they mean? (editorial) *Hepatology*, **11**, 1079–1082.
3. Michalopoulos, G. K. (1990). Liver regeneration: molecular mechanisms of growth control. *FASEB J.* **4**, 176–187.
4. LaBrecque, D. R. (1992). Liver regeneration: a picture emerges from the puzzle. *Cancer Res*, **52**, 3598–603.
5. Leffert, H. L., Koch, K. S., Lad, P. J., Shapiro, I. P., Skelly, H., d'Hemptinne, B. (1988). Hepatocyte regeneration, replication and differentiation. In: *The liver; biology and pathology*. Aris, I. M., Jacoby, W. B., Popper, H. *et al.* (eds). Raven Press, New York, 833–855.
6. Beyer, H. S., Sherman, R., Zieve, L. (1990). Aging is associated with reduced liver regeneration and diminished thymidine kinase mRNA content and enzyme activity in the rat. *J. Lab. Clin. Med.*, **117**, 101–108.
7. Tsukamoto, I., Kojo, S. (1990). The sex difference in the regulation of liver regeneration. *Biochim Biophys Acta*, **1033**, 287–290.
8. Liddle, C. (1989). Effect of liver regeneration on cytochrome P450 isoenzymes and serum sex steroids. *Gastroenterology*, **96**, 864–872.
9. Skullman, S., Ihse, I., Larsson, J. (1991). Availability of energy substrates during liver regeneration in malnourished rats. *Scand. J. Gastroenterol*, **26**, 1152–1156.
10. Ngala-Kenda, J. F., de Hemptinne, B. and Lambotte, L. (1984). Role of metabolic overload in the initiation of DNA synthesis following partial hepatectomy in the rat. *Eur Surg Res*, **16**, 294–302.
11. Himeno, Y., Engelman, R. W. and Good, R. A. (1992). Influence of calorie restriction on oncogene expression and DNA synthesis during liver regeneration. *Proc. Natl. Acad. Sci.*, **89**, 5497–5501.
12. Lindroos, P. M. and Michalopoulos, G. K. (1991). Hepatocyte Growth Factor (Hepatopoietin A) rapidly increases in plasma before DNA synthesis and liver regeneration stimulated by partial hepatectomy and carbon tetrachloride administration. *Hepatology*, **13**, 743–749.
13. Weinbren, K., Washington, S. L. A. and Smith, C. Y. (1975). The response of the rat liver to alterations in total portal blood flow. *Br. J. exp. Path.*, **56**, 148–157.
14. Higgins, G. M. and Anderson, R. M. (1931). Experimental pathology of the liver. *Arch. Pathol.*, **12**, 186–202.
15. Bucher, N. L. R., Schrock, T. R. and Moolten, F. L. (1969). An experimental view of hepatic regeneration. *Hopkins. Med. J.*, **125**, 250–257.
16. Rozga, J., Jeppsson, B. and Bengmark, S. (1986). Portal branch ligation in the rat. Reevaluation of a model. *Am. J. Pathol.*, **125**, 300–308.
17. Rozga, J., Jeppsson, B. and Bengmark, S. (1985). Hepatotrophic factors in liver growth and atrophy. *Br. J. exp. Pathol.*, **66**, 669–678.
18. Foss, A., Hochbergs, P., Ahren, B. and Bengmark, S. (1991). Computed tomography; An in vivo model for the study of liver regeneration in the rat. *Surg. Res. Comm.*, **11**, 165–168.
19. Campbell, K. A. *et al.* (1990 Sep.). *In vivo* 31P NMR spectroscopic changes during liver regeneration. *J. Surg. Res.*, **49**(3), 244–7.
20 Patrizio, G. *et al*, (1990 May). An animal model for the study of liver regeneration by NMR. *Radiol. Med. Torino.*, **79**(5), 453–7.
21. Arrive, L. *et al.*, (1989 Jun.). M. R. appearance of the liver after partial hepatectomy. *Am. J. Radiol.*, **152**(6), 1215–20.
22. Jansen, P. L. *et al.*, (1990 Feb.). Liver regeneration and restoration of liver function after partial hepatectomy in patients with liver tumours. *Scand. J. Gastroenterol.*, **25**(2), 112–8.
23. Faurous, P. *et al.*, (1989). Estimation of hepatic volume by SPECT during LR: experimental studies. *Int. J. Rad. Appl. Instrum. B.*, **16**(4), 375-9.
24. Lin, P. W. (1990 Mar.). Hemodynamic changes after hepatectomy in rats studied with radioactive microspheres. Taiwan-I-Hsuhe-Hui-Tsa-Chih., **89**(3), 177–81.
25. Furui, J. (1990 Mar.). Experimental study of changes in total bile acid concentrations in the peripheral and portal blood after partial hepatectomy. Nippon-Shakakibyo-Gakkai-Zasshi. **87**, 799–810.
26. Tsuda, H. *et al.*, (1990 Feb.). Comparative sequential changes in serum and biliary levels of bile acid components after a single dose of D-galactosamine or partial hepatectomy in the rat. *Liver*, **10**, 28–34.
27. Kwon, A.-H., Inada, Y., Uetsuji, S., Yamamura, M., Hioki, K. and Yamamoto, M. (1990). Response of fibronectin to liver regeneration after partial hepatectomy. *Hepatology*, **11**, 593–8.
28. Asahi, K. *et al.*, (1989 Feb.). Experimental study of hepatic hemodynamics and hepatic oxygen consumption after hepatectomy. Nippon-Geka-Gakkai-Zasshi **90**(2), 228–42.
29. Hanna, S. S. *et al.*, (1988 Sep.). Liver blood flow after major hepatic resection. *Can. J. Surg.*, **31**(5), 363–7.
30. Kuwano, H. *et al.*, (1988 Jul–Sep.). Healing of the resected liver: its process and the significance of the portal blood flow. *Int. Surg.*, **73**(3), 157–60.

31. Kitazawa, T. *et al.*, (1988 Jun.). Clinical and experimental studies on ligation of the draining vein in the remnant liver during hepatectomy. Nippon-Geka-Gakkai-Zasshi **89**(6), 863–70.
32. Post, J. and Hoffman, J. (1965). Dose-dependent effects of H3TDR as a DNA label on the replication of liver cells in the growing rat. *Rad. Res.*, **26**, 422–441.
33. Le Pecq, J. B. and Paoletti, C. (1966). A new fluorometric method for RNA and DNA determination. *Analyt. Biochem.*, **17**, 100–107.
34. Fausto, N. (1989). Regulation of liver growth: protooncogenes and Transforming Growth Factors. *Laboratory Invest.*, **60**(1), 4–13.
35. Koch, K. S. and Leffert, H. L. (1990). Mitogens and hepatocyte growth control *in vivo* and *in vitro*. In *Vitro Cell*, **26**, 1011–1023.
36. Ichihara, A. (1991). Mechanisms controlling growth of hepatocytes in primary culture. *Dig. Dis. and Sci.*, **36**(4), 489–493.
37. Vintermyr, O. K. (1989). Characterization of the inhibitory effect of glucocorticoids on the DNA replication of adult rat hepatocytes growing at various cell densities. *J. of Cellular Physiol*, **138**, 29–37.
38. Tomomura, A. (1987). The control of DNA synthesis in primary cultures of hepatocytes from adult and young rats: Interactions of extracellular matrix components, Epidermal Growth Factor and the cell cycle. *J. of Cellular Physiol.*, **130**, 221–227.
39. LaBrecque, D. R. (1992). Hepatocyte growth factor - how do I know thee? Let me count the ways. *Gastroenterology*, **103**, 1686–91.
40. Wu, J. C., Merlino, G. and Fausto, N. (1994). Establishment and charactherization of differentiated, nontransformed hepatocyte cell lines derived from mice transgenic for transforming growth factor a. *Proc. Natl. Acad. Sci.*, **91**, 674–8.
41. Thoreson, G. H., Refsnes, M. and Christofferson, T. (1992). Inhibition of hepatocyte DNA synthesis by transforming growth factor B, and cyclic AMP: Effect immediately before the b,/s border. *Cancer Res* **52**, 3598–603.
42. Fausto, N. (1991 May). Protooncogenes and growth factors associated with normal and abnormal liver growth. *Dig. Dis. and Sci.*, **36**(5), 653–658.
43. LaBrecque, D. R. (1991 May) Hepatic Stimulator Substance. Discovery, characteristic and mechanism of action. *Dig. Dis. and Sci.*, **36**(5), 669–673.
44. Yamamoto, N. (1989). Growth stimulation of adult rat hepatocytes in a primary culture by soluble factor(s) secreted from nonparenchymal liver cell. *Cell Structure and Function*, **14**, 217–229.
45. Katsumoto, F. and Nakayama, F. (1989). Stimulation of DNA synthesis in hepatocytes by Kupffer cells after partial hepatectomy. *Hepatology*, **9**(3), 405–410.
46. Meyer, D. H. (1991 May). Bidirectional effects of Kupffer cells on hepatocyte proliferation *in vitro*. *FEBS* **283**(1), 150–154.
47. Cruise, J. L. and Michalopoulos, G. K. (1988). Early events in the regulation of hepatocyte DNA synthesis: the role of the alfa adrenergic stimulation. *Scand J. Gastroenterol*, **23**, (suppl 151, p. 19–30.
48. Carobi, C. (1990 Jan 15). Vagal afferent innervation in regenerated rat liver. *Experientia*, **46**(1), 50–3.
49. Dubovaia, T. K., Eletskaia, IuK, and Dol'nikova, I. B. (1990 Jul.). Characteristics of liver regeneration after vagotomy. *Arkh. Anal. Gistol. Embriol.*, **99**(7), 87–91.
50. Urakawa, T. *et al.*, (1988 Aug.). Experimental study of vagotomy for prevention of gastric ulcer after hepatectomy of cirrhotic livers. Its influence on hepatic regeneration. *Scand J. Gastroenterol*, **23**(6), 705–11.

PART III

14. Methods for Studying Physiology of Bile

G. Ussia[1], R. Aldini[2] and M. Montagnani[3]

[1]*Dipartimento di Discipline Chirurgiche Anestesiologiche e dei Trapianti "A Valsalva"-Università di Bologna;*
[2]*Istituto di Scienze Cliniche - Università di Bologna;*
[3]*Dipartimento di Medicina Interna e Gastroenterologia - Università di Bologna*

Introduction

The various steps of the entero-hepatic circulation (EHC) of bile acids (BA) have been studied in the experimental animal and in man, by means of surgical experimental models. The animal models are suitable for getting such information, as they cannot be obtained from man.[1]

A major drawback of the animal model approach is the different composition of bile acids amongst animals and between animals and man: in man chenodeoxycholic is the predominant bile acid, cholic and deoxycholic being present to a lesser extent and all bile acids being mostly conjugated with glycine;[2-4] on the contrary in the rat, which is the most used species, taurocholic acid is the predominant bile acid, although all the bile acids are represented. In the rabbit, due to its particular intestinal metabolism, deoxycholate, glycine conjugated, is the major constituent of bile;[5] the animal biliary bile acid pattern which more closely resembles the human one is represented into the baboon bile.

Moreover, it appears that in the evolution of species, a number of biochemical modifications of the bile acid molecule have occurred.

The result was the formation of molecules that are so hydrophilic that they undergo little passive absorption in the biliary tree and small intestine.[6]

In general, dihydroxy bile acids are readily absorbed passively, whereas trihydroxy BA are not. Conjugation influences the physicochemical and biological properties of bile acids, since it make them much stronger BA, which, in turn, decreases their tendency to precipitate under acidic conditions.[7] The ideal animal for studying the bile acids EHC should present a biliary bile acid pattern similar to man, low cost and ease of housing; moreover, physiological studies require many experiments in order to have reproducible and reliable results.[8]

The baboon, which presents a bile acid pattern similar to that of man,[6] is highly expensive, difficult to manage and of little utility when many experiments must be performed in a non surgical department; the rat, although with a biliary bile acid pattern a little different from that of man, is most widely used, due to the feasibility of many experimental manipulations in many laboratories, without needing a great number of facilities.

The dog, rabbit and guinea-pig[9-12] have also been used, depending on: the aim of the investigation, their availability and the surgical procedure.

Methods for Studying Bile Secretion

The traditional methods for studying bile secretion are to fashion an acute or chronic fistula in an experimental model or to perfuse the isolated liver; such methods, although quite primitive in principle, continue to be used for lack of better methods.

The problem with these methods is not what is recovered in bile, is the algebraic sum of both canalicular and ductular events. In animal samples, bile is sampled at the level of the common hepatic duct or common bile duct (after cholecystectomy, in the species which have gall bladders). A variety of techniques have been described for exteriorizing the entero hepatic circulation in chronic animal preparations and stream-splitting devices that permit a controlled interruption of the EHC have been used in primates.[1]

For experiments aimed at gaining insight into bile flow or composition, it is desirable to know the composition of the sinusoidal plasma. This is achieved by means of the perfused liver.[9,13,14] The only possible limitation to this technique is the fact that the blood enters the liver only by the portal vein. Thus, it is originally considered that in the perfused liver there is no ductular function. This seems not to be the case, and ductular function seems preserved. This technique is commonly performed in the rat liver, but it has also been performed in other animals.

The dog has been perfused in vivo,[9] provided a donor liver has been obtained. The liver may be used with a continuous recycling perfusate or a single pass perfusion may be performed.[15–18] Moreover, liver perfusion is carried out in a retrograde manner, by attaching the entry cannula to the central vein and the exit cannula to the portal vein.[19,20]

With retrograde perfusion, the pericentral cells are exposed to the highest concentration of solute in the perfusate, in contrast to the prograde perfusion, in which periportal cells are exposed to the highest concentrations of solute.

Experimental Models for Studying the EHC in the Primate

Studies on chronic and acute interruption of the EHC were performed by Dowling in 1968.[1] These are basically studies which lead to the concept that the interruption of the EHC is the pathophysio-logical step of the bile acid malabsorbtion and gave the basic understanding of the interrelationship between BA pool, synthesis rate and recycling frequency, i.e. gave more insight into the dynamics of the EHC of bile acids.

Either monkeys or baboons may be studied in this way.[21–23] Monkeys are smaller, cost less and are easier to manage. However, their bile contains relatively less cholesterol than humans,[1] and, therefore, they do not develop cholesterol gallstones as humans do.

Baboons, which present a human bile and composition similar to man, and possibly develop cholesterol gallstones,[24,25] are more difficult to study by this method. Monkeys are therefore more easily used.

After anesthesia, a cannula is placed in the common bile duct and in the duodenum.

All of the bile secreted asses through an electronic drop counter; one drop every so many drops is collected; the remainder is returned to the animal duodenum. Thus, from the volume and the concentration of the total bile acids secreted, it is possible to calculate the amount of bile acid secreted by the following formula:

$$S\ (\text{secretion}) = \text{Volume} \times BA$$

In this way it is possible to extrapolate to the total amount of BA returned from the intestine, by the total amount secreted minus the amount collected. The fecal loss is also measured.[26–28]

Synthesis of newly synthesized bile acids is also measured, as it is equal to the daily loss. Bile acid pool is also measured, by creating a biliary fistula and draining all the bile acids present in the entero hepatic circulation. This model has allowed several researchers[29–32] to study the EHC in the normal state and

- in controlled portal bile diversion
- biliary fistula
- different diets
- ileal resection
- biliary obstruction.

Bile flow and bile secretion rates have been also studied by means of this model.[33–38]
The dynamics of the EHC of bile acids is therefore evaluated by the following final formulae:

$$S = BA \times V$$

$$I = S - X$$

$$A = I - F$$

$$T = F \times X$$

$$C = \frac{S}{P}$$

Where S = secretion rate; V = volume of the bile; I = BA returned from the intestine; X = collected bile acids; A = bile acid reabsorbed; F = fecal loss of BA; T = BA synthesis; P = pool; C = number of circulations per day.

Biliary Diversion Studies Investigating the Fate of A Known BA

Most biliary diversion studies are performed in the rat or hamster.[39,40] This can be as follows: 1) intravenous infusion of a bile acid; 2) intraduodenal infusion of a bile acid and its influence on choleresis and biliary lipid outputs; 3) perfusion of the distal ileum one bile acid. In all these experiments bile is collected via a bile fistula.

Rats or hamsters are usually kept fasting over twenty four hours prior to the experiments. Anesthesia is usually maintained under either or 20 mg Ketamine and 1 mg diazepam. A polyethylene tubing (Portex PP10, ID 0.28 cm) is positioned in the bile duct and another polyethylene tubing is inserted into a tail vein, after laparotomy. Bile is usually collected in hourly fractions from the start of the infusion to eighteen hours afterward.

Urine and feces may be similarly collected. Anesthesia is finally maintained throughout the study by continuous intravenous infusion of Ketamine (3mg/hour).

Intraduodenal infusion: a PVC tubing (Dural Plastics and Engineering, ID 0.5 mm) is inserted through a cut in the pyloric antrum and the tip placed at the level of the second portion of the duodenum.

For distal ileum perfusion experiments rats are operated on under ether anesthesia. Body temperature is kept at 37.5 °C. by a heating pad and a rectal thermometer. A 10 cm section of the distal ileum, approximately 5 cm proximal to the ileo-cecal valve is localized and the section is ligated and cut off, without damaging the blood vessels, leaving open ends on the 10 cm intestinal segment. A silicon rubber tubing (LKB, ID 3.0 mm) with a glass tip, is inserted in each end of the ileal section, and subsequently rinsed. In order to inspect the circulation of the perfusate and facilitate sample collection, a drop chamber may be perfusate and facilitate sample collection, a drop chamber may be included in the system. A peristaltic pump (LKB 2132 Microperpex) is set at 1.0 ml/mm and connected to the system; the temperature of the perfusate is kept at 37.5°C.

By these means, it is possible to investigate the effect of either intravenous infusion or duodenal or ileal infusion of BA on

a) bile lipid secretion
b) choleresis
c) hepatic biotransformation of bile acids
d) hepatic cholesterol metabolism.

By either the duodenal or ileal infusion of BA, it is possible to differentiate between a) passive b) active absorption of bile acids.

These experiments allow the evaluation of the influence of bile acid structure on bile flow and biliary lipid secretion: since new bile acids and bile acid analogs, with modified steroid ring or shortened side chain are now being synthesized, they are currently under study as they are por their biological properties: many of these studies are performed in the hamster, since its biliary bile acid composition resembles that of humans in the steroid structure and, as in man, the hamster conjugates bile acids both with glycine and taurine. Further, in contrast to rats, but similar to humans, the hamster does not produce 6-hydroxylate chenodeoxycholate or ursodeoxycholate appreciably, during hepatic transport.[48–54]

Experimental Systems and Methods for the Study of Canalicular Bile Flow

Canalicular bile flow may be estimated by measuring the biliary clearance of non metabolized solutes that enter the canalicular bile by passive processes[55] and are neither secreted nor reabsorbed by the ductules or the ducts. This method is similar to glomerular filtration studies by urinary clearance of substances that do not cross the renal tubular epithelium. The theoretical basis for this process has been discussed by Forker[56–59] and Wheeler.[60]

The biliary clearance C is calculated as $C = F \times (B)/(P)$, where F = bile flow and B and P are the biliary and plasma concentration of the solute.

The technique implies that the selected solute is unable to cross the ductular epithelium and that its permeability in the bile canaculi permits the diffusion.

Erythritol is used in hamsters,[61] dogs,[60] rats,[61] rabbits[62] and monkeys.[63] Mannitol equilibrates in the canalicular bile of dogs,[60] but not of hamsters[61] or rabbits.[62] In the guinea-pigs[56] the passage of these substances in the canaliculus is restricted. These observed species differences are due to the different molecular weight (MW) of these solutes, (122 MW for Erythritol and 182 MW for Mannitol).

The systems which permit these determinations are both biliary fistula and isolated and perfused livers.

Solutes such as sucrose and insulin also enter the canicular bile, but their bile to plasma ratio is such that it does not permit estimation of canicular bile flow; therefore, although they cannot be recommended in the evaluation of canicular bile flow, they provide a qualitative estimation of changes in the permeability of the biliary system.[64]

Methods for the Study of the Hepatic Uptake and Transport of the Bile Acids: the Perfused Liver

Liver perfusion in our laboratory[15–18] is performed following Martimore.[65] Male Sprague-Dawley rats (200D 250 g) are used as liver donors.

The perfusate consists of Krebs-Ringer bicarbonate, pH 7.4, containing glucose (100 mg/ml) and bovine serum albumin (3% or o.75%, w/v) (Fraction V, essentially fatty acid free, Sigma, St. Louis, Missouri, USA 6003). The perfusion is established without using any anticoagulant and the liver is washed with 40 ml of Krebs-Ringer bicarbonate solution. Complete and even blanching of all liver lobes during this washout period indicates satisfactory perfusion. Following the wash, the liver is perfused for thirty minutes with a recycling system containing either 3% or 0.75% albumin before starting the perfusion studies. Oxygenation with 95% O_2 and 5% CO_2 is carried out as described by Hamilton *et al.*,[66] using silastic tubing (Dow-Corning, Midland, Michigan, USA). Perfusate temperature and pH is monitored continuously.

Viability of the livers is assessed by light microscopy, bile output, and oxygen consumption. The outflow is collected at ten second intervals.

The liver perfusion surgical technique is hereafter detailed most of these studies have been performed in the rat, which is the classical model of the either in situ or isolated perfused organ. The surgical technique is as follows: under either ether of Ketamine (20 mg) anesthesia, a laparotomy is performed and the bile duct is cannulated with a PE10 tubing. The portal vein is then cannulated and perfusion immediately started at a mean flow rate of 10 ml/min of Krebs-Ringer bicarbonate solution containing albumin 4.0% (w/v) and gassed with 95% O_2 and 5% CO_2. Immediately after, the inferior vena cava is tied immediately above the diaphragm and cannulated and the liver is perfused from the portal vein to the inferior vena cava, throughout the parenchyma. The pH must be kept between 7.35 and 7.45 by adding small amounts of 1 M Na_2CO_2 to the perfusate.

Red blood cells must be added to the perfusate, to a final hematocrit of 35%, in order to allow a satisfactory oxygenation of the liver. Otherwise, a fluorocarbon-containing balanced electrolyte solution (Fluosol 43), supplemented with 0.5% (w/v) bovine serum albumin may be used. If the experimental conditions do not allow either of these oxygen-carrying substances to be wed, the flow rate may be increased three to four fold, by the insertion of two parallel peristaltic pumps before and behind the liver.

Oxygen consumption must be kept higher than 2.8 + 0.4 micromol/min/g liver. The system may be both recirculating (through a reservoir, by a peristaltic pump) or not recirculating. The perfusion may also be retrograde. The perfused liver allows the evaluation of:

a) liver uptake of bile acids, with
   1) single pass experiments,
   2) dose-response curve experiments
   3) competition experiments
b) secretion studies of BA and evaluation of their choleretic properties
c) effect of different BA on bile lipid composition (i.e. possible effectiveness of different BA in desaturating bile in cholesterol)
d) metabolic studies.

The classical animal, as stated above, is the rat, but studies from our laboratory showed a species difference in the hepatic uptake of BA, possibly driven by the different biliary BA composition.[15] This must be taken into account for future metabolic and physiological comparative studies.
The liver uptake of bile acids may be studied by means of both a recirculating[67] and non recirculating system.[15,16,18,68]

In a recirculating system, the bile acid is dissolved in the medium up to a known final concentration, then is allowed to recirculate throughout the liver. In this way, it is possible to evaluate a disappearance constant for each bile acid studied and to evaluate possible differences in the rate of uptake of bile acids y the liver.

A different approach to the evaluation of liver uptake is by the pulse injection of a bile acid into the liver in a non recirculating system. In the latter case, a dose-response curve may be made up. The kinetics of the uptake are evaluated by means of the relationship between dose and removal rate of the substance, according to the model of liver uptake process proposed by Goresky.[69,70] The differentiation into extravascular and intravascular spaces of the sinusoidal space may or may not be considered.[68,69] To evaluate the removal rate of the bile acid, the basic procedure is that described by Reichen and Paumgartner.[71]

The maximal uptake rate (V max) and the half saturation constant (Km) are thus evaluated according to Michaelis-Menten kinetics by linear regression analysis of the plot of the bile acid removal rate versus the ratio of BA removal rate to its dose (Eadie-Hofstee plot).[72]

The transformation of the Michealis-Menten equation is chosen because of the uniform distribution of the data in the plot and because it usually magnifies departures of data from linearity typical of kinetics more complex than those of Michaelis-Menten.

Chemical Structure of Bile Acids and Hepatobiliary Clearance

The better parameter, up until now, seems to be the hydrophilic hydrophobic balance of the substance.[18] This is also true for a series of organic cations, as has been also demonstrated in a systematic study with organic anions ranging from 74 MW to 315 MW.[73,74] The best suitable model for the study of these relationship is represented by the perfused liver, the animal of choice being the rat.

Acinar Heterogeneity in Transport

Acinar zones of the liver are unequal in transport function.[75–78] The dynamic situation may be due either to their localization in the blood stream along the sinusoidal axis or to intrinsic differences between the cells, possibly secondary to concentration gradients of CO_2 and O_2: again, a very adequate model for the purpose is the perfused liver: the isolated organ may be perfused in the normal direction or in a retrograde manner; with short term perfusion, with high extraction types of drugs, such as bile acids, at low protein medium concentrations, zone 1 and zone 3 can be loaded.

Subsequently when the normal flux is restored, the efflux to plasma and bile is monitored. The acinar distribution of taurocholate is depicted by a clear zone 1 localization.

When the perfusion is reversed, taurocholate is predominant in zone 3; this is a typical matter of presentation.

However, transport of bile is considerably slower from zone 3 and this indicates an intrinsic cellular difference in excretion efficiency for bile acid, which may even lead to intrahepatic redistribution of drugs.

The heterogeneity of liver parenchima with respect to many metabolic functions has also been investigated by the perfused liver; their preparation, in fact, permits the investigation of the tight coupling between the biosynthesis of cholesterol and its biotransformation into bile acids, or between the respective rate-limiting enzyme activities. The opposite acinar localization of the rate-limiting enzymes of the cholesterol synthesis and bile acids synthesis is best studied in the rat liver, where all bile acids are formed.

Studies of the Intestinal Absorption of Bile Acid: Intestinal Perfusion Studies

Many studies of the intestinal absorption of bile acids have been so far performed by means of intraduodenal or intestinal infusion of BA[79,80] and their recovery in bile, based on the assumption that the rate-limiting step in the overall bile acid transport is the intestinal one, the liver by far exceeding the intestinal transport capacity.[81]

Recent understandings[74] of this problem suggested that the best suitable manner for studying the intestinal transport may be obtained by direct evaluation of the intestinal transport, after direct measurement of the absorbed bile acid in the portal blood (82).

In our laboratory we have developed an experimental model in the rabbit[82] which allows: a) the infusion of a fixed concentration of a bile acid into the ileal tract and b) the complete collection of the blood draining from the perfused segment of the intestine.

Experimental Procedure and Perfusion of the Intestinal Segment

In the perfusion technique anesthesia is induced in rabbits by ketamine (Ketalar, Parke-Davis) and maintained with a gas mixture of O_2, N_2O, Fluothane.

After median longitudinal laparotomy, the intestine is exposed and kept under several gauzes soaked in 0.9% NaCl at 37° C.

In all experiments, the small intestine is ligated for a length of 50 + −7cm proximal to the ileo-cecal valve.

The perfusion technique has recently been described in detail.[82] The inlet cannula, a polyethylene catheter (I, D 5mm) is inserted into the intestinal lumen immediately distal to the ligature previously tied. The outlet cannula is positioned immediately in front of the ileo-cecal valve.

An initial perfusion to flush the intestine contents from the isolated segment is performed for at least ten minutes, using normal saline at 37° C, pH 7.2, at a mean flow rate of 10 ml/min.

The superior mesenteric vein is then isolated and the vein cannula (IV catch-18gG) inserted and secured; blood collection immediately started and the blood is not returned into the circulation. After each experiment the animal is killed by intracardiac injection of Ketamine. The flow is regulated according to protocol (see below).

After complete washing of the ileal tract, for at least ten minutes with normal saline, the perfusion solution is switched to normal saline containing a known concentration of a bile acid, at a flow rate of 4 ml/min for seven minutes. Thereafter, the perfusion solution is switched back to normal saline at a flow rate of 10 ml/min, which is continued over ten minutes. Blood is collected at sixty second intervals throughout the study, and the blood flow is measured gravimetrically into previously calibrated collecting tubes.

In this way, it is possible to extrapolate the passive and active transport mechanisms by means of a complete dose-response curve, obtained by the infusion of solutions of increasing concentrations of a known bile acid and the contemporary evaluation of concentrations of the same bile acid in the portal blood. The effect of luminal stirring may also be studied by increasing and, in turn, decreasing the flow rate of the infusate solution into the intestine.

For each bile acid the transport velocity (V uptake) is evaluated when the steady state is reached, i.e. when the BA concentration in the consecutive sixty second blood samples is constant. Bile acid absorption at the epithelium to liquid interface is considered to involve active transport and/or passive diffusion. The apparent values of maximal transport rate (T max) and the half saturation constant (Km) are evaluated according to Michaelis-Menten kinetics. For each BA studied, the following models, including active transport and/or linear (non saturated) uptake, are tested:

a) $Vupt = (T\,max \times Cep)/(km + Cep)$
b) $Vupt = (T\,max \times Cep)/(km + Cep) + b \times Cep$
c) $Vupt = b \times Cep$

Where Cep is the BA concentration at the epithelium to liquid interface, and b is the slope of the linear uptake. At a first instance, the resistance to lateral diffusion of BA is not taken into account, and BA concentration is considered to be constant throughout the lumen. A linear regression procedure (AR) allows the determination of these values.[83]

The resistance to BA diffusion from intestinal lumen to the epithelium may also be measured, assuming a lateral diffusion of BA according to a laminar flow model.[84] The following model describes the diffusion: Rt = Rl + Re + 1/b DCA where Rt is total resistance to DCA diffusion from lumen to blood, Rt and Re are liquid and epithelium resistances, respectively, and b DCA is the slope of DCA uptake rate versus infusate DCA concentration.

In this way, Re and Rl may be calculated and[85] result in Re = 0.1262 mM Min cm int/nmol and Rl = 0.06211 mM min cm int/nmol. The same value of liquid resistance is assumed to hold for the different BA.

So far, the intestinal absorption is studied by different techniques,[85, 92] mainly involving bile drainage, as stated above. Both the indication that the secretory pattern of bile acids is different when they are infused in huge amounts into bile fistula animals[74] and the existence of a cholehepatic shunt pathway,[93] may lead to an assessment of the intestinal absorption as evaluated by the biliary secretion of a bile acid infused into the intestine.

The direct transport of a bile acid from the lumen to the portal blood, seems so far to be the best approach for evaluation of intestinal absorption. This technique implies the use of rabbits, since the rat is too small an animal to allow the drainage of blood to such an extent without depleting the animal of all its blood. Since the rabbit weight is at most 3 kg, its blood volume may be approximately 200 cc: therefore, in these experiments the loss of the animal blood is not such as to impair the oxygen-carrying capacity of the blood, as well as of nutrients.

The second goal of these studies is to achieve the best discrimination of the two components of the uptake, as well as the evaluation of the effect of luminal stirring on the kinetics of absorption.[94]

The present experimental approach to the intestinal and liver transport allows a direct comparison between liver and intestine in the same animal species,[95] between liver and liver in different intestinal tracts (jejunum versus ileum) in the same animal.[97]

The latter study seems of major importance, since, besides the 99 KDa protein, which is the putative bile acid transport, and which is limited to the ileum, now acidemia is accumulating of a 87 KDa protein, which is responsible for bile acid facilitated diffusion throughout the intestine found also in the jejunum.

Therefore the jejunal perfusion studies are of increasing interest, due to the necessity of designing targets for blocking the intestinal bile acid absorption in the overall therapy of hypercholesterolemia. The liver may be either isolated as well left in situ. The studies are performed in the Rhesus Monkey.

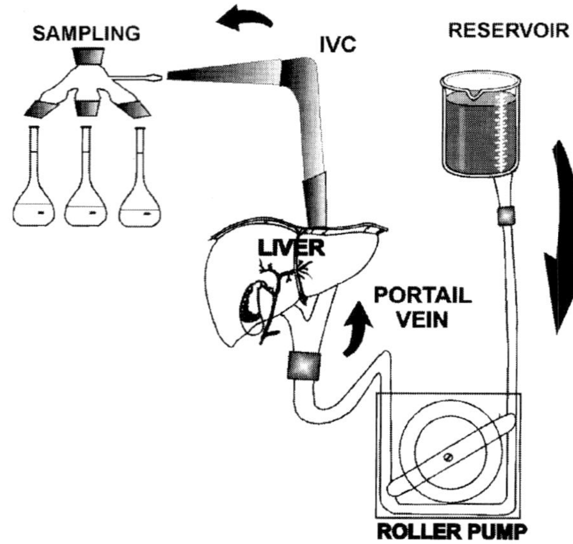

Figure 14.1. Model for the study of the hepatic extraction and biliary secretion in the perfused liver.

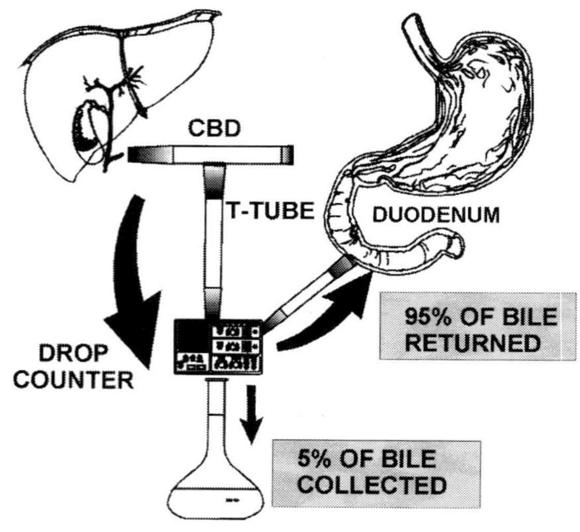

Figure 14.2. Model for a control interruption of bile acids.

A latex T-tube is inserted into the common bile distal to the hepatic duct, permitting the collection of both hepatic and gall bladder bile. The common bile duct is ligated distal to the side of insertion of the T-tube.

Table 14.3. Names and structures of common bile acids in mammals

Bile Acid	Type		Animals
lithocholic	secondary (from cheno acid)	3	many mammals
hyodeoxycholic	secondary (from hyocholic and (w-municholic)	3, 6	many mammals
hyocholic	primary	3, 6, 7	pig
a-muricholic	primary or secondary	3, 7a; 6BOH	mouse and rat
b-muricholic		3a; 6, 7, BOH	mouse and rat
w-muricholic			
murideoxycholic	secondary (from a and b muricholic)	3a, 6bOH	mouse and rat
chenodeoxycholic	primary	3a, 7aOH	many mammals
nisodeoxycholic	tertiary (hepatic or intestinal)	3a, 7bOH	many mammals
deoxycholic	secondary (from cholic)	3a, 12aOH	many mammals
lagodeoxycholic	tertiary (from deoxycholic)	3a, 12bOH	rabbit
phocecholic	primary	3a, 7a, 12aOH (but modified side chain)	seal and walrus
phocedeoxycholic	secondary (from phocecholic)	3a, 12aOH (but modified side chain)	seal and walrus

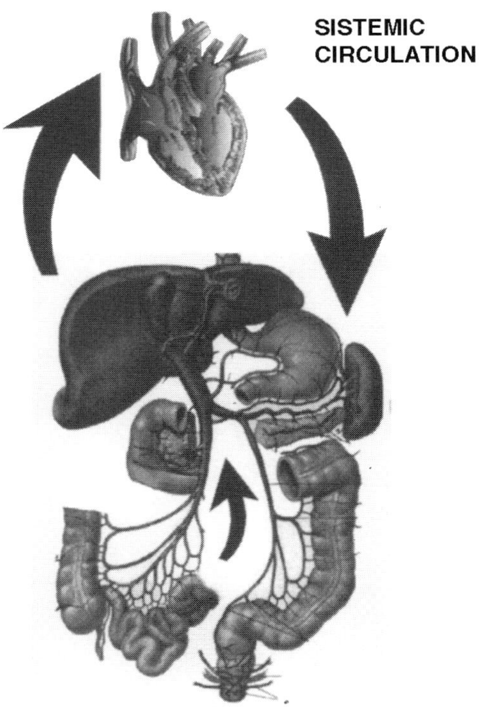

Figure 14.3. Entero-hepatic circulation of bile acids.

A second T-tube, inserted into the proximal third of the duodenum, allows the reinfusion of the bile into the intestine between experiments. In addition, a double lumen (Tygan R-3603); D1.5 lumen; OD 3.0 mm per lumen) is placed into the colon, through an incision into the anus.

The proximal infusion hole is positioned in the cecum and the aspiration site is located 10 cm distally in the transverse colon.

References

1. Dowling, R. H., Mark, E., Picott, J., Berger, J. and Small, D. M. (1968). Experimental mode for the study of the enterohepatic circulation of bile in the rhesus monkey. *J. Lab. Clin. Med.*, **72**, 169–176.
2. Carey, M. C. (1982). The enterohepatic circulation. In: The Liver: Biology and Pathobiology, I. Arias, H. Popper, D. Schachter and D. A. D. M. Shafritz, eds. New York: Raven, pp. 429–465.
3. Hofmann, A. F. (1984). The bile loss syndrome: a doubtful entity. In: Nonnsurgical Biliary Drainage, M. Classen, J. Geenen, and K. Kawai, eds. New York: Springer-Verlag, pp. 120–126.
4. MacDonald, I. A., Bokkenheuser, V. D., Winter, J., McLernon, A. M. and Mosbach, E. H. (1983). Degradation of steroids in the human gut. *J. Lipid. Res.*, **24**, 675–700.
5. Sobottka, H. (1973). Physiological Chemistry of the Bile. Baltimore MD: Williams and Wilkins.

6. Hofmann, A. F. Overview of bile secretion. Handbook of Physiology D The Gastrointestinal System III. pp. 549–566.
7. Hofmann, A. F. and Roda, A. (1984). Physiochemical properties of bile acids and their relationship to biological properties: an overview of the problem. *J. Lipid res.*, **25**, 1477–1489.
8. Hofmann, A. F., Molino, G., Milanese, M. and Belforte, G. (1983). Description and simulation of a physiological pharmacokinetic model for the metabolism and enterohepatic circulation of bile acids in man. Cholic acid in healthy man. *J. Clin. Invest.*, **71**, 1003–1022.
9. Hoffman, N. E., Donald, D. E. and Hofmann, A. F. (1975). Effect of primary bile acids on bile lipid secretion from perfused dog liver. *Am. J. Physiol.*, **229**, 714–720.
10 Gurantz, D. and Hofmann, A. F. (1984). Influence of bile acid structure on bile flow and biliary lipid secretion in the hamster. *Am. J. Physiol.*, **247**, (Gastrointest. Liver Physiol. 7): G736-G748.
11. Forker, E. L. (1967). Two sites of bile formation as determined by mannitol and erythrotol clearance in the guineas pig. *J. Clin. Invest.*, **46**, 1189–1195.
12. Eng, C. and Javitt, N. B. (1983). Chenodeoxycholic acid-3-sulfate. Metabolism and excretion in the rat and hamster and effects on hepatic transport systems. *Biochem. Pharmacol.*, **32**, 3555–558.
13. Gores, G. J., Kost, L. J. and LaRusso, N. F. (1986). The isolated perfused rat liver. Conceptual and practical considerations. *Hepatology Baltimore*, **6**, 511–517.
14. Graf, J. (1983). Canalicular bile salt-independent bile formation: concepts and clues from electrolite transport in rat liver. *Am. J. Physiol.*, **244**, (Gastrointest. Liver Physiol. 7): G233–G246.
15. Aldini, R., Roda A., Morselli Labate, A. M., Grigoli, B., Simoni, P., Roda, E. and Barbara, L. (1986). Species differences of the hepatic uptake of bile acids. *Ital. J. Gastroenterol.* **19**, 1–64.
16. Aldini, R., Roda, A., Morselli, Labate, A. M., Cappelleri, G., Roda, E. and Barbara, L. (1982). Hepatic bile acid uptake: effect of conjugation, hydroxyl and keto groups, and albumin binding. *J. Lipid res.*, **23**, 1080–1086.
17. Roda, A., Cappelleri, G., Aldini, R., Roda, E. and Barbara, L. (1982). Quantitative aspects of the interaction of bile acids with human serum albumin. *J. Lipid Res.*, **23**, 490–495.
18. Aldini, R., Roda, A., Simon, P., Roda, E. and Lenzi, P. Hepatic bile acid uptake: structure activity relationship. Hepatology.
19. Dawson, J. R., Weitering, J. G., Mulder, G. J., Stillwell, R. N. and Pang, K. S. (1985). Alteration of transit time and direction of flow to probe the heterogeneous distribution of conjugating activities for harmol in the perfused rat liver preparation. *J. Pharmacol. Exp. Ther.* **234**, 691–699.
20. Pang, K. S., Koster, H., Halsema, I. C. M., Scholtens E., Mulder, G. J. and Stillwell, R. N. (1983). Normal and retrograde perfusion to probe the zonal distribution of sulfation and glucuronidation activities of harmol in the perfused rat liver preparation. *J. Pharmacol. Exp. Ther.*, **224**, 647–653.
21. Dowling, R. H., Mack, E. and Small, D. M. (1970). Effects of controlled interruption of the enterohepatic circulation of bile salts by biliary diversion and by ileal resection on bile salt secretion, synthesis and pool size in the Rhesus monkey. *J. Clin. Invest.*, **49**, 232–242.
22. Dowling, R. H., Mack, E., Picott, J. *et al.* (1978). Experimental model for the study of the enterohepatic circulation of bile in Rhesus monkey. *J. Lab. Clin. Med.*, **72**, 169–176.
23. Berger, J., Redinger, R. and Small, D. M. (1970). Instrument for sampling and measuring bile now. *Med. Biol. Eng.*, **8**, 19–24.
24. Glenn, F. and McSherry, C. K. (1970). The baboon and experimental cholelithiasis. *Arch. Surg.* **100**, 105-108.
25. McSherry, C. K., Jevitt, N. B. and De Carvalho, J. *et al.* (1971). Cholesterol gallstones and the chemical composition of bile in baboons. *Ann. Surg.*, **173**, 569–577.
26. Dowling, R. H., Mack, E. and Small, D. M. (1971). Biliary lipid secretion and bile composition after acute and chronic interruption of the enterohepatic circulation in the Rhesus monkey: IV. Primate biliary physiology. *J. Clin. Invest.*, **50**, 1917–1926.
27. Redinger, R. N., Herman, A. R. and Small, D. M. (1971). The effects of surgery in hepatic physiology in the primate, abstracted. *Gastroenterology*, **60**, 795.
28. Herman, A. H., Redinger, R. N., Small, D. M., *et al.* (1971). The effects of surgery on bile secretion and composition. *Surg. Forum*, **22**, 378–380.
29. Redinger, R. N., Herman, A. H., and Small, D. M. (1970). Effects of biliary tract obstruction and illness on total secretion and relative composition of biliary bile salt, phospholipid and cholesterol in the rhesus monkey, abstracted. *Gastroenterology*, **58**, 1058.
30. Redinger, R. N., Herman, A. H. and Small, D. M. (1971). Effect of diet on bile composition in the rhesus monkey. *Gastroenterology*, **60**, 198.
31. Admirand, W. H. and Small, D. M. (1968). The physicochemical basis of cholesterol gallstone formation in man. *J. Clin. Invest.*, **47**, 1043–1052.
32 Dowling, R. H., White, J., Lowley, D. *et al.* (1971). The effect of dietary fat on bile composition in monkeys with intact EHCs. *Eur. J. Clin. Invest.*, **1**, 396a.
33. Wheeler, H. O. (1968). Water and electrolytes in bile, Code CC (ed.): Handbook of Physiology. Baltimore, Williams and Wilkins Co, pp. 2409–2431.
34. Wheeler, H. O. and Ramos, O. L. (1960). Determinants of the flow and composition of bile in the unanesthetised dog during constant infusion of sodium taurocholate. *J. Clin. Invest.*, **39**, 161–170.
35. Priesig, R., Cooper, H. L. and Wheeler, H. O. (1962). The relationship between taurocholate secretion rate and bile production in the unanesthetised dog during cholinergic blockage and during secretion administration. *J. Clin. Invest.*, **42**, 1152–1162.
36. Erlinger, S., Dhumeaux, D., Benhamou, J. P. *et al.* (1969). La secretion biliaire du lapin: Preuves en faveur dune important fraction independante des sels biliaires: *rev. Fr. Etud. Clin. Biol.*, **14**, 144–150.
37. Hardison, W. G. and Francis, T. I. (1969). The mechanism of cholesterol and phospholipid excretion in bile, abstracted. *Gastroenterology*, **58**, 1164.

38. Kay, R. E. and Entenman, C. (1961). Stimulation of taurocholic acid synthesis and biliary excretion of lipids. *Am. J. Physiol.*, **200**, 855.
39. Swell, L. Bell, Jr. C. C. and Entenman, C. (1968). Bile acids and lipid metabolism: III. Influence of bile acids and phospholipids in liver and bile of the isolated perfused dog liver. *Biochim.- Biophys. Acta*, **164**, 278.
40. Binet, S., Delage, Y. and Erlinger, S. (1979). Influence of taurocholate, taurochenodeoxycholate, and tauro-dehydrocholate on sulfobromophthalein transport into bile. *Am. J. Physiol.*, **236**, (Endorcinol. Metab. Gastrointest. Liver Physiol. 5): E10–E14.
41. Dumont, M., Uchman, S. and Erlinger, S. (1980). Hypercholeresis induced by ursodeoxycholic acid and 7-ketolithocholic acid in the rat. Possible role of bicarbonate transport. *Gastroenterology*, **79**, 82–89.
42. Eaton D. L. and Klaassen, C. D. (1976). Effects of acute administration of taurocholic and taurochenodeoxycholic acid on biliary lipid excretion in the rat. *Prod. Soc. Exp. Biol. Med.*, **151**, 198–202.
43. Ginsberg, R. L., Duane, W. C. and Flock, E. V. (1977). Hepatic 3-hydroxy-3-methylglutaryl CoA reductase activity in hamster on a lithogenic diet. *J. Lab. Clin. Med.*, **89**, 928–936.
44. Gregory, D. H., Vlahcevic, Z. R., Prugh, M. F. and Swell, L. (1978). Mechanism of secretion of biliary lipids: Role of a microtubular system in hepatocellular transport of biliary lipids in the rat. *Gastroenterology*, **74**, 93–100.
45. Gurantz, D., Gilmore, I. T., Hofmann, A. F. and Kozmary, S. V. (1981). Biliary lipid secretion in the hamster: Greater cholesterol secretion induced by unconjugated bile acids (Abstract). *Hepatology*, **1**, 513.
46. OMaille, E. R. L., Kozmary, S. V., Hofmann, A. F. and Gurantz, D. (1984). Differing effects of norcholate and cholate on bile flow and biliary lipid secretion in the rat. *Am. J. Physiol.* 246 (*Gastrointest. Liver Physiol.* **9**): G67-G71.
47. Paumgartner, G., Sauter, K., Schwarz, H. P. and Herz R. (1973). Hepatic excretory transport maximum for free and conjugate cholate in the rat. In: The Liver. Quantitative Aspects of Structure and Function. Basel: Karger. P. 337–344.
48. Mayer, D. (1974). Hormones and 7DalphaDhydroxylation of cholesterol. In: Matern, S., Hackenschimidt, K., back, P., Gerok, W., eds. Advances in bile acid research III: bile acid meeting, New York: Schattauer Verlag: 53–60.
49. Balasubramaniam, S., Mitropoulos, K. A. and Myant, N. B. (1974). Hormonal control of the activities of cholesterol 7-alphahydroxylase and hydroxymethylglutaryl-CoA-reductase in rats. In: Matern, S., Hachenschmidt, J., Back, P., Gerok, W., eds. Advances in bile acid research III: bile acid meeting, New York: Schattauer Verlag: 61–67.
50. Takeuchi, N., Ito, M. and Yamamura, Y. (1974). Regulation of cholesterol 7-alpha-hydroxylation by cholesterol synthesis in rat liver. *Atherosclerosis*, **20**, 481–494.
51. Kwok, C. T., Burnett, W. and Hardie, I. R. (1983). Regulation of liver microsomal cholesterol 7DalphaD-hydroxylase by Mg APT: effect of pH. *Biochem. Biophys. Res. Commun.*, **116**, 966–972.
52. Boyham, K. M. and Boyd, G. S. (1983). The effect of dibutyryladenosine 3, 5-monophosphate on the synthesis of bile salts in isolated hepatocytes from rat. *Eur. J. Biochem.*, **136**, 313–319.
53. Darielsson, H., Kalles, I., Lindstrom, B., Lundell K. and Mikvall, K. (1983). Regulation of hydroxylation by biosynthesis of bile acids:modulation of reconstituted 12-alpha-hydroxylase activity by proteins from rabbit cytosol and microsomes. *Biochem. Biophys. Res. Commun.*, **113**, 212–219.
54. Danielsson, H., Einarsson, K. and Johansson, G. (1967). Effects of biliary drainage on individual reactions in the conversion of cholesterol to taurocholic acid. *Eur. J. Biochem.*, **2**, 44–49.
55. Schanker, L. S. and Hogben, C. A. M. (1961). Biliary excretion of inulin, sucrose, and mannitol: Analysis of bile formation. *Am. J. Physiol.*, **200**, 1087–1090.
56. Forker, E. L. (1967). Two sites of bile formation as determined by mannitol and erythritol clearance in the guinea-pig. *J. Clin. Invest.*, **46**, 1189–1195.
57. Forker, E. L. (1968). Bile formation in guinea-pig: Analysis with inert solutes of graded molecular radius. *Am. J. Physiol.*, **215**, 56–62.
58. Forker, E. L. (1970). Hepatocellular uptake of inulin, sucrose and mannitol in rats. *Am. J. Physiol.*, **219**, 1568–1573.
59. Forker, E. L., Hicklin, T. and Sornson, H. (1967). The clearance of mannitol and erythritol in rat bile. *Proc. Soc. exp. Biol. Med.*, **126**, 115–119.
60. Wheeler, H. O., Ross, E. D. and Bradley, S. E. (1968). Canalicular bile production in dogs. *Am. J. Physiol.*, **214**, 866–874.
61. Boyer, J. L. (1971). Bile formation in the isolated perfused rat liver. *Am. J. Physiol.*, **221**, 1156–1163.
62. Sarfeh, I. J., Beeler, D. A., Treble, D. H. and Balint, J. A. (1974). Studies of the hepatic excretory defects ub essebtuak fatty acid deficiency. Their possible relationship to the genesis of cholesterol gallstone. *J. Clin. Invest.*, **53**, 423–430.
63. Erlinger, S., Dhumeaux, D., Berthelot, P. and Dumont, M. (1970). Effect of inhibitors of sodium transport on bile formation in the rabbit. *Am. J. Physiol.* **219**, 416–422.
64. Strasberg, S. M., Ilson, R. G., Siminovitch, K. A., Brenner, D. and Palaheimo, J. E. (1975). Analysis of the components of bile flow in the rhesus monkey. *Am. J. Physiol.*, **228**, 115–121.
65. Forker, E. L. (1969). The effect of estrogen on bile formation in the rat. *J. Clin. Invest.*, **48**, 654–663.
66. Mortimore, G. E. (1961). Effect of insulin on potassium transfer in isolated rat liver. *Am. J. Physiol.*, **200**, 1315–1319.
67. Hamilton, R. L., Berry, M. N., Williams, M. C. and Severubggays, E. M. (1974). A simple and inexpensive membrane lung for small organ perfusion. *J. Lipid res.*, **15**, 182–186.
68. Aldini, R., Roda, A., Lenzi, P. G., Festi, D., Bazzoli, F., Mozzella, G. and Contelli Forti, G. Hepatic bile acid uptake and bilancy secretain in rat. Pharmacological research, in press.
69. Aldini, R., Roda, A., Morselli Labate, A. M., Simoni, P., Roda, E. and Barbara, L. (1986). Effect of albumin on taurocholate uptake kinetics in rat liver.

70. Goresky, C. A. (1964). Initial distribution and rate of uptake of sulfobromophthalein in the liver. American Journal of Physiology. **207**, 13–26.
71. Goresky, A., Bach, G. G. and Nadeau, B. E. (1973). On the uptake of materials by intact liver: the concentrative transport of rubidium-86. Journal of Clinical Investigation. **52**, 975–990.
72. Reichen, J. and Paumgartner, G. (1976). Uptake of bile acids by perfused rat liver. *American Journal of Physiology*. **231**, 734–742.
73. Lehninger, L. A. (1955). Biochemistry. Worth Publishers Inc. New York. pp. 147–163.
74. Neef, C. and Maijer, D. K. F. (1984). Structure-pharmacokinetics relationship of quaternary ammonium compounds. Correlation of physicochemical and pharmacokinetie parameters. Naunyn-Scmiedeberg' *Arch Pharmacol* **328**, 111–118.
75. Blom, A., Scaf, A.H.J. and Meijer, D. K. F. (1982). Hepatic drug transport in the rat. A comparison between isolated hepatocytes, the isolated perfused liver and the liver in vivo. Biochem Pharmacol. **31**, 1553–1565.
76. Klaassen, C. D. and Watkins, J. B. (1984). Mechanisms of bile formation hepatic uptake and biliary excretion. *Pharmacol rev*. **36**, 1–67.
77. Groothuis, G. M. M., Hardonk, M. J. and Meijer, D. K. F. (1985). Hepatobiliary transport of drugs: do periportal and perivenous hepatocytes perform the same job? trends Pharmacol Sci. (Aug): 322–327.
78. Boyer, J. L. (1986). Mechanisms of bile secretion and hepatic transport. In: Andreoli, T. C., Hoffman, J. F., Fanestil, D. D., Schulz, S. G., eds. Physiology of Membrane Disordes 2nd ed. New York: Plenum Press, pp. 609–636.
79. Gumucio, J. J. (1983). Functional and anatomic heterogeneity in the liver acinus: impact on transport. *Am. J. Physiol.*, **244**, G578–G582.
80. Dietschy, J. M. (1986). Mechanisms for intestinal absorption of bile acids. *J. Lipid Res.*, **9**, 297–309.
81. Dietschy, J. M., Salomon, H. S. and Siperstein, M. D. (1966). Bile acid metabolism. I studies on intestinal transport. *J. Clin. Invest*. **45**, 832–846.
82. Hofmann, A. (1976). The enterohepatic circulation of bile acids. *Adv. Int. Med*. **21**, 501–534.
83. Aldini, R., Ussia, G., Roda, A., Grilli Cilioni, C., Rizzoli, R., Calcaterra, D., Roda, E., Casanova, S., Lenzi, P., Festi, D., Mazzella, G., Bazzoli, F., Vaccari, M. C. and Galletti, G. (1990). Evaluation of the ileal absorption capacity for bile acids in the rabbit. *Eur. Surg. Res.* **22**, 93–100.
84. Levitt, H. D., Fetzer, C. A., Bond, J. M., Levitt, D. G. (1987). Quantitative assessment of euminae stirring in the perfused smaee intestine of rat. *Am. J. Physioe*. **252** (Gastroint. Liver Physioe. 15): G325–32.
85. Anderson, B. W., Levine, A. S., Levitt, D. G., Kneip, J. N. and Levitt, M. D. (1988). Physiological measurement of luminal stirring in perfused rat jejunum. *Am. J. Physiol*. **254** (Gastrointest. Liver Physiol. 17): G843–G848.
86. Schiff E. R., Small, N. C. and Dietschy, J. M. (1972). Characterization of the kinetics of the passive and active transport mechanisms for bile acids absorption in the small intestine and the colon of the rat. *J. Clin. Invest*. **51**, 1351–1362.
87. Krag, E. and Phillips, S. F. (1974). Active and passive bile acid absorption in man. *J. Clin. Invest*. **53**, 1686–1694.
88. Holt, P. R. (1974). Intestinal absorption of bile salts in the rat. *Am. J. Physiol*. **207**, 1–7.
89. Lack, L. and Weiner, I. N. (1961). In vitro absorption of bile salts by small intestine of rats and guinea pigs. *Am. J. Physiol*. **200**, 313–317.
90. Wilson, F. A. and Dietschy, J. M. (1972). Characterization of bile acid absorption across the unstirred water layer and brush border of the rat jejunum. *J. Clin. Invest.*, **51**, 3015–3025.
91. Dixon, W. J. BMDP Statistical Software. Berkley. University of California Press.
92. McClintock, C. and Shian, Y. F. (1983). Jejunum is more important than terminal ileum for taurocholate absorption in rats. *Am. J. Physiol.*, **244**, G507–G514.
93. Knipers, F., Heslinga, H., Havinga, R., *et al.* (1986). Intestinal absorption of lithocholic acid sulphates in the rat: Inhibitory effect of calcium. *Am. J. Physiol.* 251/14: G189–G194.
94. De Witt, E. H. and Lack, L. (1980). Effect of sulphation patterns of intestinal transport of bile salt sulphate esters. *Am. J. Physiol.*, **1**, G34–G39.
95. Aldini, R., Roda, A., Montagnani, M., Polimeni, C., Lenzi, P., Cerrre, C., Galletti, G. and Roda, E. (1994). Hepatic uptake and intestinal absorption of bile acids in the rabbit. *Eur. S. Clin. Invest.*, **24**, 691–697.
96. Montagnani, M., Aldini, R., Roda, A., Caruso, M. L., Gioacchini, A. M. and Roda, E. (1996). Species differences in the hepatic bile acid uptake: comparative evaluation of taurocholate and tauroursodeoxycholate hepatic extraction in rat and rabbit. Comparative Biochem. And *Physiol*. **113**(2), 157–164.
97. Aldini, R., Roda, A., Montagnani, M., Hrelia, S., Biagi, P. L. and Roda, E. (1996). Intestinal absorption of bile acids in the rabbit: different transport rates in the jejunum and ileum. *Gastroenterology*, **110(2)**, 459–468.

PART IV

HAEMODYNAMICS AND VASCULAR RESEARCH

PART IV

15. Measurements of Blood Flow in Animal Research

Ib Rasmussen,*

Department of Surgery, University Hospital, Uppsala, Sweden

Introduction

An ideal method for measuring blood flow should be absolutely accurate, noninvasive, undisturbing, reproducive, inexpensive and easy to perform. No such ideal method exists. We need to know blood flow (ml/min) as related to tissue weight or volume. If the circulatory parameter cannot be expressed in basic hydraulic units referable to the mass of the organ, is it impossible to confidently relate blood flow to other measurable physiological functions of the organ or to other physiologic or pathophysiologic variables. We need to apply different blood flow measuring techniques depending on the situation. New methods should provide data which correspond closely over the widest range of conditions with accepted reference methods. In correlating the data from the method with values obtained using a simple standard, discrepancies below 1% should be the maximum allowable. Furthermore, accuracy of the used method should be independent of the flow being measured. Only if accuracy is unaffected by flow rate can one avoid the necessity of frequent calibrations or use of correction factors.

Reproducibility is a *sine qua non* for any procedure that will gain widespread acceptance and results must be reproducible over time and between laboratories. This includes that results become independent of environmental variables.

The method used for blood flow measurements must not alter any physiological function or structure in either the organ being measured or the rest of the body. The method should minimally alter the function and structure of the organism and it should not cause either pain or anxiety to the experimental animal. The ideal method should be sensitive to instant changes in blood flow. This would permit moment-to-moment measurements and would also permit correlation of rapidly changes in blood flow with other rapidly changing functions. The method should preferably permit continuous measurements. Since other physiological functions of organs also fluctuate, their correlation with changing blood flows would otherwise not be possible. The ideal technique must not require any additional measurements. The method must not itself interfere with any other method of measurement in either the organ or the body. Another important criterion for an ideal technique for blood flow measurements is that the technique should be inexpensive. It should, furthermore, be easy to perform not requiring days or even hours to conduct. This list of qualities of an ideal method for measuring blood flow is probably not complete[1]. In this chapter some of the methods for measuring blood flow will be discussed and especially judged on behalf on their advantages and disadvantages in different surgical experimental settings.

Types of Blood Flow Measurements

The term "blood flow" can be defined at at least four ways: First, flow through *large vessels* (0.5 mm and larger). Large vessel techniques include the use of venous outflow, perfusion pump, electromagnetic flowmeter, ultrasonic flowmeter, and indicator dilution. The second definition is the flow of blood through *small vessels* (precapillaries to 0.5 mm). Small vessels techniques include microsphere

*Author for correspondence.

injections and *in vivo* microscopy, flourescence microscopy, and laser doppler flowmetry. *Capillary or "exchange vessel"* flow is the third definition of blood flow and is based on the premise that rates of movement across capillary walls are affected by blood flow. Methods which measure passive movement from blood to tissue include measurement of H_2O filtration and ^{42}K or ^{86}Rb clearance. Washout from tissue to blood has been measured with H_2, $^{3}H_2O$, ^{42}K, ^{86}Rb, ^{85}Kr, ^{133}Xe, and nonradioactive xenon excitation. Some investigators have measured simultaneous clearance and washout or "exchange" of several different indicators.[2] Capillary flow has also been estimated by clearance from the lumen of the gut using such as $^{3}H_2O$, CO, K, and inert gases. Capillary flow has also been estimated by clearance of radioactive markers from the peritoneal cavity[3]. Finally, several methods have been developed to measure functional blood flow. *Functional flow* techniques includes the measurements of the clearance of nutritients, O_2, and also include measurements such as microdialysis and tonometry.

Important Questions

After considering which of the four definitions of blood flow are the most relevant in the planned study, the investigator should then ask why blood flow is being measured. The answer to this question should help to decide between a simple method or more sophisticate ones. Other questions that have to be considered include: Is the measurement important enough to justify purchase of special equipment? Can anesthetic, or surgical intervention (e.g. laparotomy) or other variables be permitted? What level of accuracy is required? Does the study requires continuous or repeated measurements or will small numbers of observations be sufficient? If the techniques requires the measurement of other parameters, can they be easily and reliably obtained? If the results are affected by factors other than blood flow can those factors be controlled? Can they be measured and their effects on the data be subtracted or otherwise standardized? If the technique itself causes a change in blood flow is this alteration kept constant during control and experimental protocols? Every method described in this chapter may be the method of choice in some set of experiments while unsuitable in others.

Large Vessel Technique

Direct Measurements

Venous Outflow Measurements

The venous outflow technique is a direct, invasive method which measures the total venous outflow from an organ or part of an organ. It does not disturb the arterial supply and is often easy to prepare. It does not contain any complicated theory or unproven assumptions. The results are extremely accurate and venous outflow can be directly measured with timed samples using flow drop counters or rate settings on pumps which reinfuses the blood. The method is often used in experimental preparations to validate other methods, or to determine the percent of total blood flow measured by capillary or functional techniques. Total blood flow and oxygen consumption measurements are easily repeatedly, and accurate obtainable. The methods has little use in chronic animal studies and it gives no clues concerning distribution of blood flow to different regions in the whole body or in an organ, although changes in distribution of flow may be more meaningful that changes in total blood flow.

The technique has changed very little in the past 50 years.[4] There are some technical problems associated with venous outflow such as increased venous blood pressure.[5] Slight increases in venous pressure can markedly reduce total flow. Probably the mechanism for reduced flow is due to precapillary sphincter response.[6,7]

Arterial Infusion Pump Technique

The Arterial infusion pump technique does not measure physiological flow rate but is limited to the monitoring of flow rates as they are controlled by the experiment. The use of this technique to measure blood flow is largely in disrepute today because the fear of the vasoactive influences by substances release by damaged blood corpuscles.[8] The technique is an obvious method of choice in isolated perfusion studies. It shares the disadvantages of venous outflow technique.

Electromagnetic Flowmetry

The electromagnetic flowmeter technique[9,10] is a direct and invasive method allowing instantaneous and continuous measurements of blood flow. The system has long stability and does not require frequent calibration. It is sensitive to the direction of the flow and the method allows *in situ* measurements. The method is based on the principle of electromagnetic induction. The flow of ions in a magnetic field induces a measurable electric current in the detecting cuff. This current is a function of flow and the electromagnetic properties of the blood. Theoretically, the results may be altered if pH, hematocrit or ion concentration are suddenly and/or significantly changed. Besides the obvious limitations imposed by the invasion to install a cuff, the major practical disadvantage of this method is the need for calibration at zero flow. If the vessel is occluded proximal to the cuff, volume in the cuff may be reduced. On the other hand, distal occlusion may not result in zero flow if vessel compliance permits a systolic-diastolic tide. A traumatic zeroing procedure may result in a longstanding alteration of flow due to e.g. spasm of the vessel wall. Even short periods of ischemia such as those following zeroing usually cause vasodilatation. During preparation there is a risk of disturbing sympathetic vascular nerves. This can have profound effect on blood flow.[7] Other disadvantages are a tendency to zero drift and perhaps the dominating disadvantage, the need for an exact probe size.[10–12] Therefore, most investigators have abandonned the electromagnetic technique in favour of ultrasound flowmetry.

Ultrasound Flowmetry

The ultrasound flowmeter technique is a direct and invasive method allowing instantaneous and continuous measurements of blood flow. The method has several significant advantages when compared with electromagnetic flowmetry. We have in our studies used the recently developed Transonic system.[13,14] The basic principle[15] of transit-time blood flow meter is two ultrasound transducers, one placed upstream and one downstream. At the opposite side of the vessel a metal acoustic reflector is positioned. The transducers function alternately as transmitter and receiver and the time-interval between the signal going downstream and that going upstream is directly proportional to the total volume flow within the acoustic window, i.e., the flow in the blood vessel positioned within the probe. Unlike with other Doppler flow probes, vessel diameter does not have to be calculated. There is no need for zero calibration and the method is rather insensitive to probe position, without a need for a tight fit on the vessel. False low flow values may be obtained if the ultrasound coupling is defective, e.g. due to letting air bubbles enter the probe. The inbuilt warning system for such an occurrence shows if measurement conditions are less than optimal. The various advantages with the Transonic system as compared with electromagnetic flowmetry suggest that it will probably by and large replace the latter system for direct measurements on blood vessels.

Indirect Methods

Indicator Dilution Technique

In 1870 Adolf Fick[16] introduced the principle upon which indicator dilution measurements are based. A fluid of different composition or temperature as the blood is introduced through a catheter and samples are taken downstream. A continuous infusion procedure was first used by Stewart in 1897,[17] and the single injection technique was validated by Hamilton and coworkers.[18] The appearance of a substance being added to the blood by a single bolus injection can be recorded distal to the injection site and used for calculation of blood flow. The substance must mix with the total blood flow. The cardiac output equals the total amount injected divided by the average concentration and the transit time determined at the recording site. In practice, a dye, for instance indocyanine green (Cardiogreen), is injected into the pulmonary artery and the passage of the dye recorded in blood continuously removed from a peripheral artery. The blood is withdrawn to pass a photometer which records light absorbed at a predetermined wavelength. The amount of absorbed light is proportional to the concentration of the dye and the system can be calibrated to give direct readings of blood flow. A major problem with the method is that recirculation of the dye occurs and it is then not possible to determine the precise transit time. Other prerequisites for the method to be applicable include that the indicator substance must be completely mixed and not lost from the circulation in the passage through the lungs. If the small volume of dye is injected into a peripheral vein or in the right atrium and blood is sampled from the pulmonary artery, mixing is often inadequate. There are limited possibilities for repeated measurements since the dye accumulates in the circulation. The accuracy of the method is least when blood flow is low. In addition, recirculation is more evident when blood flow is low and, consequently, the transit time is the very difficult to measure accurately.

Indocyanine green dye can be used for total hepatic blood flow measurements. Indocyanine green is selectively extracted by the liver. For optimal measurements, a liver vein catheter must be used. Indocyanine green dye can be infused over a longer period, allowing for the development of a steady state,[19] or as a single bolus injection.[20] One drawback with indocyanine green dye in this context is that it reflects not only liver blood flow but also liver function.[21]

Other indicators that have been used include Evans blue[22] and Fox indocyanine green[23] dyes. Other substances which change the conductivity of the blood have been used such as hypertonic saline, dextran and autologous plasma.[24–26] Isotope labelled plasma proteins and erythrocytes has also been used.[27,28]

The thermodilution method is a variant on the indicator dilution technique, in which a bolus of fluid with a temperature difference (colder or warmer than the ambient temperature) is used instead of dye.[29–31] This technique is widely used experimentally and clinically[14,32,33] for cardiac output measurements. The advantages over the use of dye include that temperature can be recorded by a thermistor incorporated in the catheter and a bridge circuit with extreme ease and accuracy, and that recirculation is unimportant since heat is dissipated in passage of the e.g. cold bolus through the body organs. Special catheters based on the same principle has been used for other purposes, e.g. measurement of portal venous blood flow.[34] We have used thermodilution technique for measurement of portal venous blood flow[13] and found it very accurate and reproducible. In a study comparing thermodilution technique with ultrasound flowmetry in the portal vein and the pulmonary artery there excellent correlations between the methods were found for both vessels.[35]

To estimate flow from dilution of an indicator several characteristics of the vascular bed and indicator must be assumed.[36] A) The vascular bed in which flow is measured must be a closed system. That is, injection and sampling sites must be located in such way that all the blood flow mixes with the indicator at some point in the circuit. B) Transit times of the indicator "particles" through the system must be the same as for fluid. C) Flow and volume in the system must be constant while the indicator concentration is measured. D) Neither the volumes of indicator injected nor the nature of the indicator should alter flow through the tissue. E) Indicator must not be lost or gained from the system between the points of infusion and measurements.[37] Small molecules which diffuse from the circulation or which are extracted for cellular metabolism are unsuitable for measuring flow by simple indicator dilution. However, such an indicator may be used for measuring flow if the extraction rate is constant and inflow and outflow concentrations are measured. In addition, any indicator which recirculates past the sampling point must be subtracted from measured concentration in order to calculate true flow by simple dilution.

The indicator technique has several advantages. It offers a particular advantage where surgical intervention must be kept to a minimum. The blood flow to a region can be measured also in situations when it is not practical to route the regional blood supply through extracorporeal measuring device. Thermodilution catheters may be positioned in regional supply vessels (position can be verified by a fluoroscope) to measure the regional blood flow in a vessel. There are, however, also limitations using this technique. Blood flow can only be measured intermittently and the total number of determinations may become limited by the buildup of an indicator dye. Specific limitations of indocyanine green dye are sedimentation in saline,[38] spectral shift on dilution,[39] and incomplete recovery from certain vascular beds.[40] Indicator dilution measurements are subject to error if blood flow changes during the measurements. Such blood flow changes may occur spontaneously in response to cardiac pulsations and respiratory measurements.[37] This may also be seen when larger amounts of fluid are given simultaneously through infusion pumps. Thermal losses from injection catheters and heat exchange with vessel walls may lead to an overestimate of blood flow.[31] Improper injection of the indicator using thermodilution technique may also give false measurements.

Small Vessel Technique

Microspheres

The injection of microspheres has been widely used to study regional perfusion and the proportions of cardiac output directed to various organs, i.e., interorgan blood flow distribution and the presence of shunting through various organs. The ability to label these particles with radionuclides to make them easily detectable or quantifiable is another major advance. Particles of several different materials, sizes, and shapes have been used previously, but currently plastic, radioactively labelled microspheres have the widest usage.

In general, accurate measurements are possible with these particles if strict standards are established and the method is appropriately applied.

The microspheres are trapped in the arteriolar or capillary vascular system on the first circulation after injection. This entrapment of microspheres within tissues essentially accomplishes the integration dilution curves by physical impaction in the microcirculation. The theoretical basis of microsphere technique is analogous to that of indicator dilution method. In order to use microspheres to measure accurately the flow distribution to or within an organ, several important factors have to be considered. Immediately after injection into the circulation the microspheres must be well mixed and evenly distributed within the blood stream so that the concentration reaching all branching sites are the same. Providing the relationship between the volume flow and sphere flow is the same at each ramification of the arterial tree, as well as fulfillment of the very important prerequisite that all the spheres reaching an organ are trapped within it, then the fraction of cardiac output directed to an organ, is identical to the fraction of the injected spheres found in that organ. If the cardiac output is also measured, the absolute values of organ flow can be calculated. Microsphere injection only gives a "snapshot" on the situation of the situation at any one time. By using different isotopes and sophisticated mathematical analyses several (approximately five) measurements can be performed, one after the other. The microspheres are trapped at the first capillary passage.This is a drawback when studying liver perfusion, as the portal blood flow has to be indirectly calculated by summing up the flow to all organs draining into the portal vein.[41]

Like most other experimental methods, determination of blood flow by microspheres is subjected to many sources of error. First, this method assumes that the spectrum of each microsphere is identical, whether it be in an isotope reference sample, a reference withdrawal sample, or a tissue sample. In practice, this is not strictly true.[42] Secondly, the microspheres may no be uniformly distributed to an organ or in an organ due to sedimentation of the microspheres in regions where flow velocity is less than a critical value,[43] axial migration of microspheres in a moving column of blood,[44] and skimming of sphere-free blood at branch points. These effects would each reduce the concentration of microspheres in blood entering small arteries branching from large ones. Each of these effects would probably be more pronounced the larger the sphere diameters. Thus, not only could there be a diminution of the concentration of spheres of a given size as blood progressed down the arterial tree, there is also a possibility that these problems could become manifest at different points in the arterial tree for spheres of differing diameters. Thus, the radial distribution of microspheres within blood vessels may not be uniform, and in addition a resulting maldistribution would likely be size-dependent, with larger spheres concentrated centrally.[43] Thirdly, there is a possibility of arteriovenous shunting or migration of the microspheres.[45] Lastly, the effects of the injection of the microspheres and their vehicles should be considered. Both dextran and Tween 80 are known to affect cardiac function. For further overview of the microspheres technique se.[41,42,46]

In vivo Microscopy

The rationale for using intravital microscopy in biomedical research is to obtain detailed information about dynamic events at the microvascular level *in vivo*. In its simplest form, direct observation of blood flow in living tissues with the aid of magnifying lenses dates back to the seventeenth century, and early use of this methodology has provided valuable, although mainly qualitative or semiquantitative, information about microvascular function.[47–49] More recently major advances in this field of research have become possible with the development of quantitative measuring techniques for use with microscopic systems.[50,51]

For the purpose of intravital microscopic observation of the microcirculation, suitable tissues are selected mainly on the condition that they can be easily exteriorized and transilluminated; frequently used preparations are the hamster cheek pouch and the mesentery of cat, rabbit, rat or mouse.[49,52,53]

Of basic concern in intravital microscopic methodology is the maintenance of physiological conditions. Thus, it is important to minimize the preparative surgical trauma, and to establish an external milieu which mimics the normal situation with respect to temperature, extracellular fluid composition, pH, gas tension, etc. Furthermore, it is important in this context to realize that different anesthetic agents can influence the physiology of the cardiovascular system.[54]

Absolute values for microvascular blood flow can be obtained by determining both vessel diameter and flow velocity in single blood vessels down to the capillary size.[55,56] A common intravital microscopic technique to study increased microvascular permeability is the use of circulating fluorescent makers for plasma proteins in combination with trans- or epi-illumination with excitatory fluorescent light.[57–59] A common intravital microscopic technique for the study of increased microvascular permeability changes during different pathological conditions is the use of circulating fluorescent makers for plasma proteins in combination with trans- or epi-illumination with excitatory fluorescent light.[57–59] This flourescence microscopy can be used when studying pathological conditions, such as shock or inflammation.

Laser Doppler Blood Flowmetry

The principle of Laser Doppler Flowmetry is a technique allowing measurement of flow in a capillary bed. Briefly, light from a heliumneon laser is guided to the tissue by a light conductor, which also serves to transmit the back-scattered light to a photodetector system.[60,61] Some of this light has undergone Doppler shifting when scattered by moving red blood cells. The number of red blood cells and their average velocity within the illuminated tissue determine the laser Doppler flow value. This value can not be defined as an absolute flow value (in ml / weight of tissue), but can be used to continuously measure relative flow changes. The method has been used on various organs with differing microvasculature.[62–64] This method allows only measurements near the surface of organs because the penetration of light is limited depending on the tissues ability to absorb light. Thus, as an example, the very dark texture of the liver absorbs light readily and the measurement depth is accordingly only approximately one millimeter.[64]

Capillary or "Exchange Vessel" Flow Techniques

The transportation between blood and tissues of various easily diffusible and more or less inert substances has been used to measure blood flow in the exchange vessel level. The radioactive isotopes of the inert gases (e.g. krypton (^{85}Kr), xenon (^{133}Xe) serves as frequently used examples of this technology. ^{85}Kr is mainly a beta-emittor and the activity of this isotope within a tissue can thus be recorded using a Geiger-Muller recorder. The penetration depth of this isotope is limited and recording from the serosal side of the bowel would only allow detection from the muscular compartment. ^{133}Xe, on the other hand, mainly emits gamma-radiation which could be recorded using e.g. a scintillation detector. The penetration depth of this radiation is such that a detector at the serosal side of the bowel would record total intestinal blood flow. Using both isotopes simultaneously would thus, allow the investigator to subtract muscularis blood flow from total and obtain a value for mucosal-submucosal blood flow.

Easily diffusible substances such as the inert gases can in principle be injected intraarterially to any organ or locally. In the first situation there is a rapid build up of local concentration of the substance in the tissue and then a slow wash-out. If the organ serves as one compartment in term of blood flow wash-out could be depicted as a straight line when plotted on a semi-logarithmic paper and from this line the concentration at time zero and the time to half concentration in the tissue could be calculated. If flow is heterogeneous the initial wash-out curve could be separated into several different curves provided the number of curves is reasonable and the difference between them high enough. By this method blood flow to the different layers of the gut has been recorded.[65]

The theoretical background for using wash-out analyses for determining regional blood flow is complicated and several prerequisites should be fullfiled to enable valid measurements. As a matter of fact there are different theoretical backgrounds which partly are in conflict with each other. For a more detailed analyses of the theoretical background the reader is referred to Lundgren and coworkers.[65,66]

Functional Flow Techniques

Oxygen Delivery and Consumption

By measuring the oxygen saturation in the blood and the blood flow coming to an organ or region the oxygen delivered can be calculated. At the same time knowing the oxygen saturation of the blood leaving the organ or region it is possible to calculate the oxygen consumed. Thus it is possible to estimate in what condition the organ or tissue is in if there are signs of increase metabolism, i.e. increased comsumption or of organ or tissue ischemia if the consumption reaches same levels as oxygen delivery.[67,13] A possible pitfall when estimating oxygen delivery and consumption is the fact that these values share the same parameters, namely blood flow, and thereby a possible false positive correlation between oxygen consumption and delivery could arise.[68]

Microdialysis

A ''biosampling'' technique called microdialysis, has recently been developed.[69] The principle of this method is to extract metabolites from extracellular fluid (ECF) without removal of fluid from the tissue. The basic device is a thin (0.65 mm diameter) double-lumen probe with a semi-permeable polycarbonate membrane at the tip. The probe is perfused with an isotonic solution (Ringer's solution). Small molecules (cut off M. W. 20,000 Dalton) cross the probe membrane by simple diffusion along the concentration gradient between the perfusate and the ECF. The dialysate microsample content of various metabolites (e.g., lactate, purines, amino acids, neurotransmitters) can be measured using sensitive analytical techniques including high-performance liquid chromatography (HPLC).[69–72] Thus, microdialysis can be used for temporal studies of regional changes in the ECF content of various metabolites. The method was originally designed for studies of brain tissue. Results from brain ischemia research indicate that the observed increased ECF levels of energy related metabolites (e.g., lactate, inosine, hypoxanthine) reflect the profound metabolic disturbance of the tissue during ischemia.[73–75] Recently, microdialysis was applied for monitoring of the disturbance in energy metabolism during kidney ischemia and recirculation[76] and in the liver during experimental sepsis.[77] Thus, the method appears to be suitable for monitoring of the energy state of tissues during ischemia. The method allows *in vivo* measurements for long periods of observations, and it can be used in tissues, cerebrospinal fluid, and in blood. A hematoma may occur when inserting the probe in tissues and a hematoma at the tip of the probe may give false or misleading results, not measuring within the tissues but rather the substances diffusing to or from the blood clot. Another concern is that the measurement only reflects the situation for a very small area when used in organ such as the kidney or the liver. This can, however, be compensated with the use of several probes simultaneously.

Tonometry

Tonometry has been used recent years to measure intraluminal carbon dioxide tension in hollow viscera. An impermeable tube with a ballon permeable to carbon dioxide on its tip is introduced to the visceral lumen. The tube and the ballon is filled with fluid containing to carbon dioxide (e.g. saline). After a period allowing equilibrium the carbon dioxide tension of the fluid in the ballon can be measured using a blood-gas analyzer. The carbon dioxide tension in the lumen is demonstrated to be in equilibrium with the carbon dioxide tension of the superficial part of the mucosa. If desired intramucosal pH can be calculated using the Henderson-Hasselbalch's equation:

$$pH = 6.1 + \log \frac{[HCO_3^-]}{PCO_2 \times 0.03}$$

assuming that simultaneously obtained bicarbonate concentrations in arterial blood equal those of the superficial mucosa. This method of determining intramucosal pH has been validated in experiments where intramucosal pH was compared with pH measured by implanted microprobes.[78] The obtained values where very closely similar in normal blood flow and in endotoxemia, a slight qualitative difference, implying an underestimation of the magnitude of the pH reduction by the tonometric technique, was found in partial regional ischemia and even more so in total intestinal ischemia. In these situations arterial bicarbonate obviously does not equal the mucosal. The value of intramucosal pH calculations is that in clinical situations this value is a better predictor of outcome than intraluminal carbon dioxide tension.[79]

The tonometrically obtained intramucosal pH value is not a measurement of blood flow. Instead intramucosal pH reflects tissue oxygenation as related to the needs of the tissue at the time of measurement. A reduced intramucosal pH could be due to insufficient regional blood flow. Overperfusion, blood flow in excess of the needs of an organ, does not provide intramucosal pH values above normal. Intramucosal pH could, thus not be considered to possibly reflect blood flow in any other instances than when blood flow is normal or reduced. Other possible mechanisms for abnormally low intramucosal pH values include impaired arterial oxygenation, impaired extraction/utilization of oxygen by the tissue and increased needs of oxygen not matched by increased regional blood flow. The two latter situations are frequently present in sepsis, at least when considering the splanchnic vascular bed.

Conclusion

Various techniques for measurement of blood flow are available. Depending on the experiment systemic or regional techniques can be addressed specifically. For most of the clinical pathophysiological situations it should be possible to combine clinical relevant models with one or several of the described methods to better understand the importance of blood flow.

References

1. Jacobson, E. (1981). Criteria for an Ideal Method of Measuring Blood Flow to a Splanchnic Organ. In *Measurement of Blood Flow. Applications to the Splanchnic Circulation*, edited by D. Granger, G. Bulkley, pp. 3–7. Baltimore, Williams and Wilkins.
2. Shepherd, A. (1981). Intestinal capillary exchange capacity and oxygen uptake rate. Applications of indicator dilution principles. In *Measurement of blood flow: Applications to the splanchnic circulation*, edited by D. Granger, G. Bulkley, pp. 289–317. Baltimore, Williams & Wilkins.
3. Bulkley, G. (1981). Washout of intraperitoneal xenon: Effective peritoneal perfusion as an estimation of splanchnic blood flow. In *Measurement of blood flow: Applications to the splanchnic circulation*, edited by D. Granger, G. Bulkley, pp. 439–453. Baltimore, Williams & Wilkins.
4. Cutting, W., Dodds, E., Noble, R. and Williams, P. (1937). Pituitary control of alimentary blood flow and secretion: The effect of alteration in blood flow on gastric secretion. *Proc R Soc Lond [Biol]*, **123**, 39–48.
5. Moody, F. (1967). Gastric blood flow and acid secretion during direct intraarterial histamine administration. *Gastroenterology*, **52**, 216–224.
6. Johnson, P. (1959). Myogenic nature of increase in intestinal vascular resistance with venous pressure elevation. *Circ. Res.*, **6**, 992–999.
7. Granger, D., Richardson, P., Kvietys, P. and Mortillaro, N. (1980). Intestinal blood flow. *Gastroenterology*, **78**, 837–863.
8. Folkow, B. (1952). A critical study of some methods used in investigation of flow circulation. *Acta Physiol Scand*, **27**, 118–129.
9. Kolin, A. (1936). An electromagnetic flowmeter: Principle of the method and its application to blood flow measurements. *Proc. Soc. Exp. Med.*, **35**, 53–56.
10. Charbon, G. and van der Mark, F. (1981). Use of Electromagnetic Flowmeters for the Study of Splanchnic Blood Flow. In *Measurement of Blood Flow. Applications to the Splanchnic Circulation*, edited by D. Granger, G. Bulkley, pp. 123–155. Baltimore, Williams and Wilkins.
11. Dobson, A., Sellers, A. and McLeod, F. (1966). Performance of a cuff-type blood flowmeter in vivo. *J. Appl. Phys.*, **21**, 1642–1648.
12. Sellers, A. and Dobson, A. (1967). Some applications and limitation of electromagnetic blood flow measurements in chronic animal preparations. *Gastroenterology*, **52**, 374–379.
13. Arvidsson, D., Rasmussen, I., Almquist, P., Niklasson, F. and Haglund U. (1991). Splanchnic oxygen consumption in septic and hemorrhagic shock. *Surgery*, **109**, 190–197.
14. Rasmussen, I., Arvidsson, D., Zak, A. and Haglund, U. (1992). Splanchnic and total body oxygen consump-

tion in experimental fecal peritonitis in pigs: Effects of dextran and iloprost. *Circ. Shock*, **36**, 299–306.
15. Drost, J. (1978). Vessel diameter – Independent volume flow measurements using ultrasound. *Proc San Diego Biomed Symp.*, **17**, 299–302.
16. Fick, A. (1870). Über die Messung des Blutquantums in den Herzventrikeln. *Sitz Physik-Med Wurzburg*, **16**.
17. Stewart, G. (1897). Researches on the circulation time and on the influences which affect it. IV. The output of the heart. *J. Physiol (Lond)*, **22**, 159–173.
18. Hamilton, W., Moore, J., Kinsman, J. and Spurling, R. (1928). Simultaneous determination of the greater and lesser circulation times, of the mean velocity of blood flow through the heart and lungs, of the cardiac output and an approximation of the amount of blood actively circulation in the heart and lungs *Am. J. Physiol.*, **85**, 377–378.
19. Caesar, J., Shaldon, S., Chiandussi, L., Guevara, L. and Sherlock, S. (1961). The use of indocyanine green in the measurement of hepatic blood flow and as a test of hepatic function. *Clin. Sci.*, **21**, 43–57.
20. Villeneuve, J., Huot, R., Marleau, D. and Huet, P. (1982). The estimation of hepatic blood flow with indocyanine green: comparison between the continuous infusion and single injection methods. *Am. J. Gastroenterol.*, **77**, 233–237.
21. Teranaka, M. and Schenk, W. (1977). Hepatic blood flow measurements. A comparison of the indocyanine green and electromagnetic techniques in normal and abnormal flow states in the dog. *Ann. Surg.*, **185**, 58–63.
22. Hamilton, W. (1962). Measurement of the cardiac output. In *Handbook of physiology*, pp. 551. Washington D.C., American Physiological Society.
23. Fox, I., Brooker, L., Heseltine, D., Essex, H. and Wood, E. (1957). A tricarbocyanine dye for continuous recording of dilution curves in whole blood dependent of variations in blood oxygen saturation. *Proc. Staff Meet Mayo Clin*, **32**, 478.
24. Geddes, L., Peery, E. and Steinberg, R. (1974). Cardiac output using an electrically calibrated flow through conductivity cell. *J. Appl. Physiol*, **37**, 972–977.
25. Goodwin, R. and Sapirstein, L. (1957). Measurement of the cardiac output in dogs by a conductivity method after single iv injections of autologenous plasma. *Circ. Res.*, **5**, 531–538.
26. Guyton, A., Lindley, J., Touchstone, R., Smith, C. and Batson, H. J. (1950). Effects of massive transfusion and hemorrhage on blood pressure and fluid shifts. *Am J. Physiol*, **163**, 525.
27. Dow, P., Hahn, P. and Hamilton, W. (1946). The simultaneous transport of T-1824 and radioactive red cells through the heart and lungs. *Am. J Physiol*, **147**, 493.
28. Kunz, A. and Smith, C. (1967). Theory and design of an online cardiac output computer. *J. Appl. Physiol*, **23**, 784.
29. Khalil, H., Ricardson, T. and Guyton, A. (1966). Measurement of cardiac output by thermal dilution and direct Fick methods in dogs. *J. Appl. Physiol*, **21**, 1131–1135.
30. Fegler, G. (1957). Measurements of blood flow and heat production in the splanchnic area. *Arch Int Physiol Biochem*, **65**, 497–501.
31. Hosie, K. (1962). Thermal-dilution techniques. *Circ Res.*, **10**, 491–504.
32. Ganz, W. (1971). A new technique for measurement of cardiac output by thermodilution in man. *Am J Cardiol*, **27**, 392–396.
33. Ganz, W. and Swan, H. (1972). Measurement of blood flow by thermodilution. *Am. J. Cardiol*, **29**, 241–246.
34. Kinoshita, H., Sakai, K., Kubo, S. and Yamazaki, O. (1989). Measurement of human portal blood flow by continuous thermodilution. *J. Surg Res.*, **46**, 235–240.
35. Rubertsson, S., Arvidsson, D., Wiklund, L. and Huglund, U. (1993). Comparison of blood flow measurement in the portal vein and pulmonary artery using transit-time ultrasound flowmetry and thermodilution techniques. *Surg. Res. Comm.*, **13**, 309–316.
36. Meier, P. and Zierler, K. (1954). On the theory of the indicator-dilution method for measurement of blood flow and volume. *J. Appl. Physiol.*, **6**, 731.
37. Hamilton, W. and Remington, J. (1949). Comparison of the time concentration curves in arterial blood of diffusible and non-diffusible substances when injected at a constant rate and when injected instantaneously. *Am. J. Physiol.*, **148**, 35–39.
38. Tripp, M., Einzig, S., Leonard, J., Gerasch, D., Swayze, C. and Manuel, M. *et al.* (1977). Recent developments in the measurement of cardiac output and regional blood flow using indocyanine green and microspheres. *Med. Instrum.*, **11**, 160–165.
39. Saunders, K., Hoffman, J., Noble, M. and Domenech, R. (1970). A source of error in measuring flow with indocyanine green. *J. Appl. Physiol.*, **28**, 190–198.
40. Donald, D. and Yipintsoi, T. (1973). Comparison of measured and indocyanine green blood flows in various organs and systems. *Mayo. Clin. Proc*, **48**, 492–500.
41. Lifton, N. (1981). Use of microspheres to measure intraorgan distribution of blood flow in the splanchnic circulation. In *Measurement of Blood Flow. Applications to the Splanchnic Circulation*, edited by D. Granger, G. Bulkley, pp. 177–194. Baltimore, Williams and Wilkins.
42. Levine, B., Sirinek, K. and Gaskill, H. (1984). The radiolabeled microsphere technique in gut blood flow measurement – current practice. *J. Surg. Res.*, **37**, 241–255.
43. Laughlin, M., Burns, J. and Loxsom, F. (1979). Use of microspheres in measurement of regional blood flow during $+G_z$ stress. *J. Appl. Physiol. Respir. Environ Excercise Physiol*, **47(6)**, 1148–1159.
44. Phibbs, R. and Dong, L. (1970). Nonuniform distribution of microspheres in blood flowing through a medium sized artery. *Can. J. Physiol. Pharmacol.*, **48**, 415–421.
45. Shepherd, A., Maxwell, L. and Jacobsson, E. (1981). Limitations of the microsphere technique to fractional intestinal blood flow. In *Measurement of blood flow. Applications to the splanchnic circulation*, edited by D. N. Granger, G. B. Bulkley, pp. 195–200. Baltimore London, Williams & Wilkins.
46. Heymann, M., Payne, B., Hoffman, J. and Rudolph, A. (1977). Blood flow measurements with radionuclide-labeled particles. *Prog. Cardiovasc. Dis.*, **20**, 55–79.
47. Grant, L. (1973). The sticking and emigration of white blood cells in inflammation. In: *The inflammatory pro-*

cess, edited by B. Zweifach, L. Grant, R. McCluskey, pp. 205–249. New York, Academic Press.
48. Zweifach, B. (1973). Microvascular aspects on tissue injury. In: *The inflammatory process*, edited by B. Zweifach, L. Grant, R. McCluskey, pp. 3–46 New York, Academic Press.
49. Wiedeman, M., Tuma, R. and Mayrovitz, H. (1981). *An introduction to microcirculation*, New York, Academic Press.
50. Messmer, K., Hammersen, F. and Zweifach, B. (1984). *Perspectives in methodology for studies of the microcirculation.*, Basel: S Karger.
51. Baker, C. and Nastuk, W. (1986). *Microcirculatory technology.*, Orlando FL: Academic Press.
52. Davis, M., Ferrer, P. and Gore, R. (1986). Vascular anatomy and hydrostatic pressure profile in the hamster check pouch. *Am. J. Physiol.*, **250**, H291–H303.
53. Lipowsky, H. and Zweifach, B. (1974). Network analysis of microcirculation of cat mesentery. *Microvasc. Res.*, **15**, 73–83.
54. Wiedeman, M. (1980). In vivo methods for study of microcirculation. Direct and indirect methods. In *Microcirculation*, edited by G. Kaley, B. Altura, pp. 277–302. Baltimore, University Park Press.
55. Wayland, H. and Johnson, P. (1967). Erythrocyte velocity measurement in microvessels by a two-slit photometric method. *J Appl Physiol*, **22**, 333–337.
56. Lipowsky, H. and Zweifach, B. (1978). Application of the "Two-Slit" photometric technique to the measurement of microvascular volumetric flow rates. *Microvasc. Res.*, **15**, 93–101.
57. Hauck, G. and Schröder, H. (1969). Vitalmikroskopische Undersuchungen zur lokalisation der Eiweiβpermeabilität an der Endstrombahn von Varmblüten, *Pflügers Arch*, **312**, 32–44.
58. Nakamura, Y. and Wayland, H. (1975). Macromolecular transport in the cat mesentery. *Microvasc Res.*, **9**, 1–21.
59. Svensjö, E., Arfors, K.-E., Arturson, G. and Rutili, G. (1978). The hamster cheek pouch preparation as a model for studies of macromolecular permeability of the microvasculature. *Upsala. J. Med. Sci.*, **83**, 71–79.
60. Stern, M. (1975). In vivo evaluation of microcirculation by coherent light scattering. *Nature*, **133**, 115–129.
61. Nilsson, G., Tenland, T. and Öberg, P. (1980). A new instrument for continuous measurement of tissue blood flow by light beating spectroscopy. *IEEE Trans Biomed. Eng.*, **BME-27**, 12–19.
62. Svensson, H., Svedman, P., Holmberg, J. and Jacobsson, S. (1983). Laser Doppler flowmetry and transcutaneous measured carbon dioxide tension for observing changes of skin blood flow in fingers. *Scand. J. Plast. Reconstr. Surg. Suppl.*, **17**, 183–186.
63. Ahn, H., Lindhagen, J., Nilsson, G., Salerud, E., Jodal, M. and Lundgren, O. (1985). Evaluation of laser Doppler flowmetry in the assessment of intestinal blood flow in the cat. *Gastroenterology*, **88**, 951–957.
64. Arvidsson, D., Svensson, H. and Haglund, U. (1988). Laser-Doppler flowmetry for estimating liver blood flow. *Am. J. Physiol.*, **254** (Gastrointest Liver Physiol 17), G471–G476.
65. Lundgren, O. (1967). Studies on blood flow distribution and countercurrent exchange in the small intestine. *Acta. Physiol. Scand. Suppl.*, **303**, 1–42.
66. Lundgren, O. (1981). Use of inert gas washout for studying blood flow and flow distribution in the intestine. In *Measurement of blood flow: Applications to the splanchnic circulation*, edited by D. Granger, G. Bulkley, pp. 227–243. Baltimore, Williams & Wilkins.
67. Arvidsson, D., Almquist, P., Rasmussen, I. and Haglund, U. (1989). Flow dependent hepatic oxygen consumption in experimental peritonitis. *Circ. Shock.*, **27**, 336–337.
68. Archie, J. (1981). Mathematic coupling of data: A common source of error. *Ann. Surg.*, **193**, 296–303.
69. Ungerstedt, U. (1984). Measurement of neurotransmitter release by intracranial dialysis. In *Measurement of neurotransmitter release in vivo*, edited by C. Marsden, pp. 81–105. New York, John Wiley & Sons Ltd.
70. Benveniste, H. (1989). Brain microdialysis. *J. Neurochem*, **52**, 1667–1679.
71. Hallström, Å., Carlsson, A., Hillered, L. and Ungersted, U. (1989). Simultaneous determination of lactate, pyruvate, and ascorbate in microdialysis samples from rat brain, blood, fat, and muscle using high-performance liquid chromatography. *J. Pharmacol Methods*, **22**, 113–124.
72. Lindefors, N., Amberg, G. and Ungerstedt, U. (1989). Intracerebral microdialysis: experimental studies of diffusion kinetics. *J. Pharmacol. Methods*, **22**, 141–156.
73. Hagberg, H., Andersson, P., Lacarewicz, J., Jacobson, I., Butcher, S. and Sandberg, M. (1987). Extracellular adenosine, inosine, hypoxanthine, and xanthine in relation to tissue nucleotides and purines in rat striatum during transient ischaemia. *J. Neurochem.*, **49**, 227–231.
74. Hillered, L., Hallström, A., Segersvärd, S., Persson, L. and Ungerstedt, U. (1989). Dynamics of extracellular metabolites in the striatum after middle cerebral artery occulsion in the rat monitored by intracerebral microdialysis. *J. Cereb. Blood. Flow. Metab.*, **9**, 607–616.
75. Hillered, L., Hallström, A., Segersvärd, S., Persson, L. and Ungerstedt, U. (1990). Dynamics of extracellular metabolites in the striatum after middle cerebral artery occulsion in the rat monitored by intracerebral microdialysis (erratum). *J. Cereb. Blood Flow Metab*, **10**, 149–151.
76. Eklund, T., Wahlberg, J., Ungerstedt, U. and Hillered, L. (1991). Interstitial lactate, inosine and hypoxanthine in rat kidney during normothermic ischaemia nad recirculation. *Acta. Physiol. Scand.*, **143**, 279–286.
77. Rasmussen, I., Hillered, L., Ungerstedt, U. and Haglund, U. (1994). Detection of liver ischemia using microdialysis during experimental peritonitis in pigs. *Shock*, **1**, 60–66.
78. Antonsson, J. B., Boyle III, C. C., Kruithoff, K. L., Wang, H. Sacristan, E. and Rothschild, H. R. *et al.* (1990). Validation of tonometric measurement of gut intramural pH during endotoxemia and mesenteric occulsion in pigs. *Am. J. Physiol.*, **259**, G519–G523.
79. Fiddian-Green, R. (1989). Studies in splanchnic ischemia and multiple organ failure. In *Splanchnic ischemia and multiple organ failure*, edited by A. Marston, G. Bulkley, R. Fiddian-Green, U. Haglund, pp. 349–363. London, Edward Arnold.

PART IV

16. Models of Venous Thrombosis in Experimental Medicine

Gottfried Harbauer

Department of Experimental Surgery, University of Cape Town Medical School, Cape Town, South Africa

Thromboembolism is still a major problem in almost all fields of clinical medicine. It is not yet possible to gain complete information, either on in vivo thrombogenesis and thrombolysis, or on the mode of operation of antithrombotic drugs merely by laboratory investigations. Thus much experimental research has been done since the investigations of Meyer and Virchow about 150 years ago roused the interest in thrombus formation and pulmonary embolism.

Historical Review

Figure 16.1 illustrates the development of experimental research on venous thrombosis during the last fifteen decades. It was impossible to gather and to evaluate all publications on this subject, but about 100 contributions can be seen as quite a representative random sample.

The first author I found is Josef Meyer. In 1843 he described flabs of "epithelial cells", circulating in frog web vessels.[1] Similar observations were made shortly afterwards by Wharton Jones. He however interpreted those "grayish-granulous-looking plugs" as being "apparently composed of colourless corpuscels and coagulated fibrin". They are sometimes seen "to block up an artery".[2]

Rudolf Virchow did many experiments during the following years, mainly in dogs. He produced thrombi by implantation of various, partially exotic, foreign bodies, such as caoutchouc or elder pith.[3] His main interest was focussed on the pulmonary embolism.

Concentrated work until 1880 uncovered the roles of fibrin,[5,12] endothelial damage[4,6,8,9,11,14] and platelets[5,7,9,10,13] in thrombogenesis. Thus almost everything was done that could be done by the means available at that time. Thrombosis research almost came to a standstill. There was little experimental work until 1900,[15,17–19] but an excellent synopsis can be found in Julius Cohnheim's *Vorlesungen über allgemeine Pathologie* 1882.[16] Even less was published up until 1916.[20,21] Then the effect of a dog liver extract onto blood clotting was observed by Jay McLean[22] and William Howell *et al.*[23] There were some experimental investigations in the following years,[24,26] but there was no general interest since the toxicity of the substance prevented its clinical use.[25,53] Many years passed until in 1936 the physiologist Charles Herbert Best – even better known because of his Nobel prize for insulin research – succeeded in purifying extracts from the lungs of cattle and in preparing a nontoxic sodium salt of heparin. In cooperation with the surgeon Gordon Murray he transferred the drugs immediately into clinical trials.[27,53] Around 1940 physicians generally began to accept the antithrombotic therapy with heparin. In 1941 Stahman published his paper on sweet clover disease in cattle[28] and in 1944 coumarin was already being used clinically.[54] Many new questions arose and great experimental effort was invested to find answers.[29–35,39,40,42,44–46,48,49,55,57,58,60–62] There were also investigations on the antithrombotic activities of plasmin,[41,43,50,56,59,63] heparinoids,[36,42,52] streptokinase[38,50] and other compounds.[37,46,47,49,51,62] Around 1960 the schedules for the clinical use of the new drugs were fairly well established.

There were few publications of experimental results with heparin during the following two decades.[73,77,81] I could find none on coumarin derivates. The interest turned mainly to platelet function inhibitors, such as acetylsalicylic acid and other anti-inflammatory agents.[69,72,76,77,80,87] There were also

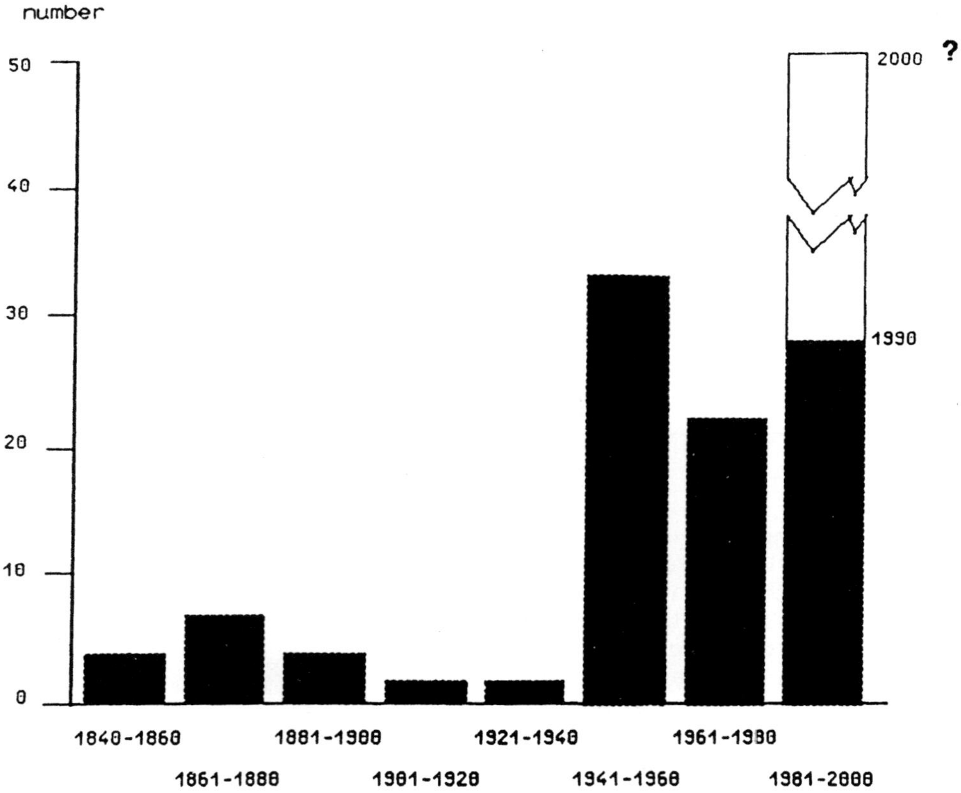

Figure 16.1. Publications of animal models of venous thrombosis.

investigations on the thrombolytic properties of streptokinase,[67,74] plasmin,[66] trypsin,[67] AMP,[72] viper venom,[73] dextran, reserpin and stypturen.[64] The influences of a fat-rich diet,[65,70] ADP,[71,84] ellagic and epsilon-aminocaproic acid[84] and prothrombin complex concentrates[86] on thrombogenesis were examined. Pathogenetic differences between thrombosis and disseminated intravascular coagulation were analysed.[83]

The variety of problems led to the development of specific experimental models both for the thrombosis production[68,77,78,80] and for the evaluation of the product.[75,82,85] On the whole, experimental work practically stagnated until, in the eighties, the low molecular weight heparins[92,111] came onto the scene. In the first ten years since 1981 there have been more publications on experimental thrombosis research than in most twenty year periods before. New or improved animal models were presented [89–91,95–97,100–106,108,109,113,114] and compared to one another.[93,94,96] The specific properties of various heparin preparations were described.[95,96,98,100,103,105–107,111,113] There were further investigations into anti-inflammatory drugs.[90,93–97,100,101,104,106,109] Enhancement or inhibition of thrombogenesis by various other compounds were examined.[90,91,95,96,100,102,104,107,108] The interest in plasmin[89,102] decreased. The new synthetic and vegetable-derived heparinoids[79,79,110,112,114] have not yet stimulated much experimental work.

Nowadays there is still a great interest in the differences in function between unfractioned heparin and its low molecular weight derivates. The significance of the heparinoids might increase during the coming years. Thus there is still a need for practicable animal models of venous thrombosis.

If one wants to develop or to adopt such a model, how can a reasonable decision be made from among the manifold possibilities? Which species and which methods have proved suitable in the past?

Components of Animal Models of Venous Thrombosis

All references for the following chapter are given in Table 16.1. Only in particular cases references are cited within the text.

The first pre-requisite for an experimental model is a suitable animal. For which species has reliable information been gathered?

Table 16.1. Components of models on venous thrombosis

a) Species	
turtle	4
frog	1, 2, 4, 7, 8, 9, 10, 15
bat	6
mouse	80
hamster	46, 64, 71, 101
rat	15, 60, 65, 68, 69, 70, 76, 77, 80, 84, 86, 87, 90, 91, 93, 94, 95, 100, 109, 106, 111, 114
guinea-pig	15, 18, 45
rabbit	6, 9, 14, 15, 16, 20, 21, 24, 31, 32, 33, 35, 36, 37, 38, 39, 42, 43, 44, 45, 48, 51, 52, 56, 57, 60, 62, 65, 67, 72, 74, 80, 81, 83, 85, 89, 96, 97, 98, 99, 103, 104, 105, 107, 112, 113
dog	3, 4, 14, 15, 18, 19, 21, 24, 26, 29, 30, 34, 37, 40, 41, 47, 49, 50, 58, 59, 61, 63, 66, 67, 73, 75, 78, 82, 102, 108
b) Location of thrombotic challenge	
systemic	24, 56, 57, 70, 80, 83, 93, 96, 100, 104, 105, 106, 109, 112
bone marraw	45
frog web	1, 2, 8
bat wing	6
mesenterium	4, 9, 10, 15, 18, 63, 69, 93, 94, 95, 100
cheek pouch	46, 71, 101
pulmon. vessels	3, 24, 74, 80
saphenous vein	9, 26, 29, 30, 49
radial vein	26, 29, 30, 34
ear vein	6, 20, 37, 38, 39, 43, 44, 51
jugular vein	3, 9, 14, 15, 19, 21, 24, 31, 32, 33, 34, 36, 42, 52, 57, 58, 59, 61, 62, 64, 68, 74, 76, 77, 78, 83, 85, 86, 90, 91, 94, 96, 97, 98, 99, 103, 104, 105, 107, 112,113, 114
femoral vein	9, 19, 21, 34, 37, 40, 41, 45, 47, 52, 61, 65, 66, 67, 72, 73, 75, 78, 82, 102, 106, 108
inf. caval vein	35, 40, 55, 60, 84, 87, 89, 106, 109, 111
c) Mode of thrombotic challenge	
foreign body	3, 4, 9, 19, 21, 26, 29, 32, 60, 78, 84, 89, 100
denervation	8, 49
electr. current	9, 40, 58, 66, 69, 75, 82, 102, 108
endothelial lesson from	
- inside	20, 21, 26, 30, 37, 38, 39, 41, 42, 43, 49, 59, 63, 67, 72, 100, 106, 109
- outside	6, 9, 10, 15, 18, 19, 26, 29, 30, 31, 32, 33, 35, 36, 37, 44, 48, 52, 55, 63, 64, 68, 76, 85, 90, 91, 93, 94, 95, 96, 97, 99, 100, 103, 104, 107, 113, 114

Table 16.1 *(Continued).*

stasis	9, 11, 16, 21, 24, 26, 33, 34, 36, 37, 38, 39, 41, 42, 43, 44, 45, 47, 48, 50, 51, 52, 57, 59, 62, 63, 65, 67, 72, 73, 74, 75, 81, 82, 83, 86, 87, 89, 96, 98, 104, 105, 106, 109, 111, 112
stenosis	31, 42, 47, 58, 59, 61, 63, 67, 72, 75, 78, 81, 99, 102, 107, 108, 113, 114
blood composition	20, 24, 32, 34, 36, 39, 43, 45, 46, 47, 48, 50, 51, 56, 57, 59, 61, 62, 63, 70, 71, 74, 80, 81, 82, 83, 84, 86, 88, 93, 98, 101, 104, 105, 106, 112
d) Endpoint methods of evaluation	
survival	24, 56, 63, 80
gross inspection	3, 4, 11, 14, 26, 29, 32, 34, 37, 38, 40, 47, 56, 60, 64, 65, 68, 70, 78, 80, 102
thrombus presence	31, 36, 72, 86, 104
thrombus size	42, 52, 55, 57, 58, 61, 83, 105, 112
thrombus weight	74, 76, 77, 81, 82, 84, 87, 91, 94, 96, 97, 104, 106, 109, 111
hb-content	76, 77, 81, 90, 100
nr. of laser shots	93, 94, 95
radioactivity	50, 75, 85, 89, 98, 102, 103, 108
phlebography	39, 41, 42, 43, 44, 45, 59, 73, 75
histology	3, 9, 14, 24, 26, 29, 30, 31, 33, 34, 35, 36, 37, 38, 40, 41, 43, 44, 45, 47, 48, 49, 58, 60, 61, 65, 68, 72, 78, 80, 102, 108
e) Dynamic evaluation	
transillumination	43, 51, 63
intravit. microscopy	1, 2, 4, 6, 7, 8, 9, 10, 18, 19, 46, 69, 71, 93, 94, 95, 101
electromagn. flowmeter	99, 107, 113
doppler device	113, 114
f) Aim of the study	
basic research	1, 2, 3, 4, 6, 7, 8, 9, 10, 11, 14, 15, 16, 18, 19, 20, 21, 40, 61, 83, 85
thrombosis model	21, 45, 55, 57, 58, 60, 61, 68, 69, 75, 78, 80, 90, 95, 99, 103, 104, 105, 113, 114
action of	
- heparin	24, 26, 31, 32, 33, 42, 44, 46, 49, 58, 62, 73, 77, 81, 96, 98, 99, 103, 105, 106, 107, 111, 112, 113, 114
action of	
- heparinoids	36, 42, 52, 105, 112
- coumarin	29, 30, 32, 34, 35, 39, 44, 58, 96
- ASA + related	72, 76, 77, 80, 87, 90, 93, 96, 97, 100, 101, 104, 106, 107, 109
- others	37, 38, 41, 43, 46, 47, 49, 50, 51, 56, 59, 62, 63, 64, 65, 66, 67, 70, 71, 72, 73, 74, 80, 82, 83, 84, 86, 88, 90, 91, 94, 96, 100, 102, 104, 105, 106, 107, 108

Experimental Animals

The turtle and the bat were mentioned in one instance, about 130 years ago. There is no special liking for the mouse, the guinea-pig nor the hamster. The frog was widely used in the early days of thrombosis research but has not been for the last 100 years. Work on dogs was quite common until 1980 and has been decreasing since, perhaps on account of the increasing sensitivity of society towards animal experiments in general. The rat was first mentioned in 1882 and has been gaining ground fast over the last twenty years. The most frequently used species is the rabbit, which has been holding its position steadily for many decades, but it seems likely that it will soon be overtaken by the rat.

The next question is: in which body-area should thrombogenesis be provoked?

Locations of Thrombotic Challenge

There is no need to discuss the bat wing, the frog web or the pulmonary vessels. The bone marrow of the calcaneus or the tibia of guinea-pigs and rabbits was used, as I found in one case, in order to produce thrombosis of the hind leg veins. Radial and saphenous veins are not favoured. The use of cheek pouch vessels is only possible in the hamster and that of the marginal ear vein is limited to the rabbit. Mesenteric veins of the frog were utilized during the last century and those of warm-blooded animals have sometimes been employed in the last thirty years. The inferior caval vein of the dog, the rabbit and the rat has occasionally been the site for thrombosis production in the last fifty years. The femoral vein of various species has been used repeatedly since the beginnings of thrombosis research. But the external jugular vein had then, as now, the greatest popularity and outrules all other sites by far.

The next problem is: how is thrombosis produced?

Modes of Thrombotic Challenge

It is a disadvantage of almost all experimental models that diseases, which occur in homo sapiens, cannot be induced in an identical shape and in a standardized manner into experimental animals, with some rare exceptions in primates. Thus, nearly all the pathological conditions we introduce into the animals – in order to study their pathophysiological basis or to find new ways to diagnosis or to therapy - remain more or less artificial.

There has been agreement for many decades that three factors are mainly responsible for thrombogenesis:

(i) alteration of the blood flow
(ii) alteration of the blood composition
(iii) alteration of the vessel wall

They are commonly referred to as Virchow's triad. I was not able to find this brilliantly simple definition in Virchow's publications. There is a similar sounding passage on page 294 of *Gesammelte Abhandlungen zur wissenschaftlichen Medizin* edited in 1856, but it relates to the consequence of pulmonary embolism.[3]

Which methods have been tried to provoke venous thrombosis? Foreign bodies were used, mainly to produce pulmonary emboli, until 1880. Since 1930 woollen threads, platinum or alloy wires have been pierced into veins as a basis for a local thrombus or else caolin suspension or gel foam have been injected for stimulation of generalized coagulation.

Vessel wall irritation by mechanical, thermal or chemical means has been performed since the beginning of thrombosis research by stretching, twisting, crushing, pressing and/or freezing the vessel or by bringing it into contact with ether, ammonia, sodium chloride crystals, croton oil, formalin or silver nitrate externally or by injection of phenol or sodium morrhuate – the sodium salt of the fatty acids of the cod fish oil – into a temporary occluded vein segment. An elegant mode of producing a generalized endothelial lesion, described by Hladovec,[106] is the intravasal injection of hypotonic saline.

Electric current was very seldom applied to achieve vessel wall lesions in the past century and has been used now and again since 1950.

Irritation of the vessel wall by laser shots onto the mesenteric vessels of the rat came into use in the early eighties.

Alteration of the blood flow has been achieved by temporary or permanent occlusion or by using non-occluding but stenosing ligatures or clips.

It has been known at least since 1857 by the investigations of Brücke[4] – confirmed by Zahn,[9] Baumgarten[11] and Senftleben[14] – that stasis alone is not sufficient for thrombosis provocation in most cases.

Most stasis and stenosis models therefore use combinations of blood flow altering measures with irritations of the vessel wall, or they influence the blood composition by injections of thrombin, thromboplastin or other substances, which trigger the coagulation cascade.

The compounds are administered intravenously or intraarterially in order to achieve systemic hypercoagulability, which starts thrombogenesis in a vein segment in which stasis is produced shortly after the injection; or they are directly injected into an isolated vein segment.

The most popular stasis model was proposed by Stanford Wessler and his coworkers[57] in 1959: In the original version homologous or heterologous serum is injected into a marginal ear vein of the rabbit. Shortly afterwards a segment of the contralateral external jugular vein is isolated by two ligatures. The segment is removed ten minutes later, opened, and the size of the thrombus (or the thrombi) is arbitrarily scored.[57]

There are various modifications of this model: Jawed Fareed *et al.* use prothrombin complex concentrates plus Russel's viper venom,[105] Giles *et al.* administer only PCC.[86] Arachidonic acid has been given in a few experiments in order to produce generalized disseminated intravascular coagulation (DIC) and to evaluate the efficiacy of antithrombotic pretreatment.[80]

The influence of fat-rich diets on the risk of thrombosis has rarely been investigated. In those studies the actual thrombogenesis is provoked by endotoxin or epinephrine.[65,70]

When thrombus formation has taken place, the results of the experiments have been evaluated and compared to those of the controls.

End Point Methods of Evaluation

Most of the evaluation methods are end point methods: thrombogenesis is started and, after a defined time, the survival rate of the animals or the presence or absence of thrombi is recorded. The thrombus volume is estimated by means of its size, its weight or its hemoglobin content. The composition of a thrombus can be analyzed by histology or its radioactivity after application of labelled fibrinogen, erythrocytes and/or platelets. The resistance to provoked thrombosis is judged by the number of laser shots which are necessary to produce an occluding thrombus. Phlebography can be used in acute and chronic experiments.

Dynamic Methods of Evaluation

Finally there are some dynamic methods which allow continuous observation in acute experiments:

Transillumination can be performed on the rabbit ear and on the mesenterium of many species. It permits gross observations as to whether the blood flows or stagnates and in which time intervals stasis or thrombolysis develop.

Intravital microscopy was widely used during the last century and has been revitalized since the sixties, nowadays mainly combined with the laser shot challenge.

Some groups have used an electromagnetic flow meter or a Doppler-device since around 1980. Decrease of the registrated flow signal or flow velocity signal means growth of a stenosing thrombus, reincrease means thrombolysis, sudden signal jumps mean break-offs of emboli and a 0-signal is interpreted as a consequence of a complete occlusion.

The huge number of different thrombosis models indicates that the ideal one has not yet been found. I will try to give some suggestions for a relatively simple and inexpensive model.

Suggestions for a Basic Model on Venous Thrombosis

Breddin gave some proposals for thrombosis models at the 15th angiological Kitzbühel Symposium in 1980: The model should be cheap and simple to handle. It should deliver reproducible results and should be reproducible in other laboratories. Antithrombotic drugs should show effects, which are comparable to those in humans.[88] Hladovec added some more suggestions at the 23rd Symposium in 1988 at the same place. Large veins should be used rather than venules, if problems of thromboembolism are under investigation. Endothelial lesion should be a component of thrombus provocation, but it should be standardized as far as possible and it should not be extensive. Systemic disturbances should be as slight as possible (Hladovec, unpublished data).

Choice of the Experimental Animal

The most frequently used species and therefore probably the most suitable are the rabbit, the dog and the rat. The dog is expensive and often not easy to handle. So this leaves the rabbit and the rat. Both are cheap to procure, house and feed and they are easy to handle. This holds true even more for the rat than for the rabbit. Moreover the rat is more stable where circulation parameters are concerned and it is available in inbred strains, which makes comparison between groups safer.

Choice of the Location of the Thrombotic Challenge

The most commonly used vessels are the mesenteric veins, the inferior caval vein, the femoral vein and the external jugular vein (Table 16.1). The preparation of the caval vein or the mesenterium means greater systemic disturbance with consequences on coagulation parameters which is not in line with Hladovec's recommendations. The femoral vein is closely adjacent to the artery, which makes preparation more difficult, and it adjoins the nerve, which is easily irritated with the consequence of some systemic disturbance. Most of the investigators prefer the external jugular vein. The preparation is simple and systemic disturbance is minimal. This vessel is even larger than the femoral vein, which meets one more of Hladovec's requirements.

Choice of the Mode of the Thrombotic Challenge

Foreign bodies should not be used according to Hladovec.

Regarding the alteration of the vessel wall: if a model, which closely imitates the thromboembolic disease, is favoured, the electric current, which is no cause for clinical thromboembolism, will not be chosen. The same is true for chemical irritation and for the laser, which is incidentally rather expensive.

But mechanical vessel lesions mimic the consequences of trauma or surgery, representing the main causes for thrombosis and embolism.

Retardation of blood flow and turbulence are important supporting factors for thrombosis in bedridden patients or in people with varicosis. On the other hand spontaneous complete stasis will be rare. Sudden imbalance of the blood composition is rare in clinical practice. Does an injection of Wessler serum, thrombin, thrombokinase or arachnidonic acid meet the reality of the thromboembolic disease? There is certainly a place for highly specified models, for instance in order to test new drugs against hyperfibrinogemia.

In my opinion a basic animal model of the ordinary venous thrombosis should be based on those two components of "Virchow's triad" which are mainly responsible for this disease: mechanical stress on the vessel and disturbance of the flow.

Choice of Evaluation Methods

For evaluation there are many end point methods. They are adequate for chronic experiments, in which the fate of a thrombus – its organization and reduction, leading to the recanalization of the vessel – are to be investigated under different therapeutic conditions.

But when the question is: how does a thrombus develop and what is the reason that none is found after a thromboprophylactic measure – then one has to keep in mind that thrombogenesis is not an unidirectional event. There is thrombus growth and simultaneously spontaneous thrombolysis; an actually existing thrombus does not exist any more in the next second, because it decided to change into an embolus.

End point methods often take their specimens too early or too late in the acute phase. The process of thrombogenesis ought to be continuously followed and recorded.

- Transillumination can only be done on the rabbit ear vein and on the mesenterium and gives little information.
- Intravital microscopy, nowadays supported by video recording, is ideal for evaluation purposes but the set-up is rather expensive.
- The electromagnetic flow meter is expensive and prone to frequent malfunction, especially in venous position.
- A cheap Doppler device, which can be connected to a DC-recorder will do fine for the rabbit. For the rat a more expensive pulsed mini-Doppler is better but not absolutely necessary.

The "Homburg model"[99,113,114] was developed in the rabbit using an electromagnetic flow meter and has been transferred onto the rat using a Doppler device.

Thrombosis is provoked in the external jugular vein by preparing a ligature around the vein plus a parallel positioned mandrin of a standardized diameter. The thread sling, with a sliding knot, is pulled on at both ends with a standardized force, for a standardized time period, and it is then secured.

Figure 16.2a. Registration of the Doppler-signal from the left external jugular vein of a rat, which had a subcutaneous injection of a heparinoid 10 hours prior to the thrombosis provocation.
a) blood flow velocity immediately after thrombosis provocation.
b) gradual decrease of the signal (growing thrombus)
c) 0-signal (complete thrombotic occlusion)
d) gradual increase of the signal (thrombolysis)
e) abrupt increase of the signal (embolic break-off)

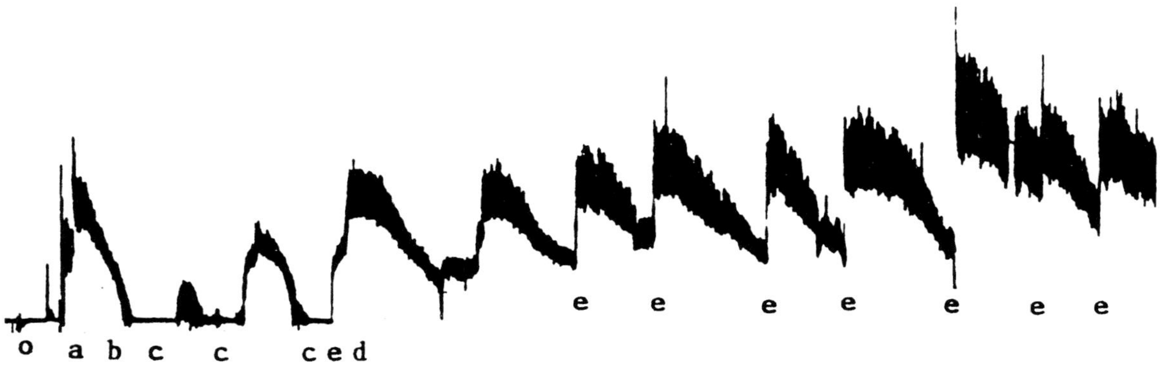

Figure 16.2b. Registration from the left external jugular vein of a rat, which had a subcutaneous injection of a heparinoid 4 hours prior to the thrombosis provocation.
o provocation of the thrombosis
a) release of the blood flow
b) gradual decrease of the signal (growing thrombus)
c) 0-signal (complete thrombotic occlusion)
d) gradual increase of the signal (thrombolysis)
e) abrupt increase of the signal (embolic break-off)

The mandrin is withdrawn. The procedure results in a standardized ringlike stenosis and in a fairly standardized ringlike lesion of the vein wall. Control animals without antithrombotic prophylaxis develop a permanent occluding thrombosis in seven to nine cases out of ten, within thirty minutes. In the rat, crushing the vein by a Mosquito forceps for three minutes (instead of the mandrin-standardized stenosis) gives comparable results.[114]

Figure 16.2a shows an original registration from a rat which had a low subcutaneous dose of a vegetable heparinoid ten hours prior to the experiment. Blood flow is present thirty minutes after the thrombosis provocation. But what had happened before? There was thrombus growth up to complete occlusion, partial reopening by thrombolysis and there were break-offs of emboli.

Figure 16.2b presents the record of a different experiment: the vein stays patent almost all the time, but small emboli go off into the lung every two minutes – is this really insignificant?

For a basis model for the evaluation of new antithrombotic drugs I would suggest:
- the rat as the first and the rabbit as the second choice
- thrombosis provocation at the external jugular vein by an, as far as possible, standardized mechanical vessel wall lesion and additionally by a standardized stenosis
- evaluation of the course of acute thrombus formation by means of a Doppler device or by video-aided intravital microscopy, if available
- follow-up of the chronic fate of the thrombus by phlebography and by autoptical and histological investigations

Specific modifications of a basic model like this one can easily be developed in order to tackle specific problems.

References

1. Meyer, J. (1843). *Frorieps Neue Notizen*, No. 560, 151; quoted by 9, pp. 82–83.
2. Jones, W. T. (1851). On the state of the blood and the blood vessels in inflammation. *Guy's Hospital Reports*, Series II, 1–100.
3. Virchow, R. (1956). *Gesammelte Abhandlungen zur wissenschaftlichen Medizin*, pp. 264–285 Frankfurt: Meidinger Sohn und Comp.
4. Brücke, E. (1857). Über die Ursache der Gerinnung des Blutes. *Archiv für pathologische Anatomie und Physiologie und für klinische Medicin*, **12**, 81–100, 172–196.
5. Schmidt, A. (1861, 1862). *Archiv für Anatomie, Physiologie und wissenschaftliche Medicin*, p. 545 (1861); pp. 428–533 (1862); quoted by 16, p. 166.
6. Samuel, S. (1867). Versuche über die Blutcirculation in der acuten Entzündung. *Archiv für pathologische Anatomie und Physiologie und für klinische Medicin*, **40**, 213–224.
7. Hering, E. Sitzungsberichte der kaiserlichen Akademie der Wissenschaften, 57, Abtlg. II, 179–180; quoted by **9**, p. 85.
8. Zielonko, J. V. (1873). Über die Entstehung der Hämorrhagien nach Verschluß der Gefäße. *Archiv für pathologische Anatomie und Physiologie und für klinische Medicin*, **57**, 436–455.

9. Zahn, F. W. (1875). Untersuchungen über Thrombose. Bildung der Thromben. *Archiv für pathologische Anatomie und Physiologie und für klinische Medicin*, **62**, 81–124.
10. Pitres, A. (1876). Recherches experimentales sur le mode de formation et sur la structure de caillots qui determinent l' hemostase. *Archives de physiologie normale et pathologique*, 2e serie, III, 230–246.
11. Baumgarten, P. (1877). *Die sogenannte Organisation des Thrombus*. Doctoral Thesis, Leipzig 1877; quoted by 16, p. 169.
12. Köhler, A. (1877). Über Thrombose und Transfusion, Eiter und septische Infection und deren Beziehung zum Fibrinferment. *Doctoral Thesis*, Dorpat 1877; quoted by 18, p. 50.
13. Hayem, G. (1878). *Archives de physiologie normale de pathologique*, 2e serie, V; quoted by **17**, p. 201.
14. Senftleben, ?. (1879). Über den Verschluß der Blutgefäße nach der Unterbindung. *Archiv für pathologische Anatomie und Physiologie und für klinische Medicin*, **77**, 421–454.
15. Bizzozero, J. (1882). Über einen neuen Formbestandteil des Blutes und dessen Rolle bei der Thrombose und Blutgerinnung. *Archiv für pathologische Anatomie und Physiologie und für klinische Medicin*, **90**, 261–332.
16. Cohnheim, J. (1882). Thrombose und Embolie. In *Vorlesungen über allgemeine Pathologie*, 2nd edn, pp. 165–231. Berlin: August Hirschwald.
17. Schimmelbusch, C. (1885). Die Blutplättchen und die Blutgerinnung. *Archiv für pathologische Anatomie und Physiologie und für klinische Medicin*, **101**, 201–244.
18. Eberth, J. C. and Schimmelbusch, C. (1886). Experimentelle Untersuchungen Über Thrombose. *Archiv für pathologische Anatomie und Physiologie und für klinische Medicin*, 103, 39–87.
19. Eberth, J. C. and Schimmelbusch, C. (1886). Experimentelle Untersuchungen der Thrombose II. *Archiv für pathologische Anatomie und Physiologie und für klinische Medicin*, **105**, 331–350.
20. Flexner, S. (1902). On thrombi, composed of agglutinated red blood corpuscels. *Journal of Medical Research*, **8**, 316–321.
21. Yatsushiro, T. (1913). Experimentelle Untersuchungen über die Thrombosefrage nebst Angabe einer einfachen Methode zur Koagulationsbestimmung des Blutes. *Deutsche Zeitschrift für Chirurgie*, **125**, 559–612.
22. McLean, J. (1916). The thromboplastic action of cephalin. *American Journal of Physiology*, **41**, 250–257.
23. Howell, W. H. and Holt, E. (1918). Two new factors in blood coagulation-heparin and pro-antithrombin. *American Journal of Physiology*, **47**, 328–341.
24. Mason, E. C. (1924). The production and prevention of experimental thrombosis and pulmonary embolism. *Surgery, Gynecology and Obstetrics*, **39**, 421–428.
25. Reed, C. I. (1925). The intravascular use of heparin. *American Journal of Physiology*, **74**, 79–81.
26. Murray, G., Jacques, L. B., Perret, T. S. and Best, C.H. (1937). Heparin and the thrombosis of veins following injury. *Surgery*, **2**, 163–187.
27. Murray, G. D. W. (1940). Heparin in thrombisis and embolism. *British Journal of Surgery*, **27**, 567–598.
28. Stahmann, M. A., Hübner, C. F. and Link, K. P. (1941). Studies on the haemorrhagic sweet clower disease; identification and synthesis of the haemorrhagic agent. *Journal of Biological Chemistry*, **138**, 513–527.
29. Dahle, D. H. and Jaques, L. B. (1942). The prevention of experimental thrombosis by dicoumarin. *Canadian Medical Association Journal*, **46**, 546–548.
30. Richards, R. K. and Cortell, R. (1942). Studies on the anticoagulant 3,3'-methhylenebis- (4-hydroxy-coumarin). *Proceedings of the Society of Experimental Biology and Medicine*, **50**, 237–242.
31. Rabinovitch, J. and Pines B. (1943). The effect of heparin on experimentally produced venous thrombosis. *Surgery*, **14**, 669–678.
32. Moses, C. (1945). Effect of heparin and dicoumarol on thrombosis induced on the presence of venous stasis. *Proceedings of the Society of Experimental Biology and Medicine*, **59**, 25–27.
33. Loewe, L., Hirsch. E. and Grayzel, D.M. (1947). The action of heparin on experimental venous thrombosis. *Surgery*, **22**, 746–760.
34. Holden, W. D., Cameron, D. B., Shea, P. C. Jr., Sun and Shaw, B. W. (1949). Trypsin and thrombin induced venous thrombosis and its prevention with 3,3 methylenebis' (4-hydroxy-coumarin). *Surgery, Gynecology and Obstetrics*, **88**, 635–638.
35. Jansen, K. F. and Tage-Hansen, E. (1949). Dicumarol in exerimental thrombosis. *Acta Chirurgica Scandinavica*, **98**, 152–156.
36. Friedrich, H. W. (1951). Experimentelle Untersuchungen Über Fibrinoyse durch Throm-bocid. *Ärztliche Wochenschrift*, **6**, 352–354.
37. Innerfield, I., Schwarz, A. and Angrist, A. (1952). Intravenous trypsinits anticoagulant, fibrinolytic and thrombolytic effects. *Journal of Clinical Investigation*, **31**, 1040–1055.
38. Johnson, A. J. and Tillet, W. E. (1952). The lysis of intravascular blood clots in rabbits by the streptococcal fibrinolytic system (streptokinase). *Journal of Experimental Medicine*. **95**, 449–464.
39. Wright, H. P., Kubik, M. M. and Hayden, M. (1952). Influence of anticoagulant administration on the rate of recanilisation of experimentally thrombosed veins. *British Journal of Surgery*, **40**, 163–166.
40. Sawyer, P. N. and Pate, I. W. (1953). Bioelertric phenomena as an ethiological factor in intravascular thrombosis. *American Journal of Physiology*, **175**, 103–107.
41. Cliffton, E. E., Grossi, C. E. and Cannamella, D. (1954). Lysis of thrombin produced by sodium morrhuate in the femoral vein of dogs by human plasmin (fibrinolysin). *Annals of Surgery*, **139**, 52–62.
42. Eysholdt, K.-G. (1954). Die experimentelle Thrombose und ihre Beeinflussung durch Heparin und Heparinoide. *Langenbecks Archiv und Deutsche Zeitschrift für Chirurgie*, **227**, 455–470.
43. Grossi, C. E., Cliffton, E. E. and Cannamella, D. A. (1954). The lysis of intravascular thrombi in rabbits with human plasmin (fibrinolysin). *Blood*, **9**, 310–320.
44. Jewell, P., Pilkington, T. and Robinson, B. (1954). Heparin and ethyl biscoumacetate in prevention of

experimental venous thrombosis. *British Medical Journal*, **1**, 1013–1016.
45. Gottlob, R. and May, R. (1954). Experimental thrombosis similar to clinical conditions on the hind-legs of minor experimental animals. *Angiology*, **5**, 307–311.
46. Berman, H. J., Fulton, G. P., Lutz, B. R. and Pierce, D. L.(1955). Susceptibility to thrombosis in normal young, aging, cortisone-treated, heparinized and x-irradiated hamsters as tested by topical application of thrombin. *Blood*, **10**, 831–840.
47. Hardy, E. G., Morris, G. C. and De Bakey, M. E. (1955). Parenteral trypsin: its effects on experimental thrombotic and inflammatory conditions. *Surgery, Gynecology and Obstetrics*, **100**, 91–96.
48. Borgström, S., Gelin, L. E., Wendeberg, B. and Zederfeldt, B. (1957). Traumatic thrombosis: Animal experiments. *Acta Chirurgica Scandinavica*, **113**, 455–458.
49. Rudowski, W. (1957). Heparin and sympathetic block in treatment of venous thrombosis. *British Medical Journal*, **1**, 626–628.
50. Back, N., Ambrus, J. L., Simpson, C. L. and Shulman, S. (1958). Effect of streptokinase activated plasmin (fibrinolysin). On clots in various stages of organization. *Journal of Clinical Investigation*, **37**, 864–871.
51. Kwaan, H. C., Lo, R. and McFadzean, A. J. S. (1958). On the lysis of thrombi experimentelly produced within veins. *British Journal of Haematolgy*, **4**, 51–62.
52. Sandritter, W., Huppert, M. and Schülter, G. (1958). Zur Frage der Fibrinolyse an experimentellen Gerinnungs- und Abscheidungsthromben. *Klinische Wochenschrift*, **36**, 651–655.
53. Best, C. H. (1959). Preparation of heparin and its use in the first clinical cases. Circulation, **19**, 79–86.
54. Link, K. P. (1959). The discovery of cumarol and its sequels. *Circulation*, **19**, 97–107.
55. Rosenberg, N., Moolten, S. E. and Vroman, L. (1959). A calibrated technique for experimental production of venous thrombosis. *Surgery*, **46**, 764–767.
56. Storm, O. (1959). Prevention of thromboplastin-induced thrombosis in rabbits by means of plasmin. *Thrombosis et Diathesis Haemorrhagica*, **3**, 21–33.
57. Wessler, S., Reimer, S. M. and Sheps, M. C. (1959). Biologic assay of a thrombosis-inducing activitiy in human serum. *Journal of Applied Physiology*, **14**, 943–946.
58. Williams, R. D. and Carey, L. C. (1959). Studies in the production of "standard" venous thrombosis. *Annals of Surgery*, **149**, 381–387.
59. Freiman, A. H., Bang, N. U. and Cliffton, E. E. (1960). Studies on the production of intravascular thrombi and their treatment with fibrinolysin. *Circulation Research*, **8**, 409–418.
60. Friedman, M., Byers, S. O. and Pearl, F. (1960). Experimental production of intraarterial and intravenous thrombi in the rabbit and rat. *American Journal of Physiology*, **199**, 770–774.
61. Marin, H. M. and Stefanini, M. (1960). Experimental production of phlebothrombosis. *Surgery, Gynecolgy and Obstetrics*, **110**, 263–268.
62. Sailer, S. and Eber, O. (1960). Über Thrombolyseversuche an Kaninchen mit einer kombinierten Nicotinsäure-Heparinbehandlung. *Klinische Wochenschrift*, **38**, 548–551.
63. Wilson, J. (1960). Fibrinolysin treatment of induced thrombosis in the dog. *Surgery*, **48**, 469–480.
64. Ashwin, J. G. and Jaques, L. B. (1961). The effect of phosphorus 32, dextran, reserpine and stypturen on thrombus formation in rats. *Thrombosis et Diathesis Haemorrhagica*, **5**, 543–552.
65. Bergentz, S. E., Gelin, L. E. and Rudenstam, C. M. (1961). Fats and thrombus formation. *Thrombosis et Diathesis Haemorrhagica*, **5**, 474–479.
66. Bradham, R. R. and Lee, W. H. Jr (1961). Dissolution effect of human fibrolysin on preformed venous thrombi in dogs. *Surgery, Gynecology and Obstetrics*, **112**, 295–298.
67. Kamiya, K. and Fukuta, K. (1961). Influences of trypsin and streptokinase on intravascular clotting. *Angiology*, **12**, 105–108.
68. Henry, R. L. (1962). Methodes for inducing experimental thrombosis. *Angiology*, **13**, 554–577.
69. Poliwoda, H., Lilli, J., Hagemann, G. and Schyma, D. (1968). Eine in-vivo-Methode zur Erzeugung und Beurteilung von plättchenreichen Abscheidungsthromben. *Zeitschrift für die Gesamte Experimentelle Medizin*, **145**, 252–259.
70. Renaud, S. and Godu, J. (1969). Induction of large thrombi in hyperlipemic rats by epinephrine and endotoxin. *Laboratory Investigation*, **21**, 512–518.
71. Begent, N. and Born, G. V. R. (1970). Growth rate in vivo of platelet thrombi, produced by iontophoresis of ADP as a function of mean blood flow velocity. *Nature*, **227**, 926–930.
72. Peterson, J. and Zucker, M. B. (1970). The effect of adenosine monophosphate, arcaine and antiinflammatory agents on thrombosis and platelet function in rabbits. *Thrombosis et Diathesis Haemorrhagica*, **23**, 148–158.
73. Rahimtoola, S. H., Raphael, M. J., Pitney, W. R., Olsen, E. J. G. and Webb-Peploe, M. (1970). Therapeutic defibrination and heparin therapy in the prevention and resolution of experimental venous thrombosis. *Circulation*, **42**, 729–737.
74. Nowak, A. and Gurewich, V. (1974). Thrombolysis with streptokinase in rabbits. Dose response, fibrinclot specifity and laboratory investigation of fibrinolytic effect. *Thrombosis et Diathesis Haemorrhagica*, **31**, 265–272.
75. Henkin, R. E., Martonffy, K., Kenfield, K., Yao, J. S. T. and Quinn, J. L. (1975). An animal model of radionuclide Venography. *Radiology*, **116**, 211–213.
76. Meng, K. (1975). Tierexperimentelle Untersuchungen zur antithrombotischen Wirkung von Acetylsalicylsäure. *Therapeutische Berichte*, **47**, 69–79.
77. Schmidt, R. (1975). Eine neue Methode zur Erzeugung von Thromben durch Unterühlung der Gefäßwand und ihre Anwendung zur Prüfung von Acetylsalicylsäure und Heparin. *Inauguraldissertation*, Universität Giessen, FRG.
78. Turnipseed, W. D. and Evans, W. E. (1975). A model of ileofemoral thrombophlebitis. *Journal of Surgical Research*, **19**, 397–401.
79. Joffe, S. (1976). Drug prevention of postoperative deep vein thrombosis. A comperative study of cal-

cium heparinate and sodium pentosan polysulfate. *Archives of Surgery*, **111**, 37–40.

80. Kohler, C., Wooding, W. and Ellenbogen, L. (1976). Intravenous arachidonate in the mouse: a model for the evaluation of antithrombotic drugs. *Thrombosis Research*, **9**, 67–80.
81. Chiu, H. M., Hirsch, J., Yung, W. L., Regoeczi, E. and Gent, M. (1977). Relationship between the anticoagulant and antithrombotic effects of heparin in experimental venous thrombosis. *Blood*, **49**, 171–184.
82. Francis, P. B. and Wilson, J. E. (1977). Correlation of the serial dilution protamine sulfate test with models of hypercoagulability and thrombosis in the dogs. *American Journal of Pathology*, **67**, 356–361.
83. Gurewich, V., Nunn, T. and Lipinski, B. (1979). Activation of instrisic or extrinsic blood coagulation in experimental venous thrombosis and disseminated intravascular coagulation: pathogenetic differences. *Thrombosis Research*, **14**, 931–940.
84. MacJomhair, M. and Lavelle, S. M. (1979). Thrombus weight as a measure of hypercoagulability induced by drugs. *Thrombosis and Haemostasis*, **42**, 1018–1021.
85. Zimmermann, R., Zeltsch, C. and Lange, D. (1979). Estimation of thrombus formation by labeling of platelets, red cells and fibrinogen in experimental thrombosis. *Thrombosis Research*, **16**, 147–158.
86. Giles, A. R., Johnston, M., Hoogendoorn, H., Blajchman, M. and Hirsh, J. (1980). The thrombogenicity of prothrombin complex concentrates: I. The relationship between in vitro characteristics and in vivo thrombogenicity in rabbits. *Thrombosis Research*, **17**, 353–366.
87. Reyers, I., Mussoni, L, Donati, M. B. and de Gaetano, G. (1980). Failure of aspirin at different dosis to modify experimental thrombosis in rats. *Thrombosis Research*, **18**, 669–674.
88. Breddin, K. (1981). Voraussetzungen und Anforderungen für Thrombosemodelle am Tier. In *15. Angiologisches Symposium in Kitzbühel* (1980). Verhandlungsbericht herausgegeben von K. Breddin, D. Gross, W. Rotter, pp. 25–29. Stuttgart-New York: Schattauer.
89. Dupe, R. J., English, P. D., Smith, R. A. G. and Green, J. (1981). The evaluation of plasmin and streptokinase activator complexes in a new rabbit model of venous thrombosis. *Thrombosis and Haemostasis*, **46**, 528–534.
90. Grünwald, M., Schlag, M. and Redl, H. (1981). Venöse Thrombose der Ratte durch lokale Unterkühlung. In *15. Angiologisches Symposium in Kitzbühel* (1980). Verhandlungsbericht herausgegeben von K. Breddin, D. Gross, W. Rotter, pp. 125–131. Stuttgart-New York: Schattauer.
91. Haas, S., Söhnchen, R., Geng, R. and Blümel, G. (1981). Erfahrungen mit der kältein-duzierten experimentellen Thrombose. In *15. Angiologisches Symposium in Kitzbühel* (1980). Verhandlungsbericht herausgegeben von K. Breddin, D. Gross, W. Rotter, pp. 133–138. Stuttgart-New York: Schattauer.
92. Holmer, E., Kurachi, K. and Soderstrom, G. (1981). The molecular weight dependence of the rate-enhancing effect of heparin on the inhibition of thrombin, factor Xa, factor Ixa, factor Xia, factor XIIa and kallikrein by antithrombin. *Biochemical Journal*, **193**, 395–400.
93. Lehmann, H. D. (1981). Vergleich der thrombozytenaggregationshemmenden Wirkung von Acetylsalicylsäure, Dipyridalmol und Sulfinpyrazon in verschiedenen Tiermodellen. In *15. Angiologisches Symposium in Kitzbühel* (1980). Verhandlungsbericht herausgegeben von K. Breddin, D. Gross, W. Rotter, pp. 189–192. Stuttgart-New York: Schattauer.
94. Melzer, G. (1981). Erfahrungen mit Acetylsalicylsäure, Dipyridamol und Sulfinpyrazon in verschiedenen Tiermodellen. In *15. Angiologisches Symposium in Kitzbühel* (1980). Verhandlungsbericht herausgegeben von K. Breddin, D. Gross, W. Rotter, pp. 193–199. Stuttgart-New York: Schattauer.
95. Weichert, W., Wiedemann, R. and Breddin, K. (1981). Versuche zur Standardisierung eines Tiermodells: Erzeugung laserinduzierter Thromben im Rattenmesenterium. In *15. Angiologisches Symposium in Kitzbühel* (1980). Verhandlungsbericht herausgegeben von K. Breddin, D. Gross, W. Rotter, pp. 117–124. Stuttgart-New York: Schattauer.
96. Zimmermann, R. (1981). Wirksamkeit antithrombotischer Substanzen an Modellen arterieller und venöser Thrombosen an Kaninchen. In *15. Angiologisches Symposium in Kitzbühel* (1980). Verhandlungsbericht herausgegeben von K. Breddin, D. Gross, W. Rotter, pp. 99–104. Stuttgart-New York: Schattauer.
97. Zimmermann, R., Thiessen, M., Mörl, H., Harenberg, J. and Weckesser, G. (1981). Erfahrungen mit der Acetylsalicylsäure im Tiermodell. In *15. Angiologisches Symposium in Kitzbühel* (1980). Verhandlungsbericht herausgegeben von K. Breddin, D. Gross, W. Rotter, pp. 211–214. Stuttgart-New York: Schattauer.
98. Carter, C. J., Kelton, J. G., Hirsh, J., Cerskus, A., Santos, A. V. and Gent, M. (1982). The relationship between the haemorrhagic and antithrombotic properties of low molecular weight heparins in rabbits. *Blood*, **59**, 1239–1245.
99. Harbauer, G., Hiller, W., Uhl, U. J. and Wenzel, E. (1983). A venous thrombosis model in the rabbit. In *17. Angiologisches Symposium in Kitzbühel* (1982). Edited by K. Breddin, R. Zimmermann, pp. 115–119. Stuttgart-New York: Schattauer.
100. Hladovec, J. (1983). Induction of venous thrombosis in the rat. In *17. Angiologisches Symposium in Kitzbühel* (1982). Edited by K. Breddin, R. Zimmermann, pp. 123–132. Stuttgart-New York: Schattauer.
101. Michal, M. and Giessinger, N. (1983). Induction of thrombi by the iontophoretic application of ADP to venules in the hamster cheek pouch (1983). In *17. Angiologisches Symposium in Kitzbühel* (1982). Edited by K. Breddin, R. Zimmermann, pp. 133–145. Stuttgart-New York: Schattauer.
102. Ulutin, O. N. and Balkuv-Ulutin, S. (1983). The dog venous thrombosis model using electrical stimulation and some results using this model. In *17. Angiologisches Symposium in Kitzbühel* (1982). Edited by K. Breddin, R. Zimmermann, pp. 83–90. Stuttgart-New York: Schattauer.

103. Zekert, F. (1983). Silver nitrate induced venous thrombosis in the rabbit. In *17. Angiologisches Symposium in Kitzbühel* (1982). Edited by K. Breddin, R. Zimmermann, pp. 129–132. Stuttgart-New York: Schattauer.
104. Zimmermann, R., Harenberg, J., Jung, G. and Peter, J. (1983). Experimental venous thrombosis in the rabbit. In *17. Angiologisches Symposium in Kitzbühel* (1982). Edited by K. Breddin, R. Zimmermann, pp. 91–96. Stuttgart-New York: Schattauer.
105. Fareed, J., Walenga, J. M., Kumar, A. and Rock, A. (1985). A modified stasis thrombosis model to study the antithrombotic actions of heparin and its fractions. *Seminars in Thrombosis and Hemostasis*, **11**, 155–175 (1985).
106. Hladovec, J. (1986). A sensitive model of venous thrombosis in rats. *Thrombosis Research*, **43**, 539–544.
107. Just, M. (1986). Pharmakologische Beeinflussung einer experimentellen Thrombose beim Kaninchen. In *Verhandlungsbericht der 29. Tagung der Deutschen Arbeitsgemeinschaft für Blutgerinnungsforschung* (Saarbrücken, 1985). Herausgegeben von E. Wenzel, P. Hellstern, E. Morgenstern, M. Köhler und G. von Blohn, pp. 4.98–4.98. Stuttgart-New York: Schattauer.
108. Ulutin, O. N., Tunali, H., Ugur, M. S., Aytis, S., Erbengi, T. and Balkuv-Ulutin, S. (1986). Effect of defibrotide in electrically induced thrombosis in dogs. *Haemostasis*, **16**; suppl. **1**, 9–12.
109. Hladovec, J. (1988). Models of thrombosis. *General Pharmacology*, **19**, 171-175.
110. Sie, P. (1988). Les heparinoides. *La Revue du Practicicien*, **38**, 952–956.
111. Doutremepuich, C., Guyot, M., Lalanne, M., Deharo, E., Walenga, J. M. and Fareed, J. (1989). Comparative Pharmacology of low molecular weight heparins; implications of manufacturing process on biological effects. Thrombosis Research, 55, 419–426.
112. Hoppensteadt, D., Racanelli A., Walenga, J. M. and Fareed, J. (1989). Comparative antithrombotic and hemorrhagic effects of dermatan sulfate, heparan sulfate and heparin. *Seminars in Thrombosis and Hemostasis*, **15**, 378–385.
113. Harbauer, G., Hiller, W., Hermes, M., Kiefer, S., and Hellstern, P. (1988). Heparin in einem Tiermodell der venösen Thrombose. In *Angiologie und Hämostaseologie;* 18–20. *Kitzbühler Symposium (1983*–1985), herausgegeben von K. Breddin, D. Gross und H. Rieger, pp. 325–334. Stuttgart-New York: Gustav Fischer.
114. Kiesel, J., Harbauer, G., Wenzel, E., Pindur, G., and Bohnert, St. (1991). Comparison of the effects of sodium pentosanpolysulfate and unfractioned heparin on venous thrombosis; an experimental study in rats. *Thrombosis Research*, **64**, 301–308.

PART V

CANCER

PART V

17. Animal Models for Colorectal Liver Metastases

Yves Panis[1] and Bernard Nordlinger[2]

[1]*Digestive surgery unit (Professor Valleur), Hôpital Lariboisière, 2 rue Ambroise Paré, 75475 Paris cedex 10, France*
[2]*Laboratory of experimental hepatic surgery (INSERM U181) and digestive surgery unit (Professor Parc), hôpital Saint-Antoine, 184 rue du faubourg Saint-Antoine, 75012, France*

Introduction

The prognosis of colorectal cancer has not improved in recent years, and the five-year survival rate remains close to 40 percent after surgical excision.[1] The liver is the most frequent site of metastases from colorectal cancer. Liver metastases (L.M.) are present in 20 % of cases at the time of excision of primary colon cancer,[2] and appear later in a further 30 to 50% of patients.[3–6] Once L.M. have occurred, only surgical excision has proven able to prolong survival in some patients.[3,5,7] Results of chemotherapy for unresectable colorectal L.M. have been disappointing with few exceptions which remain to be confirmed.[8]

Although many experimental models exist for other sites of metastatic cancer, such as the lungs,[9] very few have been devoted to the development of experimental L.M. from colorectal cancers.[10,11] In recent years, the interest in the development of such experimental models of L.M. has increased due to the need for new adjuvant therapies to improve the treatment of colorectal cancers. At present, the aims of developing an ideal experimental model of colorectal L.M. are two-fold: first, to help improve the treatment of patent L.M., and perhaps more importantly, to develop adjuvant treatments in order to prevent the development of L.M. after surgical excision of the primary cancer. Consequently, a good animal model should mimic the natural history of human colon adenocarcinomas and induce colon cancers and L.M. with a good yield.

We will study the different procedures which yield experimental L.M. of colorectal origin, the factors which influence their growth, and finally the methods of treatment which can be applied to them.

Spontaneous Liver Metastases from Chemically-Induced Colon Cancer in Rats

Experimental colorectal cancers can be induced in rats by chemical carcinogens. Dimethylhydralazin and its derived compounds are those most often used.[12,13] The factors that can modulate the effect of the carcinogen are the dose and the route of administration, the age and sex of the animal[12,14] and exogen factors. Carcinogens are most often administered subcutaneously by weekly injections of 10 to 20 mg/kg over several weeks. When high doses of carcinogen are used, a good yield of colon cancer is obtained, but the specificity is reduced since they are often associated with small bowel tumors, particularly on the duodenum (10, 12). After administration of azoxymethane at the dose of 10 mg/week for twelve weeks, gastrointestinal tract tumors were observed in 89% of the rats six months after beginning the administration of the carcinogen, but only 35 % of the animals had isolated colon cancers.[15] With lower doses, the rate of digestive tumours was reduced but the proportion of isolated colon cancers was increased: after administration of azoxymethane at the dose of 8 mg/kg/week for ten weeks (total dose of 80 mg vs 120 mg in the previous study), 38 % of the rats had digestive tract tumors, but 70 % of them were isolated colon cancers.[16]

Several studies have observed a higher efficiency of carcinogens in male than in female rats, but the mechanisms remain unknown.[12 14]

Pathological examination of the colonic tumors shows that adenocarcinoma is the most often observed lesion; in rats, it seems that approximately 75% of the colon cancers develop *de novo* from dysplastic lesions on normal colonic mucosa.[17]

There are only a few studies on the spontaneous liver metastases obtained with these models.[12,17] After subcutaneous administration of azoxymethane at the dose of 10 mg/kg/week for twelve weeks, L.M. were observed in 14% of the rats and peritoneal carcinomatosis in 24% (associated with L.M. in half of the cases).[15] With higher doses of azoxymethane (up to 20 mg/kg/week), spontaneous L.M. were observed in 20–25% of rats in six months.[18,19] L.M. had the same pathological characteristics (i.e. differentiation) than the primary tumor.[20]

Several means have been used to increase the yield of colonic cancers and L.M. in these models. Surgical colonic trauma induced by caecotomy and its suture has increased the rate of chemically-induced colonic cancers from 26% to 50%.[16] Similar results were obtained after partial colectomy or creation of a colonic diverticula.[21,22] The inflammatory reaction associated with cellular proliferation, secondary to the colonic suture could enhance the effect of carcinogens.[16]

Immunosuppression induced by Cyclosporin A (CSA) promotes metastatic dissemination in several experimental models of tumors in rats.[23] CSA administration at a dose of 10 mg/kg/28 days did not result in an increased number or size of chemically-induced colonic tumors, but in an increased number of tumors with serosal infiltration (36% vs 7 in the control group; $p < 0.05$) and of spontaneous L.M. (25% vs 12.5% in the control group; N.S.).[16] Martin *et al.* have shown that chemically-induced colon cancers in rats developed in 70–80 % of cases within the lymphoid formations of the colon, confirming the relation between tumor development and the status of immunological defense mechanisms.[17] One of the mechanisms to explain this relation could be the presence on lymphoid formations (Peyer's patches) of immunosuppressive cells (i.e. T-cells) which inhibit the immune antitumoral response.[24] CSA could stimulate the infiltrating capacity of cancer cells by diminution of local immune defences and perhaps by stimulation of immunosuppressive cells present within the lymphoid formations.[16] Dawiskiba has developed an experimental model of L.M. using a colon carcinoma cell line injected into lymphoid formation of the caecum: 100% of caecal cancers and 70% of L.M. were observed after eight weeks, but the drawback of this model is the presence of peritoneal carcinomatosis due to local dissemination in almost half of the rats[25] and approximately 40% of rats presented early occlusion due to the caecal tumour.[26–29]

In conclusion, this experimental model with chemically-induced primary colon cancer and spontaneous liver metastases mimic quite well the natural history of the development of colorectal L.M. in humans; however, a low yield of both colonic cancers (isolated in less than 50% of rats) and L.M. (20–25% of rats) is obtain after six months of latency. For these reasons, this model is now seldom used, although it seems the most appropriate for the development of new adjuvant therapies for the prevention of L.M. of colorectal origin.

Liver Metastases Induced by Direct Intraportal Injection of Cancer Cells from a Colon Carcinoma Cell Line

At the present time, this experimental model is the one most often used for both biological and therapeutic studies. Cancer cells originate from an established colon carcinoma cell line developed in animals,[30–34], or sometimes in humans.[35] Cells injection can be performed through the portal,[32,34,36] splenic,[35,37–41] or ileocolic.[30,33,42–46] This kind of model allows a higher yield of L.M. to be obtain with a shorter period of latency than the previous model.

The type of colon carcinoma cell line chosen and the number of cells injected modulate the rate of L.M.[47,48] One of the most frequently used in recent experimental studies is DHDK 12 cell line, developed by Martin. It was obtained from an established transplantable carcinoma cell line originating from a 1,2-dimethylhydralazine-induced colon cancer in syngeneic BDIX rats.[42–53]

Intraportal injection of approximately 10 to 20 millions of cancer cells into the rat allows 80 to 100% of rats with L.M. to be obtain three to six weeks later.[30,36,44,54] Associated extrahepatic metastases (i.e. lungs), due to cells escaping from the liver sinusoids, can be observed with a high rate;[49] this drawback can be minimized by using voluminous cell aggregates[34] instead of single-cell suspensions: using this model, we observed only 1/19 rats with associated lung metastases.[55] Cell aggregates are obtained from incubation of single cells on a giratory shacker for forty eight hours.[34] Injecting cell aggregates (150–200 μm in diameter) rather that single cells may also mimic what occurs in humans where the great majority of cancer cells in the venous blood are aggregated in clumps.[56] Furthermore, when single cells are used in animal models, a greater number must be injected (1 to 10 million cells/rat); with cell aggregates, the same L.M. rate can be obtained with only 0.5 million cells/rat.[34,55] A similar model, using synthetic microspheres containing cancer cells was developed with a high yield of L.M. and without associated lungs metastases.[57]

Several factors have been shown to increase the growth of experimental L.M. After partial liver resection, the regenerating liver releases growth factors which promotes hepatic mitotic activity followed by liver regeneration.[58] These factors have been shown to increase the growth rate of subcutaneously implanted tumors and L.M. of some tumors.[59–64] During the regeneration process, the tumor growth is enhanced in proportion to the regenerative stimuli and in contrast in atrophied lobes the tumor growth is partially inhibited.[62] However, data regarding the effect of partial hepatectomy on experimental tumors are controversial. Regression of tumors or even disappearance of L.M. have been reported following partial hepatectomy in rats.[42,43] Partial hepatectomy could augment the liver's host defence by enhancing cytostatic or cytotoxic antitumor activity of nonparenchymal liver cells.[43] However, most of of the experimental studies showed a tumor growth stimulation after partial hepatectomy.[58–60,62] Fisher observed that a 70% hepatectomy performed in the rat within twenty four hours before or after the intraportal injection of cancer cells promoted the development of L.M., with the greatest effect occuring when the resection was performed five minutes after the inoculation of the cancer cells.[59] On the other hand, if hepatectomy is performed 5 min. before intraportal injection of the cells rather than after, in order to avoid losing a large proportion of the injected cells in the resected specimen, the number of L.M. is not increased compared to control rats. This may be explained by the liver sinusoidal obstruction and acute portal hypertension resulting from the migration of all the injected cell aggregates to the 70% reduced liver. It may have responsible for early postoperative deaths observed and the reduced liver regeneration and tumor promotion in the surviving rats.[55]

In this model, we and others have shown that immunosuppression by Cyclosporin A increase the number of L.M., provided CSA is given at the time of cancer cells injection (80 to 100% of rats with L.M. vs 30% in the control group.[32,34,55]. The effect of CSA, which inhibits T-lymphocytes confirms the role of cellular immunity in metastatic spread and growth. In the rats receiving CSA, L.M. were larger and more numerous than in control group.[34,55] We used, as others, subcutaneous administration of CSA, because it is well tolerated, easy to administer, and yields reproducible CSA plasma levels with an adequate peak and few variations with time.[65] The degree of host cell infiltration around the liver metastases, due to the high immunogenicity of the DHDK12 cell line, seems to be related to the metastatic potential of the tumor. The host cell infiltration which surrounded experimental L.M. was reduced in previously CsA-immunosuppressed rats.[34] Martin has observed in an histoimmunological study of the cellular reaction induced in the liver by metastatic tumor cells that a contrast exists between the early metastatic foci which are surrounded by an accumulation of inflammatory cells (i.e. T lymphocytes and NK cells) and the larger, late metastases which no longer induce a significant cellular reaction.[44] The stimulation of metastases growth by CsA could be explained by the reduction of the host cell infiltration around the L.M.[32,55]

Other factors have been shown to modulate the growth of experimental L.M.: Cyclosphosphamide,[49,66] Kupffer cell inhibition,[38,67,68] splenectomy[31] and repeated surgical procedures.[60] It is well known that the operative period is critical for the development of metastases, due to profound immunodepression both in animals[54,60,69 70] and in humans.[71–73] Animals undergoing intraportal injection of tumor cells develop more tumors when the injection is followed by several exploratory laparotomies.[60]

The theory of "dormant cells" was first introduced by Fisher in 1959.[48] These cells may continue to survive indefinitely in a dormant state. Some studies in humans have suggested the existence of two different tumor cell populations at the site of secondary tumor growth within the liver:[74,75] tumor cells growing to form metastasis and tumor cells which appear to be, according to their morphology, in a resting non-proliferating state. The presence of two tumor cell types at the secondary site is compatible with the existence of two genetically different tumor cell subpopulations with, eventually, a different latent period before growing as a metastatic tumor or with a different degree of repression of the metastatic gene.[76] In a rat model, when a small number of cells are injected intraportally, no tumor was apparent five months later; however, all rats were subjected to repeated laparotomies at seven-days intervals, three months after injection of the cells, 100% developed a tumor within a few weeks.[60] Using the model of colorectal L.M. induced by intraportal cell aggregates, we observed after stimulation by hepatectomy or CSA administration, the presence of macroscopic L.M. twelve weeks after injection of cells, although non apparent liver metastases, were observed at 8 weeks.[32] It can be hypothesized that undetectable liver micrometastases were present at eight weeks and were probably stimulated by CsA-immunosuppression and by hepatectomy. These findings support the existence of "dormant cells" in the liver which remain in a dormant state and are not destroyed by the host, and begin to proliferate when immunosuppression is induced by CsA, by repeated surgical procedures or by liver resections.[32] This would also explain why some supposedly curative liver resections performed for liver metastases in humans are followed by early recurrence in the remaining liver.[3]

LaFrenière *et al.* have developed an original model of L.M., by injection of cancer cells into the splenic vein of rats, prior to splenectomy and observed a high rate (80 to 100%) of L.M. This can be explained by the great number of cells injected 10^6 and by the immunosuppressive effect of splenectomy.[39]

Using a similar model, but without splenectomy, Giavazzi *et al.* obtained an equivalent yield of L.M. but with associated splenic tumors in all cases.[77] Wu et al. have developed an original model using DHDK 12 cell line, with selective injection of cancer cells in only three segments of the liver, to obtain a model of localized L.M.[78]

These models allowed several studies concerning the vascularization of experimental L.M. Ackerman *et al.* have demonstrate that micrometastases have both portal and arterial vascularization, when larger metastases have a predominant arterial vascularization.[79–82] These findings could have some therapeutic implications: in the case of preventive adjuvant treatment of colorectal L.M. after resection of the primary colonic tumor in order to prevent L.M. in humans, the portal route should be prefered for administration of chemotherapy; on the other hand, palliative chemotherapy for unresectable L.M. should be administered by the arterial route. Some authors have studied the interaction between liver microcirculation and L.M.[83] and it was demonstrated, using an experimental model with intraportal injection of cancer cells, that early detection of occult L.M. can be made by the study of the hepatic perfusion index (decreased portal venous outflow without modification of hepatic arterial outflow)[84] or even by CT-scan in rats.[85] Using monoclonal antibodies against cancer cells and nonparenchymal liver cells, it is possible to study the different phases of metastase development: in the early phase, L.M. is surrounded by inflammatory cells (which could theoritically destroy the cancer cells); later on, L.M. began larger, and inflammatory cells disappear, probably eliminated by the immunosuppressive tumor effect.[44]

In conclusion, this model, using direct intraportal injection of cancer cells originating from a colon carcinoma cell line is the one most often used in recent experimental studies. Although it bypasses the natural history of colon cancer, this model is easily reproducible, and rapidly produces L.M. with a high yield. The absence of associated extrahepatic tumors permits specific studies of L.M. (i.e. modulating factors of L.M. growth). Furthermore, it seems very indicated for the development of curative treatent of L.M.

Treatments of Experimental Colorectal Liver Metastases

Most of the experimental therapeutic studies have been performed with the model of direct intraportal cancer cell injection. This model is better adapted for the study of curative treatments of patent L.M., than to test the efficacy of preventive treatment modalities of L.M., since their natural history is completely bypassed. The model of chemically-induced colon cancer with spontaneous L.M. has the advantage of respecting the spontaneous step-by-step development of L.M. from colorectal cancers. Thus, it appears well suited for the study of new preventive adjuvant chemotherapy of colorectal cancers. Unfortunately the yield is low. We studied an original model reproducing the conditions of curative resections in humans:[15] after six months of administration of carcinogen in 236 rats, 83 rats (35%) with isolated colon cancers underwent total colectomy and ileorectal anastomosis; forty of these rats (without L.M. and peritoneal carcinomatosis) were randomized into two groups: surgery alone or surgery lus intraperitoneal 5-FU (5mg/kg/day for five days). The good tolerance of intraperitoneal chemotherapy particularly concerning the healing of recent colonic anastomosis had been demonstrated in a previous work in rats.[86] Peritoneal carcinomatosis and L.M. were observed only in the control group respectively in 27.7% and 22.2% of the animals. Using a similar model, Ross *et al.* have shown the preventive effect of specific immunotherapy against L.M. and peritoneal carcinomatosis.[87]

Several recent studies have used direct intraportal injection of cancer cells to evaluate different treatment modalities of L.M., particularly immunologic. The effect of hepatic artery ligation on the growth of L.M. was demonstrated.[18,80,88,89] The efficiency of anticoagulant therapy (warfarin or heparin) on L.M. was reported.[19,90] Successful immunologic treatments against experimental L.M. were reported using interleukine 2,[31,37,39,91,92] interferon,[91] LAK cells[31,92] lymphokines within liposomes,[93] specific monoclonal antibodies[94] tumour necrosis factor,[45] levamisole[95] and other immunostimulating factors.[95,96] The effect of chemotherapy with doxorubicine within liposomes[97] or nanoparticles,[98] chemoembolization,[46] or new drugs such as flavone acetic acid[35] has been studied on the same type of model.

Although only few experimental studies have been devoted, until now, to the treatment of L.M. by gene therapy, it has been already shown that gene therapy could be a potential treatment for L.M.[99] Selective gene delivery directly within the liver tumor[100] or through the portal vein have been shown feasible under total vascular exclusion of the liver.[99] Furthermore, L.M. grow within a quiescent extratumoral liver. This particularity allows to expect high selectivity for strategies using either vectors or gene products whose functionalities are restricted to dividing cells. This is the case of recombinant retroviruses which can only integrate and be expressed in dividing cells, and of suicide gene such as *Herpes Simplex* type 1 thymidine kinase. Yet, significant antitumoral effect against L.M. have been reported after both retroviral[100] and adenoviral mediated gene-transfer.[101]

Conclusion

Over recent years, the interest in the development of experimental models of colorectal liver metastases has increased due to the need for new adjuvant therapies to improve the treatment of both colorectal cancer and liver metastases. Two kinds of experimental model can be used:

(a) the induction of colon cancer by carcinogens into rats with spontaneous liver metastases fairly closely mimics the natural history of colon cancer but a low yield of both colonic cancer (less than 50%) and liver metastases (approx. 25%) is obtained after six months of latency;

(b) direct intraportal injection of cancer cells is the model most often used. Although it bypasses the natural history of cancer, this simple model produces 100% of liver metastases 6 weeks after injection of cells.

These two models have been used for several studies of liver metastases concerning their morphology, their modulating factors and their vascularization. They have been used to test new adjuvant therapies for colorectal cancer.

References

1. Kune, G. A., Kune, S., Field, B., White, R., Brough, W., Schellenberger, R. and Watson, L. F. (1990). Survival in patients with large bowel cancer. A population-based investigation from the Melbourne Colorectal Cancer Study. *Dis. Colon. Rectum*, **33**, 938–946.
2. Finlay, I. G. and McArdle C.S. (1986). Occult hepatic metastases in colorectal carcinoma. *Br. J. Surg.*, **73**, 732–735.
3. Nordlinger, B., Quilichini, A., Parc, R., Hannoun, L., Delva, E., Huguet, C. and Hepatic, (1987). resection for colorectal liver metastases : influence on survival of preoperative factors and surgery for recurrences in 80 patients. *Ann. Surg.*, **205**, 256–263.
4. Cady, B. (1983). Natural history of primary and secondary tumours of the liver. *Sem. Oncol.*, **86**, 550–555.
5. Steele, G. and Ravikumar, T. (1989). Resection of hepatic metastases from colorectal cancer. Biologic perspectives. *Ann. Surg.*, **210**, 127–138.
6. Woods, C. B. (1984). Natural history of liver metastases. In : Van de Velde, C. J. H, Sugarbaker, P. M. Eds. Liver metastases. Amsterdam: *Martinus Nishoff*, 47–54.
7. Scheele, J., Stangl, R. and Altendorf-Hofmann, G. (1990). Hepatic metastases from colorectal carcinoma: impact of surgical resection on the natural history. *Br. J. Surg.*, **77**, 1241–1246.
8. Patt, Y. Z., Chuang, V. P., Wallace, S., Hersh, E. M, Freireich, E. J. and Mavlight, G. M. (1981). The palliative role of hepatic arterial infusion and arterial occlusion in colorectal carcinoma metastatic to the liver. *Lancet*, **i**, 349–350.
9. Wexter, H. (1966). Accurate identification of experimental pulmonary metastases. *J. Natl. Cancer Inst.*, **36**, 641–645.
10. Panis, Y. and Nordlinger, B. (1991). Les modèles expérimentaux de métastases hépatiques d'origine colorectale. *Ann. Chir.*, **45**, 222–228.
11. Prat, F., Chauffert, B. and Martin, M. (1991). Les modèles animaux de métastases hépatiques. *Gastro-entérol. Clin. Biol.*, **15**, 689–696.
12. Lamont, J. R. (1978). Experimental colon cancer. *Gastroenterology*, **75**, 1157–1169.
13. Ward, J., Yamamoto, R. and Brown, C. (1973). Pathology of intestinal neoplasmas and other lesions in rats exposed to Azoxymethane. *J. Natl. Cancer. Inst.*, **51**, 1029–1039.
14. Balish, E., Shih C. X, Croft, W., Pamucku, A., Lower, G., Bryan, G. and Yale, C. (1977). Effect of age, sex and intestinal flora on the induction of colon tumors in rat. *J. Natl. Cancer. Inst.*, **58**, 1103–1106.
15. Nordlinger, B., Panis, Y., Puts, J. P., Hervé, J. P., Delelo, R. and Ballet, F. (1991). An experimental model of colon cancer: recurrences after surgery alone or associated with intraperitoneal 5-fluorouracile chemotherapy. *Dis Colon Rectum*, **34**, 658–663.
16. Panis, Y., Ballet, F., Oliveri, J. M., Hervé, J. P., Delelo, R. and Nordlinger, B. (1991). An experimental model of colon cancer in the rat : effects of surgical trauma and immunosuppression on colonic carcinogenesis and spontaneous liver metastases. *Surg. Res. Commun.*, **11**, 99–105.
17. Martin M. S. (1982). Cancérogènése colique expérimentale. *Gastroentérol. Clin. Biol.*, **6**, 382–391.
18. Jenkins, S. A., Day, D.W., Mooney, B., Devitt, P., Taylor, I. and Shields, R. (1984). The effect of vasopressin and hepatic artery ligation on the blood supply to normal and metastatic liver tissue. *Br. J. Cancer*, **50**, 785–791.
19. Mooney, B., Serlin, M. and Taylor, I. (1982). The effect of warfarin on spontaneously metastasing colorectal cancer in the rat. *Clin. Oncol.*, **8**, 55–59.
20. Shamsuddin, A. M. (1984). Comparative studies of primary, metastatic and transplanted colon adenocarcinomas of Fisher 344 rats. *J. Submicrosc. Cytol.*, **16**, 327–339.
21. Pozharisski, K. (1975). The significance of nonspecific injury for colon carcinogenesis in rats. *Cancer Res.*, **35**, 3824–3830.
22. Williamson, R., Bauer, F., Terpstra, O., Ross, J. and Malt, R. (1980). Contrasting effects of subtotal enteric bypass, enterectomy and colectomy on azoxymethane-induced intestinal carcinogenesis. *Cancer Res.*, **40**, 538–543.
23. Eccles, S. A., Heckford, S. E. and Alexander, P. (1980). Effect of Cyclosporin A on the growth and spontaneous metastasis of syngeneic animal tumours. *Br. J. Cancer*, **42**, 252–259.
24. Martin, M. S. (1987). Défense immunitaire anti-tumorale dans les cancers colorectaux. Acquisitions récentes dans les cancers colorectaux, congrès, Paris.
25. Dawiskiba, J., Jeppson, B., Perssòn, G., Hägerstrand, I. and Bengmark, S. A. (1983). model of colonic carcinoma with liver metastases in the rat. *J. Exp. Clin. Cancer Res.*, **3**, 269–272.
26. Bresalier, R., Raper, S., Hujanen, E. and Kim, Y. (1987). A new animal model for human colon cancer metastasis. *Int. J. Cancer*, **39**, 625–630.

27. Goldrosen, M. H. (1980). Murine colon adenocarcinoma. Immunobiology of metastases. *Cancer*, **45**, 1223–1228.
28. Goldrosen, M. H., Paolini, N. and Holyoke, E. Description of a murine model of experimental hepatic metastases. *J. Natl. Cancer Inst.*, **77**, 823–828.
29. Tan, M. H., Holyoke, E. D. and Goldrosen, M. H. (1977). Murine colon adenocarcinoma : syngeneic orthotopic transplantation and subsequent hepatic metastases. *J. Natl. Cancer Inst.*, **59**, 1537–1544.
30. Chauffert, B., Shimizu, T. and Caignard, A. (1988). A simple model of liver metastases from rat colon cancer; an histological study with a monoclonal antibody. *In vivo*, **2**, 301–306.
31. LaFrenière, R. and Rosenberg, S. (1986). A novel approach to the generation and identification of experimental hepatic metastases in a murine model. *J. Natl. Cancer Inst.*, **76**, 309–322.
32. Panis, Y., Ribeiro, Y., Chrétien, Y., Ballet, F. and Nordlinger, B. (1992). Dormant liver metastases: an experimental study. *Br. J. Surg.*, **79**, 221–223.
33. Tominaga, T., Yoshida, Y., Kitamura, M. and Kosaki, G. (1987). Liver metastasis of colon 26 cells implanted into the superior mesenteric vein in mice. *Jpn. J. Cancer Res.* (Gann), **78**, 846–850.
34. Van der Elst, J., De Greeve, J., Geerts, F., De Neve, W., Storme, G. and Willems, G. (1986). Quantitative study of liver metastases from colon cancer in rats after treatment with Cyclosporine A. *J. Natl. Cancer Inst.*, **77**, 227–231.
35. Giavazzi, R., Garofalo, A., Damia, G., Garattini, S. and D'Incalci, M. (1988). Response to flavone acetic acid (NSC 347512) of primary and metastatic human colorectal carcinoma xenografts. *Br. J. Cancer*, **57**, 277–280.
36. Weese, H. K., Emoto, S. E. and Sondel, P. M. (1987). Reduced incidence of hepatic metastases by perioperative treatment with recombinant human interleukin-2. *Dis. Colon Rectum*, **30**, 503–507.
37. Barnaron, Y., Iwaki, H., Bash, J., Brettschreider, F., Milsenbeck, S., Darnell, E. and Wallack, M. K. (1988). Treatment of murine hepatic metastases with vaccinia colon oncolysates and IL-2. *J. Surg. Res.*, **45**, 523–530.
38. Cohen, S. and Goldrosen, M. (1989). Modulation of a colon-derived experimental hepatic metastasis by murine non-parenchymal liver cells. *Immunol. Invest.*, **18**, 351–363.
39. LaFrenière, R., Borkenhagen, K., Bryant, L. and Muchcroft, S. (1989). Immunotherapy of murine hepatic metastases with lymphokine-activated killer cells expanded in serum-free media and recombinant interleukin-2. Cancer Res., **49**, 2409–2414.
40. Timar, J., Kopper, L., Pal, K. and Lapis, K. (1983). Transmission and scanning-electron microscopy of experimental liver metastases derived from intrasplenically growing primary tumor. *Path. Res. Pract.*, **177**, 47–59.
41. Yeatman, T., Bland, K., Copeland, E. and Kimura, A. (1989). Tumor cell-surface galactose correlates with the degree of colorectal liver metastasis. *J. Surg. Res.*, **46**, 567–571.
42. Castillo, M., Doevr, R., Paolini, N., Cohen, S. and Goldrosen, M. (1989). Hepatectomy prolongs survival of mice with induced liver metastases. *Arch. Surg.*, **124**, 167–169.
43. Doevr, R., Castillo, M., Evans, P., Paolini, N., Goldrosen, M. and Cohen, S. (1989). Partial hepatectomy augments the liver's antitumor response. *Arch. Surg.*, **124**, 170–174.
44 Martin, M. S., Chauffert, B., Caignard, A., Pelletier, M., Hammann, A. and Martin, F. (1989). Histoimmunological characterization of the cellular reaction to liver metastases induced by colon cancer cells in syngeneic rats. *Invasion Metastasis*, **9**, 216–230.
45. Nishiyama, Y., Fuchimoto, S. and Orita, K. (1989). Preventive and antiproliferative effects of tumor necrosis factor against experimental hepatic metastases of mouse colon-26 tumor. *Jpn. J. Cancer Res.*, **80**, 366–372.
46. Yamasaki, K., Konno, T., Mizauchi, Y. and Maeda, H. (1987). Reduction of hepatic metastases in rabbits by administration of an oily anticancer agent into the portal vein. *Cancer Res.*, **47**, 852–855.
47. Fisher, E. R. and Fischer, B. (1959). Experimental studies of factors influencing hepatic metastases I. Effect of number of tumor cells injected and time of growth. *Cancer*, **12**, 926–928.
48. Fisher, B. and Fisher, E. R. (1959). Experimental evidence in support of the dormant tumour cell. *Science*, **130**, 918–919.
49. Mendes da Costa, P., DeFleur, V. and Lejeune, F. J. (1978). Le carcinome colique transplanté chez le rat BD IX: un nouveau modèle pour l'étude de la dissémination métastatique et de thérapeutiques expérimentales. *Acta. Zool. Pathol.*, **72**, 65–76.
50. Dunnington, D., Buscarino, C., Gennaro, D., Greig, R. and Poste G. (1987). Characterization of an animal model of metastatic colon carcinoma. *Int. J. Cancer*, **39**, 248–254.
51. Martin, M. S., Bastien, H., Martin, F., Michiels, R., Martin, M. R. Justrabo, E. (1973). Transplantation of intestinal carcinoma in inbred rats. *Biomedicine*, **19**, 555–558.
52. Caignard, A., Martin, M. S. and Michel M. F. (1985). Interaction between two cellular subpopulations of a rat colonic carcinoma when inoculated to the syngeneic host. *Int. J. Cancer*, **36**, 273–279.
53. Martin, M. S., Caignard, A. and Hamman, A. (1987). An immunohistological study of cells infiltrating progressive and regressive tumors induced by two variant subpopulations of a rat colon cancer cell line. *Int. J. Cancer*, **40**, 87–93.
54. Weese, J. L, Ottery, F. D. Emoto, S. E. (1986). Do operation facilitate tumor growth ? An experimental model in rats. *Surgery*, **100**, 273–277.
55. Panis, Y., Nordlinger, B., Ballet, F. and Delelo, R. (1990). Infante J. Experimental colorectal liver metastases: influence of sex, immunological status and liver regeneration. *J. Hepatol*, **11**, 53–57.
56. Roberts, S., Watne, A., McGrath, R., Mc, Crew, E. and Cole, W. H. (1958). Technique and results of isolation of cancer cells from the circulating blood. *Arch. Surg.*, **76**, 334–346.

57. Archer, S. and Gray, B. (1988). A new reproductible model of hepatic and peritoneal metastases from colonic carcinoma. *Eur. J. Cancer Clin. Oncol.*, **24**, 1623–1632.
58. Naniemo, T., Takeichi, N., Hata, Y., Kobayashi, H. and Uchino, J. (1989). Analysis of mechanism of the effect of partial hepatectomy on enhanced growth of liver cancers in rats: Experimental analysis of the mechanism of recurrences after liver surgery in human liver cancers. *Acta. Hepatol. Jap.*, **30**, 889–897.
59. Fisher, B. and Fischer, E. R. (1959). Experimental studies of factors influencing hepatic metastases. II. Effect of partial hepatectomy. *Cancer*, 12, 929–932.
60. Fischer, B. and Fisher, E. R. (1959). Experimental studies of factors influencing hepatic metastases. III. Effect of surgical trauma with special reference to liver injury. *Ann. Surg.*, **150**, 731–744.
61. Gershbein, L. (1963). Transplantable tumor growth and liver regeneration in the rat. *J. Natl. Cancer Inst.*, **31**, 521.
62. Rozga, J., Tanaka, N., Jeppsson, B., Hargerstrand, I. and Bengmark, S. (1985). Tumor growth in liver atrophy and growth. An experimental study in rats. *Eur. J. Cancer Clin. Oncol*, **21**, 35–40.
63. Lee, J. (1971). Effects of partial hepatectomy in rats on two transplantable hepatomas. *Am. J. Pathol.*, **65**, 347–355.
64. Trotter, N. C. (1961). The effect of partial hepatectomy on subcutaneously transplanted hepatomas in mice. *Cancer Res.*, **21**, 778.
65. Wassef, R., Cohen, Z. and Langer, B. (1985). Pharmacokinetic profiles of cyclosporine in rats. *Transplantation*, **40**, 489–493..
66. Van Dale, P., Penasse, W., Popoxski, A., Jacobowitz, D. and Galana, P. (1984). An experimental model of liver metastases from rat colorectal adenocarcinoma. Treatment of metastasis: problems and prospects. Proc, Taylor–Francis, London, Philadelphia 1986.
67. Pearson, M. J., Anderson, J., Chamberlain, J. and Bell, P. (1986). The effect of Kupffer cell stimulation or depression on the development of liver metastases in the rat. *Cancer Immunol. Immunother*, **23**, 214–216.
68. Gjoen, T., Seljelid, R. and Kolset, S. (1989). Binding of metastatic colon carcinoma cells to liver macrophages. *J. Leuko Biol.*, **45**, 362–369.
69. Shapiro, J., Jersky, J. and Katzav, S. (1981). Anesthetic drugs accelerate the progression of post-operative metastases of mouse tumors. *J. Clin. Invest.*, 68, 678–685.
70. Weese, J. L., Gilbertson, E. M., Syrjala, S. E., Whitney, P. D. and Starling, J. R. (1985). Reduced incidence of rat colon cancer metastases by perioperative immunostimulation with maleic anhydride divinyl ether-2 (MVE-2). *Dis. Colon Rectum*, **28**, 217–221.
71. Park, S., Brody, J. I., Wallace, H. A. and Blakemore, W. S. (1971). Immunosuppressive effect of surgery. *L ancet*, **1**, 53–55.
72. Roth, J. A., Golub, S. H., Grimm, E. A., Eilber, F. R. and Morton, D. C. (1976). Effects of operation on immune response in cancer patients: sequential evaluation of in vitro lymphocyte function. *Surgery*, **79**, 46–51.
73. Slade, M. S., Simmons, J. L, Yunis, E. and Greenberg, J. L. (1975). Immunodepression after major surgery in normal patients. Surgery, **78**, 363–372.
74. Noltenius, C. and Noltenius, H. (1985). Dormant tumor cells in liver and brain. *Path. Res. Pract.*, **179**, 504–11.
75. Brattain, M. G., Fine, W. D., Khaled, F. M., Thompson, J. and Brattain, D. E. (1981). Heterogeneity of malignant cells from a human colonic carcinoma. *Cancer Res.*, **41**, 1751–6.
76. Ramshaw, I. A., Carlsen, S., Wang, H. C. and Badenoch-Jones, P. (1983). The use of cell fusion to analyse factors involved in tumor cell metastasis. *Int. J. Cancer*, **32**, 471–8..
77. Giavazzi, R., Jessap, J. M., Campbell, D. E., Walker, S. and Fidler, I. (1986). Experimental nude mouse model of human colorectal cancer liver metastases. *J. Natl. Cancer Inst.*, **77**, 1303–1308.
78. Wu, A., Nicolet, J., Morin, J., Kemeny, F., Smadja, C. and Franco, D. (1989). A simple experimental model of localised hepatic metastases of colonic origin by intraportal injection. *Surg. Res. Commun.*, **6**, 205–211.
79. Ackerman, N. B., Lien, W. M., Kondi, E. S. and Silverman, N. A. (1969). The blood supply of experimental liver metastases. I. The distribution of hepatic artery and portal vein blood to "small" and "large" tumors. *Surgery*, **66**, 1067–1072.
80. Ackerman, N. B., Lien, W. M. and Silverman, N. A. (1972). The blood supply of experimental liver metastases. III. The effects of acute ligation of the hepatic artery or portal vein. *Surgery*, **71**, 636–641.
81 Ackerman, N. B. (1974). The blood supply of experimental liver metastases. IV. Changes in vascularity with increasing tumor growth. *Surgery*, **75**, 589–596.
82. Lien, W. M. and Ackerman, N. B. (1970). The blood supply of experimental liver metastases. II. A microcirculatory study of the normal and tumor vessels of the liver with the use of perfused silicone rubber. *Surgery*, **68**, 334–340.
83 Bagge, U., Skolnik, G. and Ericson, L. E. (1983). The arrest of circulating tumor cells in the liver microcirculation. A vital fluorescence microscopic, electron microscopic and isotope study in the rat. *J. Cancer Res. Clin. Oncol.*, **105**, 134–140.
84. Nott, D., Crime, S., Yates, J., Day, D., Buxter, N., Jenkins, S. and Cooke, T. (1989). Changes in the hepatic perfusion index during the development of experimental hepatic tumours. *Br. J. Surg.*, **76**, 259–267.
85. Qin, Y., Koen, D. and Willems, G. (1990). Easy production and early detection of liver metastases of colon cancer in rats. *Surg. Res. Commun.*, **9**, 99–105.
86. Hillan, K., Nordlinger, B., Ballet, F., Puts, J. P. and Infante, R. (1988). The healing of colonic anastomoses after early intraperitoneal chemotherapy : an experimental study in rats. *J. Surg. Res.*, **44**, 176–177.
87. Ross, D., Steele, G., Madara, J., Lahey, S., Ravikumar, T., Wilson, R., Munroe, A., Wright, D. and Kingt, V. (1984). Effects of specific active immunization on tumor recurrence following primary tumor resection in WF rats with 1,2 dimethylhydralazine-induced bowel cancer. *J. Natl. Cancer Inst.*, **73**, 731–735.
88. Mooney, B. and Taylor, I. (1982). The effect of hepatic

arterial ligation on spontaneously developing colorectal liver metastases in the rat. *Clin. Oncol*, **8**, 231–235.

89. Hafstrsm, L. and Carlsson, G. (1984). Influence of hepatic artery ligation on survival. An experimental study in rats. *Am. J. Surg.*, **147**, 688–691.
90. Paris, P., Nagawa, H., Chauffert, B. and Martin, F. (1989). Les anticoagulants peuvent-ils prévenir les métastases hépatiques d'origine colorectale ? *Gastroentérol. Clin. Biol.*, **A52** (abs).
91. Cameron, R., Mc Intosh, J. and Rosenberg, S. (1988). Synergistic antitumor effects of combination immunotherapy with recombinant interleukin-2 and a recombinant hybrid a-interferon in the treatment of established murine hepatic metastases. *Cancer Res.*, **48**, 5810–5817.
92. LaFrenière, R. and Rosenberg, S. (1985). Adoptive immunotherapy of murine hepatic metastases with lymphokine activated killer (LAK). cells and recombinant interleukin-2 (RIL-2) can mediate the regression of both immunogenic and non-immunogenic sarcomas and an adenocarcinoma. *J. Immunol.*, **135**, 4273–4280.
93. Thrombe, P. S. and Deodnar, S. D. (1984). Inhibition of liver metastases in murine colon adenocarcinoma by liposomes containing human C reactive protein or crude lymphokine. Cancer Immunol. *Immunother.*, **16**, 145–150.
94. Otaka, M., Singhal, A. and Hakomori, S. (1989). Antibody-mediated targeting of differentiation inducers to tumor cells : inhibition of colonic cancer growth in vitro and in vivo. A preliminary note. *Biochem. Biophys. Res. Commun.*, **158**, 202–208.
95. Weese, J. L., Gilbertson, E. M., Syrjala, S. E. and Starling, J. R. (1984). Prevention of rat colon cancer metastases by perioperative stimulation. *Surgery*, **96**, 420–426.
96. Sukumar, S., Hunter, J. T. and Terata, N. (1983). Eradication of microscopic hepatic metastases by active specific immunization. *Cancer Immunol. Immunother.*, **14**, 151–154.
97. Mayhew, E., Goldrosen, M. H., Vaage, J. and Rustrum, Y. (1987). Effects of liposome-entrapped Doxorubicin on liver metastases of mouse colon carcinoma 26 and 38. *J. Natl. Cancer Inst.*, **78**, 707–713.
98. Chiannilkuchai, N., Driovich, A., Benoit, J. P., Parodi, A. L. and Couvreur, P. (1985). Doxorubicin-loaded nanoparticles : increased efficiency in murine hepatic metastases. Selective Cancer Therapeut, **5**, 1–10.
99. Panis, Y., Kienmanesh Rad, A. R., Boyer, O., Houssin, D., Salzmann, J. L. and Klatzmann, D. (1996). Gene therapy for liver tumors. *Surg. Oncol. Clin. North Am.*, **5**, 461–473.
100. Caruso, M., Panis, Y., Gagandeep, S., Houssin, D., Salzmann, J. L. and Klatzmann, D. (1993). Regression of an established macroscopic liver metastases after in situ transduction of a suicide gene. *Proc. Natl. Ac. Sci. USA*, **90**, 7024–7028.
101. Chen, S. H., Chen, X. H., Wang, Y., Kosai, K. I., Finegold, M. J. and Rich, S. (1995). Combination gene therapy for liver metastasis of colon carcinoma in vivo. *Proc. Natl. Ac. Sci. USA*, **92**, 2577–2581.

PART V

18. Animal Models in Breast Cancer Research

Jan Hein van Dierendonck and Henk-Jan van Slooten

Department of Surgery, Leiden University Medical Center, The Netherlands

Introduction

The mammary gland is a phylogenetically young organ, unique to the class Mammalia. Its complex development is the result of the different growth phases of the organism, initiation of hormonal changes at puberty, and changes induced by hormonal stimulation during pregnancy. In some domestic species, such as dog and cat, and in women of the industrialized Western World it is a relatively high-risk site for development of (pre)neoplastic lesions. As a result, the mammary gland is one of the most extensively studied organs, of interest to researchers in the field of developmental biology and experimental pathobiology.

Animal model systems are required to study (a) developmental aspects: hormonal and growth factor control on the natural developmental cycles of growth, morphogenesis, differentiation and involution of the mammary gland, and various aspects of stromal-parenchymal interactions, (b) carcinogenesis: the influence of inherited genetic factors, exposure to chemical carcinogens, damaging radiation, oncogenic virusses, mitogenic hormones and various other host factors (including age and reproductive history) on mammary cancer development, (c) tumor-host interactions: effects of (neuro)endocrine, immunological, and dietary factors and chronobiological phenomena on the progression of primary and metastatic mammary tumors, (d) prevention an treatment: the biological principles, feasibility and efficacy of different treatment modalities.

Because of the complexity of the entire organism, *in vitro* methods are widely applied to address specific questions on cellular and molecular biology of growth regulation and antitumor treatment and therefore serve as a valuable complement to *in vivo* research. A major challenge of present research is to establish the physiological significance of *in vitro* observations and their potential relevance to the prevention and treatment of human breast cancer. Knowledge of comparative aspects of mammary gland development and cancer in different mammalian species is of utmost importance to interpret data obtained from *in vivo* models in order to extrapolate them to the human species. This chapter focusses on the origin and characteristics of a number of animal models frequently used in breast cancer research and discusses their relevance for the study of human disease.

Comparatative Aspects of Mammary Gland Development

Major progress in understanding of the biology of breast cancer has involved research on inbred mice and rat strains, in which cancers of the mammary gland are quite common. There are many excellent recent reviews dealing with various aspects of rodent mammary biology, covering gland architecture, embryology, endocrine requirements for growth and differentiation and factors involved in neoplastic transformation.[1–5]

Studies in mice revealed that mammary epithelium initially develops similar in males and females, with the formation of five pairs of ectodermal bulb-shaped structures. In males, these structures regress under the influence of androgen (acting on the surrounding mesenchyme). In females, gland development proceeds by outgrowth of mammary sprouts penetrating the mammary fat pad precursor tissue and ramifying to form a system of branching ducts. In the prepuberal period, this allometric growth may vary considerably among different strains of mice and rats; in mice it strongly depends on the presence of the ovaries, whereas in rats it continues even after ovariectomy, probably due to the action of adrenal sex steroids.[4] The end of the ducts consists of a specialized group of cells, the terminal end bud (TEB); in rats each TEB can progressively differentiate into 3 to 5 smaller alveolar buds (ABs), a process accentuated by each estrus cycle which starts when animals are about 40 days old.[6] Each AB, surrounded by connective tissue, forms a lobular structure; this structure is found in human and rat mammary glands, but not in mice.[3]

The degree of differentiation of the mammary gland is important in the initiation of carcinogenesis,[7] examplified by the protective effect against breast cancer of pregnancy at young age in humans. However, it is not clear whether the intralobular site in human breasts is equivalent to the TEB, the most suspectible site in rodents.[6] Also the precise role and interspecies differences of various endocrine control mechanisms on mammary gland development and on the initiation, development and growth of its neoplasms needs to be determined.

A basic morphological difference between rodent and human mammary glands is that in humans, ductal structures grow along connective tissue septa and that lobular structures rarely grow into adjacent fat tissue. This difference might be particularly of relevance in studies on stroma-paranchyma interactions. Another difference is that in human females the relationship between age and gland development is less well defined than in rodents, probably due to a combination of endocrinologic, genetic and socioeconomic differences.[6]

The mammary parenchyme is composed of two major cell types: the contractile myoepithelial cells, arranged parallel to the basement membrane, and centrally located secretory epithelial cells. It has been hypothesized that an intermediate cell may represent a common developmental stage of epithelial and myoepithelial cells.[8]

Epithelial cells can be further divided on basis of topography and function into ductal and (milk secreting) alveolar luminal cells. A basic question with respect to the development of different histological types of mammary cancer is whether they originate from these different cell types or from a pluripotent stem cell. Both in rats[9,10] and mice[11–12] potential stem cell candidates have been described. From an *in vivo* transplantation model in mice, Smith *et al.* deduced that mammary epithelial committed stem cells are present within the glandular population throughout the adult life.[13] These cells divide and give rise to daughter cells with specific developmental potential and a limited proliferative life span, i.e., one cell remains a stem cell whereas the other cell will activate a specific differentiation program. In the prepuberal gland, stem cells will provide ductal progenitors; at the onset of puberty the ovarian hormones activate local growth promotors, leading to an increased activity of ductal extension. During this period some of the stem cell progeny receive signals for lobular (i.e. secretary) differentiative programs; at the onset of pregnancy and during secretory development, the mammary stem cells produce primarily lobular committed progeny.[13]

Mammary Tumors in the Mouse

Preneoplastic Dysplasia and the Role of Tumor Virus

The incidence of mammary tumors in inbred laboratory mice depends upon genetic background, endocrine factors, presence of mouse mammary tumor retrovirus (MMTV) and a variety of other factors.[14] Spontaneous mouse mammary tumors were the first tumors continuously transplanted and studied with respect to hereditary factors and influence of hormones on tumor development.[15] The mouse is also the most extensively studied model for preneoplastic dysplasia and neoplastic progression. Its gland may contain a distinctive focal epithelial hyperplasia, the hyperplastic alveolar nodule (HAN), which is likely comprised of mammary stem cells and their lobular-committed daughters[13] and is prone to develop into mammary carcinoma.[16] HANs can be indefinitely maintained in syngeneic hosts as serial transplants, but no growth is observed outside the mammary fat pad.[17] MMTV seems to play a major role in HAN formation: in a vast majority of mouse strains comparable MMTV copy numbers have been detected.[14] This endogenous MMTV is a proviral form, associated with a relatively late onset of mammary tumor development, in contrast to the exogenous MMTV, which is transmitted via the milk and is associated with earlier oncogenesis.

The mutagenic potential of MMTV is involved in the origin of HANs as well as their progression towards invasive and metastatic lesions. The genes at MMTV insertion sites were originally collectively called *int* genes.[18] *Int*-1 appeared to be homologous to the *wingless* gene in Drosophila, and became the first member of a family of highly conserved developmental genes, (*Wnt* genes) encoding secretory glycoproteins involved in cell-cell signaling events.[19] Several of these genes were found to induce morphological transformation and altered growth characteristics in an epithelial cell line derived from normal mouse mammary tissue.[20]

Another gene in which MMTV insertion frequently occurs, *int*-2, encodes for a member of the fibroblast growth factor (*Fgf*) gene family[21]. Until now, eight different genes (*Wnt*-1, *Wnt*-3, *Wnt*-10b, *Fgf*-3, *Fgf*-4, *Fgf*-8, *Int*-3 and *Int*-6) have been shown to be altered in multiple mammary tumors as a consequence of MMTV integration. Some MMTV-infected mice develop tumors in which both *Wnt*-1 and *Fgf*-3 are activated by the viral genome, suggesting cooperating effects in tumorigenesis (discussed in 22). However, no proof has been found that these genes play a significant role in human breast cancer.[22,23]

It is known for a long time that most HANs and HAN-derived mammary neoplasms are immunogenic by virtue of the virus-encoded tumor antigens as well as viral-induced nonvirus-associated antigens.[24] Especially tumors arising from HANs after treatment with carcinogens in general are strongly immunogenic.[25]

Another virus known for its oncogenic properties in the induction of mammary tumors in mice is the murine polyoma virus (MPyV): ductal adenocarcinomas develop at high frequency and with short latency in infected immuno-competent adult female or normal neonatal mice or in transgenic mice expressing the viral oncogene middle T.[26]

Carcinogen Induced Tumors

Many mouse strains are highly susceptable to carcinogens like 7,12-dimethylbenzanthracene (DMBA), methylcholanthrene (MCA) and urethan, leading to various mammary dysplasias, most frequently ductal hyperplasia. The type of mammary tumors induced are most often adenoacanthomas and type-B mammary adenocarcinomas.[27] In contrast to the rat model (discussed below), multiple carcinogen injections are required for optimal tumor yields, (interfering with animal health and the possibility to properly study the process of tumor progression). Also the latency period is relatively long, although in Balb/c mice this period can be significantly shortened by prolonged hormonal stimulation.27

A transplantable tumor line derived from a carcinogen-treated BD2F1 mouse is the ovarian-dependent MXT ductal carcinoma;[28] this tumor line has been frequently used for endocrine-related studies.

Carcinogen-induced mouse mammary tumors are often hormone-responsive, but the relative importance of pituitary and ovarian hormones for development and growth of these tumors is not completely understood.

Endocrine Factors

In several strains such as DBA, CBA, and C3H, pituitary isografting or multiple pregnancies stimulate the development of spontaneous mammary tumors, whereas certain substrains of BALB/c and C57BL are relatively resistant to these endocrine effects[29] and other strains, such as TSI and STS, are almost or completely refractory.[30]

Even the 'high-incidence' mouse strains do not develop mammary tumors in the absence of ovarian steroids and prolactin, but once formed, they are mainly hormone independent, i.e. hormones are not required for continued growth. In GR/A mice however, in which mammary tumors are readily induced by the administration of estrogen and progesteron or by repeated pregnancies, resulting ductal carcinomas most often retain their hormone responsiveness upon transplantation and tumor growth is inhibited by drugs that suppress prolactin secretion.[31] Also mammary tumors in the above mentioned CH3 mice can be transplanted to syngeneic hosts, but these tumors are neither pituitary nor ovary dependent.[32]

The transplantable ductal TPDMT-4 carcinoma, derived from a pregnant DDD mouse, is not strictly hormone-dependent, but its growth is stimulated by prolactin and a combination of estrogen and progesteron[32]. Another interesting transplantable model is the Shionogi Carcinoma 115 tumor, a medullary carcinoma that arose spontaneously in a DD/S mouse and is dependent on androgens for growth.[34]

Parity has been shown to have an important effect on mouse mammary tumorigenesis: breeders generally show a higher incidence than virgins.[35] The stimulating effect of pregnancies is not found in humans: parity before 30 years of age has a risk-decreasing effect that is maximal when first child is born before the age of 20.[36]

Transgenic Mice

In the study of tumorigenesis and metasatasis at the genetic level, transgenic technology has been used to modify the mammalian genome by introducing foreign DNA into the germ line in a stable fashion. With respect to mammary epithelium, two related approaches have been pursued: the development of oncogen-containing transgenic mice[37] and transplantation of oncogene- containing mammary epithelial cells into mice from which the endogenous epithelium had been removed early in development.[38] These systems have been used to study the influence of *myc*, *ras*, *neu*, *src*, *cyclin* D1, int-1 and *int*-2 oncogenes, TGF-a and PyV middle T protein on mammary tumorigenesis (reference 39 and references therein).

Targeted expression of genes to the mammary epithelium has been accomplished by linking them to either the MMTV long terminal repeat or to the wey acidic protein promotor (WAP), the latter being transcriptionally active exclusively in the mid-pregnant mammary gland.[40]

It has been found that induction of tumors in most of these transgenic models is a multi-step process where transgene expression, although required, is not sufficient for tumorigenesis. In hormone-stimulated mice, c-*myc* causes mild hyperplasia, but introduction of v-Ha-*ras* into c-*myc* expressing cells causes tumor development.[41] Analyses of the offspring from interbreeding between the MMTV/*int*-2 and MMTV/*Wnt*-1 transgenic mice provided direct evidence of the cooperation of *int*-2 and *Wnt*-1 in mouse mammary tumorigenesis.[42]

Also MMTV/TGF a mice develop mammary gland hyperplasias and require additional genetic events to transform the primary epithelial cell.[43] Interestingly, mice carying both the MMTV/activated *ras* and MMTV/activated *neu* transgenes have been reported to show a reduced incidence of mammary tumors by comparison to monogenic animals.[44]

Both *neu* and PyV middle T (both encoding for potent tyrosine kinases) seem to be exceptions to the multi-step tumor progression model: gland-specific expression of either Neu or PyV middle T protein has been correlated with the rapid and synchronous appearance of multifocal mammary tumors.[45,46] The effect of the latter protein has been reported to be mediated by the product of the c-*src*-proto-oncogene.[46]

Studies on the histology of mammary tumors arising in transgenic mice containing *ras*, *myc* or *neu* have suggested that the phenotype of these tumors is related to their particular genotype. Although most murine mammary tumors do not seem to have morphologic counter-parts amongst the common types of human breast cancer, the pale intermediate-cell tumor associated with the *neu* transgene seems to have considerable resemblance to human comedo-carcinomas.[44]

Tumor growth results from the simple fact that proliferation rate of tumor cells exceeds their rate of death; a reduction of the latter may therefore produce the same net result as elevated growth rate. In response to a variety of extracellular signals, cells can initiate a series of molecular events leading to their own self-destruction. This active process, known as 'programmed cell death' or 'apoptosis', is distinguished from generalized necrosis resulting from environmental insults, such as lack of oxygen or nutrients.[47] An interesting observation in this context is that although MMTV-*ras*, MMTV-*myc* and MMTV-*myc/ras* tumors were recently reported to have similar growth rates, MMTV-*myc* tumors showed much higher levels of both mitosis and apoptosis than MMTV-*ras* tumors, whereas intermediate levels were observed in MMTV-myc/*ras* tumors.[48]

Apart from (proto) oncogenes much interest is focussed nowadays on tumor suppressor genes.[49] E.g., the p53 tumor suppressor gene is the gene known to be mutated most frequently in human cancer, with point mutations and deletions observed in about one quarter of all sporadic breast cancers.[50] The P53 protein is a transcription factor that probably plays a central role in the surveillance of gene amplification, abnormal recombination and the control over ploidy. It has been proposed that p53-mediated apoptosis is an important part of the tumor suppressor phenotype and subsequent selection for mutant p53 genes in cancers. However, p53 also acts in the cell cycle checkpoints and is responsible for cell cycle arrest or senescence.[51]

Metastatic Behavior

Tumor metastasis remains the major cause of morbidity and death in breast cancer patients. A metastatic outgrowth is the result of a complex series of tumor-host interactions that follow primary tumor initiation and progression. Metastatic cells must invade host tissue adjacent to the primary tumor, intravasate blood vessel and/or lymphatics, extravasate at a distant site and interact with it in such a way that the colony can grow beyond a certain size. The acquirement of these metastatic properties is probably the result of an accumulation of both genetic and epigenetic events resulting in a growth-dominant phenotype and increased tumor cell hetero geneity within the primary tumor.[55]

A major reason why relatively few rodent mammary tumor models with evidence of dissemination to other organs have been described is probably related to the limited life span of rodents.

However, in CD8F1 mice, a relatively high incidence of pulmonary metastases has been reported for spontaneous mammary carcinomas.[56] In SHN mice, which have a very high incidence of spontaneous mammary tumors at a relatively early age, repeated surgical reduction of the primary tumor increases the incidence of lung metastases.[57] Lung colony formation after intravenous injection of tumor cells is often used in studies on metastasis, but this method has been repeatedly questioned as a reliable measure of metastatic capacity.[58,59]

The transgenic mouse has emerged as an important model to study the progression towards a metastatic phenotype. E.g., a remarkable feature of the above described MMTV/PyV middle T and Neu tumors is that they metastasize with very high frequency to the lungs.[45] Recent findings suggest that enhanced expression of PEA3, a member of the *ets* oncogene family, may be required to facilitate the metastatic behavior of these tumors.[53]

Frequently transplanted mouse mammary tumors often show a progression towards a metastatic phenotype. However, whereas in these tumor models blood borne metastases are frequently observed, relatively few models have been described showing dissemination along lymphatic pathways.[61,62]

Mammary Tumors in the Rat

Spontaneous Tumors and the Role of Endocrine Factors

Mammary tumors are the most common type of neoplasm in female laboratory rats. The frequency depends on strain and period of observation.[63] In rats that are allowed to live a full life time, at least 85% of all mammary neoplasms are benign (fibroma, fibroadenoma, adenoma).[64] Malignant neoplasms are the invasive and non-invasive ductal carcinoma (with cribriform, comedo, solid, papillary or not 'otherwise specified' growth patterns), followed by the fibrosarcoma and the carcinosarcoma, all with their counterpart in human disease.[6] No evidence for the involvement of a viral factor has been found.[65]

As in mice, endocrine factors play an important role in carcinogenesis; tumors arise more frequently in multiparous than in nulliparous rats[66] and early ovariectomy-adrenalectomy or hypophysectomy completely blocks this process.[67] Chronic treatment with moderate levels of estrogen, e.g. by implantation of an estrogen containing pellet (EP), enhances the incidence[64,68] and this enhancement is prevented by hypophysectomy.[69] The latency period of tumor development in EP-bearing rats is increased by ovariectomy, adrenalectomy, treatment with progesterone or testosterone.[69]

Also treatment with prolactin (e.g. by implantation of pituitary homografts[66] or induced hypothalamic lesions[70]), or with growth hormone[71] increases mammary tumor incidence, but this concerns predominantly benign tumors.[66,70]

In general, tumors arising in estrogenized rats remain highly dependent on estrogen for their growth and regress upon EP removal.[69,72,73] The effect of estrogen on tumor growth is mainly mediated by stimulation of pituitary prolactin secretion,[73] a major growth-promoting agent in most hormone-dependent rat mammary cancers.[74]

In tumors which are induced and maintained under supraphysiological hormone levels, progression towards autonomous growth is a relatively slow process,[75] even after many serial transplantations in EP-bearing recipients. This relative stability with respect to hormone-dependency is probably related to selection of the most rapidly growing hormone-dependent tumor cell subpopulation from transplants in EP-bearing hosts.[76] A remarkable feature of these carcinomas is that after regression of primary or transplanted tumor or after transplantation to untreated females, a certain population of residual tumor cells persists, which (even after many months of 'dormancy') retains the capacity to resume growth upon hormonal stimulation. Occasionally, these highly hormone-dependent growth characteristics are combined with a truly malignant phenotype. This is demonstrated by the recently described transplantable EMR-86 adenocarcinoma, which metastasizes to lymph nodes and lungs in estrogenized WAG/Olac rats.[77] After EP-removal, tumor cell division decreases simultaneously with serum prolactin levels and is followed by a transient increase in apoptosis;[78] within 10 days, this results in elimination of the majority of tumor cells without affecting the gross histology of the tumor. This orderly regression greatly resembles involution of lactating mammary glands following weaning, a process thought to be initiated by prolactin depletion and which involves apoptotic cell death.[79]

After prolonged hormonal withdrawal, within the residual dormant EMR-86 tumors a small fraction of cells continue to proliferate, which is not further reduced by additional treatment either with bromocriptin or by ovariectomy.[77] Recent data on cell cycle kinetics of hormonal recruitment of dormant EMR-86 tumors is consistent with the concept that during dormancy, proliferation of cells with pluripotent self-renewal capacity ('stem cells') is balanced by loss of the (non-proliferating) stem cell daughter cells; upon stem cell division, one cell enters a resting state and eventually dies. When serum prolactin levels are increased above a critical treshold, cells that had only recently entered the non-proliferating state are stimulated to divide and rapidly repopulate the tumor (J.H. Wijsman *et al.*, submitted for publication).

Radiation- and Carcinogen-Induced Rat Mammary Tumors

Sublethal doses of different types of whole body irradiation have been shown to induce mammary tumors in rats.[80] Fractionated schemes did not differ in that respect from single doses, but the hormonal status appeared of paramount importance, because ovariectomy completely prevents and estrogen treatment enhances tumor formation. In EP-bearing rats, the latency period is shortened, the fraction of rats with carcinomas, (especially of the cribriform type), is increased, as is the number of tumors in tumor bearing rats.[81] A striking difference between radiation- and chemical carcinogen-induced rat mammary tumors is that the former show a random anatomical distribution, do not show an age-dependent induction and virgin, pregnant and lactating rats appear to be equally sensitive.[82]

In 1959, Huggins reported on the optimal conditions for induction of mammary carcinomas in Sprague-Dawley (SD) rats by gastric intubations of MCA.[83] Depending on optimal dosage, timing and hormonal status in the recipient animal, multiple mammary carcinomas were induced within 60 days. In the same rat strain, an even better result was obtained with DMBA; given by gavage in a single dose of 2.5 to 20 mg this compound induces tumors with latencies that generally range between 8 and 21 weeks with a final incidence close to 100%.[84] This 'Huggins tumor' probably represents the most intensely investigated animal model for human breast cancer.

In 1975, Bots and Willighagen reported that monthly i.v. injections of NMU into female Lewis rats resulted in a high yield (65%) of multiple mammary carcinomas 6 months after administration[85] and Gullino *et al.* presented similar results for SD and Fischer 344 rats (yield: 73 and 89% respectively).[86]

Russo and Russo have extensively investigated histogenesis of DMBA carcinomas.[87–89] They observed that administration of DMBA to virgin rats of different ages induces tumors with an incidence proportional to the density of highly proliferating TEBs. A sharp decrease in number of TEBs, observed in animals older than 55 days, resulted in a lower incidence as well as a lower number of tumors per animal. As a result of DMBA treatment, many TEBs become larger, the stage at which they are called intraductal proliferations (IDPs), and their confluence leads to the formation of microtumors. These intraductal carcinomas show various patterns such as cribriform, comedo or papillary type. ABs do not produce carcinomas, but may give rise to tubular adenomas, cystic dilatations or hyperplastic alveolar nodules.

The process of differentiation of TEBs into ABs does not occur simultaneously in all glands but depends upon the topographic location of each gland. The observed differences in degree of tumor development between thoracic and abdominal glands explain the higher incidence of ductal carcinomas originating from the former.[89] It was concluded that differentiation modifies gland structure and cell kinetics, decreasing the growth fraction and lengthening the cell cycle. Pregnancy occuring at a young age of 45 to 55 days, induces rapid differentiation of TEBs into ABs and of ABs into lobules, which soon show secretory activity. Decreased tumor incidence has been observed when DMBA administration occurred after pregnancy and lactation or when mammary growth had been prestimulated by hypothalamic lesions or pituitary grafting.

MNU preferentially induces guanine-adenine transitions in the cellular DNA, responsible for mutation of the H-*ras*-1 locus, which becomes activated in a vast majority of MNU-induced rat mammary carcinomas.[90] DMBA-induced tumors have a relatively low incidence of activated *ras* expression ($\sim 25\%$)[90] and in human breast tumors mutation of ras is extremely rare, with activation through amplification occurring in less than 25% of cases.[91] It has been postulated that ras oncogenes may play a role in tumor promotion or progression if they become activated in proliferating cells that have been initiated by other mechanisms involving chemical carcinogens or activation of other types of oncogenes.

Isaacs found that Copenhagen (COP) strains did not develop mammary tumors after treatment with DMBA or MNU. Analysis of breed produced by crossing COP with a highly carcinogen susceptible Osborne-Mendel (OM) strain revealed that this resistance is due to the presence of a single dominant autosome allele in the COP germ line. Moreover, this autosomal dominant gene appears to be specifically a mammary cancer suppressor gene, which is present in feral rats, but can be functionally inactivated by mutation or deletion in the germ line of certain strains like the OM and inbred SD.[92]

Metastatic Behavior

Although the most reliable criterion of malignancy, few authors reported the occurence of metastases from spontaneous or experimentally induced rat mammary tumors. Therefore, the parameters generally applied to discriminate malignant from benign are the histopathologic features that are used in human pathology, e.g., loss of tubuloalveolar growth patterns, increased nucleo-cytoplasmic ratio, prominent nucleoli, lack of myoepithelium, invasion of stroma and neighbouring fat, and stromal response.[81]

In general, carcinogen-induced rat mammary carcinomas appear to have antigenic determinants not observed in normal mammae. It is still not understood how these neoplasms escape immunological response or whether this antigen expression is related to their lack of metastatic behavior.[6,81,89]

Kim and colleagues compared a series of nonmetastasizing and metastasizing transplantable rat mammary carcinomas derived from MCA- or DMBA-treated female Wistar-Furth rats and reported that the metastasizing carcinomas possessed little or no membrane glycocalix, whereas all nonmetastasizing tumors had a thick glycocalix.[93] Moreover, metastasizing variants had significantly increased glycoprotein metabolism and shedding of glycoprotein membrane components compared with nonmetastasizing cells.[94] In the transplantable 13762 rat mammary carcinoma model a correlation has been reported between the metastatic potential of several clones, their expression of sialogalactoprotein and their sensivity to macrophage-mediated cytolysis.[95]

On a genetic level, increased expression and/or mutation of particular genes may result in an increased metastatic ability, but also the decrease of expression or total elimination of so called 'metastasis suppressor genes' has been hypothesized to be an important factor.[96] E.g., expression of the NM23 gene is associated with low metastatic ability of a variety of cells, including NMU-induced cancer cells.[97]

Mammary Tumors in Cats and Dogs

Canine Mammary Cancer

Mammary tumors represent the most common form of spontaneous neoplastic disease in dogs, with an incidence approximately three times higher than observed in women.[98,99] This incidence increases significantly with age, probably because ovarian activity does not stop in this species.[100]

A majority of all canine mammary tumors is benign and these tumors often contain a prominent myoepithelial cell component, which may undergo metaplasia to cartilage and bone (i.e. mixed tumors). This histological characteristic distinguishes this species from many others, including man, where such a change occurs rarely.[81] True malignant mixed tumors, i.e. carcinosarcomas, are relatively uncommon.[101]

About 30% of the carcinomas and 75% of the sarcomas give rise to metastases; lymph nodes and lungs are the most frequently affected metastatic sites.[101] It is known for long that progesterone and many of its derivatives promote mammary tumor development,[102] and that progestin can induce overproduction of growth hormone (GH).[103] Recently it was shown that progestins stimulate GH synthesis in the mammary glands of a variety of species, but it needs to be proven whether this local biosynthesis is the cause of mitogenic effects of progestins on mammary epithelium.[104,105] Also the role of prolactin in mammary tumorigenesis remains uncertain[106] and young age at first pregnancy was not found to be protective.[107]

As in humans, in a large proportion of canine mammary cancers estrogen and progestin receptors have been found.[108] However, in dogs steroid receptor positivity of metastases is less common, and receptor levels in both primary cancers and metastases are often lower than in man.[108] Autonomously growing canine mammary tumor cell lines derived from spleen, lung and lymph node metastases were found to be estrogen receptor negative, whereas high levels of progesterone receptors were present.[109]

In women with breast cancer, DNA aneuploidy as defined by flow cytometric DNA analysis has been associated with an unfavorable prognosis.[110] Although also in dogs a high frequency of aneuploidy is observed, a correlation with clinical stage and histological malignancy could not be established.[111] A remarkable finding is the presence of aneuploidy in non-malignant mammary cancers and the relative high frequency of hypodiploid tumors, being most prominent in anaplastic carcinomas.[112] This hypodiploidy might be related to the larger number of chromosomes in the canine karyotype, which reduces the chance of losing vital genes with chromosomal loss. It was hypothesized that this chromosomal loss may be a factor in etiology and metastatic capacity of canine mammary tumors, as it may imply the loss of tumor suppressor genes.[111]

The potential advantage of domestic dogs as a model for human disease is that relatively large numbers of dogs can be easily recuited for trials on mammary cancer prevention.[113] Moreover, female pet dogs presenting with advanced metastatic mammary cancer and considered for euthanasia may provide valuable autopsy material, e.g. to study pharmacokinetics of immunodiagnostic and antineoplastic compounds.

Feline Mammary Cancer

Mammary neoplasms occur less commonly in cats than in dogs.[101] The majority (85%) of these tumors are carcinomas and most of these metastasize to regional lymph nodes or distant sites.[114–116] Histological features of feline mammary carcinomas are more similar to those of human carcinoma than to those of either murine or canine mammary carcinoma.[114]

As in dogs, there is no definitive evidence for a role of viruses in the etiology of feline mammary neoplasms and although viral particles (Feline Leukemia Virus) were described,[117] according to the current opinion these should be regarded as passenger viruses.[118,119] At an early age, ovariectomy reduces risk of tumor development seven-fold, whereas it does not significantly affect growth of established tumors.[114] At clinical presentation, tumors were frequently found to be negative for estrogen receptors, and (in contrast to dogs) also for progesterone receptors.[118]

Little is known about genetic factors involved in feline mammary carcinomas. Flow cytometric DNA-ploidy analysis revaled that half of the malignancies contain aueuploid stemlines, clustering in the near-diploid range and, as in dogs, with a relatively high percentage of hypodiploid stemlines.[120]

In experimental setting, allotransplantation of feline mammary carcinoma cell lines in fetal cats resulted in tumor growth at the site of injection, metastatic spread of tumor cells to various organs, and death from progression 2-4 months after birth. It has been proposed that this model system may offer an attactive alternative to rodent models for experimental diagnostic and therapeutic studies.[121]

Mammary Xenografts in Immune-Deficient Mice

A last model that deserves to be discussed is the immune-deficient mouse. The congenitally athymic ('nude') mouse was discovered by chance[122] and heterotransplantation of malignant human tumors was first reported in 1969 by Rygaard and Povlsen.[123] Since then, congenitally athymic mice have been widely used as hosts for human tumor xenografts to explore different treatment modalities. A number of other mouse strains with varying defects in their immune-system were also developed and are now used in breast cancer research. Mice homozygous for the nude (nu) mutation are congenitally athymic and consequently have a lack of T-cells, whereas mice with X-chromosome-linked immunodeficiency lack T-cell independent B-cells.[124] Natural killer cell activity is strongly reduced in mice carrying the beige (bg) mutation.[125] Crossing of strains carrying these different mutations has resulted in a number of strains carrying one or more of these mutations, like the Bg-nu-XID mice, which carries all three; however, little advantage has been found above the conventional nude mouse.[126]

SCID (sever combined immune deficient) mice lack both functional B- and T-cells because of nonfunctional rearrangements of their immunoglobulin and T-cell receptor genes.[127]

Because of their deficient immune response, immune-deficient mice seldom reject human or other xenografts, thereby providing a valuable *in vivo* model system for the study of normal and neoplastic tissues. E.g. with respect to normal mammary tissue, athymic mice were used to study mitogenic effects of various hormones on ductal epithelium.[128] Grafts can be maintained this way for the entire life-span of the animal.[129] Xenotransplantation of human cancers into immune-deficientmice has been used to test various anti-cancer therapies.[130,131] Moreover, the effect of a certain treatment or gene-transfer on a cell-line *in vitro* can be compared to the effect in the *in vivo* situation. An advantage of using established human cancer cell lines for that purpose is that they usually have been extensively characterized, allowing more accurate interpretation of experimental findings. However, because these cells have adapted to growth *in vitro*, molecular, cellular and metabolic changes in these cells lines may or may not fully reflect the human disease.[132]

Although immune-deficient mice cannot raise a very effective immune-response to xenografted tissues, not all xenogeneic breast tumors or breast tumor cell lines show progressive growth after transplantation. It has been hypothesized that the residual immunoreactivity may negatively affect transplantability of some xenografts in immune-deficient mice.[133] This is supported by the finding that the more severely immune-compromised SCID mice are better recipients of xenografts than nude mice and that further immunosuppression by whole body irradiation improves transplantability.[134]

For some tumor types, such as colon,[135] melanoma,[136] and pancreatic cancer,[137] orthotopic implantation in these animals has been reported to significantly improve both take rate and expression of metastatic phenotype. Breast cancer, however, shows low take rates, even when implanted orthotopically, and the majority of primary tumors that do grow out are hormone-independent.[138,139] Furthermore, only a limited number of hormone-dependent breast cancer cell lines have been established *in vitro*. Therefore, *in vivo/in vitro* studies of hormone-dependent breast cancer mainly depend on the MCF-7, T-47D and ZR-75-1 cell lines; these cell lines were established from pleural effusions and are tumorigenic in estrogen-supplemented immune-deficient mice.[140–142]

A number of recent publications demonstrated that co-inoculations of tumor cells with basement membrane mix Matrigel or fibroblasts dramatically improved tumor take rate of various types of cancer, including breast cancer.[143–145] After co-inoculation with Matrigel and estrogen supplementation, the BT-474 breast cancer cell line forms solid tumors and lymphogenic metastases. This cell line, which was (in contrast to other hormone-dependent cell lines) isolated from a primary breast carcinoma, strongly overexpress the HER-2/*neu* oncogene, making it a useful model for immuno-therapy.[146]

In conclusion, immune-deficient mice offer the opportunity to grow human breast cancer xenografts *in vivo*, making it an exceptionally useful model for the study of breast cancer growth regulation and dissemination and for testing various treatment modalities. SCID mice that have received whole body irradiation have the lowest level of immunoreactivity against human tumor xenografts and coinoculation with Matrigel dramatically improves take rate of human breast tumor cells.

Concluding Remarks

Animal models are required to increase our understanding of certain clinical observations and serve as a necessary step in sequences of investigations aimed at clinical applications. Many factors can influence selection of a particular model for experiments, but a critical factor should always be the specific problem to be investigated; evidently, the choice for experimentation on animals should be very much determined by the relative importance of the question involved and one should always be aware of the possibility that certain questions can be properly addressed in *in vitro* systems.

Within this context, it seems important to realize that cells in a solid tumor may show entirely different responses than the same tumor cells growing as monolayers *in vitro*. An example of this may be related to the phenomenon that *in vitro* drug response by human cancer cells removed from treated patients not always correlates with clinical response. Recently, Teicher and coworkers derived stable drug-resistant variants of EMT-6 mouse mammary tumors by *in vivo* drug administration to tumor-bearing mice; surprisingly, when plated as monolayer *in vitro* these cells were as chemosensitive as the parental EMR-6 cell line.[147] Additional studies have confirmed that when mammary tumor cells grow as solid tumors or under three-dimensional conditions in vitro, they rapidly acquire a type of multicellular (or 'tissue') drug resistance that is not observed in monolayers.[148,149]

The available types of animal models include spontaneous and induced tumors, early-generation transplanted tumors, established transplanted tumor lines and human xenografts. In this overview, special attention was paid to fundamental interspecies differences in development and structure of the mammary gland and in endocrine, genetic and other factors involved in its neoplastic diseases.

Extensive characterization of a model seems essential in order to interpret experimental findings. Primary tumors or first generation transplants in rats, mice or other species are far less characterized than established transplanted tumors and reproducibility of experiments is compromized by variability. As a result of extensive cell selection, established tumor cell lines may not only have lost various differentiation characteristics, but also proliferate much faster than their clinical counterparts (which is, on the other hand, a major advantage in terms of experimental design). Tumor models like the above described EMR-86 rat mammary carcinoma may therefore be particularly relevant to questions related to treatment of very slowly growing tumor cell populations, since this tumor remains dormant when transplanted into untreated females, but shows a rapid outgrowth after proper hormonal stimulation.[77] This type of model seems useful to test *in vivo* whether the sensitivity of tumor stem cells to chemotherapeutic drugs can be enhanced by hormonal 'recruitment' or whether these cells can be eliminated by immunotherapeutic strategies.[78]

Sex hormones may have both a direct effect on mammary tumor growth and induce a host of regulatory factors.[150,151] *In vitro* studies have provided evidence for intimate or paracrine communications between stromal and epithelial cells: both cell types seem to be required for growth in response to estrogens.[150] As stated in the introduction paragraph, a major challenge of present research is to elucidate whether these *in vitro* observations have their counterpart in appropriate *in vivo* models. Immune deficient mice may be helpful to evaluate the physiological significance of *in vitro* studies by comparing both systems with respect to patterns of cytokinetics and gene expression.[146]

Another challenge is to elucidate whether experimentally defined host factors affecting tumor growth and metastatic behaviour[152,153] are significant to clinical practice; as long as animal models have been used for cancer research, interesting observations were made from which it is still uncertain if similar phenomena occur in man. Examples are the observations that a tumor mass can inhibit the outgrowth of remote metastases or secondary tumor implants[154–156] and that surgical removal of tumors can modulate the cytokinetics of distant metastases, secondary tumor implants and local residual disease.[157–159]

An important consideration with respect to the clinical relevance of these observations in tumor-bearing laboratory animals is that the ratio between tumor size and body weight may differ considerably from that observed in man. Nevertheless, identification of the systemic factors responsible for stimulation or inhibition of metastastic tumor growth may lead to a specific search of these activities in the blood or urine of cancer patients. In this respect the recent discovery in tumor-bearing mice of a highly stable 38 kDa plasminogen fragment ('angiostatin') that potently blocks neovascularization and growth of metastases may provide interesting oportunities for further research.[160]

It is clear that this type of animal studies is still indispensable for the progress of our understanding of a large variety of biological phenomena and for the proper testing of new drugs and treatment modalities. But as was recently emphasized by Robert Clarke, "The decision to use animal models in no way abrogates the investigators' obligation to responsibly use animals and many governments already have enacted legislation that accepts the principle that the use of animals requires specific justification, a demonstration that *ex vivo* techniques are inapplicable, that studies are not unnecessarily duplicative, that appropriate anesthesia and analgesia are provided to limit pain and stress, and that the number of animals used should be clearly justified and minimized".[161]

References

1. Kratochwil, K. (1987). Epithelium-mesenchyme interaction in the fetal mammary gland. In: *Cellular and Molecular Biology of Mammary Cancer*. Edited by D. Medina, W. Kidwell, G. Heppner and E. Anderson, pp. 67–79, New York and London: Plenum Press.
2. Daniel, C. W. and Silberstein, G. B. (1987). Postnatal development of the rodent mammary gland. In: *The Mammary Gland*. Edited by M. C. Neville and C. W. Daniels, p. 3, New York: Plenum Press.
3. Sakakura, T. (1991). New aspects of stroma-parenchyma relations in mammary gland differentiation. *International Reviews in Cytology*, **125**, 165–202.
4. Imagawa, W., Bandyopadhyay, G. K. and Nandi, S. (1990). Regulation of mammary epithelial cell growth in mice and rats. *Endocrine Reviews*, **11**, 494–523
5. Rønnov-Jessen, L., Petersen, O. W. and Bissell, M. J. (1996). Cellular changes involved in conversion of normal to malignant breast: importance of the stromal reaction. *Physiological Reviews*, **76**, 69–125.
6. Russo, J., Gusterson, B. A., Rogers, A. E., Russo, I. H., Wellings, S. R. and Van Zwieten, M. J. (1990). Comparative study of human and rat mammary tumorigenesis. *Laboratory Investigation*, **62**, 244–278.
7. Russo, J. and Russo, I. H. (1996). Experimentally induced mammary tumors in rats. *Breast Cancer Research and Treatment*, **39**, 7–20.
8. Rudland, P. S. (1991). Histochemical organization and cellular composition of ductal buds in developing human breast: evidence of cytochemical intermediates between epithelial and myoepithelial cells. *Journal of Histochemistry and Cytochemistry*, **39**, 1471–1484.
9. Dulbecco, R., Hanahan, M. and Armstrong, B. (1982). Cell types and morphogenesis in the mammary gland. *Proceedings of the National Academy of Science, U.S.A.*, **79**, 7346–7350.
10. Williams, J. M. and Daniel, C. W. (1983). Mammary ductal elongation: differentiation of myoepithelium and basal lamina during branching morphogenesis. *Developmental Biology*, **97**, 274–290.
11. Sonnenberg, A., Daams, H., Van der Valk, M., Hilkens, J. and Hilgers, J. (1986). Development of the mouse mammary gland: identification of stages in differentiation of luminal and myoepithelial cells using monoclonal antibodies and polyvalent antiserum against keratin. *Journal of Histochemistry and cytochemistry*, **34**, 1037–1046.
12. Smith, G. H. and Medina, D. (1988). A morphological distinct candidate for an epithelial stem cell in the mouse mammary gland. *Journal of Cell Science*, **89**, 173–183.
13. Smith, G. H. (1996). Experimental mammary epithelial morphogenesis in an *in vivo* model: evidence for distinct cellular progenitors of the ductal and lobular phenotype. *Breast cancer Research and Treatment*, **39**, 21–31.
14. Cardiff, R. D. (1984). Protoneoplasia: the molecular biology of murine mammary hyperplasia. *Advances in Cancer Research*, **42**, 167–190.
15. Lathrop, A. E. C. and Loeb, L. (1913). The influence of pregnancies on the incidence of cancer in mice. *Proceedings of the Society of Experimental Biology and Medicine*, **11**, 38–40.
16. Nandi, S. and McGrath, C. M. (1973). Mammary neoplasia in mice. *Advances in Cancer Research*, **17**, 353–414.
17. De Ome, K. B., Faulklin, L. J., Bern, H. A. and Blair, P. B. (1959). Development of mammary tumors from hyperplastic alveolar nodules transplanted into gland-free mammary fat pads of female CH3 mice. *Cancer Research*, **19**, 515–520.
18. Nusse, R., Van Ooyen, D., Cox, D., Funf, Y. K. T. and Varmus, H. E. (1984). Mode of proviral activation of a putative mammary oncogene (*int*-1) on mouse chromosome 15. *Nature*, **307**, 131–136.
19. Nusse, R. and Varmus, H. E. (1992). *Wnt genes. Cell*, **69**, 1073–1088.
20. Wong, G. T., Gavin B. J. and McMahon, A. P. (1994). Differential transformation of mammary epithelial cells by Wnt genes. *Molecular and Cellular Biology*, **14**, 6278–6286.
21. Van Leeuwen, F. and Nusse, R. (1995). Oncogene activation and oncogene cooperation in MMTV-induced mouse mammary cancer. *Seminars in cancer Biology*, **6**, 127–133.
22. Callahan, R. (1996). MMTV-induced muatations in mouse mammary tumors: their potential relevance to human breast cancer. *Breast Cancer Research and Treatment*, **39**, 33–44.
23. Van de Vijver, M. J., Petersen, J., Mooi, R., Lomans, J., Devilee, P., Cornelisse, C. J., Bos, J. L. and Nusse, R. (1989). Oncogene activation in human breast cancer. *Cancer Cells*, **7**, 385–391.
24. Vaage, J. (1968). Nonvirus-associated antigens in virus-induced mouse mammary tumors. *Cancer Research*, **28**, 2477–2483.
25. Ruppert, B., Wei, W., Medina, D. and Heppner, G. H. (1978). Effect of chemical carcinogen treatment on the immunogenicity of mouse mammary tumors arising from hyperplastic alveolar nodule outgrowth. *Journal of the National Cancer Institute*, **61**, 1165–1169.
26. Fluck, M. M. and Haslam, S. Z. (1996). Mammary tumors induced by polyomavirus. *Breast Cancer Research and Treatment*, **39**, 45–56.
27. Medina, D. (1974). Mammary carcinogenesis in chemical carcinogen-treated mice. I. Incidence in BALB/c and C57BL mice. II. Dependence on hormone stimulation for tumorigenesis. *Journal of the National Cancer Institute*, **53**, 213–226.
28. Watson, W. S., Medina, D. and Clark, J. H. (1977). Estrogen receptor characterization in a transplantable mouse mammary tumor. *Cancer Research*, **37**, 3344–3348.
29. Van der Valk, M. A. (1981). Survival, tumor incidence and gross pathology in 33 mouse strains. In: *Mammary Tumours in the Mouse*. Edited by J. Hilgers and M. Sluyser. pp. 45–115. Amsterdam: Elsevier/Noth Holland Biomedical Press.
30. Van der Gugten, A. A., Ropcke, G., Van Nie, R. and Hilgers, J. (1985). Mouse strain (STS/A) resistant to

mammary tumor induction by hypophysal isografts. *Cancer Research*, **45**, 3448–3453.
31. Van Nie, R. and Dux, A. (1971). Biological and morphological characterization of mammary tumors in GR mice. *Journal of the National Cancer Institute*, **46**, 885–897.
32. Braunschweiger, P. G. and Schiffer, L. M. (1981). Antiproliferative effects of corticosteroids in CH3/HeJ mammary tumors and implications for sequential combination chemotherapy. *Cancer Research*, **41**, 3324–3330.
33. Matsuzawa, A. (1982). Progesterone effect on tumor growth and development. In: *Hormonal Regulation of Mammary Tumors*. Edited by B. S. Leung. Vol I, pp. 183–215. Montreal: Eden Press Inc.
34. Minesita, T. and Yamaguchi, T. (1965). An androgen-dependent mouse mammary tumor. *Cancer Research*, **25**, 1168–1175.
35. Moore, D. H. and Holben, J. A. (1979). Effect of parity regimen on the rate of occurence of mammary tumors in A, C3H, and RIII mice. *International Journal of Cancer*, **24**, 161–164.
36. Trapido, E. J. (1983). Age at first birth, parity, and breast cancer risk. *Cancer*, **51**, 946–948.
37. Steward, T. A., Pattengale, P. K. and Leder, P. (1984). Spontaneous mammary adenocarci nomas in transgenic mice that cary and express MMTV/*myc* fusion genes. *Cell*, **38**, 627–637.
38. Edwards, P. A. W., Ward, J. L. and Bradbury, J. M. (1988). Alteration of morphogenesis by the v-myc oncogene in transplants of mammary gland. *Oncogene*, **2**, 407–412.
39. Munn, R. J., Webster, M., Muller, W. J. and Cardiff, R. D. (1995). Histopathology of transgenic mouse mammary tumors (a short atlas). *Seminars in Cancer Biology*, **6**, 153–158.
40. Webster, M. A. and Muller, W. J. (1994). Mammary tumorigenesis and metastasis in transgenic mice. *Seminars in Cancer Biology*, **5**, 69–76.
41. Sinn, E., Muller, W. J., Pattengale, P., Tepler, I., Wallace, R. and Leder, P. (1987). Coexpression of MMTV/v-Ha-*ras* and MMTV/c-*myc* genes in transgenic mice: synergistic action of oncogenes *in vivo*. *Cell*, **49**, 565–575.
42. Kwan, H., Pecenka, V., Tsukamoto, A., Parslow, T. G., Guzman, R., Lin, T. -P., Muller, W. J., Lee, F. S., Leder, P. and Varmus, H.E. (1992). Transgenes expressing the *Wnt*-1 and *int*-2 proto-oncogenes cooperate during mammary carcinogenesis in doubly transgenic mice. *Molecular Cell Biology*, **12**, 147–154.
43. Matsui, Y., Halter, S. A., Holt, J. T., Hogan, B. L. and Coffey, R. J. (1990). Development of mammary hyperplasia and neoplasia in MMTV-TGFα transgenic mice. *Cell*, **61**, 1147–1155.
44. Cardiff, R. D., Sinn, E., Muller, W. and Leder, P. (1991). Transgenic oncogene mice: tumor phenotype predicts genotype. *American Journal of Pathology*, **139**, 495–501.
45. Muller, W. J., Sinn, E., Pattengale, P. K., Wallace, R. and Leder, P. (1988). Single-step induction of mammary adenocarcinoma in transgenic mice bearing the activation c-*neu* oncogene. *Cell*, **54**, 105–115.
46. Guy, C. T., Muthuswamy, S. K., Cardiff, R. D., Soriano, P. and Muller, W. J. (1994). Activation of the c-*Src* tyrosine kinase is required for the induction of metastatic mammary tumors in transgenic mice. *Genes & Development*, **8**, 23–32.
47. Kerr, J. F. R., Wyllie, A. H. and Currie, A. R. (1972). Apoptosis: a basic biological phenomenon with wide-ranging implications in tissue kinetics. *British Journal of Cancer*, **26**, 239–257.
48. Hundley, J. E., Koester, S. K., Troyer, D. A., Hilsenbeck, S. G., Barrington, R. E. and Windle, J. J. (1997). Differential regulation of cell cycle characteristics and apoptosis in MMTV-myc and MMTV-ras mouse mammary tumors. *Cancer Research*, **57**, 600–603.
49. Lee, E. Y. H. P. (1995) Tumor suppressor genes and their alterations in breast cancer. *Seminars in Cancer Biology*, **6**, 119–125.
50. Greenblatt, M. S., Bennet, W. P., Hollstein, M. and Harris, C. C. (1994). Mutation in p53 tumor suppressor gene: clues to cancer etiology and molecular pathogenesis. *Cancer Research*, **54**, 4855–4878.
51. Levine, A. J. (1997). P53, the cellular gatekeeper for growth and division. *Cell*, **88**, 323–331.
52. Donehower, L. A., Godely, L. A., Aldaz, C. M., Pyle, R., Shi, Y.-P., Pinkel, D., Gray, J., Bradley, A., Medina, D. and Varmus, H. E. (1995). Deficiency of p53 accelerates mammary tumorigenesis in *Wnt*-1 transgenic mice and promotes chromosomal instability. *Genes & Development*, **9**, 882–895.
53. Hundley, J. E., Koester, S. K., Troyer, D. A., Hilsenbeck, S. G., Subler, M. A. and Windle, J. J. (1997). Increased tumor proliferation and genomic instability without decreased apoptosis in MMTV-*ras* mice deficient in p53. *Molecular and Cellular Biology*, **17**, 723–731.
54. Li, M., Hu, J., Heermeier, K., Hennighausen, L. and Furth, P. A. (1996). Apoptosis and remodeling of mammary gland tissue during involution proceeds through p53-independent pathways. *Cell Growth & Differentiation*, **7**, 13–20.
55. Kerbel, R. S. (1990). Growth dominance of the metastatic cancer cell: cellular and molecular aspects. *Advances in Cancer Research*, **55**, 87–132.
56. Anderson, J. C., Fugman, R. A., Stolfi, R. L. and Martin, D. S. (1974). Metastatic incidence of a spontaneous murine mammary carcinoma. *Cancer Research*, **34**, 1916–1920.
57. Nagasawa, H., Yanai, R., Taniguchi, H., Tokuzen, R. and Nakahara, W. (1976). Two-way selection of a stock of Swiss albino mice for mammary tumorigenesis: establishment of two new strains (SHN and SLN). *Journal of the National Cancer Institute*, **57**, 425–430.
58. Stackpole, C. W. (1981). Distinct lung colonizing and lung metastasizing cell populations in B16 mouse melanoma. Nature, **289**, 789–799.
59. Price, J. E., Carr, D., Jones, L. D., Messer, P. and Tarin, D. (1982). Experimental analysis of factors affecting metastatic spread using naturally occuring tumors. *Invasion and Metastasis*, **2**, 77.
60. Trimble, M. S., Xin, J. H., Guy, C. T., Muller, W. J. and Hassel, J. A. (1993). PEA3 is overexpressed in

mouse metastatic mammary adenocarcinomas. *Oncogene*, **8**, 3037–3042.
61. Van de Velde, C. J. H., Van Putten, L. M. and Zwaveling, A. (1977). A new metastasizing mammary carcinoma model in mice: model characteristics and applications. *European Journal of Cancer*, **13**, 555–565.
62. Eccles, S. A. (1983). Differentiation and neoplasia. Invasion and metastasis: experimental systems. *Journal of Pathology*, **141**, 333–353.
63. Sher, S. P. (1972). Mammary tumors in control rats: literature tabulation. *Toxicology and Applied Pharmacology*, **22**, 562–588.
64. Noble, R. L. and Cutts, J. H. (1959). Mammary tumors of the rat: a review. *Cancer Research*, **19**, 1125–1139.
65. Bentvelzen, P., Westenbrink, F., Broerse, J. J. and Van Zwieten, M. J. (1981). Absence of antigens related to murine mammary tumour virus polypeptides in rat mammary tumours. *European Journal of Cancer*, **17**, 407–410.
66. Welsch, C. W., Jenkins, T. W. and Meites, J. (1970). Increased incidence of mammary tumors in the female rat grafted with multiple pituitaries. *Cancer Research*, **30**, 1024–1029.
67. Durbin, P. W., Williams, M. H., Jeung, N. and Arnold, J. S. (1966). Development of spontaneous mammary tumors over the life-span of the female Charles River (Sprague Dawley) rat; the influence of ovariectomy, thyroidectomy, and adrenalectomy-ovariectomy. *Cancer Research*, **26**, 400–411.
68. Cutts, J. H. and Noble, R. L. (1964). Estrone-induced mammary tumors in thee rat. I. Induction and behavior of tumors. *Cancer Research*, **24**, 1116–1123.
69. Cutts, J. H. (1964). Estrone-induced mammary tumors in the rat. II. Effect of alterations in the hormonal environment on tumor induction, behavior, and growth. *Cancer Research*, **24**, 1124–1130.
70. Welsch, C. W., Nagasawa, H. and Meites, J. (1970). Increased incidence of spontaneous mammary tumors in female rats with induced hypothalamic lesions. *Cancer Research*, **35**, 766–780.
71. Moon, H. D., Simpson, M. E., Li, C. H. and Evans, H. M. (1951). Neoplasms in rats treated with pituitary growth hormone. V. Absence of neplasms in hypophysectomized rats. *Cancer Research*, **11**, 535–539.
72. Noble, R. L., Hochachka, B. C. and King, D. (1975). Spontaneous and estrogen-produced tumors in Nb rats and their behavior after transplantation. *Cancer Research*, **35**, 766–780
73. Van Dierendonck, J. H., Cornelisse, C. J., Van der Linden, P. W. G., Van Putten, L. M. and Van de Velde, C. J. H. (1987). Characterization of a slow-growing, transplantable rat mammary tumor (MCR-83): a model for endocrine-related cell kinetic studies. *Cancer Research*, **47**, 4093–4098.
74. Welsch, C. W. and Nagasawa, H. (1977). Prolactin and murine mammary tumorigenesis: A review. *Cancer Research*, **37**, 951–963.
75. Noble, R. L. and Hoover, L. A. (1975). A classification of transplantable tumors in Nb rats controlled by estrogen from dormancy to autonomy. *Cancer Research*, **35**, 2935–2941.
76. Noble, R. L. (1977). Hormonal control of growth and progression in tumors of Nb rats and a theory of action. *Cancer Research*, **37**, 82–94.
77. Wijsman, J. H., Cornelisse, C. J., Keijzer, R., Van de Velde, C. J. H. and Van Dierendonck, J.H. (1991). A prolactin-dependent, metastasizing rat mammary carcinoma as a model for endocrine-related tumour dormancy. *British Journal of Cancer*, **64**, 463–468.
78. Wijsman, J. H., Cornelisse, C. J., Keijzer, R., Van de Velde, C. J. H., Elvers, B. and Van Dierendonck, J. H. (1996). Effect of hormone depletion on cell survival in the EMR-86 rat mammary carcinoma. *British Journal of Cancer*, **73**, 1210–1215.
79. Walker, N. I., Bennet, R. E. and Kerr, J. F. R. (1989). Cell death by apoptosis during involution of the lactating breast in mice and rats. *The American Journal of Anatomy*, **185**, 19–32.
80. Shellabarger, C. J. (1981). Pituitary and steroid hormones in radiation-induced mammary tumors. In : *Hormones and Breast Cancer*. Edited by M. C. Pike, P. K. Siiteri and C. W. Welsch. Cold Spring Harbor Laboratory.
81. Van Zwieten, M. J. (1984). In: *The Rat as Animal Model in Breast Cancer Research*. Thesis, Boston: Martinus Nijhoff Publishers.
82. Welsch, C. W. (1987). Rodent models to examine *in vivo* hormonal regulation of mammary gland tumorigenesis. In: *Cellular and Molecular Biology of Mammary Cancer*. Edited by D. Medina, W. Kidwell, G. Heppner, and E. Anderson. pp. 163–179, New York: Plenum Press.
83. Huggins, C. Briziarelli, G. and Sutton, H. (1959). Rapid induction of mammary carcinoma in the rat and the influence of hormones on the tumors. *Journal of Experimental Medicine*, **109**, 25–42.
84. Huggins, C., Grand, L. C. and Brillantes, F. P. (1961). Mammary cancer induced by a single feeding of polynuclear hydrocarbons and its suppression. *Nature*, **189**, 204–207.
85. Bots, G. T. A. M. and Willighagen, R. G. J. (1975). Tumours in the mammary gland induced in Lewis rats by intravenous methylnitrosurea. *The British Journal of Cancer*, **31**, 372–374.
86. Gullino, P. M., Pettigrew, H. M. and Grantham, F. H. (1975). N-Nitrosomethylurea as mammary gland carcinogen in rats. *Journal of the National Cancer Institute*, **54**, 401–414.
87. Russo, J. and Russo, I. H. (1978). DNA-labeling index and structure of the rat mammary gland as determinants of susceptibility to carcinogenesis. *Journal of the National Cancer Institute*, **61**, 1451–1459.
88. Russo, J. and Russo, I. H. (1980). Influence of differentiation and cell kinetics on the susceptibility of the rat mammary gland to carcinogenesis. *Cancer Research*, **40**, 2677–2687.
89. Russo, J. and Russo, I. H. (1987). Biological and molecular bases of mammary carcinoge nesis. *Laboratory Investigations*, **57**, 112–136.
90. Zhang, R., Haag, J. D. and Gould, M. N. (1990). Reduction in the frequency of activated *ras* oncogenes in rat mammary carcinomas with increasing N-methyl-

N-nitrosourea dose or increasing prolactin levels. *Cancer Research*, **50**, 4286–4290.

91. Bos, J. L. (1989). *Ras* oncogenes in human cancer: a review. *Cancer Research*, **49**, 4682–4689.
92. Isaacs, J. T. (1991). A mammary cancer suppressor gene and its site of action in the rat. *Cancer Research*, **51**, 1591–1595.
93. Kim, U., Baumler, A., Carruthers, C. and Beelat, K. (1975). Immunological escape mechanism in spontaneously metastasizing mammary tumors. *Proceedings of the National Academy of Sciences, U.S.A.*, **72**, 1012–1016.
94. Gosh, S. K., Grossberg, A. L., Kim, U. and Pressman, D. (1979). A tumor-associated organ specific antigen characteristic of spontaneously metastatic rat mammary carcinoma. *Journal of the National Cancer Institute*, **62**, 1229–1233.
95. North, S. M. and Nicholson, G. L. (1985). Heterogeneity in the sensitivities of the 13762NF mammary adenocarcinoma cell clones to cytolysis mediated by extra- and intratumoral macrophages. *Cancer Research*, **45**, 1453–1458.
96. Sobel, M. E. (1990). Metastasis suppressor genes. *Journal of the National Cancer Institute*, **82**, 267–276.
97. Steeg, P. S., Bevilacqua, G., Kopper, L., Thorgeirsson, U. P., Talmadge, J. E., Liotta, L. A. and Sobel, M. E. (1988). Evidence for a novel gene associated with low tumor metastatic potential. *Journal of the National Cancer Institute*, **80**, 200–204.
98. Schneider, R. (1970). Comparison of age, sex, and incidence rates in human and canine breast cancer. *Cancer*, **26**, 419–426.
99. Raynaud, J. P., Cotard, M., Andre, F., Mialot, J. P., Rolland, P. H. and Martin, P. M. (1981). Spontaneous canine mammary tumors: a model for endocrine therapy? *Journal of Steroid Biochemistry*, **15**, 201–207.
100. Rutteman, G. R. (1990). Hormones and mammary tumour disease in the bitch: an update. *In Vitro*, **4**, 33–40.
101. Dorn, C., Taylor, D. O. N., Schneider, R., Hibbord, H. H. and Klauber, M. R. (1986). Survey of animal neoplasms in Alameda and Contra Costa Countries, California. II. Cancer morbidity in dogs and cats from Alameda Country. *Journal of the National Cancer Institute*, **40**, 307–318.
102. Owen, L. N. (1979). A comparative study of canine and human breast cancer. *Investigations in Cellular Pathology*, **2**, 257–275.
103. Concannon, P. W., Altszuler, N., Hampshire, J., Butler, J. and Hansel, W. (1980). Growth hormone, prolactin and cortosol in dogs developing mammary nodules and an acromegaly-like appearance during treatment with medroxyprogesterone acetate. *Endocrinology*, **106**, 1173–1177.
104. Mol, J. A., Van Garderen, E., Selman, P. J., Wolfswinkel, J., Rijnberk, A. and Rutteman, G. R. (1995). Growth hormone mRNA in mammary gland tumors of dogs and cats. *Journal of Clinical Investigation*, **95**, 2028–2034.
105. Mol, J. A., Van Garderen E., Rutteman, G. R. and Rijnberk, A. (1996). New insights in the molecular mechanism of progestin-induced proliferation of mammary epithelium: induction of the local biosynthesis of growth hormone (GH) in the mammary glands of dogs, cats and humans. *Journal of Steroid Biochemistry & Molecular Biology*, **57**, 67–71.
106. Rutteman, G. R., Bevers, M. M., Misdorp, W. and Van den Brom, W.E. (1989). Anterior pituitary function in female dogs with mammary tumors; II. Prolactin. Anticancer Research, **9**, 241–245.
107. Fidler, I. J., Abt, D. A. and Brodey, R. S. (1967). The biological behaviour of canine mammary neoplasms. *Journal of the American Veterinary Association*, **151**, 1311–1318.
108. Rutteman, G. R., Misdorp, W., Blankenstein, M. A. and Van den Brom, W. E. (1988). Oestrogen (ER) and progestin (PR) receptors in mammary tissue of the female dog: different receptor profiles in non-malignant and malignant states. *British Journal of cancer*, **58**, 594–599.
109. Van den Burg, B., Van Selm-Miltenburg, A.J,P., Van Maurik, P., Rutteman, G. R., Misdorp, W., De Laat, S. W. and Van Zoelen, E. J. J. (1989). Isolation of autonomously growing dog mammary tumor cell lines using medium supplemented with serum treated to inactivate growth factors. *Journal of the National Cancer Institute*, **81**, 1545–1551.
110. Cornelisse, C. J., Capers, R., Hermans, J., Molenaar, A.J. and Van de Velde, C. J. H. (1987). DNA ploidy and survival in 565 breast cancer patients. *Cytometry*, **8**, 225–234.
111. Rutteman, G. R., Cornelisse, C. J., Dijkshoorn, N. J., Poortman, J. and Misdorp, W. (1988). Flow cytometric analysis of DNA ploidy in canine mammary tumors. *Cancer Research*, **48**, 3411–3417.
112. Hellmen, E. and Svensson, S. (1995). Progression of canine mammary tumours as reflected by DNA ploidy in primary tumours and their metastases. *Journal of Comparative Pathology*, **113**, 327–342.
113. Shofer, F. S., Sonnenschein, E. G., Goldschmidt, M. H., Lester, L. L. and Glickman, L. T. (1989). Histopathologic and dietary factors for canine mammary carcinoma. *Breast Cancer Research and Treatment*, **13**, 49–60.
114. Misdorp, W. and Weijer, K. (1980). Animal model: feline mammary carcinoma. *American Journal of Pathology*, **98**, 573–576.
115. Weijer, K. and Hart, A. A. M. (1983). Prognostic factors in feline mammary carcinoma. *Journal of the National Cancer Institute*, **70**, 709–716.
116. Hahn, K. A., Bravo, L. and Avenell, J. S. (1994). Feline breast carcinoma as a pathologic and therapeutic model for human breast cancer. *In Vivo*, **8**, 825–828.
117. Calafat, J., Weijer, K. and Daams, H. (1977). Feline malignant tumors. III. Presence of C-particles and intracisternal A-particles and their relationship with feline leukemia virus antigens and RD-114 virus antigens. *International Journal of Cancer*, **20**, 759–767.
118. Rutteman, G. R., Blankenstein, A., Minke, J, M. and Misdorp, W. (1991). Steroid receptors in mammary tumors of the cat. *Acta Endocrinologica*, **125**, Supplement 1, 32–37.
119. Minke, J. M. (1990). Feline Mammary Carcinoma. *Thesis*, University of Utrecht, The Netherlands.

120. Minke, J. M., Cornelisse, C. J., Stolwijk, J. A. M., Kuipers-Dijkshoorn, J., Rutteman, G. R. and Misdorp, W. (1990). Flow cytometric DNA ploidy analysis of feline mammary tumors. *Cancer Research*, **50**, 4003–4007.
121. Minke, J. M., Hensen, E. J. and Misdorp, W. (1995). Uterine carcinomas in mother cats after intraferal inoculation of allogeneic tumor cells (K248 C and P). *Veterinary Immunology & Immunopathology*, **46**, 361–366.
122. Pantelouris, E. M. (1968). Absence of the thymus in a mouse mutant. *Nature*, **217**, 370–371.
123. Rygaard, J. and Povlsen, C. O. (1969). Heterotransplantation of a human malignant tumor to nude mice. *Acta Pathologica Microbiologica Scandinavia*, **77**, 758–760.
124. Karagogeos, D., Rosenberg, N. and Wortis, H. (1986). Early arrest of B-cell development in nude, X-linked immune-deficient mice. *Europeian Journal of Immunology*, **16**, 1125–1130.
125. Roder, J. and Duwe, A. (1979). The beige mutation in the mouse selectively impairs natural-killer function. *Nature*, **278**, 451–453.
126. Zietman, A., Sugiyama, E., Ramsay, J., Silobrcic, E. Sedlacek, R. and Suit, H. A. (1991). A comparative study of xenotransplantability of human solid tumors into mice with different genetic immune deficiencies. *International Journal of Cancer*, **47**, 755–759.
127. Bosma, G. C., Custer, R. P. and Bosma, M. J. A. (1983). Severe combined immunodeficiency mutation in the mouse. *Nature*, **30**, 527–530.
128. Manus, M. J. and Welsch, C. W. (1990). The effect of estrogen, progesterone, thyroxine, and human placental lactogen on DNA synthesis of human breast ductal epithelium maintained in athymic nude mice. *Cancer*, **54**, 1920–1927.
129. Hillman, E. A., Valerio, M. G., Halter, S. A., Barrett-Boone, L. A. and Trump, B. F. (1983). Long-term explant culture of normal mammary epithelium. *Cancer Research*, **43**, 245–257.
130. Giovanella, B. C., Stehlin, J. S., Williams, L. J., Lee, S. S. and Stephard, R. C. (1978). Hetero-transplantation of human cancers into nude mice: a model system for human cancer chemotherapy. *Cancer*, **42**, 2269–2281.
131. Brünner, N., Spang-Thomsen, M. and Cullen, K. (1996). The T61 human breast cancer xenograft: an experimental model of estrogen therapy of breast cancer. *Breast cancer Research & Treatment*, **39**, 87–92.
132. Clarke, R. (1996). Human breast cancer cell line xenografts as models of breast cancer – the immunobiologies of recipient mice and the characteristics of several tumorigenic cell lines. *Breast Cancer Research & Treatment*, **39**, 69–86.
133. Rofstad, E. K. (1989). Local tumor control following single dose irradiation of human melanoma xenografts: relationship to cellular radiosensitivity and influence of an immune response by the athymic mouse. *Cancer Research*, **49**, 3163–3167.
134. Taghian, A., Budach, W., Zietman, A., Freeman, J., Gioioso, D., Ruka, W. and Suit, H. D. (1993). Quantitative comparison between the transplantability of human and murine tumors into the subcutaneous tissue of NCr/Sed-*nu/nu* nude and severe combined immunodeficient mice. *Cancer Research*, **53**, 5012–5017.
135 Fu, X., Besterman, J. M. M., Monosov, A. and Hoffman, R.M. (1991). Models of human metastatic colon cancer in nude mide orthotopically constructed by using histologically intact patient specimens. *Proceedings of the National Academy of Sciences, U.S.A.*, **88**, 9345–9349.
136. Cornil, I., Man, S., Fernandez, B. and Kerbel, R. S. (1989). Enhanced tumorigenicity, melanogenesis, and metastases of a human malignant melanoma after subdermal implantation in nude mice. *Journal of the National Cancer Institute*, **81**, 938–944.
137. Fu, X., Guadagni, F. and Hoffman, R. M. (1992). A metastatic nude-mouse model of human pancreatic cancer constructed orthopically with histologically intact specimens. *Proceedings of the National Academy of Sciences, U.S.A.*, **89**, 5645–5649.
138. Giovanella, B. C. and Fogh, J. (1985). The nude mouse in cancer research. *Advances in Cancer Research*, **44**, 69–120.
139. Price, J. E., Polyzos A., Zhang, R. D. and Daniels, L. M. (1990). Tumorigenicity and metastasis of human breast carcinoma cell lines in nude mice. *Cancer Research*, **50**, 717–721.
140. Leung, C. K. and Shiu, R. P. (1981). Required presence of both estrogen and pituitary factors for the growth of human breast cancer cells in athymic nude mice. *Cancer Research*, **41**, 546–551.
141. Yue, W. and Brodie, A. (1993). MCF-7 human breast carcinomas in nude mide as a model for evaluating aromatase inhibitors. *Journal of Steroid Biochemistry and Molecular Biology*, **44**, 671–673.
142. Weckbecker, G., Liu, R., Tolcsvai, L. and Bruns, C. (1992). Antiproliferative effects of somatostatin analogue octrotide (SMS 201–995) on ZR-75-1 human breast cancer cells *in vivo* and *in vitro*. *Cancer Research*, **52**, 4973–4978.
143. Fridman, R, Kibbey, M. C., Royce, L. S., Zain, M., Sweeney, M., Jicha, D. L., Yanelli, J. R., Martin, G. R. and Kleinman, H. K. (1991). Enhanced tumor growth of both primary and established human and murine tumor cells in athymic mice after coinjection with Matrigel. *Journal of the National Cancer Institute*, **83**, 169–774.
144. Mehta, R. R., Graves, J. M., Hart, G. D., Shilkaitis, A. and Das Gupta, T. K. (1993). Growth and metastasis of human breast carcinomas with Matrigel in athymic mice. *Breast Cancer Research & Treatment*, **25**, 65–71.
145. Price, J. E. (1996). Metastasis from human breast cancer cell lines. *Breast Cancer Research & Treatment*, **39**, 93–102.
146. Van Slooten, H. -J., Bonsing, B. A., Hiller, A. J., Colbern, G., Van Dierendonck, J. H., Cornelisse, C. J. and Smith, H. S. (1995). Outgrowth of BT-474 human breast cancer cells in immune deficient mice: a new model for hormone-dependent breast cancer. *British Journal of Cancer*, **72**, 22–30.
147. Teicher, B. A., Herman, T. S., Holden, S. A., Wang, Y., Pfeffer, R., Crawford, J. W. and Frei, E. III. (1990).

Tumor resistance to alkylating agents conferred by mechanisms operative only *in vivo*. Science, **247**, 1457–1461.

148. Kobayashi, H., Man, S., Graham, C. H., Kapitain, S. J., Teicher, B. A. and Kerbel, R. S. (1993). Acquired multicellular-mediated resistance to alkylating agents in cancer. *Proceedings of the National Academy of Sciences U.S.A.*, **90**, 3294–3298.
149. Graham, C. H., Kobayashi, H., Stankiewicz, K. S., Man, S., Kapitain, S. J. and Kerbel, R. S. (1994). Rapid acquisition of multicellular drug resistance after a single exposure of mammary tumor cells to antitumor alkylating agents. *Journal of the National Cancer Institute*, **86**, 975–982.
150. Lippman, M. E. and Dickson, R. B. (1989). Mechanisms of growth in normal and malignant breast epithelium. *Recent Progress in Hormone Research*, **45**, 383–440.
151. Van den Burg, B., De Groot, R. P., Isbrücker, L., Kruijer, W. And De Laat, S. W. (1992). Direct stimulation by estrogen of growth factor signal transduction pathways in human breast cancer cells. *Journal of Steroid Biochemistry and Molecular Biology*, **43**, 111–115.
152. Shapot, V. S. (1979). On the multiform relationships between the tumor and the host. *Advances in Cancer Research*, **30**, 89–150.
153. Welsch, C. W. (1985). Host factors affecting the growth of carcinogen-induced rat mammary carcinomas: a review and tribute to Charles Brenton Higgins. *Cancer Research*, **45**, 3415–3443.
154. Tyzzer, E. E. (1913). Factors in the production and growth of tumor metastases. *Journal of Medical Research*, **28**, 309–333.
155. Gorelik, E., Segal, S. and Feldman, M. (1981). On the mechanism of tumor concomitant immunity. *International Journal of Cancer*, **27**, 847–856.
156. Prehn, R.T. (1991). The inhibition of tumor growth by tumor mass. *Cancer Research*, **53**, 3266–3269.
157. Simpson-Herren, L., Sanford, A.H. and Holmquist, J.P. (1979). Effect of surgery on the cell kinetics of residual tumor. *Cancer Treatment Reports*, **60**, 1749–1760.
158. Fisher, B., Gunduz, N. and Saffer, E.A. (1983). Influence of the interval between primary tumor removal and chemotherapy on kinetics and growth of metastases. *Cancer Research*, **43**, 1488–1492.
159. Van Dierendonck, J. H., Keijzer, R., Cornelisse, C. J. and Van de Velde, C. J. H. (1991). Surgically induced cytokinetic responses in experimental rat mammary tumor models. *Cancer*, **68**, 759–767.
160. O'Reilly, M. S., Holmgren, L., Shing, Y., Chen, C., Rosenthal, R. A., Moses, M., Lane, W. S., Cao, Y., Sage, E. H. and Folkman, J. (1994). Angiostatin: a novel angiogenesid inhibitor that mediates the suppression of metastases by Lewis lung carcinoma. *Cell*, **79**, 315–328.
161. Clarke, R. (1996). Animal models in breast cancer: their diversity and role in biomedical research. *Breast Cancer Research & Treatment*, **39**, 1–6.

PART VI

TRANSPLANTATION

PART VI

19. Surgical Techniques in Transplantation Biology Allograft Models and Manipulations of the Laboratory Rat

David Whitley and Nicholas L. Tilney

From the Surgical Research Laboratory, Harvard Medical School, and the Department of Surgery, Brigham and Women's Hospital, 75 Francis Street, Boston, Mass 02115, USA

Introduction

Since Medawar's first description of the "homograft response" in 1944, considerable energy has been expended by workers in a plethora of disciplines analyzing the host immunological responses activated by the introduction of transplantation antigens. Although he initially described the phenomenon of rejection of skin, it has become clear subsequently that many of the original rules governing the process have had to be redefined in the context of vascularized organ allograft models, rescuing us, as Medawar himself put it, "from the tyranny of the skin graft". Developments contributing to current understanding of the mechanisms of this remarkable host process have particularly involved the use of inbred strains of laboratory animals and the perfection of microvascular surgical techniques which have allowed reproducible models for experiments in transplantation biology. In addition, the more recent availability of fluorescence activated cell sorting, immunohistology and techniques for gene detection and amplification are being used to produce a clearer understanding of the cellular and extra-cellular interactions between the host and a foreign graft.

It is obvious that many technical and pharmacological successes in clinical transplantation over the past several years have been a direct result of knowledge gained by exploiting both in vitro and in vivo experimental surgical models. Using these models, we have been able to dissect the selective and interrelated roles of specific cell subsets and their soluble products which appear in the rejection cascade as well as to begin to develop appropriate therapeutic protocols based on these events.

Transplantation biologists have typically chosen the rat for their studies, as they are inbred, several strains are available, and the animals are relatively inexpensive and easy to handle. Skin engraftment can be performed quickly and with facility. In addition laboratory rats are large enough to allow satisfactory transplantation of vascularized organ grafts at relatively low magnification (0–2.5 ×) with typical microsurgical instruments and techniques, and after moderate training and perseverance on the part of the investigator. Such operations now have become so standardized that increasing numbers of individuals are performing similar procedures on mice, an advantage to those performing experiments based on murine genetics. As the majority of experimental transplant models still involve the rat, however, this chapter will concentrate primarily upon the various surgical techniques used widely on that species. It is generally helpful to employ male rats weighing between 150–200 gm as recipients. The vessels of older and larger rats are not proportionally greater in size; in addition, increased amounts of retroperitoneal fat, both in older animals and in female rats, may hamper easy dissection.

Microvascular Instruments

These are a standard part of the surgical armamentarium (Table 19.1). Fine tipped watchmarker forceps are especially useful as are fine sharp scissors. Cork boards and a focused light source are crucial. Loupes (2.5 × magnification) are often used.

Skin Transplantation

Initially the primary model for studies in transplant biology, skin grafting remains an important experimental tool. A large piece of abdominal skin is

Table 19.1. Microvascular instruments used in described procedures

Capsulotomy scissors	8.12 cm
Iris forceps	10.8 cm
Jeweller's forceps	10.9 cm
Jaffe lid retractor	–
Mosquito forceps (curved)	7.5 cm
Schwartz temporary vessel clip (smooth jawed)	2.5 cm
Castroviejo needle holder (straight)	14.6 cm
Metzenbaum scissors	18 cm
Micro dissecting forceps (serrated, full curve)	10.2 cm

usually used. The donor animal can be salvaged for later use if desired by repairing the resultant defect in an oblique fashion. The skin is laid on a damp gauze and tacked to a cork board, hair side down. All areolar and subcutaneous tissues are removed with scissors until the pearly gray color of the deep dermis is uniform. The skin is then cut to the desired size (usually c 2 × 2cm) and stored in a saline soaked sponge.

The anesthetized recipient animal is laid on his side and his flank shaved. A piece of skin over the lower thorax is excised, being careful to leave the panniculus carnosus muscle intact with its blood supply. The skin can often be dissected off sharply via a cleavage layer immediately between dermis and panniculus. The graft is then sutured into place. The animal is wrapped circumferentially with gauze, then with tape, making sure not to restrict his thoracic excursions. It is easier to keep the dressing in place when the graft is on the chest wall, not the abdomen. The dressing is held by extending the tape beyond the edges of the gauze and sticking it to the fur. It can be changed and the graft examined as desired. After a week or so, it may be removed as the rat will no longer scratch at the operative site. Alternatively, the graft can be placed high on the back and left unbandaged, as it is difficult for the animal to dislodge it in that position.

Cardiac Transplantation

Recipient Preparation

This Procedure has been Well Described (1–3). The rat is fastened supine to a cork board and a long midline abdominal incision made. The abdominal wall is held open with curved stainless steel retractors attached to the dissecting board with tacks. The bowel is then placed on the left side of the animal and wrapped in a saline soaked sponge. The infrarenal abdominal aorta and vena cava are exposed to their bifurcations by gently opening the overlying retroperitoneum with damp cotton tipped swabs. A short length of aorta is then isolated from the vena cava by dissecting into the small space immediately below the renal arteries; the closely adherent distal aorta and vena cava are not separated. Removal of longitudinal nerves running anteriorly between these vessels allows cleaner dissection. The lumbar veins, including a large one usually entering the back wall of the cava, are ligated. Loops of 4-0 silk are then passed individually around proximal aorta and vena cava; a single loop is then passed around the distal great vessels. These vascular segments are occluded by placing the loops under tension by clipping them to the abdominal wall with small hemostats or tacking the clamp to the cork board. An arteriotomy is made by passing a 7-0 proline suture through the anterior wall of the aorta; traction is placed on the suture and a small (approx. 3mm) oval fashioned by cutting beneath the suture with small iris scissors. Alternately, the anterior aortic wall is picked up with watch makers forceps, a tiny cross incision made with sharp scissors, a point of the scissors placed within the hole and the arteriotomy extended longitudinally proximally and distally. The distal component is crucial to prevent aortic occlusion at the time of anastomosis with resultant hind limb paralysis and recipient death. The open aorta is flushed with saline and a 7-0 or 8-0 proline suture (depending upon preference) placed out-to-in at each end of the arteriotomy. Preparation of the recipient takes about 4 minutes.

Donor Preparation

The abdomen is opened in the midline, the bowel mobilized to the left, the aorta and vena cava incised and the animal exsanguinated. The thoracic cavity is opened by extending the abdominal incision subcostally in a T shape. The anterior chest wall is mobilized cephalad by incising the diaphragm and dividing the ribs bilateraly. The pericardium is opened. The superior and inferior vena cavae are isolated and ligated. The blade of a small scissors is introduced into the space beneath the ascending aorta and pulmonary artery, and the vessels divided. The heart is irrigated gently with cold saline via a blunt 16 gauge needle inserted into the aorta. The pulmonary veins are then ligated by passing a long tie beneath both superior vena cavae, around the heart and beneath the inferior vena cava, being careful not to include the atria. The cavae are then divided distal to their ties, and while lifting up on the ties the pulmonary veins are divided distal to their surrounding ligature. It is unnecessary to perfuse out all coronary blood although the small clots in the aortic root and proximal pulmonary artery must be removed. Alternatively the isolated heart may be flushed with cold saline via the aorta until the effluent from the pulmonary artery is clear. Care must be taken not to overdistend the heart during flushing or to introduce air bubbles into the coronary arteries.

Transplantation

End-to-side anastomoses are performed between donor and recipient aortas and donor pulmonary artery and host vena cava. The heart graft is placed on the right side of the recipient and the sutures previously placed in the recipient aorta are brought through the donor aorta and tied. The ends of the ligatures around the pulmonary veins can be held under slight tension for optimal exposure of the aortic wall. The left wall of the aortic anastomosis is completed in a running fashion using no more than 5 sutures. After tying this suture to the short end of the opposite suture, the heart is then flipped to the left and the right wall of the aortic anastomosis completed. For the venous anastomosis the sharp point of a small iris scissors is inserted into the blood filled recipient vena cava laterally (not anteriorly) and the vein incised approximately 4mm. Alternatively, a small elipse can be cut formally from the wall. The vena cava is now flushed with saline to remove clots. Two 7-0 proline sutures are placed in the proximal and distal ends of the venotomy, passed out through the donor pulmonary artery, and ligated. The anastomosis of the back wall is run from the inside then the suture is passed outside the vessels and tied to the opposite suture. The outside wall of the venous anastomoses (that nearest the surgeon) is then run and tied to the opposite suture tail. Care must be taken not to narrow the anastomosis when tying. The occlusive loops around the proximal inferior vena cava, distal vena cava and aorta, and proximal aorta are slowly removed in that order. Bleeding is controlled carefully using finger pressure and opening and closing the snares. The small bronchial artery running between donor aorta and vena cava is suture ligated with a 7-0 proline separately as this can hemorrhage later. The anastomosis takes about 10 minues (Figure 19.1). Some magnification (2 ×) with loupes is often helpful in this procedure.

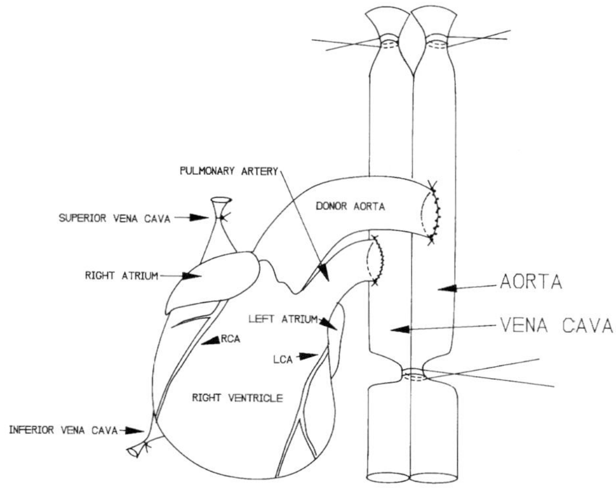

Figure 19.1. Heterotopic cardiac transplant utilizing end-to-side donor aorta and pulmonary artery to recipient aorta and vena caval anastomoses, respectively.

Kidney Transplantation

Two techniques ae used for transplanting the rat kidney: heterotopic end-to-side vascular anastomoses and a silastic stent for the uretero-vesical anastomosis, or orthotopic end-to-end anastomoss of vessels and ureter. The first technique, using aortic and vena caval patches surrounding the donor renal vessels, is technically easier than the second as it employs microsurgical techniques similar to those used in the heart graft model. The disadvantage is that multiple arterial branches adjacent to the donor renal artery require careful and time consuming dissection and ligation. In addition, as these kidneys tend to develop hydronephrosis, the method should not be used for long-term studies. The second technique requires much higher magnification (5–10 ×) under a dissecting microscope and needs a longer learning curve; however, once learned it is faster with fewer late ureteric complications, particularly stenosis and hydronephrosis. In addition, both donor kidneys can be used for separate recipients. In brief, the second technique is superior to the first.

First Technique

Recipient Preparation

The abdomen is opened via the midline. The right renal artery and vein are dissected free, doubly ligated and transected. The right ureter is then ligated, transected and the kidney removed. The infrarenal portions of aorta and vena cava are isolated.

Donor Preparation

The abdomen is opened in the midline and the abdominal aorta and vena cava exposed by blunt dissection with a cotton swab. The entire ureter is isolated, ligated distally and allowed to dilate with urine to facilitate later stent insertion. The right renal arery and aorta are isolated. The multiple small vessels arising from the aorta adjacent to the renal artery are doubly ligated and divided. The renal vein is dissected to the level of the vena cava; it is not necessary to divide the small branches coming off of the cava at this level. A small opening is then made in the distal ureter from which urine flows. A silastic stent 0.012 inch I.D. and approximately 2cm in length is inserted into the ureter and secured with an 8-0 double armed nylon suture passed through the adventitia and tied around the ureter and indwelling stent. Urine should then be seen draining from the end of the stent. This suture should be retained as it will be used later to join ureter to recipient bladder. The aorta is then ligated approximately 2mm proximal to the renal artery orifice with 5-0 silk. The renal vein is transected at its origin including a small vena caval cuff. A catheter is then inserted into the aortic bifurcation and the segment flushed with heparinized saline until the kidney pales and the venous effluent becoems clear. The aorta is transected at a 45 degree angle approximately 4mm below the renal artery, and a small eliptical patch surrounding the renal artery orifice made. The freed kidney is then placed in cold saline.

Transplantation

The recipient bowel has been retracted to the left. The donor organ is then placed in the right abdominal cavity in an anatomical position. Stay sutures of 8-0 proline, placed at each end of the aortotomy, made as described for heat transplantation, are passed through the donor aorta from inside to outside and the tissue approximated. The medial and lateral walls of the aortic patch are run using approximately 6-10 sutures on each side. The lateral wall of the vena cava is then grasped with forceps and a small elipse removed. Superior and inferior stay sutures are placed in the cava, passed inside to outside the renal vein patch and the veins approximated. The back wall is anastomosed from the inside (approx 10 sutures/wall), the suture passed outside and tied to the stay suture, then the outside wall anastomosed. The snare loop around the proximal vena cava is removed. The proximal aortic loop is slowly loosened and bleeding controlled with finger pressure or rarely, interrupted 8-0 patching sutures. The inferior snare is removed once the organ is well perfused. The ureter is then anastomosed to the bladder. A 18 gauge needle is passed through the anterior inferior aspect of the bladder and out through the apex. The silastic stent in the donor ueter is inserted into the needle and pulled through the bladder via the needle. The donor ureter is now closely opposed to the bladder wall. Two purse strings around the bladder opening are made using the double armed 8-0 nylon suture tied around ureter and stent. These are ligated, being careful not to stenose the soft stent. The stent end protruding out of the anterior inferior wall of the bladder is trimmed to the level of the bladder wall and allowed to fall back into the bladder. The small cystostomy is then closed using interrupted 8-0 nylon sutures. The abdomen is then rinsed and closed (Figure 19.2).

An alternative method of ureteric reconstruction involves dissecting out the entire, donor ureter, plus a surrounding patch of bladder. This bladder patch is then anastomosed using 7–0 absorbable suture to a defect made on the dome of the receipt bladder. Although the method was originally for kidney transplation in rats, it should not be used for other than short studies because of the formation of stones. These lead to hydronephrosis and/or pyelonephritis.

Second Technique

Recipient Preparation

After opening the abdomen and retracting the bowel, the abdominal aorta and vena cava are exposed. The left renal artery and vein are dissected free to their origins and small vascular clips placed on their proximal segments. The ureter is isolated from the retroperitoneum and transected approximately 1cm from the kidney. The organ is then removed by dividing the artery and vein distal to the clips.

Donor Preparation

Approximately 100 units is of heparin in saline is given I.V. to the donor. Five minutes later the left kidney is freed from the surrounding ret-

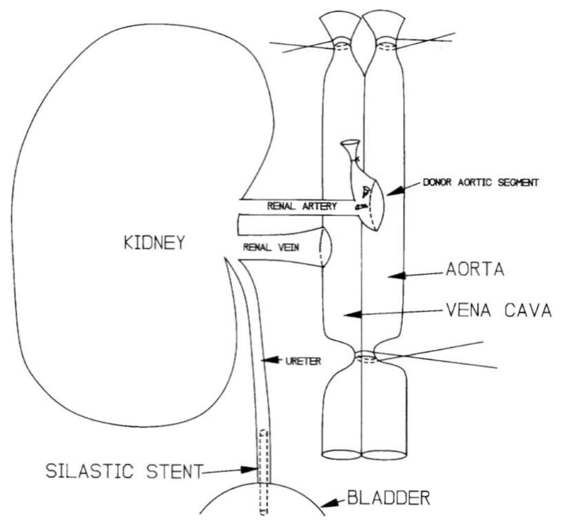

Figure 19.2. Heterotopic kidney transplant utilizing end-to-side anastomosis of donor aorta and renal vein with patch to recipient aorta and vena cava. Ureter anastomosis is end-to-bladder using a silastic stent.

roperitoneum, and the renal artery and vein dissected to their origins. The ureter is then isolated and transected just above the bladder. The animal is then exsanguinated by transecting the distal great vessels, and the renal artery and vein divided close to their origins. The kidney is removed and placed in iced saline without flushing.

Transplantation

The isolated organ is then placed into the recipient abdomen in anatomical position and under the dissecting microscope the renal artery anastomosed directly to recipient renal artery using interrupted 10-0 proline sutures (approx 5). The vein is run using 10-0 proline (approx. 8-10 sutures) after placing superior and inferior stay sutures. The vascular clip can then be taken off the renal vein. Once venous bleeding is controlled the arterial clip is slowly removed. The kidney should become pink and start producing urine within a few minutes. The donor ureter is then trimmed of any excess length and the anastomosis carried out in an interrupted fashion using 10-0 proline (approx. 4 sutures) It is not necessary to anchor the kidney to the abdominal wall (Figure 19.3).

Small Intestinal Transplantation

Recipient Preparation

The abdominal incision, isolation and preparation of the great vessels are identical to those described in the heart graft model.[5]

Donor Preparation

The small intestine from proximal jejunum to mid-ileum is isolated on a vascular pedicle consisting of the superior mesenteric artery with an aortic patch

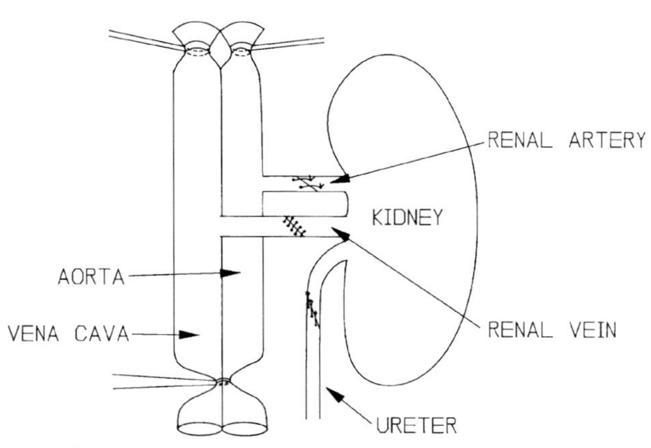

Figure 19.3. Orthotopic kidney transplant utilizing end-to-end anastomosis of renal artery, vein and ureter.

and a patch segment of mesenteric and portal veins. The donor is then heparinized with 150 units of heparin in saline. The intestinal lumen is flushed with cold saline and the isolated bowel placed in cold saline.

Anstomosis

The donor aortic patch is anastomosed end-to-side to recipient aorta and the donor vein with surrounding patch joined in similar fashion to the recipient vena cava using 7-0 proline sutures. Both ends of the small intestine are brought out through the abdominal wall as stomas. The bowel can be later reapproximated if desired.[3]

Liver Transplantation

The following technique for orthotopic liver transplantation in rats utilizes a cuff technique for anastomosis of the bile duct and protal vein. This has been found efficacious because of the fragility of these structures as well as shortening the anastomosis time considerably.[6]

Donor Preparation

The abdomen is entered through a transverse incision. The lumbar veins and right suprarenal vein are ligated and divided, as are the pyloric, splenic and inferior mesenteric and left phrenic veins. The right phrenic vein is left intact. The bile duct is cannulated using a 1 cm. nylon tube (O.D 1.2 mm) inserted into its distal lumen and secured with a silk tie. The animal is heparinized with 30 units of heparin in 2 cc saline via the penile vein. The hepatic artery and right renal artery and vein are ligated and divided. The infrahepatic vena cava is clipped and divided at the level of the left renal vein.

The liver is perfused via the portal vein with 4 cc iced saline to which 50 units of heparin have been added then flushed with 5cc of human plasma protein (HPP). The suprahepatic vena cava is transected with a surrounding cuff of diaphragm, and the portal vein divided at the level of the inferior mesenteric vein. The clip on the infrahepatic vena cava is removed and the liver placed in iced saline. The cuff of donor portal vein is fashoned after isolating the structure from the adjacent bile duct. A 7-0 silk suture is then placed through its tip and passed through the lumen of a 0.4 cm polyethylene tube (O.D. 2.1 mm). The vein is everted over the plastic cuff and secured into position with a circumferential 7-0 silk ligature (Figure 19.4). A cannula is then inserted into the cuff end of the portal vein and flushed with 2 cc of iced HPP, assuring patency.

Recipient Preparation

An I.V. is established in the anesthetized recipient via the femoral or jugular vein and 30 mg. of hydro-

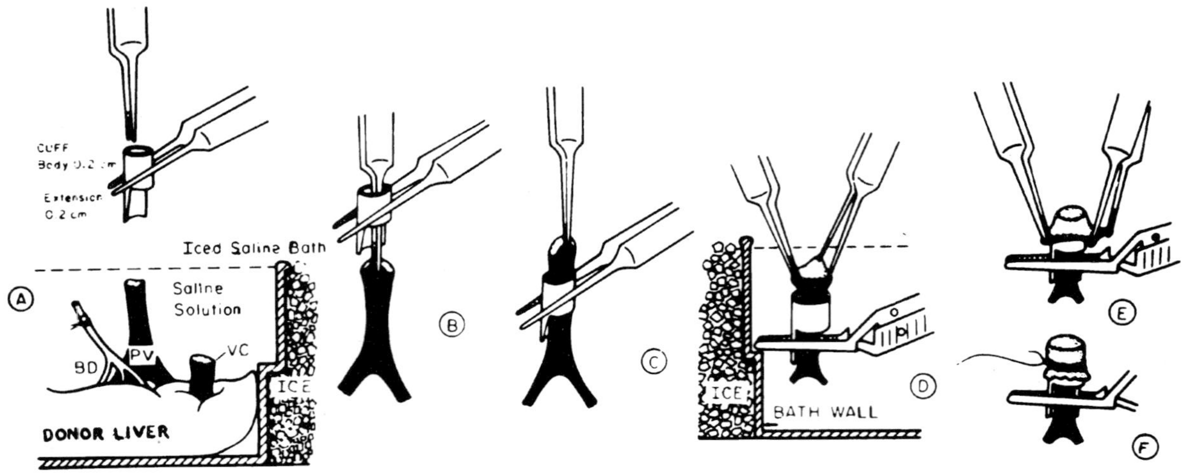

Figure 19.4. Liver transplantation using cuff technique for anastomosis. Cuff preparation is in a saline bath. (A) The cuff is held with forceps; (B) another pair of forceps is passed through the lumen of the cuff to grasp the portal vein; (C) the cuff is slipped over the portal vein; (D) The cuff and vein are secured with a bulldog clamp; the distal end of the vein is spread with two forceps and then (E) turned back over the cuff body; (F) a circumferential suture secures the cuff. PV = portal vein; VC = vena cava; BD = bile duct (from Kamada, N. ref. 4, with permission).

cortisone administered. A midline abdominal incision is made. The right suprarenal, lumbar and left phrenic veins and the hepatic artery are ligated and divided. The recipient bile duct is then cannulated by dissecting it free from the portal vein and a small opening made in its anterior wall. A 0.5 cm long polyethylene cannula (O.D. 1.34 mm) is inserted and secured with a circumferential 7-0 silk ligature. The bile duct is then transected distal to the tie. The infrahepatic vena cava and portal vein are clipped. After the liver pales, the suprahepatic vena cava is clamped with a small Satinsky clamp. The vessels are divided being careful to retain as much portal vein as possible. The recipient liver is then removed (Figure 19.5).

Liver Implantation

The donor liver is placed in an orthotopic position and covered with a cold gauze. Donor and recipient suprahepatic vena cavae are anastomosed end-to-end using 7-0 proline continuous suture. An incision is made in the anterior wall of the recipient portal vein which is irrigated free of blood. The previously prepared cuff of donor portal vein is then inserted through the venotomy. A circumferential silk tied around the recipient portal vein, completes the anastomosis. After release of the clips on portal vein and suprahepatic vena cava, the liver should regain its normal color and begin to produce bile. The infrahepatic vena cava is anastomosed end-to-end using a continuous 8-0 proline suture. The hepatic artery remains ligated. The polythylene tube in the donor bile duct is telescoped into the larger diameter tube in the recipient bile duct. The anastomosis is completed by tying together the ligatures around each segment of bile duct (Figure 19.6) . The animals receive an averge of 3 to 4 cc HPP I.V during the operation.

Spleen Transplantation

Recipient Preparation

The recipient great vessels are prepared as described.

Donor Preparation

The upper abdominal aorta and celiac axis of the heparinized donor are isolated, and the hepatic, and right and left gastric arteries ligated. The splenic pedicle is separted from omentum and pancreas, and the junction of portal and splenic veins cleared of surrounding tissues. Care is taken to ligate all small veins entering the pancreas from the splenic pedicle to obviate venous ooze and hematoma formation upon restoration of the circulation. A segment of aorta at the celiac axis is removed following ligation of its distal portion. An ellipse of portal vein surrounding the splenic vein is fashioned, and the spleen and its vascular pedicle removed and placed in a cold gauze.

Transplantation

The aortic segment of the splenic pedicle is sutured end-to-side to the infrarenal aorta of the recipient with 8-0 monofilament suture and the portal vein

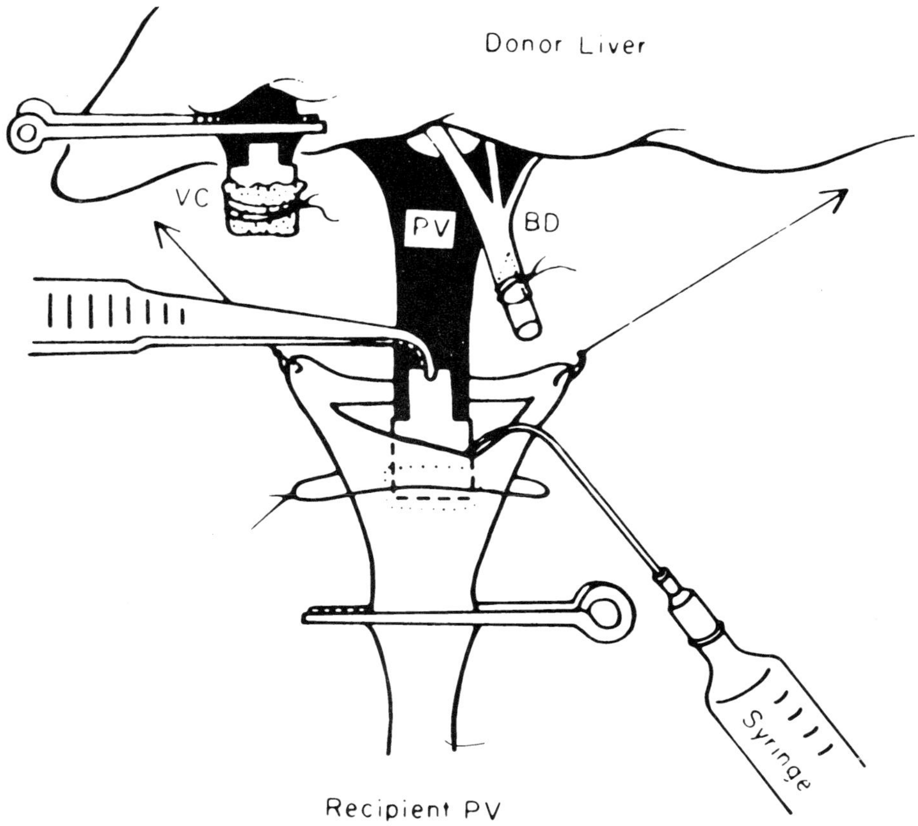

Figure 19.5. Preparation of the portal vein anastomosis (from Kamada, N. ref 4, with permission).

anastomosed in similar fashion as a patch graft to the inferior vena cava. Ischemic time is 30–40 minutes. Arterial pulsations appearing in the celiac axis and splenic artery immediately upon release of the vascular clamps, gradually spread to vessels in the splenic hilum. Veins fill with blood and the flaccid and cyanotic spleen slowly regains its normal tone and color. The vascularized graft is placed along the right perinephric and psoas gutter of the recipient, where it remains firm, compact and barely palpable through the flank. Infrequently in the post-operative period, the organ obviously enlarges and becomes tender, a consequence of vascular accident or torsion of the pedicle. The spleen can be kept in position by a few small sutures between its capsule and the recipient tissues.[7]

Pancreatic Transplantation

The following technique for pancreas transplantation in rats utilizes a cuff technique for anastomosis of the aorta and portal-renal vein. This technique shortens the anastomotic and warm ischemic time considerably.[8]

Donor Preparation

The pancreas and spleen are removed from the donor with the segment of portal vein and aorta which contains celiac and supeior mesenteric arteries. The organ is then placed in cold saline and the pancreatic ducts are ligated or left open to drain into the peritoneal cavity. The distal donor aorta is clamped and the proximal aorta gently perfused with 0.5 cc cold plasma protein. The splenic vessels are ligated at the hilum but not divided. A 3 mm long cuff of polyethylene tube (6FG) is slipped over the vein and the ends are everted over the cuff and secured with a circumferential 8-0 silk suture.

Recipient Preparation

A midline vertical incision is made and the left renal vein isolated. The suprarenal and testicular veins are ligated. A 1.5 cm segment of recipient infrarenal aorta is isolated and vascular clamps applied, the aorta transected between the clamps. A 3 mm long polyethylene tube (5FG) is slipped over the proximal and distal ends of the transected recipient aorta, the

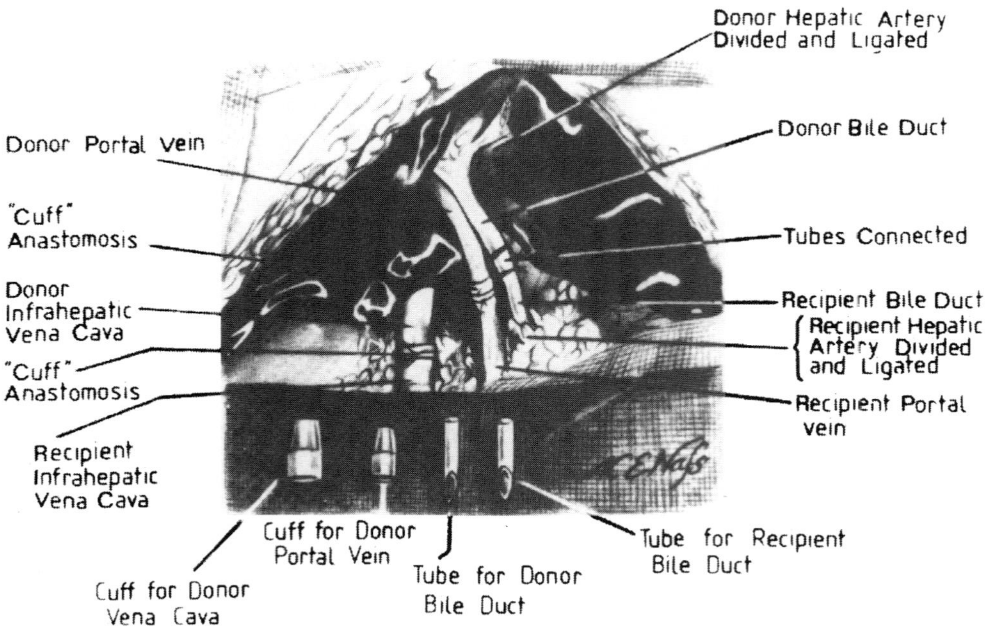

Figure 19.6. Completed anastomoses (from Kamada, N. ref 4, with permission).

ends are everted over the cuff and secured with a 8-0 circumferential silk suture. This leaves a 1.5 cm. gap in the aorta.

Anastomosis

The proximal and distal ends of the donor aorta is pulled over the proximal and distal cuffs of the recipient aorta. Both anastomoses are then secured with a 8-0 silk circumferential suture. Two vascular clips are then applied to the celiac and mesenteric arteries on the donor segment of aorta and the clamps are removed from the proximal and distal recipient aorta, restoring blood flow to the lower extremities and pelvis of the recipient. A left nephrectomy is then performed as close to the kidney as possible. A vascular clip is placed on the left renal vein at its origin. The cuff of donor portal vein is inserted into the recipient renal vein and secured with an 8-0 circumferential silk suture. The venous clamps and those occluding the celiac and superior mesenteric artery are removed restoring blood flow to the pancreas. The donor spleen is then removed (Figure 19.7).

The B Rat

The B rat (T cell depleted rat) is a model in which animals have undergone thymectomy, whole body x-radiation and repletion with marrow cells from a syngeneic, thymectomized and thoracic duct drained donor.[9] Organ allografts survive indefinitely in this iatrogenic equivalent of the "nude" mouse, a preparation useful in reconstitution and cell transfer experiments.

Thymectomy

Thymectomy is performed in 6 week old, 80 gm rats through a middline cervico-thoracic incision with splitting of the upper third of the sternum. The thymus is exposed, teased away from the surrounding tissues and removed intact. The strap muscles and sternum are reapproximated quickly using a single 4-0 silk suture. As the chest is closed, it should be compressed transiently to express residual air. The animals should be examined after sacrifice for remnants of thymic tissue. If the thymus itself is to be used for cell transfer, it is crucial that the parathymic lymph nodes are removed from the lateral lobes before preparation of the thymocytes.

X-Radiation

Approximately 3–4 weeks after thymectomy, rats are placed in a plastic carrier and x-radiated with a ^{137}Cs small animal x-radiator to deliver a sublethal total body 750 rads.

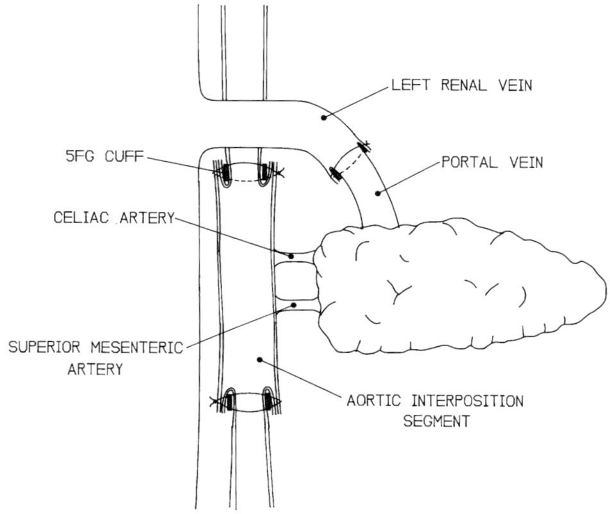

Figure 19.7. Pancreas transplantation utilizing a plastic cuff anastomosis technique. The donor aortic interposition segment containing the superior mesenteric and celiac artery is anastomosed to the recipient aorta using a 5 FG cuff while the portal vein is anastomosed to the left renal vein.

Thoracic Duct Cannulation and Drainage

Potential marrow donors are thymectomized, then 3–5 weeks later undergo thoracic duct drainage. The anesthetized rat is positioned supine with both front legs extended to his right. A short left subcostal incision is made, the kidney retracted medially and the abdominal portion of the thoracic duct isolated circumferentially by blunt dissection immediately cephalad to the cisterna chyli. The duct is opened and cannulated with polyethylene tubing (0.61 mm O.D.) which is tied in place and secured with 4-0 silk sutures to the muscle layers and skin. The distal end is led out the dorsal flank through a large bore needle and placed dependently in a 50 cc collection tube. The femoral vein is then cannulated with medical grade silicon tubing (0.020 in. I.D), and the rat placed in a restraining cage and given a continuous infusion of D_5 Ringers lactate at 3 ml/hr via an infusion pump. These animals routinely drain over 50 ml cell-rich lymph/24 hours. After 48 hours of drainage, by which time the output of circulating lymphocytes has decreased to low and stable levels, the rat is exsanguinated and bone marrow harvested.

Bone Marrow Harvest and Reconstitution

The humerus, femur and tibia of the thymectomized and thoracic duct drained rats are then flushed with RPMI 1640 medium and the cells dispersed by repeated expression through an 18 gauge needle. These are then filtered through a nylon wool column, washed twice and counted in a hemocytometer. The T cell depleted bone marrow cells (1×10^7) are then transfused I.V. into the syngeneic thymectomized and radiated animals which may be grafted 3-5 weeks later.

Lymphocyte Harvesting

Lymphocyte suspensions are useful in may experimental protocols in transplantation biology, particularly mixed lymphocyte reaction (MLR), assays for cytotoxicity, adoptive transfer, and fluorescence activated cytometry studies (FACS). The spleen can be removed separately. Commonly harvested peripheral lymph nodes include superficial and deep cervival, axillary and brachial nodes. Femoral lymph nodes are usually tiny and not worth using. Mesentonic lymph nodes are large and easily obtained. The nodes are cleaned of surrounding areolar tissue and placed into RPMI 1640 media on ice. They are expressed through 60 gauge stainless steel mesh using a rubber stopper, the dispersed cells resuspended in approximately 10 ml of media and washed twice by centrifugation at 1000 RPM $\times$ 10 minutes. The supernatant is removed without disturbing the cell pellet and the cells resuspended in complete media (RPMI 1640, 5mM HEPES buffer, 5×10-5 M 2-ME, 100 U/ml penicillin, 100 ug/ml streptomycin, and 10% fetal calf serum). Washed splenic lymphocytes are resuspended in 0.83% hypotonic (pH 7.21) ammonium chloride and allowed to stand at room temperature for 10 minutes to lyse erythrocyts. The erythrocyte free lymphocytes are then washed three times and resuspended in complete medium as above. The cell number is then determined using an hemocytometer and adjusted as necessary. Cells are stored on ice until use.

Removing Graft Infiltrating Cells

To examine mechanisms of rejection and host unresponsiveness at the graft level, it may be desirable to isolate the infiltrating lymphocyte populations. The organ is removed, diced into small pieces in media with scissors and incubated for 1 hour at 37 C in medium containing dispase (0.6 U/ml, Boehringer Mannheim GmbH, FRG). The fragments are then expressed through a 60 gauge stainless mesh with complete medium. The cell suspension is then washed three times and the mononuclear interphase collected after Ficoll-Hypaque gradient centrifugation.

Cell Separation

Discrete populations and sub-populations of lymphocytes may be identified and isolated by phenotype (CD4-, CD8- etc.) using monoclonal antibodies.

These cells may be harvested from host lymphoid compartments for comparison with graft infiltrating cells. The suspension may be enriched for T cells by resuspension in complete media and slow filtering (1 hour, 37 C) through a nylon wool column (loose nylon wool placed in the barrel of a 20 ml syringe, Fenwal Lab Inc., Deerfield, IL). B cells and macrophages adhere to the nylon wool and are left behind. The non-adherant T cells in the effluent are then labelled with the appropriate monoclonal antibody for 1 hour at 4 C (to remove th CD8 + population, for instance, cells are labelled with a monoclonal antibody directed against the CD8 receptor). They are then washed three times with media and incubated with goat-antimouse IgG on magnetic beads (Dynabeads M-450, Dinal Inc., Great Neck, NY) for 1 hour at 4 C. The antibody positive cells, now attached to the magnetic beads, are removed with a magnetic particle concentrator (Dinal). The purity of the negatively selected population should be confirmed by FACS analysis. The cells are then counted and numbers adjusted for use (Figure 19.8).

Lymphocyte Migration

Unstimulated lymphocytes recirculate continuously in normal animals by leaving the blood stream to enter various organs and tissus, then return to the blood.[10,11] As the circulating lymphocyte is critical to host immunological responses including delayed-type hypersensitivity and allograft rejection, its migration patterns have been investigated in many species including man.[12]

Lymphocyte Labelling

The use of radioisotopes provides a conenient and objective means of cell labelling in vivo or in situ, directly by autoradiography or quantitatively by measuring radioactivity in particular tissues. Available markers include ^{3}H, ^{14}C. ^{51}Cr, ^{125}I, ^{99}mTc, ^{75}Se and ^{111}In.[13] For both clinical and experimenal migration studies ^{111}In oxyquinoline seems to be the optimal marker. Its half life of (67.2 hours) is short and its two high gamma photoemissions of 171.3 KeV and 245.4 KeV ae advantageous for external cameral imaging.[14] The neutral and lipid-soluble oxyquinoline complex allows the label to penetrate the cell membrane and dissociate into free oxyquinoline and ^{111}In which attaches to cytoplasmic components.[15]

Migration of peripheral lymphocytes has been examined in transplant models,[16] however, almost any cell population or subpopulation can be studied. Lymphocyte suspensions are prepared. The cells are incubated with 10-15 uCi/10^{8} cells for 15 minutes at room temperature using the modified method of Thakur et al[14], washed three times, counted and transferred I.V. in a concentration of $3–5 \times 10^{7}$ cells/ml (1 ml injection). Approximatley 100 ul of the cell suspension should be counted in the gamma counter to determine the administered dose of radiation necessary for future calculations. Typically, over 80% of the radioactivity will be incorporated into the cell fraction which retains greater than 90% viability as determined by trypan blue exclusion. Twenty four hours following transfer of the labelled cells, the animals are sacrificed, their tissues harvested and placed in preweighted tubes. Tissues and organs usually analyzed include graft, whole blood, serum, peripheral and mesenteric lymph nodes, Peyers patches, small intestine, liver, spleen, thymus, parathymic lymph nodes, lung, heart and femur (marrow).[17] The radiation in each tissue is ascertained in a well type gamma counter. The data of lymphocyte incorporation can be expressed as percentage of total dose injected per gram of tissue weight or average of counts per gram of tissue weight. The large amounts of data generated are best handled by use of commercially available computer spreadsheets.

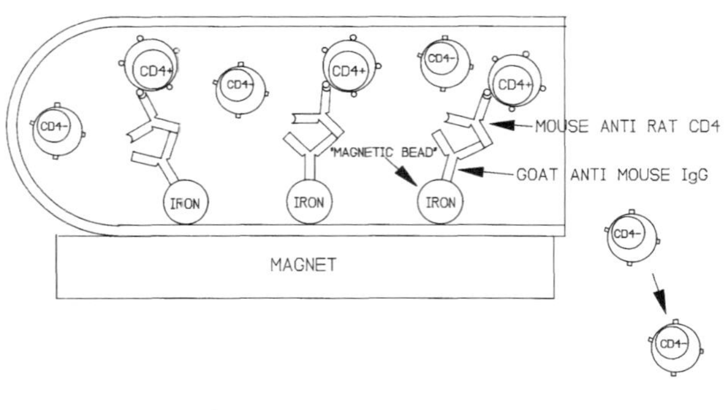

Figure 19.8. Magnetic beads coated with anti-mouse antibody are attached to mouse anti-rat antibody cell surface receptor complex. This technique renders a negatively selected cell population.

Routes of Intravenous Access and Drug Administration

Intravenous access is most easily obtained via the penile vein in the male rat, a site useful for daily injections using a 25 gauge steel needle. Alternatively, the subglossal veins can be used. The tail veins can be used alternatively, but are more difficult to cannulate. Animals with indwelling femoral vein catheters should be placed in a restraining cage. The femoral vein of the anesthetized rat is exposed through a vertical incision made approximately 1 cm below the inguinal ligament. The subcutaneous tissues are gently teased apart using cotton swabs, the femoral vein exposed for approximately 1 cm of its length and its distal portion ligated with a 4-0 silk suture. Vascular control is obtained by passing a 4-0 silk suture around the proximal vein and fixing it under tension to the abdominal wall with a hemostat. A transverse venotomy is then made in the vein and a soft polythylene tube (OD. 0.965 mm I.D. 0.28 mm) introduced. The proximal tie is then loosened and the catheter inserted approximately 1 cm into the vein and tied into position, being careful not to occlude the catheter which is fixed to the abdominal wall at at least 2 places or led out though the buttock (Figure 19.9). It may be connected to an external pump by inserting a blunt 23 gauge needle into its distal end, or connected to an implantable osmotic pump.

Osmotic pumps are devices which allow continuous delivery of compounds at predetermined rates. Implanted under the skin or into the peritoneal cavity of the animal, these preclude the need for complex and cumbersome restraining devices. They may be modified to allow delivery of compounds subcutaneously, intravenously by attaching a catheter tunneled under the skin and inserted in the femoral vein, or intraarterially. Intra-arterial delivery of agents is a very useful tool in some transplantation models as a cannula may be inserted into an arterial branch on the donor organ to allow continuous "in situ" perfusion of the organ with small doses of the compound being tested. If cardiac allografts are to be perfused, for instance, an extra long segment of the donor aorta including the right carotid artery is excised. While the donor heart is still in cold saline, a 0.061 mm O.D. cannula which has been

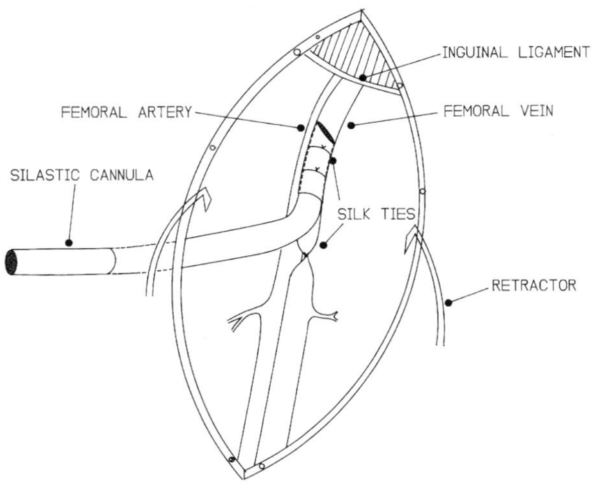

Figure 19.9. The soft silastic catheter is fixed in the femoral vein. This catheter is tunneled out of the skin and fixed at multiple sites to the abdominal wall.

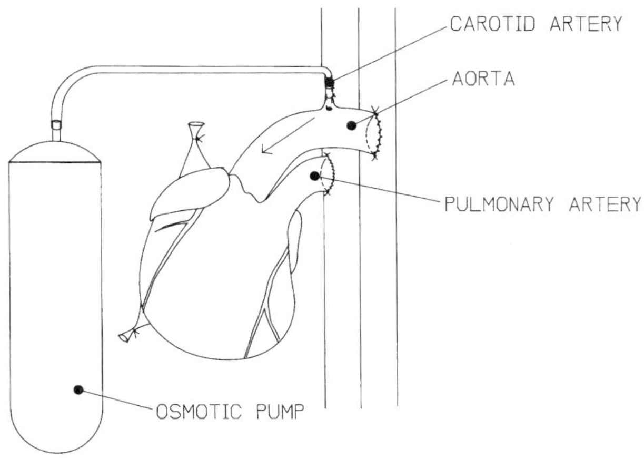

Figure 19.10. A catheter connected to a mechanical or osmotic (implantable) pump is fixed into the carotid artery on the donor aorta. This catheter can be used to deliver minute quantities of compounds to the heart *in situ*. Arrow indicates direction of blood flow.

joined to a 0.76 mm I.D., 1.22 mm O.D. catheter (1 cm length) is inserted into the right carotid artery of the donor aortic segment until the tip lies at the carotid orifice (Figure 19.10). This is then secured using 5-0 silk ligatures and the heart transplanted as usual. For intra-arterial infusion, the core temperature of the animal must be maintained by heat lamp or a heating pad. If the osmotic pump is inserted into a cool peritoneal cavity, it cannot pump against arterial pressure and will clot (this is not a problem when the catheter is inserted in the low pressure venous system). The proximal catheter is then connected to the osmotic pump and allowed to begin pumping before the occlusive loops are removed from the abdominal vessels and the heart perfused. For "in-situ" perfusion of renal allografts the distal catheter should be inserted into the adrenal artery coming off the renal artery and secured in similar fashion.

References

1. Ono, K. and Lindsey, E. S. (1969). Improved technique of heart transplantation in rats. *Journal of T horacic and Cardiovascular Surgery*, **57**, 225–229.
2. Araneda, D., Padberg, W. M. and Tilney, N. L. (1989). Refined techniques for heterotopic cardiac transplantation in the rat. *T ransplantation Proceedings*, **21**, 2562.
3. Schmid, C., Heemann, U. W., Azuma, H. and Tilney, N. L. (1994). Successful heterotopic heart Transplantation in the rat. *Microsurg.*, **15**, 279.
4. Feldman, J. D. and Lee, S. (1967). Renal Transplantation in rats. I. Allogenic recipients. *J. Expt. Med.*, **126**, 783.

5. Monchik, G. J. and Russell, P. S. (1971). Transplantation of small bowel in rat: technical and immunological considerations. *Surgery*, **70**, 693.
6. Kamada, N. and Calne, R. Y. (1979). Orthotopic Liver Transplantation in the Rat. *T ransplantation*, **28**, 47–50.
7. Parrott, D. M. V., Tilney, N. L. and Shess, F. (1975). The different migratory characteristics of lymphocyte populations in a whole spleen transplant. *Experimetnal Immunology*, **19**, 459.
8. Nolan, M. S., Lindsey, N. J., Savas, C. P., Slater, D. N. and Fox, M. (1982). Pancreatic transplantation in the rat, a simplified method using aortic interposition and cuff techniques. *Transplantation*, **33**, 327–329.
9. Tilney, N. L., MacLennon, I., Strom, T. B. and Baldwin, W. M. III. (1981). Effects of thymectomy, thoracic duct drainage and radiation of T and B lymphocyte distribution and cardiac allograft survival in rats. *Transplantation*, **31**, 26.
10. Gowans, J. L. (1957). The effect of the continuous re-infusion of lymph and lymphocytes from the thoracic duct of unanaesthetized rats. *British Journal of Experimental Pathology*, **38**, 67.
11. Gowans, J. L. and Knight, E. J. (1964). The route of recirculation of lymphocytes in the rat. *Proceedings Royal Society of London*, **159**, 257.
12. Kupiec-Weglinski, J. W. and Tilney, N. L. (1989). Lymphocyte migration patterns in organ allograft recipients. *Immunological Reviews*, **108**, 63–82.
13. Rannie, G. H. and Donald, K. J. (1977). Estimation of the migration of thoracic duct lymphocytes to non-lymphoid tissues. A comparison of the distribution of radioactivity at intervals following i.v. transfusion of cells labelled with ^{3}H, ^{14}C, ^{75}Se, ^{99m}Tc, ^{125}I and ^{53}Cr in the rat. *Cell Tissue kinetics*, **10**, 523.
14. Thakur, M. L., Coleman, R. E., Mayhill, G. G. and Welch, M. J. (1975). ^{11}indium labelled leukocytes: Preparation and evaluation as an abscess imaging agent. *Radiology*, **26**, 175.
15. Dewanjee, M. K., Rao, S. A. and Didesheim, P. J. (1981) Indium-111 tropolone, a new high affinity platelet agent: preparation and evaluation of labelling parameters. *Journal of Nuclear Medicine*, **22**, 981.
16. Kupiec-Weglinski, J. W., de Sousa, M. and Tilney, N. L. (1985). Importance of lymphocyte migration patterns in experimental organ transplantation. *Transplantation*, **40**, 1.
17. Tilney, N. L. (1971). Patterns of lymphatic drainage in the adult laboratory rat. *Journal of Anatomy*, **109**, 369.

PART VI

20. Methods in Experimental Xenogeneic Transplantation State of the Art

Claus Hammer[1] and Gunnar Tufveson[2]

[1]*Institute for Surgical Research, University of Munich, Germany*
[2]*Department of Transplant, University Hospital Uppsala, Sweden*

The increasing shortage of allografts for clinical transplantation stimulated the idea to use xenografts at least in desperate situations. Historical trials to use animal organs were never successful. Observations, that organs of monkeys do not undergo fatal destruction has lead to the first attempt to investigate organs from closely related species to man like chimpanzees and baboons. Kidney grafts and hearts did work for weeks and months.[1–3]

These promising data on the other hand have opened an unlimited field for experimentation. The estimated 4070 species of mammals on earth and the wide variety of different organs would allow endless series of experiments. Even human recipients have been used in these experiments. Thus, livers from baboons[4] and islets from piglets[5] have been transplanted. Human kidneys have also been connected to the circulation.[6,7]

In order to obtain in a relatively short time relevant data for clinical use, the models have to be reduced to suitable species with special regard to man. The optimal approach would be to use a few combinations reflecting the biochemical, physiological and genetical as well as immunological human situation.[8] Other animal experimentation is only justified if a wished biological mechanism is studied, e.g. test of immunosuppressant for a defined immunological reaction.

These investigations have to deal with surgical methods, in vivo modulation and manipulation of the rejection mechanisms as well as in vitro tests which clarify the rather complex features of xenogeneic rejection.

Whether these modifications should be applied to donor or recipients depends basically on the zoological i.e. evolutionary relationship of the species involved.

Classification of Xenografts

Like in allogeneic transplantation models, the success of xenografts depends very much on the histoincompatibility of the two individuals involved. Extensive reviews were published in two recent comprehensive books on xenotransplantation.[9,10]

Systematic evaluation of xenogeneic graft rejection has shown, that two major donor recipient combinations exist (Table 20.1).

Organs which survive for days and weeks without immunosuppression originate from closely related species. The cellular type of rejection in concordant systems can to some extent be modulated by immunosuppressive treatment. They might be of clinical interest.

Grafts which are destroyed within minutes or hours in a hyperacute, humoral type of rejection are usually taken from zoologically widely divergent species. The antibody mediated rejection in these discordant models cannot be influenced by immunosuppressive regimens known at the present time. The response may however be modulated by plasma exchange and/or complement inactivation.[11–13]

Various combinations have shown, that intermediate combinations exist in which a primarily cellular type of rejection is overlapped by humoral mechanisms.

Still the most reliable classification is of Darwin and Linne who classified the members of the animal kingdom according to the evolutionary aspects.

Table 20.1

- one zoological family	Closely related cellular rejection
- one zoological order	Moderately related cellular/humoral rejection
- different zoological orders/families	Widely divergent humoral rejection (can be followed by a cellular rejection)

Tissue Differences in Xenografting

Vascularized and non vascularized xenogeneic tissues of organs behave different in terms of rejection.

Parenchymal organs like liver and kidney are more vulnerable than muscular organs like hearts. In both cases the vascular system including endothelium are destroyed first. However, it seems as if hepatocytes and tubular cells suffer more rapidly and severely from ischemia. Myocytes also show in in-vitro tests a higher resistance to the xenogeneic rejection mechanisms.

Bradytrophic tissues like cartilage, bone and skin undergo rapid deterioration. They are resorbed sooner, often after severe reactions, similar to what is called "Arthus Phenomenon". Single cells when unprotected and not encapsulated induce immunological reactions. Bone marrow, endocrine cells and hepatocytes or islets are destroyed by antibody mediated cellular reaction in which phagocytic cells play a leading role. If the antibody mediated rejection is prevented macrophages controlled by T-cells may reject the graft.[14]

In vitro single hepatocytes, beating myocytes and endothelial cells behave differently. After adding PNAB (Performed Natural Antibodies) all cells show short intervals of activation with following decrease of function. They usually recover to survive even under the influence of PNAB until the culture is too old or they are overgrown by fibroblasts.[15]

Carbohydrate Antigens

It is now clear that carbohydrate antigens are important for hyperacute rejection of grafts from a distantly related species. Humans and old world monkeys lack the Gal-alfa-1, 3-Gal antigen and possess preformed IgM antibodies to this structure.[16,17] Recently a mouse with a gene knock out for the enzymes synthesising this antigen has been produced.[18] One carbohydrate antigen (Forssman) is also identified in the mouse to rat heart model.[19]

Evolutionary Aspects

The most important factor of xenotransplantation seems today still to be the genetic background. The genetic relationship between species underlies the similar structural organizational development of the DNA. In reassociation experiments and DNA hybridisation it was shown that 98% of human DNA can be found in the genes of chimpanzees, 92% in other old world monkeys and 85% in Asiatic primates. Similar studies have been done in small rodents, with almost the same results. However, while the numerical differences in nucleic acids may be small, differences in genes and genetic control mechanisms are so obvious that they distinguish clearly between closely related species like man and ape. Karyological approaches to taxonomy have shown that fusion and fission of chromosomes are the driving force in speciation.

While some loci stay however very conservative and carry the same gene for 400 million years, others become silenced or reactivated. A third group may vary fast, some times from one species to the next related one. Some characters shift fast and unpredictable, others for example albumin share great structural resemblance, so that the aminoacid sequence changes can be used as "index of dissimilarity" or "immunological distance". This regularity has been called "immunological clock".

The characters of each species change however to a different rate. This leads to the problem, that enzymes or protein hormones show considerable immunological variations. While porcine and human insulin differ in one amino acid, the C-peptide of preproinsulin, the connecting part of the two insulin chains is so different, that monoclonal antibodies do not even recognize the two species specific molecules. This shows, that even functionally compatible molecules, handling the same substrate perfectly are highly immunogenic.

Other situations exist where substrate specificity disappeared to various extends as shown for angiotensin II and renin or kallikreins, hormones and complement. These variations in comparative biochemistry and physiology seem to be endless and might only be overcome, when applying transgenic techniques.[20–23]

Surgical Techniques

In principle the same surgical techniques like in allogeneic transplantation are useful.

Kidneys

Renal grafts in closely divergent species are connected heterotopically to the iliac vessels. In case small organs had to be used, "en bloc" transplantation was described.[24]

Organs in widely divergent combinations were hemoperfused "ex vivo" i.e. connected by tubing to the vessels of a heparinized perfusing "recipient".

Various perfusion models, pulsatile or non pulsatile for in vitro hemoperfusion were described.[25] All of the perfusion models show that normotherm autologous whole blood perfusion damages the graft. Complement activation is partially responsible for this problem and should therefore be taken into account and tubing, oxygenators etc. should be carefully chosen.[26,27]

Heart

Cardiac transplantation was performed ortho- and heterotopically. For immunological investigations transplantation to the cervical vessels is preformed. Intensive observations are possible in terms of ECG, flow measurement or biopsies. Langendorff models and working heart devices were developed for rat, rabbit and pig hearts (Fig.1).[28]

Liver

Hepatic grafts were transplanted in a few cases orthotopically. Some efforts have been made to place hepatocytes on the other side of a dialysis membrane and thereby creating a protective barrier to immunological damage.[29,30]

Small Animal Models

Connection of rodent organs has been described extensively. Suturing and cuff-techniques are performed. The most common donors are hamsters and guinea-pigs where the former is concordant to rats and the latter discordant. Of some interest is mouse to rats as the mouse can be genetically altered by transgenic or knock out techniques. This combination is concordant but can be converted to mimic a discordant situtation with immunisation.[31]

Skin Transplantation

Transplantation of xenogeneic skin was described in dogs/fox, rat/mouse, guinea pig etc.[32]

Perfusion Models

Advantages of perfusion models in xenotransplantation are, that all parameters can be measured without disturbance by secreted or liberated inhibitors or factors from the perfusing animal. Collection of urine and blood samples is possible at each time required. Oxygen consumption and other physiological parameters can be monitored in line. Addition of metabolites is guarantied in exact dosages.

Absorption or mediator release, complement consumption can be correlated at a given time intervals.

Models for the Concordant Systems

Clinically interesting survival times of 3–9 months were achieved, when kidneys from baboon or chimpanzee were grafted to patients. 9 kidneys were the only organs yet, that established satisfactory function. Hearts were less successful.

These concordant combinations undergo primarily a cellular type of rejection. CD-antigens, interleukins and complement are compatible so that antigen presentation, interleukin interaction and proliferation as well as cytotoxicity are comparable to what is known from allogeneic immunreactions. Mixed lymphocyte reactions using cells from closely related species are described and the ^{3}H-thymidin uptake is comparable to that of an alloresponse. However proliferation rate and survival time i.e. severity of rejection showed no correlation. Mixed leukocyte reactions can also be obtained between more distant species provided that adequate cofactors are added.[33]

Primates were tested for blood groups and isoantibodies, which do not exist in other animal species. Secretory blood groups like CEA (canine erythrocyte antigens) in dogs or pigs had also no detectable effect on the survival time. Preformed natural antibodies do not exist in closely related species. Transplantation however leads to secondary IgG type antibodies, which were shown to be directed against class I and possibly class II antigens.

These antibodies are capable to induce the typical second set, humoral rejection. They even reject a following allograft (bridging) in accelerated or hyperacute fashion. This seems to be different in rodents.[13,34,35]

Inflammation during rejection is comparable to the allogeneic type. Fine-needle aspiration and corebiopsy i.e. endomyocardial biopsy show, that T-lymphocytes migrate into the graft followed by B-cells and macrophages. Without immunosuppression B-cells and their products, mainly IgM-type antibodies attack the graft.

Even extreme immunosuppression is unable to suppress the switch from T-cell dominated rejection into B-cell reactions.[36]

In summary several mechanisms for xenograft rejection has been described. These mechanisms are summarised in table two. Chimerism, mixed chimerism and tolerance are achieved in rodent models. Various class I and class II deficient strains as well as thymusless (nude) mouse strains and the beige mouse (T- and B-cell deficient) exists.

Using biological, chemical or physical immunosuppression regimens long term survival can be achieved. Of particular interest are some of the new immunosuppressants, which when combined with Cyclosporine (Sandimmune) or FK 506 (Prograf) yield substantially prolonged graft survival. The anti B-cell properties of Deoxyspergualin,[38–40] Mycophenolate mofetil (Cellcept);[41,42] Leflunomide[43,44] and Brequinar sodium[45,46] should be kept

Table 20.2 Mechanism of xenograft rejection

Designation of phenomenon	*Time to appear*	*Effector Mechanism*
Hyperacute rejection	seconds - minutes	antibodies
Hyperacute rejection	seconds - minutes	complement
Accelerated acute rejection	1–3 days	Macrophages
Acute rejection	3–7 days	T-cells
Induced antibody mediated rejection	2–5 days	B-cells
Chronic rejection	Weeks-months	?

in mind. Likewise compounds inhibiting the B7/CD28 interaction[47,48] have to be considered. The latter may be of particular interest to xenografts where large quantities of preformed B-cells may act as antigen pre-senters according to a novel mechanism described by Kelly *et al.*[49]

Moderately Related Species

Experiments using moderately related animals have been rarely described. In this combination using animals of 2 zoological families, but one zoological order like cat/dog, guineapig/rat or sheep/pig low titers of preformed natural antibodies exist. The delayed humoral rejection allows some cellular components to take part in this type of rejection. Survival times measured were hours to days. Conventional immunosuppression had no significant impact on survival time. However second set reactions of hyperacute fashion can be induced.[18]

Rather strong activation of complement was seen, using both the alternative and the classical pathway. Activation of endothelial cells in this model was shown, suggesting release of cytokines and mediators which lead to the typical accelerated blood coagulation, vasoconstriction and extravasation of blood.

Marked differences exist in isoenzymes, blood proteins, blood groups and possibly hormones and the cytokine systems.

Widely Divergent Systems

Suitable organs in large numbers could only be harvested from species widely divergent to man, here especially domestic animals. These animals belong to different zoological orders and show total histoincompatibility. No enzyme triggered cascade, like blood clotting, complement or other protein metabolism do find immunological and substrate specific identical homologues. Therefore components destroying the xenograft have to be split in imunological, physiological and biochemical compartments.

All widely divergent species possess PNAB against each other. Fixation of these highly avid, rather low specific IgM antibodies to their epitope on the endothelial cells lead immediately, within minutes to less than 2 hours to what is called "hyperacute xenogeneic rejection" (HXAR).

The characteristics of the HXAR like blood clotting, vasoconstriction, cessation of blood flow and cessation of function are typical. The immediate binding of PNAB is followed by activation of the alternative and classical complement pathway. This mechanism recruits thrombocytes and granulocytes to the graft. Lymphoid cells are not involved in this early phase of HXAR. Other mediators like thromboxan from platelets, prostaglandin and prostacyclin from endothelial cells together with other mediators like platelet activation factor (PAF) etc. are released. Thus humoral factors activate blood coagulation, fibrin deposition and lead to irreversible slugging.

It is not known, whether enzyme systems interact in these combinations. Long term survival time of disparate grafts have been shown in immunologically deficient mice. Transgenic animals might be produced which carry one or two donor specific antigens on the cell surface. The transfection of the whole recipient characteristics seems to be impossible in the next future.

In case these organs would survive, their products and metabolites are expected to be highly antigenic. It is not sure if their metabolites are capable to support xenogeneic organisms.

Experiments with transgenic animals suggest that hormones and their mediators or enzymes and their inactivators would not interact. It is totally unknown, whether such organs would work in this strange environment.

Preformed Natural Antibodies

The rejection of xenografts is linked with the existence of so called PNAB. Lack of these IgM-type gamma globulins is connected with a long term survival time. With increasing titres of PNAB the survival time becomes shortened. The antibodies are directed against virtually all xeno antigens available. The amount of antibodies seems to vary in each individual. Total lack of PNAB was not found, not even in immunosuppressed or immunodeficient (AIDS) persons. PNAB have agglutinating, complement fixing and lysing as well as opsonizing properties.

Life without PNAB seems not to be possible. PNAB therefore are transferred via placenta from mother to offspring. Large amounts are absorbed from colostrum or milk. Under gnotobiotic conditions the own antibody production is delayed until the time of first contamination.

Bacteria seem to be the strongest stimulants. There seems to be an archaic recognition mechanism which induces the production of a broad spectrum of PNAB even when one single antigen is applied. Non pathogenic germs are as potent as pathogenic bacteria.

After reaching an individual titre of PNAB, additional contact with food antigens, immunisation, vaccination or blood transfusion etc. does not increase the titre of polyclonal IgM-PNAB significantly.[50] It only increases the amount of specific IgG antibodies.[51]

Absorption studies have shown, that PNAB can be highly specific in small amounts. Cross reactivity is closely linked to the zoological disparity. Cleavage of the sulfur bond of IgM parameters reduces or eliminates the antibody action.[52]

Even if the PNAB would be an epiphenomenon in xenografting, their estimation in in vitro tests, like binding hemagglutinins by endothelial cells in ELISA or absorption studies would help to assess the evolutionary distance of the species involved.

Concordant Animals Models

From numerous experiments it seems that xenogeneic models using rodents have to be distinguished from other zoological species. While in rodents long term survival time can easily be achieved this is not true for other mammals, especially for canines, which seem to attack extremely aggressive other mammalian antigens. Pigs on the other hand express rather few class I antigens and might therefore be advantageous as donors.

Primates Models

Xenotransplantation using non human primates was described. Serological criteria of donor selection seem to be very important. Especially the ABO blood group system existing in part in primates has to be taken into the consideration.

Still, the survival time of heart grafts from velvet monkeys transplantated into baboons function not longer than 10.3 ± 5.2 days. ABO incompatible pairs were rejected within 7.3 ± 5.6 days with 30% still ending in HXAR. Immunosuppressive therapy using Cyclosporin A (CSA), Methylprednisolone and Azathioprine did not significantly prolong the survival time.[53,54]

Other immunosuppressive strategies achieved interesting survival times often, however, with unacceptable high mortality and morbidity. Even the application of highly specific monoclonal antibodies was not able to control these severe xenogeneic cellular rejections possibly because of their own species specific restriction.

Distantly related donor species result with very few exceptions in HXAR.[55] Best results were found after absorption of PNAB on absorption columns[56] or by application of plasmapheresis together with maximal immnunosuppression.[57]

In anticipation to the clinical situation pigs were used as heart and kidney donors. The histopathological features especially in the long term surviving kidneys and hearts were still oedema, interstitial haemorrhage and intravascular coagulation reflecting a delayed mode of HXAR, 3 out of 9 pig kidneys survived with satisfactory function for 13–23 days in baboons.[57]

It should be emphasized that the cardiac models used were only heterotopic grafts perfused and not orthotopic working one's.

Other conventional immunosuppressive therapies including splenectomy were not found to be of advantage as monotherapy.

Canines

Investigating groups lacking the access to primates have used canine species as preclinical models. Immunogenetic markers and analysis of biochemical and physiological characteristics have proven that this surgically important model is able to mimic the preclinical situation in man.

Interestingly enough very similar results were found as compared to the primate system. Wolf and fox kidneys, hearts and skin showed survival times identical to those in the system velvet monkey/baboon. Chemotherapy and antilymphocyte globulin was able to prolong the concordant vascularized and non vascularized grafts up to a month. Variable results after treatrnent with a low titer anti-macrophage-serum could give interesting hints on the mechanism of this acute xenogeneic rejection. The irreversibility of the mechanism was found to be due to B-cell action and antibodies at the end of the experiment.[27]

Studies using organs from moderately related species like felines show a different behaviour. Perfusion and function time of cat kidneys transplanted into mongrel dogs did not exceed 24 hours.

Concentration of urine for up to 12 hours proves a delayed type of HXAR. It was found that the low titers of PNAB triggered the rejection mechanism and treatment with ALG was not able to improve the results. In contrast, a second transplant after 3 weeks ended in hyperacute rejection due to the extremely elevated antibody titers of the recipient dog.[59]

The histology showed slight congestion of veins and capillaries and platelet aggregates, vacuolisation and pycnosis of the nuclei of tubular cells, supporting that mechanisms of HXAR are governing the rejection in these organs. Organs of widely divergent species mainly from pig and sheep, rabbit and rat connected extracorporally to the dogs circulation were destroyed within minutes. HXAR as such was probably first described due to observations in these experiments.

No major difference exists between the results known from primate experiments.

Ungulates

Despite the favorable experimental situation within the order of ungulates, where many large species exist as domestic animals (e.g., sheep, goats, cattle and pigs) and many species are under domestication (e.g., deer and antelopes) few attempts at xenotransplantation have been ventured. In an effort to find a model for separation of humoral and cellular mechanisms, pig kidneys were transplanted into goats.

Untreated animals rejected the xenografts within 3 days and ALG-treated individuals after 11 days.

The kidneys functioned and produced concentrated urine. The rejection was of both cellular and humoral type. Since ALG did eliminate mononuclear cells from the circulation and prolong survival time, a cellular mechanism was presumed to be responsible for the accelerated rejection seen in untreated animals.

Survival time of sheep kidneys in unsensitized goats was 11.2 days and almost identical to that of allografts (11.8 days). Sensitization reduced the survival time to 2.7 days compared with 2.9 days in allografts. Similar findings were reported concerning skin transplantation where a survival time of 7 days was found as compared with 9 days in allografts. Intensive studies in this interesting experimental model have not been carried out. Cardiac and skin transplantation between calves and goats resulted in respective survival times of 5 and 7 days. Antilymphocyte serum (ALS) was able to prolong skin graft survival to between 28 and 79 days, but prolonged heart grafts only to 14–30 days.

Orthotopic cardiac xenografting using donor lambs in the newborn goat resulted in mean graft survival of 32 days (range 13–72 days) when treated with cylosporine. Survival of the controls was only 6 days. The longest surviving animal died after 165 days from anemia (28, 29).

Rodents

While rodents especially rats and mice are the most important species to reveal immunological and genetical problems in the field of allogeneic transplantation they play still a minor role in xenotransplantation. This might be due to a lack of suitable size donors outside the order of rodents i.e. family of rats. In addition it is almost impossible to run the extensive biochemical and hematological studies in the short time intervals because of the small amount of blood of these animals.

Another disadvantage is the limited access for invasive or non-invasive methods and the rather weak response to xenogeneic antigens. The first extensive reviews in this field[61,62] are thus restricted to data about survival time and effect of chemotherapy –and antibody on models using hamster, mouse and guinea pig as donors and rat as recipient.

Plasma exchange and splenectomy in xenotransplantation in rodents and its effect on mean SVT has been described exhaustively for various organs and immunosuppressive regimens.[63] Similar detailed data comparing skin versus heart grafts gives insight in the different types of reaction of primary and non primary vascularized grafts.[64] Treatment strategies including total body irradiation and chemotherapy as well as monoclonal antibody therapy indicate that the behaviour of xenografts in rodents differs considerably from those in large animals including primates. The advantage in this model are the genetically modified and immunodeficient strains of mice and rats, which have become available for experimental studies and do not exist in other animals.[65]

Bridging with Xenografts

One of the future tasks of xenotransplantation could be the bridging of time, when desperately needed organs are not available. In rat models 'bridging' with organs from either rat, hamster and Guinea pig had no impact on the following allograft. Sensitisation with primary vascularized grafts or non-vascularized grafts showed no difference in the outcome, suggesting that a sensitization is rather specific and the species specific and organ specific antigens might be not cross reacting.

In contrast, combinations of large animals including primates rejected the following allograft in an accelerated if not hyperacute fashion.[66] These types of rejection were reflected by different release of cytokines and the production of species specific antibodies directed also against cross reacting histocompatibility antigens. The latter could be used for xenogeneic tissue typing.[67] This field needs extensive investigation in the near future.

Ex Vivo Perfusion of Xenogeneic Organs

In order to receive accurate data from xenogeneic events occurring in the hemoperfusate (blood and its components) and the graft itself, ex vivo models have been developed.[68] This experimental approach achieves results which are not disturbed by changes in the perfusing animal, which suffers from all types of multiorgan failures under such exhausting procedures.

The collection of urine, bile, pancreas juice etc. allow further measurements of products released under the different types of rejection. On line measurement of ion changes, oxygen and oxygen radicals are under investigation. Multiple modifications of the hemoperfusate have been described from complement depletion to antibody absorption.[69]

In Vitro Testing of Xenogeneic Reactions

Alternative methods for xenogeneic investigations are in vitro tests. Mixed lymphocyte reactions for immunogenetic and cellular studies of xenogeneic responses have been used. Solid phase tests with endothelial cells have been developed recently.[70] They allow intermediate measurement of antibody fixing, complement activation and mediator release. Delicate variations of the xenoagents used indicate the impact on endothelial cell membrane molecules and might thus open a new field of xenoimmunochemistry.

Encapsulation and organ protection using membranes, alginate and biohybrids could protect small organs or single cells from the attack of cytotoxic cells or antibodies. This approach is not only of clinical interest since it would allow to measure biochemical and physiological mediators in the metabolism of organs, which are exposed to a totally foreign environment.

Summary

This comprehensive report is based on the most recent reviews in the field of xenotransplantation. Numerous experimental models and approaches exist already for xenogeneic transplantation research. Almost every question that had to be answered in allografting reappears in this field with the additional aspect of the difference in the animal disparity. Since xenografting should aim towards a fast solution of the organ shortage, experiments should concentrate on the finding of an optimal donor for man, its suitable size and easy access.

It seems to be a waste of time and money to develop methods or recipient treatments which are connected with unbearable problems for human beings, with high mortality and morbidity or a low quality of life after having received a xenograft.

Its advantage would be however that in contrast to allografts, organs from healthy, selected donors could be harvested. Animals raised in special laboratories could be modified immunologically, genetically or both.[71]

Minimal ethical, religious and moral problems would appear. And the logistic and financial problems could be reduced to a minimum.

At the time of writing of this revision of this chapter (Febr 1997), xenografts to humans has been subjected to a moratorium in the United Kindgdom in the fear of retroviral transmission from pig to humans.

References

1. Reemtsma, K. (1964). Renal heterotransplantation from non-human primates to man. *Ann. N.Y. Acad. Sci.*, **162**, 412–438.
2. Starzl, T., Marchioro, T., Peters, G., Kirkpatrick, C., Wilson, W., Porter, K., Rifkind, D., Ogden, D., Hitchcock, C. and Waddell, W. (1964). Renal heterotransplantation from baboon to man: experience with six cases *Transplantation*, **2**, 752–776.
3. Bailey L, Nehlsen-Cannarella, S. Concepcion W. and Jolley, W. (1985). Baboon-to-human cardiac xenotransplantation in a neonate *J.A.M.A.* **254**, 3321–3326.
4. Starzl, T. E., Fung, J., Tzakis, A., Todo, S., Demetris, A. J., Marino, I. R. *et al.* (1993). Baboon-to-human liver transplantation. *Lancet* i, 65–71
5. Groth, C. G., Korsgren, O., Tibell, A., Tollemar, J., Moller, and E. Bolinder, J. *et al.* (1994). Transplantation of porcine fetal pancreas to diabetic patients. *Lancet* ii, 1402–1404
6. Cairns, T. D., Taube, D. H., Stevens, N., Binns, R. and Welsh, K. I. (1991). Xenografts future prospects for clinical transplantation. *Immunol-Lett.* **29**, 167–170
7. Breimer, M. E., Björk, S., Svalander, C. T., Bengtsson, A., Rydberg, L. Lie-Karlsen L. *et al.* (1996). Extracorporeal ("ex vivo"). connection of pig kidneys to humans. 1. Clinical data and studies of platelet destruction. Xenotransplantation **3**, 328–339
8. Hammer, C. (1991). Evolutionary, physiological, and immunological considerations in defining a suitable donor for man. Xenotransplantation, edited by D. Cooper, E. Kemp, K. Reemtsma and D. White, pp. 429–438. Berlin: Springer Verlag.
9. Hardy, M. (1989). Xenograft 25. New York, Oxford: Elsevier
10. Cooper, D., Kemp, E., Reemtsma, K. and White, D. (1997). Xenotransplantation. Berlin: Springer Verlag.
11. Besse, T., Duck, L., Latinne, D., Gianello, P., Squifflet, J. P., Lievens, M. *et al.* (1994). Effect of plasmapheresis and splenectomy on parameters involved in vascular rejection of discordant xenografts in the swine to baboon model. *Transplant Proc.*, **26**, 1042–1044
12. Forty, J., White, D. G. and Wallwork J. (1993). A technique for perfusion of an isolated working heart to investigate hyper acute discordant xenograft rejection. *J. Thorac. Cardiovasc. Surg,.* **106**, 308–316
13. Rosengard, A. M., Cary, N. R. Langford, G. A., Tucker, A. W., Wallwork, J. and White, D. J. (1995). Tissue expression of human complement inhibitor, decay-accelerating factor, in transgenic pigs. A potential approach for preventing xenograft rejection. *Transplantation*, **59**, 1325–1333

14. Wallgren, A. C., Karlsson-Parra, A. and Korsgren, O. (1995). The main infiltrating cell in xenograft rejection is a CD4+ macrophage and not a T lymphocyte. *Transplantation*, **60**, 594–601
15. Hammer, C., Csapo, C., Höbel, G., Müller, U. and Werdan, K. (1992). Preformed natural antibodies: Action on Endothelial cells. *Transpl. Proc.*, **24**, 590–591
16. Galili, U. (1993). Interaction of the natural anti-Gal antibody with alpha-galactosyl epitopes: a major obstacle for xenotransplantation in humans [see comments]. *Immunol. Today*. **14**, 480–482
17. Samuelsson, B. E., Rydberg, L., Breimer, M. E., Backer, A., Gustavsson, M., Holgersson, J. *et al.* (1994). Natural antibodies and human xenotransplantation. *Immunol. Rev.* **141**, 151–168
18. Tearle, T. G., Tange, M. K., Zannettino, Z. L., Katerelos, M., Shinkel, T. A., Van-Denderen, B. J. *et al.* (1996). The alpha-1,3-galactosyltransferese knockout mouse. Implications for xenotransplantation. *Transplantation*, **61**, 13–19
19. Gustavsson, M. L., Gannedahl, G., Backer, A. E., Larson, G., Olling, A., Tufveson, G. and Samuelsson, B. E. (1996). Anti-carbohydrate antibodies associated with hyperacute rejection in a vascularized mouse heart-to-rat xenotransplantation model. *Transplantation*, **61**, 957–963
20. Subha, S., Fhrasawa, K., Smeby, R. and Bumpus, F. (1971). Measurement of plasma renin substrate using homologous and heterologous renin. *Am. J. Physiol*,. **221**, 1476–1480
21. Macdonald, R., Margolius, H. and Erdös, E. (1988). Molecular biology of tissue kallikrein *Biochem. J.*, **253**, 313–321
22. Wani, R., Hermanns, E. W., Folger, S., Wolf, E. and Brem, G. (1991). Accelerated growth and visceral lesions in transgenic mice expressing foreign genes of the growth hormone family *Ped. Nephrol*,. **5**, 513–521
23. Rollins, S., Ji, Z., Haruhiko, N. and Sims, P. (1991). Inhibition of homologous complement by CD59 is mediated by a species selective recognition conferred through binding to C8 within C5b-8 or C9 within C5b-9 *J. Immunol.*, **146**, 2345–2351
24. Chaussy, C., Hammer, C., Scheel, J. V., Eisenberger, F., Pielsticker, K. and Brendel W. (1973). Experimental xenogeneic kidney transplantation in closely related species. *Res. Exp. Med.*, **159**, 266–275
25. Otte, K., Steinbrüchel, D. and Kew, E (1991). Ex vivo organ perfusion studies in xenograft research. Xenotransplantation pp. 395–403. Berlin: Springer Verlag
26. Fiane, A. E., Videm, V., Scholz, T., Karlsen, H., Geiran, O. *et al.* (1995). Ex vivo model for xenotransplantation. *Transplant Proc.*, **27**, 3560–3561
27. Forty, J., White, D. J. and Wallwork, J. (1993). Activation of the alternative pathway of complement in hyperacute xenograft rejection of rabbit hearts by human blood. *J. Heart. Lung Transplant*, **12**, 283–286; discussion 286–287
28. Schütz, A., Engelhardt, M., Brandl, U., Breuer, M., Hammer, C., Reichart, B. and Kemkes, B. (1992). Current monitoring methods in heart transplantation, pp. 186–206. In Transplantation Monitoring, edited by S. Shapira, A. Yussim, C. Hammer, Pabst Verlag.
29. Fuchs, M., Gerlach, J., Unger, J., Encke, J., Smith, M., Neuhaus, P., Nundel, M. and Riedel, E. (1994). Alpha-keto acid metabolism by hepatocytes cultured in a hybrid liver support bioreactor. *Int. J. Artif. Organs*,. **17**, 554–558
30. Demetriou Rozga, J., Podesta, L., Lee Page, E., Hoffman, A., Morsiani, E., Sher, L., Woolf, G. M. and Makowka, L. (1993). Control of cerebral oedema by total hepatectomy and extracorporeal liver support in fulminant hepatic failure. *Lancet* ii, **342**, 898–899
31. Gannedahl, G., Karlsson-Parra, A., Wallgren, A., Roos-Engstrand, E., Nilsson, B., Totterman, T. H. and Tufveson, G. (1994). Role of antibody synthesis and complement activation in concordant xenograft retransplantation. *Transplantation*, **58**, 337–344
32. Hammer, C., Chaussy, C., Welter, H., Wembacher, J., Höbel, G. and Brendel, W. (1981). Exceptionally long survival time in xenogeneic organ transplantation *Transpl. Proc.*, **13**, 81–884
33. Lee, L. A., Gritsch, H. A., Sergio, J. J., Arn, J. S., Glaser, R. M., Sablinski, T. and Sachs, D. H. (1994). Sykes-M Specific tolerance across a discordant xenogeneic transplantation barrier. *Proc. Natl. Acad. Sci.*, **91**, 10864–10867
34. Auchincloss, H. (1988). Xenogeneic Transplantation. *Rev. Transplant*, **46**, 1–20
35. Pratschke, J., Breuer, M., Hammer, C., Schütz, A., Kemkes, B., Engelhart, M. *et al.* (1992). Allogeneic heart transplantation following xenografting. *Transpl. Proc.* **24**, 688–690
36. Reichenspurner, H., Ertel, W., Reichart, B., Peters, D., Welz, A., Überfuhr, P. *et al.* (1986). Xenogeneic and allogenic canine heart transplantation: A model for cytologic and immunologic monitoring of rejection mechanisms. *J. Heart. Transpl.*, **5**, 471–476
37. Sykes, M., Lee, L. A. and Sachs, D. H. (1994). Xenograft tolerance. *Immunol. Rev.*, **141**, 245–276
38. Valdivia, L. A., Monden, M. Gotoh, M., Nakano, Y, Tono, T. and Mori, T. (1990). Evidence that deoxyspergualin prevents sensitization and first-set cardiac xenograft rejection in rats by suppression on antibody formation. *Transplantation*, **50**, 132–136
39. Tufveson, G. and Gannedahl, G. (1994). Deoxyspergualin - a different and intriguing immunosuppressant. *Transplant Proc.*, **26**, 3029–3039
40. Gannedahl, G., Tufveson, G., Sundberg, B. and Groth, C. G. (1996). The effect of plasmapheresis and deoxyspergualin or cyclophosphamide treatment on anti-porcine Gal-alpha (1,3). Gal antibody titres in humans. *Transplant Proc.*, **28**, 551
41. Murase, N., Starzl, T. E., Demetris, A. J., Valdivia, L., Tanabe, M., Cramer, D. and Makowka, L. (1993). Hamster-to-rat heart and liver xenotransplantation with FK506 plus antiproliferative drugs. *Transplantation*, **55**, 701–707; discussion 707–708
42. Fujino, Y., Kawamura, T., Hullett, D. A. and Sollinger, H. W. (1994). Evaluation of cyclosporine, mycophenolate mofetil, and brequinar sodium combination therapy on hamster-to-rat cardiac xenotransplantation. *Transplantation*, **57**, 41–46
43. Siemasko, K. F., Chong, A. S., Williams, J. W., Bremer, E. G. and Finnegan, A. (1996). Regulation of B cell

function by the immunosuppressive agent leflunomide. *Transplantation*, **61,** 635–642
44. Wennberg, L., Wallgren, A. C., Sundberg, B., Rafael, E., Zhu, S., Liu, J. *et al.* (1995). Immunosupression with leflunomide and cyclosporine prevents pig-to-rat islet xenograft rejection. *Transplant Proc.*, **27**, 3314–3314
45. Hullett, D. A., Kawamura, T., Fujino, Y., Allison, A. M. and Sollinger, H. W. (1993). Prolongation of allograft and xenograft survival with mycophenolate mofetil (RS-61443) and brequinar sodium (DUP-785). *Transplant Proc.*, **25**, 700–701
46. Cosenza, C. A., Cramer, D. V., Tuso, P. J., Chapman, F. A., Wang H. K. and Makowka, L. (1994). Combination therapy with brequinar sodium and cyclosporine synergistically prolongs hamster-to-rat cardia xenograft survival. *J Heart Lung-transplant* **13**, 489–497
47. Sayegh, M. H., Akalin, E., Hancock, W. W., Russell, M. E., Carpenter, C. B., Linsley, P. S. and Turka, L. A. (1995). CD28-B7 blockade after alloantigenic challenge in vivo inhibits Th1 cytokines but spares Th2. *J. Exp. Med.*, **181**, 1869–1874
48. Bluestone, J. A. (1996). Costimulation and its role in organ transplantation. *Clin Transplant*, **10**, 104–109
49. Kelly, C. M., Benham, A. M., Sawyer, G. J., Dalchau, R. and Fabre, J. W. (1996). A three cell cluster hypothesis for noncognate T-B collaboration via direct T cell recognition of allogeneic dendritic cells. *Transplantation*, **61**, 1094–1099
50. Hammer, C. and Hingerle, M. (1992). Development of preformed natural antibodies in gnotobiotic dogs and pigs, impact of food antigens and antibody specificity. *Transpl. Proc.*, **24**, 707–709
51. Hammer, C., Höbel, G. and Csapo, C. (1992). Fixation of mammalian natural antibodies to porcine endothelial cefis. *Transpl. Proc.*, **24**, 588–589
52. Hammer, C. (1987). Isohemagglutinins and preformed natural antibodies in xenogeneic organ transplantation. *Transpl. Proc.*, **19**, 4443–4447
53. Sanchez, J., Michler, R., Rose, E. and Cooper, D. (1991). Experimental xenotransplantation between closely related primate species. Xenotransplantation, edited by D. Cooper, E. Kemp, K. Reemtsma, D. White, Berlin: Springer Verlag
54. Ye, Y. and Cooper, D. (1991). Experimental xenotransplantation in non human primates, using distantly related donor species. Xenotransplantation, edited by D. Cooper, E. Kemp, K. Reemtsma, D. White, Berlin: Springer Verlag.
55. Kemp, E. (1991). Experimental xenotransplantation between distantly related non human species. Xenotransplantation, edited by D. Cooper, E. Kemp, K. Reemtsma, D. White, Berlin: Springer Verlag.
56. Bach, F., Platt, F. and Cooper, D. (1991). Accomodation, the role of natural antibody and complement in discordant xenograft rejection. Xenotransplantation, edited by D. Cooper, E. Kemp, K. Reemtsma, D. White, Berlin: Springer Verlag
57. Alexandre, G., Gianello, P., Latinne, D., Carlier, M. and Dewalle, A. and Obberch, L. V. (1989). Plasmaphereses and splenectomy in experimental renal xenotransplantation. Xenograft 25, Edited by M. Hardy, Amsterdam, New York, Oxford: Excerpta Medica
58. Hammer, C. (1991). Experimental xenotransplantation between closely related species. In: Xenotransplantation edited by D. Cooper, E. Kemp, K. Reemtsma, D. White. Berlin: Springer Verlag
59. Perper, R. and Najarian, J. (1966). Experimental renal heterotransplantation II, closely related species. *Transplantation* **4**, 700–709
60. Baily, L. and Niehlssen-Cannarella, S. (1986). Observation of cardiac xenotransplants *Transplant Proc.*, **18**, 88–91
61. Reding, R. (1989). Organ xenografts in rodents. In: Xenograft 25, edited by M. Hardy Amsterdam, New York, Oxford: Excerpta Medica.
62. Bollinger, R., Knechtle, S., Harland, R., Jamaguchi, Y. and Halperin E. (1989). New approches to immunosuppression for xenotransplantation. In: Xenograft 25, edited by M. Hardy, Amsterdam, New York, Oxford: Excerpta Medica
63. Reding, R. (1991). Experimental xenotransplantation in rodents. Plasma exchange and splenectomy. In: Xenotransplantation, edited by D. Cooper, E. Kemp, K. Reemtsma, D. White, Berlin: Springer Verlag
64. Bowman, G., Wolvekamp, M., Debruin, R., Jeekel, J. and Marquet, R. (1991). Experimental xenotransplantation in rodents II. Skin and heart grafts In: Xenotransplantation, edited by D. Cooper, E. Kemp, K. Reemtsma, D. White, Berlin: Springer Verlag
65. Steinbrüchel, D., Nielson, B., Kemp, E. (1991). Experimental xenotransplantation in rodents III. Total lymphoid irradiation, Cyclosporin and monoclonal antibody. In: Xenotransplantation edited by D. Cooper, E. Kemp, K. Reemtsma, D. White. Berlin: Springer Verlag.
66. Schütz, A., Breuer, M., Engelhart, M., Brandl, U., Hatz, R., Hammer, C. and Kemkes, B. (1992). Heterotopic cervical heart transplantation with unsensitized vs. presensitized donors ("Domino"). Comparison of kinetics in acute rejection. *Texas Heart Inst. J.*, **18**, 286–292.
67. Vriesendorp, H., Westbroek, D., A'amaro, J., v.d. Steen, J., v. Rood, J., Albert, E., *et al.* (1973). Joint report of the 1st International Workshop on Canine Immunogenetic. *Tissue Antigens*, **3**, 145–172
68. Forty, J., Metcalfe, M., White, D. and Wallwork J. (1991). Development and use of an ex vivo heart model to study hyperacute xenograft rejection. *Minerva Chir* **46**, 123–127
69. Bach, F (1989). Immunogenetic and cellular studies of xenogeneic responses. In Xenograft 25, edited by M. Hardy. Amsterdam, New York, Oxford: Excerpta Medica
70. Platt, J., Turman, M., Noreen, H., Fischel, I., Bolman, R. and Bach, F. (1990). An ELISA assay for xenoreactive natural antibody. *Transplantation* **49**, 1000–1001
71. Hammer, C. (1991). Xenografting, its future role in clinical organ transplantation. Ethics, Justice and Commerce, edited by W. Land, J. Dossetor. Berlin: Springer Verlag.

PART VII

ENDOCRINOLOGY

PART VII

21. Animal Models in Hormone Research

Gerhard Rümenapf, Paul Otto Schwille and Angelika Schmiedl

Division of Experimental Surgery and Endocrine Research Laboratory, Department of Surgery, University of Erlangen, FRG

1. Introduction

1.1. General Remarks

In this chapter we discuss some of our scientific work in the surgical field, most of which is based on animal studies, and the pertinent literature. Surgery owes much of its present high level of achievement to prior extensive experimental work on animals. Such work has aided

(i) the development of new surgical techniques
(ii) the acquisition of (micro)surgical skills
(iii) the study of adverse effects of surgery in humans
(iv) the systematic utilization of surgery-induced metabolic changes as a tool for physiological and pathophysiological research.

In recent years, animal experiments have come to be regarded with repugnance, and public opinion is negative. In response to this a critical analysis of the need for animal trials has stimulated the search for alternatives.

As a result the use of animals for the development of new surgical techniques has declined, in part owing to the high technical standard of surgery already attained. Moreover, the training of (micro)surgical skills can be done on cadavers. However, the study of metabolic and functional side effects of surgery systematically appplied in physiological experiments requires the intact organism of a living animal. Much experimental work is already being done *in vitro*, but the relevance of the data thus obtained to physiology, and their comparability with the results from *in vivo* studies remain uncertain.

1.2. Clinical Background

The digestion and absorption of nutrients involves an extremely complex interaction of the various organs of the digestive system. In the regulation of digestive processes gastroentero-pancreatic hormones, local paracrine factors and (peptidergic) neurotransmitters all play an important role. Our present understanding of these interrelations is still limited, although new hormonal factors are constantly being discovered.[1] Although gastrointestinal surgery grossly disturbs physiological functions, the body is well able to cope with these massive changes over the short-term, for example after Whipple's operation, or after total gastrectomy. The surgeon is concerned with the patient mainly during the peri- and immediate postoperative period (thirty days), but rarely beyond this. Long-term metabolic changes after surgical interventions therefore often go unnoticed, and comparatively little scientific work has been devoted to these points.

One of the long-term disturbances after gastrointestinal surgery concerns the metabolism of bone and minerals. Below, we outline some such changes following gastric surgery, bariatric jejunoileal bypass surgery and porta-caval shunt operations, in man and the rat, with special regard to endocrine changes and their repercussions on bone and mineral metabolism. Owing to its versatility as an experimental animal, we have used the rat exclusively. While the results obtained underscore the need for animal models, it remains our intention to make more use of *in vitro* studies on organs and cells, as based on the results presented below, in our future work.

2. Influence of Gastric Surgery on Bone and Mineral Metabolism

2.1. "Postgastrectomy Bone Disease"

The large number of gastric resections performed in the past decades have resulted in many physiological disturbances and nutritional deficiencies, such as iron, vitamin B12, and folate deficiencies.[2,3] One of the late complications is metabolic bone disease which develops after various types of gastric surgery.[4–11] It was first reported in gastrectomized puppies.[12] Subsequently, a number of clinical reports were published on small numbers of patients[13–16] who developed osteopenia after certain types of gastric resection. With increasing interest in this disease, larger numbers of gastrectomy patients were evaluated.[4–9] British and Australian authors reported that "postgastrectomy bone disease"[7] developed in 10–30% of patients with partial gastrectomy, especially of the Billroth II type.[4,7–9] In the earlier reports, postgastrectomy bone disease was mainly regarded as osteomalacia, and a number of unclear and less well characterized forms of osteopenia have also been described.[7,8,11,17,18] Many hypotheses of the pathogenesis of postgastrectomy bone disease have been proposed, such as deficient calcium absorption due to decreased intake of calcium, maldigestion, gastric achlorhydria, gastrointestinal hurry, small gastric reservoir, defective stimulation of pancreatic secretion, defective gall bladder emptying, malabsorption due to morphological alterations of the gastrointestinal mucosa, steatorrhea, vitamin D deficiency, lactose malabsorption, increased endogenous fecal calcium loss, inadequate urinary calcium losses, disturbances of vitamin D metabolism, alterations in the interplay of calciotropic hormones, and changes in the gastroentero-pancreatic hormones with dysfunction of a hypothetical gastroentero-thyroid-parathyroid axis. No final consensus has yet been achieved with respect to the pathophysiology of the disease, and the clinical relevance of the disease has even been doubted.[4,5]

It would apppear that the changes in bone metabolism remain subclinical for a long period after gastric surgery. Only extreme cases of osteopenia are recognized sequelae of gastric resection. However, in view of the increasing prevalence of osteoporosis in the elderly, interest has again focused on gastric surgery as an important risk factor. This implies that postgastrectomy bone disease is considered to be secondary osteoporosis. However, osteopenia in this setting may be specific[18] and morphologically different from either osteoporosis or classical osteomalacia traditionally diagnosed on the basis of histology.

Nevertheless, the evaluation of its pathomechanisms and the development of prophylactic measures might be as rewarding as in the osteoporosis field.

The clinical evaluation of early postgastrectomy bone disease is difficult. In many clinical studies populations heterogeneous with respect to age, sex, and nature of gastric surgery were investigated; no case-control studies have been performed to date. Other diseases which might give rise to osteopenia were not excluded, and an accurate histological evaluation of the skeleton is mostly lacking. Diagnostic procedures such as skeletal computed tomography, dual photon absorptiometry or nuclear magnetic resonance are expensive, not available on a larger scale, indirect, and often not accurate enough. Biochemical markers of bone and mineral metabolism, such as bone alkaline phosphatase, osteocalcin, urinary hydroxyproline, are routinely available and can be determined repeatedly. However, their diagnostic validity still remains uncertain.

In view of the many clinical reports which have appeared over the past fifty years, it is surprising how few relevant animal studies have been performed.[12,19–22] In order to reevaluate the pathogenesis of postgastrectomy bone disease, and in particular to determine whether different types of the disease relative to the different types of surgery performed might present and to avoid the drawbacks frequently associated with clinical trials, we have performed a number of studies in the rat. Several types of gastric surgery were investigated, including

a) highly selective vagotomy
b) fundectomy
c) antrectomy
d) total gastrectomy (Longmire or Roux-en-Y reconstruction).

These interventions can be roughly divided into those leading to hypogastrinemia (c, d) and those resulting in hypergastrinemia (a, b). Additionally, different functional parts of the stomach with some endocrine role in the interplay between the stomach and the thyroid and parathyroid gland were resected.

2.2. The Gastroentero-Thyroid-Parathyroid Hormonal Axis

For an understanding of our animal models, it is important to outline a comprehensive model of the role played by the stomach in calcium metabolism (see Figure 1).

2.2.1. Gastro-parathyroid Axis

A relationship between gastrin and calcium metabolism has long been suspected,[23] owing to the in-

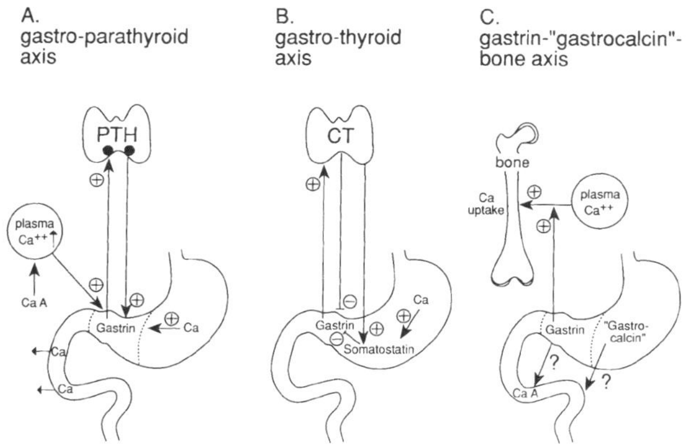

Figure 21.1 Diagrammatic representation of the (partly hypothetical) interrelations between the stomach, bone, and the thyroid and parathyroid glands.
→ : stimulation; —| : suppression; CaA: intestinal calcium absorption; CT: calcitonin; PTH: parathyroid hormone.

creased prevalence of duodenal ulcers in patients with primary hyperparathyroidism,[24] in whom high blood gastrin levels are detectable.[25] Graded hypercalcemia as achieved by i.v. calcium infusions,[26] as well as the oral administration of calcium carbonate loads not inducing hypercalcemia,[27] caused increases in serum gastrin concentrations in man. The latter effect was explained by local stimulation of antral and duodenal mucosa by the ingested calcium.[28] Also, parathyroid hormone was shown to be a gastrin secretagogue.[29,30] While in the pig, gastrin stimulation by parathyroid hormone was independent of the serum calcium level,[29,30] clinical studies in patients with chronic hypercalcemia of parathyroid or non-parathyroid origin revealed a direct interrelation only between hypercalcemia and elevated serum gastrin levels, i.e. independent of parathyroid hormone.[31] Conversely, gastrin stimulates parathyroid hormone secretion *in vivo* in man[32] and *in vitro* from isolated bovine parathyroid cells,[33] and endogenous hypergastrinemia in antral excluded rats was followed by hyperplasia of the parathyroid glands.[34]

2.2.2. Gastro-thyroid Axis

Gastrin acts as a calcitonin secretagogue in man and other species.[35,36] On the other hand, antagonism between calcitonin and gastrin has been reported.[37,38] Gastrin release from G cells was shown to be directly inhibited by calcitonin in duodenal ulcer patients.[39] In the rat, acute thyroidectomy with subsequent lack of calcitonin deinhibits gastrin secretion,[40] and intracerebroventricular administration of calcitonin suppresses gastric acid secretion.[41] In man, calcitonin stimulates the secretion of somatostatin, which is associated with a reduction in gastrin secretion.[42] Calcitonin might therefore reduce gastrin secretion by way of somatostatin.

Although the physiological significance of these findings has been questioned, a recent study in patients with partial gastrectomy of the Billroth I and II type[43] reports decreased postprandial calcitonin levels in these patients that correlate with the low postprandial serum gastrin levels as well as the degree of bone loss. These findings support the existence of the gastrothyroid axis, and suggest that reduction of calcitonin release may contribute to gastric surgery osteopenia.[43]

2.2.3. Hypothetical Gastrin-Gastrocalcin-Bone Axis

Pharmacological doses of gastrin induce hypocalcemia in man and other species.[44] While this effect was long thought to result from gastrin-induced calcitonin release from thyroid,[44,45] other workers have shown that it continues to be observed in thyroparathyroidectomized animals, suggesting that the hypocalcemic action of gastrin is calcitonin-independent.[46] However, the effect is abolished either by fundectomy or by total gastrectomy.[46–48] In this context, a gastric factor was thought to mediate the hypocalcemic action of endogenous or exogenous hypergastrinemia.[46–48] Recently, this factor has been reported to be a peptide hormone.[48] It may be synthesized in the gastric oxyntic mucosa and has been given the name "gastrocalcin". It has been shown to stimulate calcium uptake into bone as well as bone turnover, and its action appears to be dependent on gastrin.[47] Removal of either the gastrin-producing antrum, or the "gastrocalcin"-producing fundus, may therefore induce changes in calcium metabolism on an endocrine basis (see below).

The positive and negative feedback mechanisms involved between calcium, gastrin, calcitonin, parathyroid hormone, "gastrocalcin" and bone, are explained by a hypothetical gastroentero-thyroid-parathyroid axis.[29] This complex endocrine interrelationship between the stomach and the thyroid and parathyroid glands, acting either directly or via changes in plasma calcium, might serve to optimize the organism's calcium balance after calcium ingestion, as following:

(i) Ingested calcium may stimulate gastrin directly or via hypercalcemia. Either of these factors may stimulate calcitonin, possibly resulting in a shift of calcium to bone, thereby attenuating postprandial hypercalcemia.
(ii) Postprandial hypergastrinemia stimulates parathyroid hormone secretion, and this in turn may accelerate gastrin secretion. In the early phase of food ingestion this positive-posi-

tive feedback mechanism may result in rapid parathyroid hormone secretion. This sequence of events may also be associated with a postprandial rise in calcitonin, which in turn ameliorates postprandial hypercalcemia, thereby also stimulating parathyroid hormone secretion. Sufficiently high blood levels of parathyroid hormone serve to reduce postprandial renal calcium losses via stimulation of renal tubular calcium reabsorption. In the long term, renal 1a-hydroxylase will be stimulated, thus maintaining the production of 1,25-dihydroxyvitamin D at the order of magnitude needed to keep intestinal calcium absorption at a high level.

The rationale behind all these interrelationships is the conservation or positivation of the bone balance, especially bone calcium. Disturbances of any of these steps might result in disturbances of bone metabolism.

Additionally, gastrin has been reported to affect calcium absorption directly; however, recent short-term physiological studies do not provide evidence for such a role. Nevertheless, long-standing hypergastrinemia might alter the capacity of the bowel to absorb calcium via trophic changes in the epithelial tissues.[49]

2.3. Effects of Highly Selective Vagotomy (HSV) on Bone and Mineral Metabolism

Vagotomy for the treatment of duodenal ulcer disease was introduced in 1943. The initial procedure – supradiaphragmatic bilateral truncal vagotomy[50] - was later combined with a drainage procedure aimed at facilitating gastric emptying.[51] Since 1970 highly selective (parietal cell) vagotomy (HSV) has been performed.[52,53] This preserves the antral fibers (N. Latarjet), also the hepatic and celiac vagal fibers and does not require drainage. Its development was based on the clinical observation that sacrifice of the extragastric and antral vagal fibers adversely affects gastric emptying, small bowel, pancreatic and gallbladder function, with the frequent sequelae of diarrhea, steatorrhea and gall stone formation. While the frequency of vagotomy diarrhea and dumping syndrome was largely reduced by the latter procedure,[54] HSV still has certain distinct influences on gastric emptying[55] as well as on extragastric functions, such as dietary habits[56] and exocrine pancreatic function.[57] Additionally, it may deprive central regulatory mechanisms of afferent data pertinent to eating behavior.[58] Although HSV is aimed at protecting the stomach from efferent vagal stimulation, it has to be borne in mind that the vagus is predominantly an afferent nerve,[59] and that the transmission of signals from the periphery to the central nervous system is of major importance.[1]

Additionally, HSV induces certain endocrine changes, the most important being chronic hypergastrinemia induced by the decrease in the acid inhibition of the antral gastrin-producing G cells and the elimination of gastrin release inhibiting vagal signals.

2.3.1 Data from clinical studies

2.3.1.1. Influence of vagotomy on bone metabolism. Gastric resections may result in bone disease (see above). Many resection procedures were combined with some kind of vagotomy, so that specific influences of the surgical interventions on bone could not be differentiated. Also, a number of side effects of vagotomy, such as a massive reduction of gastric acid production, intestinal hurry, diarrhea and steatorrhea, were comparable to those regarded as crucial for the development of bone disease after gastric resection. Vagotomy was therefore suspected to alter bone and mineral metabolism per se. However, published data are sparse. More than 10 % of patients with vagotomy and drainage had elevated serum alkaline phosphatase levels, but this was not found to correlate with any sign of bone disease.[60] Furthermore, the mean daily food consumption, the vitamin D intake, the body weight distribution[60] and the incidence of "thin" bones[61] one to five years after vagotomy and drainage were not found to differ significantly vis-a-vis unoperated ulcer patients. In contrast to these findings, Imamura *et al.*[62] demonstrated that 40 % of patients who had undergone vagotomy with pyloroplasty three to seventeen years previously had evidence of thin bones. The authors concluded that vagotomy itself can induce osteopenia. They attributed bone disease to the high incidence of diarrhea after vagotomy. Data from our own laboratory[63–65] indicate that five years after HSV, there are no changes in the bone mineral content of these patients. At the same time, intestinal calcium absorption is significantly elevated (see below), while total serum calcium is decreased, serum calcitonin, parathyroid hormone, 24,25(OH)2D and urinary nephrogenous cyclic AMP excretion are unchanged, and the preoperatively low 25-hydroxyvitamin D levels are normalized. In conclusion, there were no indications for any pathological changes in mineral and bone metabolism after HSV in man, although the observation period was presumably too short to detect clinically important changes that might ensue from the observed minor changes in calcium metabolism after HSV.

2.3.1.2. Influence of HSV on calcium absorption

After administration of pharmacological doses of pentagastrin, intestinal absorption of calcium was shown to be decreased in man.[66,67] Neither calcitonin nor parathyroid hormone was altered by these infusions, indicating that this effect of gastrin was not dependent on the two hormones.[67] The only available data on the effects of chronic hypergastrinemia on intestinal calcium absorption of man have been reported in HSV patients ([65] see above). Five years after HSV, three hour-cumulative intestinal calcium absorption as evaluated by a double isotope deconvolution method was elevated significantly above normal,[65] in the presence of unchanged calcitonin and parathyroid hormone levels, normal urinary calcium excretion and a normal bone mass. However, the influence of physiological plasma levels of gastrin on intestinal calcium absorption in healthy man has never been investigated.

In conclusion, the clinical data on the influence of vagotomy on calcium and bone metabolism are contradictory and inconclusive with respect to the existence of mere postvagotomy bone disease. The clinical data do not throw light on the influences on bone and mineral metabolism of vagotomy-induced endocrine changes, such as that of hypergastrinemia, disturbances of the gastroentero-thyroid-parathyroid axis, etc. In addition, bone composition and turnover after vagotomy remain unknown.

2.3.2. Data from Animal Experiments

For the above reasons, we have performed several studies on the influence of HSV on bone and mineral metabolism in the rat. Our working hypothesis was that long-standing hypergastrinemia, as induced by HSV, might alter certain steps of calcium metabolism (see above).

2.3.2.1. Influence of Hypergastrinemia and HSV on Intestinal Calcium Absorption. Elevated intestinal calcium absorption was found in rats after administration of pharmacological doses of tetragastrin.[68] This effect was thought to be mediated by parathyroid hyperactivity. On the basis of these findings, we evaluated the effect of HSV on intestinal calcium absorption of the rat. Using an *in situ* loop technique, we found that duodenal intestinal calcium absorption was not abnormally altered in either intact or thyroidectomized HSV rats, while it was decreased in parathyroidectomized as well as thyroparathyroidectomized HSV rats.[69] Acute or chronic pretreatment of thyroparathyroidectomized rats with pentagastrin did not alter intestinal calcium absorption. We concluded that HSV lowers intestinal calcium absorption only in the absence of the parathyroid glands, and that this effect is presumably not related to endogenous postvagotomy hypergastrinemia.

In a second trial, we studied the influence of HSV on the net absorption and the bidirectional fluxes of calcium in the duodenum and ileum of intact or parathyroidectomized rats.[70] After HSV, net intestinal calcium absorption was increased in the duodenum and decreased in the ileum, while after additional parathyroidectomy, net intestinal calcium absorption was decreased in both gut segments. We concluded that after parathyroidectomy, HSV is potentially harmful to the calcium balance of the body. In order to investigate whether physiological hypergastrinemia might be responsible for the above changes in intestinal calcium absorption in HSV rats, we have recently performed a dose-response study in intact rats on the influence of graded doses of iv gastrin 17-I on duodenal calcium transport, using *in situ* perfusion for the determination of intestinal calcium absorption.[71] While plasma gastrin immunoreactivity rose to levels seen in the postprandial period in a dose-dependent manner, the bidirectional calcium fluxes, net intestinal calcium absorption, serum calcium, parathyroid hormone and calcitonin as well as the urinary excretion of calcium remained unchanged.[71] From these observations we concluded that a direct physiological, short-term regulating role of gastrin in intestinal calcium absorption or calcium metabolism of the rat is unlikely. However, long-term effects of hypergastrinemia on the above parameters have not yet been studied. Also, gastrin might modify the small intestinal calcium transport capacity on a long-term basis via its trophic action on the gut mucosa ([72] see below).

2.3.2.2. Influence of HSV on Bone Metabolism Since the available clinical data neither rule out nor suggest[62,63] certain long-term influences of HSV on bone metabolism, we recently performed long-term studies in the rat aimed at clarifying this point.[73]

Study design Male growing Sprague Dawley rats were either subjected to microsurgical HSV (for details see[69,70]) or to sham operation (for numbers of animals and general data see Table 21.1). The completeness of HSV was confirmed by largely reduced gastric acid secretion during hypoglycemia. They received normal rat chow and tap water ad lib. Approximately fifteen weeks after HSV, all rats were housed in individual metabolic cages. On three consecutive days, the intake of minerals as well as their excretion in the feces and urine were assessed. The daily mineral balances, as well as fractional intestinal mineral absorption were calculated on the basis of these data. In the urine, cyclic AMP and hydroxyproline were determined. Thereafter, the rats were starved for twenty hours, anesthetized (for

Table 21.1 Influence of highly selective vagotomy (HSV), fundectomy (FX) or sham operation (SHAM) on bone and mineral metabolism in the rat, as assessed 7–8 weeks postoperatively. (a) general data and markers data from serum and urine; (b) data from bone, and bone markers.
Mean values $\pm$ 1 SEM. °: $p < 0.06$; *: $p < 0.05$; •: $p < 0.01$; ••: $p < 0.001$ vs SHAM

a	SHAM ($n = 12$)	HSV ($n = 12$)	SHAM ($n = 12$)	FX ($n = 14$)
General data				
Body weight (g);				
Initial	243 ± 3	244 ± 2	240 ± 10	228 ± 6
Final	506 ± 13	491 ± 9	518 ± 18	417 ± 14••
Food consumption; g/day	11.8 ± 0.5	13.2 ± 0.3	19 ± 0.5	15 ± 0.6••
Fractional absorption of calcium; %	21 ± 4	29 ± 2	41 ± 2	43 ± 3
Fractional absorption of magnesium; %	33 ± 5.2	33 ± 3.4	14 ± 3	7 ± 4
Calcium balance; mg	29 ± 6	39 ± 5	69 ± 4	59 ± 6
Data from serum and urine				
Calcium; mg/dl	9.05 ± 0.09	8.98 ± 0.07	9.7 ± 0.2	9.8 ± 0.2
Gastrin; pg-equiv/ml	12 ± 3	33 ± 7	37 ± 4	565 ± 86••
Mid-regional PTH; pg-equiv/ml	143 ± 45	328 ± 58*		
N-terminal PTH; pg-equiv/ml	22 ± 3	48 ± 14°	31 ± 2	34 ± 3
Calcitonin; pg-equiv/ml	38 ± 5	48 ± 6	33 ± 10	31 ± 7
25-Hydroxyvitamin D; ng/ml	11 ± 1	10 ± 1	9 ± 1	7 ± 1
1,25-Dihydroxyvitamin D; pg/ml	26 ± 7	33 ± 7	48 ± 23	114 ± 22°
Urinary calcium; μg/mg creatinine	53 ± 7	49 ± 3	50 ± 1	20 ± 0.1••
Urinary cyclic AMP; ng/mg creatinine	18.4 ± 1.6	17.4 ± 1	16 ± 1	18 ± 2
Specific activity (serum) calcium^{45}; (cpm/mg) × 10^{-4}			5.02 ± 0.2	6.92 ± 0.3••
Specific activity (urine) calcium^{45}; (cpm/mg) × 10^{-4}			8.4 ± 0.3	10.5 ± 0.6*

b	SHAM ($n = 11$)	HSV ($n = 12$)	SHAM ($n = 12$)	FX ($n = 14$)
Data from bone				
Volume; μl	533 ± 12	523 ± 15	795 ± 29	740 ± 20
Dry weight/100g body weight; mg	186 ± 4.2	187 ± 3	162 ± 5	166 ± 3
Mean specific density; g/ml	1.77 ± 0.02	1.75 ± 0.01	1.52 ± 0.01	1.44 ± 0.02
Ash weight/volume; g/ml	1.05 ± 0.02	1.04 ± 0.01	0.65 ± 0.01	0.57 ± 0.01•
Calcium/volume; mg/ml	455 ± 16	449 ± 15	278 ± 7	239 ± 13•
Calcium^{45}/volume; (cpm/ml) × 10^{-6}			4.91 ± 0.2	6.41 ± 0.3•
Specific activity calcium^{45}; (cpm/mg) × 10^{4}			1.74 ± 0.1	2.81 ± 0.2•
Magnesium/volume; mg/ml	7.6 ± 0.2	8.6 ± 0.6	4.2 ± 0.1	3.7 ± 0.1•
Phosphorus/volume; mg/ml	270 ± 24	227 ± 10	141 ± 5.3	119 ± 5.2•
Bone markers				
Bone alkaline phosphatase; U/l	30 ± 8	36 ± 11	73 ± 5	98 ± 8*
Osteocalcin; ng/ml	38 ± 7	35 ± 3	49 ± 2	60 ± 3•
Urinary hydroxyproline/creatinine; μg/mg	23 ± 2	34 ± 3*	48 ± 5	59 ± 6

details, see[70]), laparotomized and bled from the abdominal aorta. Blood was sampled for the determination of minerals, gastrin, calciotropic hormones (Table 21.1) and biochemical markers of the activity of bone metabolism, such as bone alkaline phosphatase and osteocalcin. The right femur was excised and its wet, dry, and ash weight, as well as mean specific density were determined. The gastric antrum and fundus, as well as the pancreas, were excised and their wet weight recorded.

Results HSV induced mild hypergastrinemia (see Table 21.1). The final body weight of the HSV rats was no different from that of the controls. Interestingly, HSV rats consumed more food than control rats. However, calculated food efficiency was low but not statistically different. Fractional intestinal calcium absorption as well as the calcium balance tended to higher values, but the difference did not attain significance. Except for a significant increase in serum parathyroid hormone (mid-regional, i.e.

non-bioactive), there were no changes in serum minerals, other calciotropic hormones or markers of bone turnover. While the urinary excretion of minerals and cyclic AMP was no different from that in controls, HSV rats excreted significantly more hydroxyproline. There were no significant changes in the bone composition and mineral content between HSV and controls; however, bone histomorphometry was not performed. The wet weight of the gastric fundus and antrum increased by approximately 40 %, that of the pancreas by approximately 13 %, as compared with controls.

Discussion of the Results Apparently, HSV induces some tendency in the rats to ingest more than the control rats. This might be in compensation for a lack of weight gain in the early phase after HSV which escaped our observation, since at the end of the trial the HSV rats had a similar weight to the controls. Malabsorption or maldigestion as a cause of increased food consumption can be ruled out, since there were no signs of steatorrhea, and fecal weight was similar. In contrast, intestinal calcium absorption as well as the calcium balance tended to higher than normal values. With the simple methods employed in our study, no HSV-related differences were detectable with respect to bone mass or mineral content, as compared with controls. Also, bone alkaline phosphatase (indicating osteoblastic activity) as well as osteocalcin (indicating osteoclastic activity) were unchanged by HSV as compared with controls. Our results are thus in line with the majority of the clinical data which indicate that significant bone disease may rarely develop after HSV.[60,63] However, we cannot completely rule out certain influences on bone metabolism, since the postoperative observation period was not very long (approximately 100 days), and we did not perform bone histomorphometry which might have revealed changes in bone turnover with no repercussion on the bone mass, as indicated by the elevated urinary excretion of hydroxyproline (an indicator of osteoclastic activity). Recently, an increased bone turnover rate in the presence of unchanged bone mass has been described in rats which had been made chronically hypergastrinemic via the subcutaneous delivery of gastrin 17 from osmotic minipumps.[47] Thus, further studies would be worthwhile and are presently performed in our laboratory, involving the use of adult rats, a one year postoperative observation period, and bone histomorphometry with the aim of studying the dynamics of bone turnover. An unexpected finding was the increase in the weight of the pancreas and the stomach of the HSV rats. This finding is in line with the trophic action of gastrin which is apparently not confined to the gastric mucosa. One might speculate that the small intestinal mucosa has also gained weight, a process which might compensate for gastrin-induced reductions in the intestinal capacity to absorb substances, e.g. water, electrolytes,[74] and calcium.[66,67] Another compensation of hypergastrinemia-induced reductions in the intestinal transport capacity for calcium might be the slight elevation of parathyroid hormone after HSV; however, this increase is not paralleled by concomitant increases in the urinary excretion of cyclic AMP. On the other hand, this elevation of parathyroid hormone may result from gastrin stimulation of the parathyroid glands, i.e. HSV influences the hypothetical gastro-entero-thyroid-parathyroid axis (see above).

HSV rats have lost the capacity to produce large amounts of gastric acid. Nevertheless, their intestinal calcium absorption is normal. This indicates that gastric acid production is not a major determinant of calcium absorption, which is compatible with recent results from clinical studies.[75,76]

In conclusion, the use of HSV in rats for the study of influences on bone and mineral metabolism has confirmed the general clinical observation that bone disease must be rare after HSV. However, since minor changes in mineral metabolism have been found after HSV, we hope that modifications of the study design may offer an opportunity to detect as yet unknown mechanisms in the interrelations between vagal denervation, gastrointestinal hormones and mineral metabolism.

2.4. Effects of Fundectomy (FX) on Bone and Mineral Metabolism

FX was formerly used in the treatment of both duodenal and proximal gastric ulcer disease.[77,78] It was designed to reduce gastric acid output by removing the acid-producing portion of the stomach. The FX procedure was also used in the treatment of proximal gastric cancer. The various types of FX were subsequently abandoned because of their negative side effects, such as recurrent (anastomotic) ulcers, which developed when residual oxyntic mucosa was stimulated by the concomitant hypergastrinemia (see below), insufficiency of the oesophago-gastrostomy, reflux esophagitis, and problems of gastric emptying associated with concomitant vagotomy. In tumor surgery, FX was replaced by total gastrectomy, in ulcer surgery mainly by highly selective vagotomy.

2.4.1. FX-Effects on Mineral and Bone Metabolism – Current State of our Knowledge

2.4.1.1. Data from the Rat. Although FX is not of clinical interest, it serves as a valuable animal model for the induction of extreme endogenous hypergastrinemia and G cell hyperplasia.[79] This is thought to result from the complete abolition of the acid feedback inhibition of antral gastrin release, in conjunction with the preserved passage of ingested food.[80] Vagal innervation appears to have no influence in this context.[80] Additionally, FX is thought to remove some fundic hormone-like agent that normally helps inhibit gastrin release.[81]

Interrelations between gastrin and calcium metabolism have been discussed in the respective section of this chapter. Pharmacological doses of gastrin induce hypocalcemia in the rat which is independent of calcitonin release, since it also occurs in thyroidectomized rats.[82,83] However, it may be linked to the release of some fundic factor ("gastrocalcin") which is eliminated by FX or gastrectomy, since after the latter procedures hypergastrinemia-induced hypocalcemia is absent.[48] During hypergastrinemia, bone turnover was found to be considerably accelerated, while the bone mass and calcium content remained unchanged.[47] The latter two parameters were found to be decreased after FX or gastrectomy, while bone turnover remained normal.[47]

2.4.1.2. Data from Man. In analogy to these animal experimental data, changes in bone mass, calcium content and turnover might also be expected in gastrectomized patients, since in total gastrectomy the fundus is completely removed, while in partial distal gastrectomies the oxyntic fundic surface is reduced to various degrees. In fact, postgastrectomy bone disease is observed after each of the latter conditions (see above). Unfortunately, only few clinical data are available on the effects of FX on bone and mineral metabolism, which is presumably due to the small numbers of patients involved. So, Imamura *et al.*[62] reported finding pathologically thin bone in five patients who had undergone proximal gastrectomy for gastric cancer two to nine years previously. Serum levels of calcium, phosphorus and alkaline phosphatase were normal in these patients. However, the relative contribution of concomitant partial fundic resections to postgastrectomy bone disease after partial distal gastrectomies remains unknown. Also unclear is the effect of persistent excessive hypergastrinemia with its trophic[72] and possibly tumor-inducing capacity.

2.4.2. FX and Bone – Data from our own Studies in the Rat

In order to evaluate the relative importance of changes in the serum gastrin levels and of resection of the fundus, in the pathogenesis of postgastrectomy bone disease, we have performed long-term studies on the influence of FX on bone and mineral metabolism in the rat.[84]

Study Design Growing male Sprague Dawley rats were subjected either to microsurgical FX with end-to-end anastomosis of rumen and antrum, or to sham operation (for numbers of animals and general data see Table 21.1). The study design was similar to that described for highly selective vagotomy (see above). In addition, FX rats received a subcutaneous injection of 45-calcium fifteen days before sacrifice, with the aim of studying the dynamics of calcium metabolism. Furthermore, in a second trial, an additional group of FX rats received 50 mg of tetracycline via intraperitoneal injection fourteen and five days before being killed. Immediately after sacrifice, their femurs were excised and kept in Schaffer's solution prior to histomorphometric evaluation.

Results FX induced extreme hypergastrinemia (see Table 21.1). Although the initial body weight of FX and SHAM rats was similar, the final body weight was lower in FX than in SHAM. Food intake decreased in FX, while food efficiency was roughly unchanged as compared with SHAM. Calcium balance as well as the fractional absorption of calcium and magnesium did not change vis-a-vis SHAM. In FX, serum levels of 1,25-dihydroxyvitamin D tended toward higher values. Serum activity of 45-calcium was significantly elevated, while urinary excretion of calcium and magnesium was decreased. FX reduced bone specific density and ash weight, as well as calcium, magnesium and phosphorus content. In contrast, 45-calcium/volume was elevated as compared with SHAM. Both total alkaline phosphatase and osteocalcin levels were elevated in FX, while the urinary excretion of hydroxyproline was not different from that in SHAM. In FX rats, the wet weight of the liver, pancreas, gastric antrum, duodenum, cecum, colon and kidneys was significantly elevated as compared with SHAM.

Comments and Conclusions Our data indicate that FX induces osteopenia over the long term. Apparently, normocalcemia is maintained at the expense of the bone mass. Elevated 1,25-dihydroxyvitamin D, increased serum and bone 45-calcium specific activity, decreased urinary excretion of calcium, hyperphosphaturia (data not shown) and normal intestinal calcium absorption may indicate that secondary hyperparathyroidism has been developed. Bone turnover is apparently elevated, as indicated by elevated alkaline phosphatase and osteocalcin, and the higher specific activity of bone with respect to 45-calcium. However, not in line with the assumption of increased bone turnover in the presence of secondary hyperparathyroidism is the normal urinary excretion of hydroxyproline and cyclic AMP. The diagnostic value of the markers of bone turnover is therefore doubtful in the rat.

Unfortunately, published clinical data on post-FX bone disease are sparse,[62] and no information is available on biochemical parameters of bone and mineral metabolism other than serum calcium, phosphorus and alkaline phosphatase. The latter parameters were all reported to be normal in FX patients.

Our findings differ from those of recent reports of osteopenia in conjunction with normal bone turnover in rats that had undergone FX one month previously.[47] This discrepancy may be due to the difference in the time lapse between the FX procedure and the analysis of bone changes, since in the present study the postoperative observation period was three months.

An important finding is that FX rats develop osteopenia in the presence of normal gastric emptying and a preserved duodenal food passage, while signs of malabsorption or maldigestion are absent (no steatorrhea). Pilot studies have indicated that the stomach of FX rats re-attains a normal volume by two weeks after the operation. This favors the assumption that other than functional factors such as loss of gastric reservoir, gastrointestinal hurry, or steatorrhea, may contribute to the development of postfundectomy (and possibly postgastrectomy) bone disease. Our findings may therefore open up new aspects in the elucidation of the pathomechanisms of postgastrectomy bone disease, and may indicate that our animal studies represent a useful and valuable supplement to the clinical studies on the subject. An alternative explanation for the decreased bone mass in FX rats might be the loss of gastric acid production, which may be important for the maintenance of normal intestinal calcium absorption, since ingested calcium salts are more soluble in an acid medium. However, in man an important contribution of gastric acid to intestinal calcium absorption has been ruled out.[75,76] Also, in HSV rats (see above) intestinal calcium absorption is practically normal, in the absence of secondary hyperparathyroidism, although gastric acid production is largely reduced.

The role of gastrin is unclear in the present context, but the recent reports on "gastrocalcin", an as yet poorly characterized fundic calciotropic peptide hormone which is thought to mediate the hypergastrinemia-induced shift of absorbed calcium to the bone[47,48] may be of interest. The normal intestinal calcium absorption in the FX rats might indicate that FX induces some intrinsic tendency toward decreased intestinal calcium absorption which is compensated by slight secondary hyperparathyroidism and a tendency toward small-intestinal hyperplasia (see below). This tendency toward low intestinal calcium absorption after FX may either be related to hypergastrinemia, or may be due to the loss of some fundic calciotropic factor which normally stimulates intestinal calcium absorption. As is seen in gastrectomized rats, which are hypogastrinemic and have low intestinal calcium absorption but still show all the signs of secondary hyperparathyroidism in the presence of massive osteopenia, FX-induced hypergastrinemia is presumably not the factor decisive for bone disease. However, it may explain the general tendency of FX rats to develop visceromegaly.[85]

In conclusion, FX induces osteopenia, with elevated bone turnover in the rat, in the presence of excessive hypergastrinemia. Since FX does not lead to significant functional disturbances, the changes in bone and mineral metabolism may possibly result from endocrine changes. Further detailed animal studies may help clarify these aspects - with possible repercussions for the understanding of postgastrectomy bone disease in man.

2.5. Effects of Antrectomy (BI) on Bone and Mineral Metabolism

Antrectomy with end-to-end gastroduodenostomy was first performed by Billroth in 1881 for the treatment of a distal gastric cancer. Later, this surgical intervention (which is generally known as the Billroth I procedure) was adopted for the treatment of distal gastric or duodenal ulcers. Since the recurrence rates were high, it was subsequently usually combined with some kind of vagotomy.

Antrectomy removes the gastrin-producing part of the stomach and thus eliminates gastrin, the most efficient stimulus of the gastric oxyntic glands, with the result that gastric acid output is greatly reduced. Hypogastrinemia persists after intake of meals. Since the duodenal passage of food is preserved, mixing of the gastric chyme with the pancreatic and biliary secretions is guaranteed, and bilio-pancreatico-cibal asynchrony is avoided. Functionally, BI diminishes the gastric reservoir and abolishes the gastric mill mechanism. Gastric emptying is accelerated and no longer controlled by the pylorus. Maldigestion and malabsorption occur with varying frequency, however, steatorrhea is less frequently found than after BII or total gastrectomy.[86]

2.5.1. BI and Bone Disease – Current Knowledge

2.5.1.1. Data from clinical studies Although numerous reports on postgastrectomy bone disease have been published, most did not evaluate BI strictly separately from BII patients[8,11,87]. Most reports on postgastrectomy bone disease dealt predominantly or exclusively with BII patients[18,88,89]. In general, changes in bone metabolism have been reported to be less important after BI as compared with BII[43,62,90,91]. However, in the majority of reports BI was combined with some sort of vagotomy, making it difficult to differentiate between the specific effects of the two procedures on bone and mineral metabolism.

The absorption and retention of 47calcium was found to be decreased in BII patients, but increased in BI patients.[92] Other studies revealed that intestinal calcium absorption from calcium chloride is higher in BII than in BI patients, which was thought to reflect a higher calcium deficiency of the former patients.[6] In man, even a case of nutritional secondary hyperparathyroidism has been observed, which was attributed to a previous BII resection that was later modified to BI.[93] In a recent study, however, intestinal calcium absorption was normal in both BI and BII patients.[11]

Alhava *et al.* found a significant decrease in bone mineral density in women 3.3–8.5 years after BI for duodenal ulcer disease.[94] This was accompanied by significantly elevated serum alkaline phosphatase levels. However, the incidence of elevated serum alkaline phosphatase levels has been shown to be higher in BII than in BI patients.[11]

In both, BI and BII patients, reduced 25-hydroxyvitamin D levels and raised 1,25-dihydroxyvitamin D levels have been found.[87,95] It has been suggested that the former represents vitamin D deficiency,[87] while the latter is the result of a compensatory process aimed at preventing hypocalcemia.[95] Other authors found decreased levels of 25-hydroxyvitamin D only in BI patients, while 1,25-dihydroxyvitamin D levels were elevated in BI as well as BII patients.[43] In addition, decreased levels of vitamin D-binding protein were found in BI.[96]

The changes in bone structure and composition after BI have rarely been accurately characterized. They have mostly been classified as osteomalacia on the basis of the frequent finding of elevated alkaline phosphatase and elevated calcium retention during a calcium infusion test.[6] In the few studies where histomorphometry was performed on bone biopsy specimens obtained from iliac crest, BII-related data are mostly reported, or BI data are not differentiated from BII data.[11,87] Only one study mentions an elevated volumetric density of osteoid and trabecular bone surface covered with osteoid in BI patients.[97] However, trabecular bone volume was also elevated in BI patients. It was concluded that the changes might constitute a mild form of osteomalacia.[97]

Specific influences of hypogastrinemia on the development of gastric surgery osteopenia have been suspected,[43] since in response to a meal the increases in both gastrin and calcitonin were lower in BI patients than in controls. In BII patients, the reduction in the blood levels of these hormones correlated with the decrease in bone mass. This finding might indicate that antrectomy may possibly affect bone mass through a disturbance of the gastrothyroid axis, since calcitonin is thought to have a bone-conserving effect (see above).

2.5.1.2. Data from Animal Studies Unfortunately, no data from animal studies have so far been published on specific effects of BI on bone and mineral metabolism. BI in the rat may serve as a model for endogenous hypogastrinemia. While hypergastrinemia was found to accelerate bone turnover as long as the gastric fundus is preserved,[47] little is known about the effect on bone of long-term hypogastrinemia as produced by antrectomy. In the present context mention must be made of a study on the effect of hypogastrinemia on the trophic status of the pancreas and gastrointestinal mucosa.[85] The authors were able to demonstrate that antrectomy in the rat significantly reduced the weight of the pancreas and the colon one month after the operation, concomitantly with a decrease in the RNA and DNA content of the pancreas.[85] These changes were reversed by the administration of pentagastrin. According to these findings, changes in the trophic status of the absorptive intestinal mucosa must be taken into account when interpreting bone and mineral metabolism after BI in man and the rat. Several histological studies on small intestinal biopsy specimens have indicated that some atrophy of the jejunal mucosa is a frequent finding in gastrectomy patients.[98]

2.5.2. Data from Our own Animal Studies

In order to better understand the influences of antrectomy and antrectomy-related endocrine changes on bone and mineral metabolism, we have carried out long-term studies on the influence of BI on bone and mineral metabolism in the rat.[99]

Study Design Growing male Sprague Dawley rats were either subjected to microsurgical antrectomy with end-to-end gastroduodenostomy (= BI) or to sham operation (for numbers of animals and general data, see Table 21.2). The study design was similar to that described for fundectomy (see above).

Results (see Tables 21.2 a and b) BI induced sustained hypogastrinemia. At the end of the trial, the body weight of BI rats was similar to that of SHAM rats, as was the daily food intake. There were no changes in serum calcium, magnesium, phosphorus, nor did the levels of parathyroid hormone, calcitonin, 25-hydroxyvitamin D and 1,25-dihydroxyvitamin D differ from those in SHAM animals. However, both intestinal absorption and urinary excretion of calcium and magnesium was significantly decreased. The dry, specific and ash weights of bone, as well as bone calcium and magnesium content, were significantly decreased as compared with SHAM.

Table 21.2 Influence of antrectomy (BI), total gastrectomy (Longmire, Roux-Y) and sham operation (SHAM) on bone and mineral metabolism in the rat, as assessed 7–8 weeks postoperatively. (a) general data and data from serum and urine; (b) data from bone, and bone markers. ( ): numbers of animals.
Mean values ± 1SEM. °: $p < 0.06$;*: $p < 0.05$;•: $p < 0.01$;••: $p < 0.001$ vs SHAM

a				*Total gastrectomy*	
	SHAM (*n* = 11)	BI (*n* = 11)	SHAM (*n* = 10)	Longmire (*n* = 12)	Roux-Y (*n* = 1[illegible])
General data					
Body weight (g);					
Initial	250 ± 3	250 ± 3	283 ± 7	273 ± 5	266 ± 3
Final	496 ± 8	500 ± 7	505 ± 12	412 ± 10••	336 ± 14••
Food consumption; g/day	13.6 ± 0.8	13.1 ± 0.7	16.5 ± 1.1	13.9 ± 0.6	11.9 ± 0.41
Fractional absorption of calcium; %	32.5 ± 3.3	21.5 ± 3.9*	18.6 ± 3.1	15.1 ± 2.1	14 ± 4.2
Fractional absorption of magnesium, %	35 ± 4.4	17.2 ± 3.7*	5 ± 1.6	4.3 ± 1.7•	4 ± 1.9*
Calcium balance; mg	45 ± 7	34 ± 6	21 ± 4.5	8 ± 2.5*	4 ± 2•
Data from serum and urine					
Calcium; mg/dl	9.2 ± 0.08	9.2 ± 0.05	9.10 ± 0.43	8.97 ± 0.36	9.7 ± 0.51
Gastrin; pg-equiv/ml	14 ± 3	6 ± 2*	19 ± 4	6 ± 1•	5 ± 2•
Mid-regional PTH; pg-equiv/ml	174 ± 18	162 ± 20			
N-terminal PTH; pg-equiv/ml	21 ± 2	28 ± 8	55 ± 11	52 ± 7	43 ± 8
Calcitonin; pg-equiv/ml	30 ± 7	24 ± 5	77 ± 18	66 ± 27	47 ± 4
25-Hydroxyvitamin D; ng/ml	27 ± 4	28 ± 5	18 ± 2	10 ± 2•	12 ± 2•
1,25-Dihydroxyvitamin D; pg/ml	31 ± 3	28 ± 5	38 ± 6	60 ± 14•	52 ± 16•
Urinary calcium; μg/mg creatinine	44 ± 3	30 ± 2*	48 ± 7	33 ± 2	49 ± 11
Urinary cyclic AMP; ng/mg creatine	9.1 ± 0.4	9.4 ± 0.4	11.3 ± 1.9	18.2 ± 1.4•	18.2 ± 1.6••
Specific activity (serum) calcium^{45}; (cpm/mg) $\times 10^{-4}$	3.9 ± 0.1	4.3 ± 0.2			
Specific activity (urine) calcium^{45}; (cpm/mg) $\times 10^{-4}$	3.7 ± 3	4.5 ± 4			

b				*Total gastrectomy*	
	SHAM (*n* = 11)	BI (*n* = 11)	SHAM (*n* = 10)	Longmire (*n* = 12)	Roux-Y (*n* = 11)
Data from bone					
Volume; μl	706 ± 13	716 ± 26	508 ± 12	466 ± 17•	423 ± 13••
Dry weight/100 g body weight; mg	170 ± 3	160 ± 2*	172 ± 4	175 ± 4	188 ± 8
Mean specific density; g/ml	1.61 ± 0.01	1.54 ± 0.04*	1.71 ± 0.02	1.47 ± 0.03••	1.39 ± 0.05••
Ash weight/volume; g/ml	0.77 ± 0.01	0.73 ± 0.02*	1.07 ± 0.01	0.85 ± 0.02••	0.78 ± 0.03••
Calcium/volume; mg/ml	317 ± 5	294 ± 9*	444 ± 13	356 ± 12••	319 ± 14••
Calcium^{45}/volume; (cpm/ml) $\times 10^{-6}$	4.88 ± 0.2 (10)	4.89 ± 0.3 (12)			
Specific activity calcium^{45}; (cpm/mg) $\times 10^{4}$	1.59 ± 0.03 (10)	1.68 ± 0.05(12)			
Magnesium/volume; mg/ml	5.6 ± 0.1	5.2 ± 0.1*	7.8 ± 0.2	6.2 ± 0.2••	5.8 − 0.3••
Phosphorus/volume; mg/ml	194 ± 12	170 ± 12	202 ± 11	171 ± 6*	163 ± 7•
Bone markers					
Bone alkaline phosphates; U/l	25 ± 3 (11)	23 ± 2 (11)	26 ± 3	41 ± 5*	78 ± 17•
Osteocalcin; ng/ml	42 ± 1 (11)	45 ± 1* (11)	43 ± 3	46 ± 3	42 ± 3
Urinary hydroxyproline/creatinine; μg/mg	60 ± 10 (11)	90 ± 10* (11)	29 ± 3	38 ± 3	35 ± 7

However, both serum and bone specific activities with respect to 45-calcium were no different from those of SHAM. While serum osteocalcin levels and urinary excretion of hydroxyproline were increased, the activity of serum alkaline phosphatase was decreased, and that of bone isoenzyme was normal.

Comments and Conclusions BI in the rat induces a decrease in bone mineral content as well as a decrease in the intestinal absorption of calcium and magnesium. The normal body weight and the absence of steatorrhea in the BI rats rules out a key role for a general intestinal malabsorption syndrome involving malabsorption of calcium in the development of post-antrectomy osteopenia. Apparently, no counterregulation in terms of the activation of the parathyroid glands has occurred. The BI rats have attained a new equilibrium which is characterized by diminished absorption and urinary excretion of calcium and magnesium, and the maintenance of normocalcemia at the expense of bone mass. The concomitant changes in the biochemical bone markers are contradictory. While elevated osteocalcin and urinary hydroxyproline in the presence of normal parathyroid hormone suggest parathyroid-independent enhanced bone turnover, the normal bone isoenzyme activity of alkaline phosphatase militates against abnormal bone turnover. However, since total alkaline is decreased after BI, this may argue in favor of the above mentioned possibility that hypogastrinemia has induced some trophic change in the intestinal mucosa.[85,98] This is supported by the fact that calcium and magnesium absorption were decreased in our rats despite normal levels of 1,25-dihydroxyvitamin D, which indicates a certain resistance of the enterocytes to the latter calciotropic hormone. While in fundectomy rats, the loss of the speculated "gastrocalcin" hormone[47] might be responsible for decreased bone mass, in BI rats hypogastrinemia might possibly have induced osteopenia. Since gastrocalcin is apparently effective only in the presence of normal or elevated gastrin levels,[47] hypogastrinemia might possibly be responsible for the loss of its action. The result might be a decrease in bone mass. In contrast to the findings in humans,[6,11,87] our BI rats revealed no biochemical signs of osteomalacia. Thus, alkaline phosphatase was low, serum calcium and phosphorus were normal, and there was no sign of secondary hyperparathyroidism. These differences do not militate against the animal model, on the contrary may even favor it inasmuch as certain pathomechanisms have been focused on which, in man, are usually masked by a number of concomitant disturbances. In addition, low magnesium absorption after BI is a new finding which has not previously been observed. Its consequence might be some magnesium deficiency at the bone level and the development of osteopenia.[100] Magnesium deficiency has been suspected of playing a key role in the pathogenesis of human osteoporosis.[101] Apparently, the loss of a regulated gastric emptying mechanism, as well as of the antral mill, is of minor importance in the rat, since both body weight and food efficiency remained normal. This may indicate that functional disturbances may only play a minor role in the development of post-antrectomy osteopenia. The minor importance of decreased gastric acid secretion has been discussed in the section on fundectomy. So the striking new finding of our studies is that both fundectomy with hypergastrinemia and antrectomy with hypogastrinemia may induce osteopenia. This indicates either that the pathomechanisms of osteopenia differ after the two interventions, or that the interplay of two factors, namely normo- or hypergastrinemia and the presence of a fundic calciotropic factor, may be responsible for bone mass conservation. After both, fundectomy and antrectomy, one of these two factors is lacking.

2.6. Effects of Total Gastrectomy (GX) on Bone and Mineral Metabolism

Total gastrectomy was first performed successfully by Schlatter in 1897. In recent years total gastrectomy (GX) has been used to an increasing extent in the treatment of gastric cancer, since the primary mortality is acceptably low and the five-year survival rate high enough to justify this extensive surgical procedure. Two methods of re-establishing continuity have proven successful: esophago-intestinal passage is re-established either by a Roux-Y anastomosis, or by interposing an intestinal segment between the esophagus and the duodenum. According to Longmire *et al.*,[102] a jejunal segment is interposed, while State *et al.* used a segment of the transverse colon.[103] One of the invariable sequelae of GX is steatorrhea[104] which is due to reduced pancreatic function, intestinal hurry, insufficient mixture of food and digestive juices, and altered bacterial flora of the small intestine. Additionally, since in most cases truncal vagotomy is performed concomitantly with GX, post-vagotomy diarrhea may aggravate steatorrhea.

2.6.1 Mineral and Bone Metabolism after GX – Current Knowledge

2.6.1.1. Data from Clinical Studies. Nicolaysen and Ragard[105] in 1955 were the first to report on severely disturbed calcium metabolism in patients with total gastrectomy. They found highly negative calcium and phosphorus balances in these patients, the net intestinal calcium absorption being negative in more than half of them; however, their mean calcium intake of approximately 550 mg was abnormally low. Low intestinal calcium absorption was attributed to the concomitant steatorrhea as well as to vitamin D deficiency. The authors concluded that "such negative balances as here reported cannot continue for years without serious consequences".[105] After total gastrectomy, the changes in bone metabolism have been reported to occur earlier, more often and to be more severe than after partial gastrectomy.[62] However, in the series reported by Nishimura *et al.*[106] there was no difference in bone mineral content between the partial distal gastrectomy and total gastrectomy groups. In patients with total gastrectomy, decreased levels of 25-hydroxyvitamin D have been found concomitantly with decreased levels of vitamin D-binding protein.[96] In these patients, serum calcium as corrected for serum protein was significantly decreased, while total alkaline phosphatase was elevated. The decreased bone mass was attributed to low intestinal calcium absorption as induced by vitamin D deficiency, the latter being the sequela of impaired intestinal vitamin D absorption as well as of a failure of hepatic protein synthesis.[96] In four gastrectomized patients with a Roux-en-Y reconstruction and eight with a Longmire interposition, Fischermann *et al.*[107] failed to find major abnormalities in calcium metabolism. Only one of the Roux-en-Y patients had radiologically apparent osteoporosis, while intestinal calcium absorption was practically normal in all but two patients. However, in this study, the mean follow-up period was rather short.

2.6.1.2. Data from Animal Studies. Skeletal growth is impaired after GX in young pigs, dogs and monkeys.[12,108,109] Bussabarger *et al.*[12] were the first to detect osteopenia in dogs that had been gastrectomized when they were puppies. They had less radiopaque bones, wider medullary spaces, thinner cortices and fewer trabeculae, and bony deformities and spontaneous fractures were common. Parathyroid gland structure was generally normal. At the same time, serum calcium and phosphorus, as well as alkaline phosphatase, were usually within the normal range, while intestinal calcium absorption was abnormally low and calcium balance was often negative. The authors concluded that after GX, the bones do not normally ossify on a diet adequate for normal puppies, and that a kind of osteoporosis develops. They attributed the changes in bone structure to the absence of gastric acidity and reservoir function, the increased velocity of intestinal transport of food, and the presence of postcibal acidosis ("acid tide"). However, neither was quantitative determination of the changes in bone metabolism performed, nor was information on the histological structure of the bone after GX provided.

In growing Rhesus monkeys, Freeman *et al.* [109] demonstrated that GX prevented further weight gain, reduced skeletal growth, and induced osteopenia resembling rickets and osteitis fibrosa on histology. At the same time, parathyroid hyperplasia developed, and serum calcium was decreased, while serum alkaline phosphatase was increased. The emptying time of the upper small intestine was enhanced, and signs of intestinal malabsorption were often present. After GX, rats are known to gain less body weight,[110] which was suggested to be due to inadequate food consumption, whereas impaired intestinal absorption of nitrogen and fat did not play a part. Johnston *et al.*[21] discovered that GX in the rat caused a general deficiency syndrome with weight loss. However, the metabolic rate in GX rats was normal, and was therefore not crucial for their postoperative weight loss. They also found that skeletal growth is diminished in GX rats, an effect which was not reversible by the administration of growth hormone.[21] However, the oral administration of whole rat stomachs or of human gastric mucosa produced a temporary weight gain in one-third of the adult rats which had developed the deficiency syndrome.[21] Unfortunately, the state of bone metabolism and the quality of bone were not further investigated in their study.

2.6.2. Data from Our Own Animal Studies

With the aim of better understanding the pathomechanisms of osteopenia following total gastrectomy, we have performed long-term studies in the rat. In addition, the role of preserved duodenal passage in this context was investigated.[111]

Study design Growing male Sprague Dawley rats were either subjected to microsurgical total gastrectomy with Roux-en-Y gastrojejunostomy (GXR), total gastrectomy with interposition of a jejunal segment according to Longmire (GXL), or to sham-operation (for numbers of animals and general data, see Table 21.2). The study design was similar to that described for highly selective vagotomy (see above).
Results (see Tables 21.2 a and b)
Both GXR and GXL induced sustained hypogastrinemia. At the end of the trial, the body weight of the GXR rats was roughly two-thirds, that of GXL approximately 80 per cent of the body weight of SHAM rats. The daily food intake was decreased by the same proportions. There were no changes in serum calcium, magnesium and phosphorus. The levels of parathyroid hormone and calcitonin were no different from those in SHAM animals, while serum 25-hydroxyvitamin D was significantly decreased and 1,25-dihydroxyvitamin D was increased in the GX groups. The urinary excretion of cyclic AMP was significantly elevated in both GXL and GXR. The intestinal absorption and the urinary excretion of calcium tended toward lower values, while the absorption of magnesium was significantly decreased. However, calcium balance was drastically decreased in GXL, and even lower in GXR. The bone volume was significantly decreased in both GX groups as compared with the SHAM group, while bone dry weight when factorized for the body

weight was similar in all three groups. The mean specific density of bone, ash weight per volume, and the factors calcium, magnesium and phosphorus per volume were significantly decreased in both GX groups, the lowest value being encountered in GXR. While the serum osteocalcin levels as well as the urinary excretion of hydroxyproline were normal, the serum levels of both total alkaline phosphatase and its bone isoenzyme were increased.

Discussion of the Results Both GXL and GXR rats had a final body weight far below that of the control rats. However, the food consumption of GX rats was decreased by the same proportion, and signs of malabsorption or maldigestion were absent, although the food efficiency was slightly lower and the fecal dry weight somewhat higher than in SHAM. No steatorrhea was present in GX rats. Some change in the satiety control may therefore have occurred in response to GX; alternatively the animals had learned to consume only as much as they were able to digest without getting diarrhea. The latter mechanisms may be comparable with the situation in gastrectomized humans who mostly lose weight in the postoperative course without becoming hungrier, and who often unconsciously avoid certain nutrients, such as dairy products.

GX induced a drastic decrease in the specific weight of bone as well as bone calcium and magnesium, and to a minor degree phosphorus, per bone volume. In other words: even when corrected for the body weight, which was dramatically lower than in SHAM animals, GX rats had a decrease in bone mineral content exceeding that encountered in antrectomy and fundectomy rats (see above). The most important finding in this context, which underlines the usefulness of our animal trials, is the fact that all the changes in bone and mineral metabolism brought about by GX were more pronounced in the Roux-en-Y group than in the Longmire-type reconstruction. Thus, our trials have proven for the first time that duodenal passage with all its theoretical advantages, such as pancreatico-bilio-cibal synchrony and early mixing of the ingesta, pancreatic juice and bile, is indeed important for the conservation of the bone mass after total gastrectomy. One might speculate that it also plays an important role in partial gastrectomy. The results of our trials may therefore be proof of the theory that the higher frequency and severity of osteopenia after BII as compared with BI-antrectomy are due to the fact that in BI the duodenal passage is preserved. Nevertheless, additional factors must be operative in inducing osteopenia after GX, since the changes in bone mineral content were more pronounced in the Longmire-type GX group than in BI-antrectomy or fundectomy, despite the preservation of the duodenal passage in all. We can only speculate about the reasons for this osteopenia. The most probable cause is the loss of the gastric reservoir. Also, changes in the endocrine regulation of calcium metabolism may be important. GX removes both gastrin and a calciotropic factor generated in the gastric fundus ("gastrocalcin"; [112]). As discussed in the section on antrectomy, the deficiency of either of the two factors may be a prerequisite for the induction of osteopenia. Thus, after removal of both factors, even more severe osteopenia should be expected to occur. While N-terminal parathyroid hormone levels were unchanged, the urinary excretion of cyclic AMP was largely increased, leaving insolved whether, secondary hyperparathyroidism had developed.[109] This might be in line with the finding of normal intestinal calcium absorption in the presence of osteopenia, suggesting some counterregulation which developed during the postoperative period, but which was not able to restore bone mass to normal. The loss of gastric acid, however, has recently been shown not to be of any importance for the development of postgastrectomy bone disease.[113]

The data on the markers of bone metabolism are contradictory, which again makes their diagnostic value questionable in the context of gastric surgery osteopenia. While total alkaline phosphatase and its bone isoenzyme were significantly elevated, indicating increased osteoblastic activity, both osteocalcin and urinary hydroxyproline were unchanged as compared with SHAM, which militates against increased bone turnover. Histomorphometric evaluation of the bones of GX rats is therefore the only way to elucidate the dynamics of bone turnover, and is presently being undertaken in our laboratory.

3. Influence of Jejunoileal Bypass (JIB) on Bone and Mineral Metabolism

After animal experiments[114] had suggested their usefulness, intestinal bypass procedures were introduced into clinical surgery for the treatment of morbid obesity. In 1963 Payne described jejunocolic bypass.[115] This intervention was later replaced by jejunoileal bypass[116] which proved to reduce body weight more effectively than conservative methods of treatment.[117] In recent years JIB has attracted renewed interest as a means of reducing blood cholesterol levels.[118,119] However, JIB has been abandoned in the treatment of morbid obesity, owing to serious side effects, such as malabsorption, steatorrhea and diarrhea,[120] liver failure due to diffuse fatty infiltration,[121] rheumatoid-like polyarthritis,[122] hypokalemia,[123] vitamin B 12 depletion.[124]

3.1. Influences of JIB on Bone and Mineral Metabolism – Current Knowledge

3.1.1. Data from Man

In addition to the above mentioned side effects, various disturbances of bone and mineral metabolism have been observed after JIB in man, such as metabolic bone disease with osteopenia, disturbed calcium and magnesium absorption, alterations in the blood levels of calciotropic hormones, as well as calcium oxalate urolithiasis.

Several studies have indicated that the bone mineral content is reduced in 20–40% of patients who have undergone JIB for morbid obesity.[119,125–135] The changes in bone structure were mostly interpreted as osteomalacia, and some workers have confirmed the diagnosis by histomorphometry.[126–128,130,133] However, signs of low bone turnover, defective osteoblastic function, decreased matrix synthesis, as well as a reduced bone formation rate were also found on histomorphometry.[127] In some cases, the surface extent of Howship's lacunae was increased, indicating accelerated bone resorption.[127] Several workers, however, were not able to find significant changes in the bone mass after JIB in man.[136–139]

In general, the bone disease after JIB was considered to result from calcium malabsorption. The latter was proven by several workers,[127,128,133,139–141] but not confirmed by others who found normal calcium absorption after JIB.[134] The possible causes of calcium malabsorption include precipitation of insoluble calcium soaps within the intestinal lumen, reduced small intestinal length (calcium absorption was shown to correlate with the extent of ileum that is resected or excluded; 139), intestinal hurry, diseased intestinal mucosa, as well as malabsorption and deficiency of vitamin D. In favour of the latter assumption is the frequent finding of decreased serum levels of 25-hydroxyvitamin D in JIB patients,[126–129,131,133,135,142] although normal 25-hydroxyvitamin D levels have also been found.[137] Despite decreased 25-hydroxyvitamin D, 1,25-dihydroxyvitamin D levels are apparently normal,[138] but decreased levels of this calciotropic hormone have also been described.[130] Although small intestinal resections or exclusions induce adaptive hypertrophy of the remaining mucosa,[143] the levels of small intestinal calcium-binding protein (CaBP; 144) are not changed by JIB in man. This indicates some resistance of the mucosa against 1,25-dihydroxyvitamin D, as well as some 1,25-dihydroxyvitamin D and CaBP-independent mechanism of compensatory stimulation of intestinal calcium absorption.[139]

These changes in the serum levels of the vitamin D metabolites were often accompanied by elevations of the alkaline phosphatase[135,137] as well as elevated,[127,128,131,145] unchanged[126,128,135,137] or even decreased[146] levels of parathyroid hormone. Rarely, however, were the changes interpreted as secondary hyperparathyroidism,[131] which is astonishing in a situation where intestinal calcium absorption is largely reduced (see above). The levels of serum calcium generally decrease after JIB, which is presumably due to the accompanying decrease in the serum protein levels.[127,135,137] Also, serum magnesium levels were reported to be decreased after JIB.[127,135,137]

Calcium oxalate (CaOX) urolithiasis occurs in approximately 20–30 % of patients after small intestinal resections or JIB.[147,148] It is mainly due to hyperoxaluria, but other factors such as decreased urinary excretion of inhibitors of stone formation, such as citrate, may also play an important role.[137] Hyperoxaluria after JIB is thought to reflect intestinal hyperabsorption of oxalate. The latter was attributed to either decreased intraluminal binding of calcium to oxalate, owing to binding of calcium to high amounts of free fatty acids (so-called "solubility theory"), or to increased permeability of the colonic mucosa under the influence of high concentrations of unabsorbed fatty acids and bile salts.[149]

In conclusion, metabolic bone disease, disturbances of calcium metabolism, as well as CaOX urolithiasis are a frequent finding after JIB in man. While the pathophysiology of the latter seems to be elucidated, the pathomechanisms resulting in bone disease are as unclear as the histological classification of the type of osteopenia. None of the above-mentioned findings has stood uncontradicted. Furthermore, the sound identification and interpretation of the nature of such disturbances is hampered by the fact that most clinical studies on JIB have dealt with patients of both sexes, who in addition to morbid obesity, suffered from a number of associated abnormalities such as hypercholesterolemia, diabetes mellitus and hypertension.

3.1.2. Data from Animal Studies

Kuhlencordt and Jowsey[22] were the first to describe osteoporosis-like changes in the bone structure of dogs after various small intestinal resections. Increased bone resorption, a tendency towards a loss of bone mass, and the absence of histological signs of osteomalacia were the most prominent findings. The authors concluded that the changes in mineral metabolism and bone turnover after the operations reflected secondary intestinal hyperparathyroidism.

In adult female rats, jejunoileal end-to-end bypass with either resection or jejunocolic shunting of the remaining small intestine led to the development of progressively more radiolucent bones within five weeks after the operations (150). On histology, there was a loss of trabecular bone, along with a loss in histochemically identifiable osteoid. Changes in blood calcium, and magnesium as well as biochemical parameters of bone metabolism in resected rats were indicative of hyperparathyroidism secondary to impaired intestinal calcium absorption. The existence of some endotoxin-like "bone resorptive factor" generating from excluded gut segments was speculated on. However, in these studies no calciotropic hormones were determined, and bone histomorphometry was not performed.

3.2. Data from Our Own Animal Studies

In order to evaluate the nature and causes of bone disease after JIB and whether enteric hyperoxaluria after JIB might serve as an animal model of CaOX urolithiasis (which would be extremely useful for the study of inhibitor substances), we have performed JIB in growing male rats.[151]

Study design Male growing Sprague-Dawley rats were subjected to end-to-side jejunoileal bypass (JIB) excluding 90 % of the small intestinal length. SHAM-operated rats served as controls (for numbers and body weight see Table 21.3).

They received normal rat chow and tap water ad lib. Approximately ten weeks later, all rats were housed in individual metabolic cages. On three consecutive days, the intake as well as fecal and urinary excretion of minerals and oxalate was assessed. The daily mineral balances as well as their fractional intestinal absorption were calculated on the basis of these data. At the end of the trial, the right femur was excised and evaluated in terms of bone volume, density, and mineral content. In an additional setting, the small intestinal absorption as well as the colonic disappearance of 14C-oxalate were determined.

Results (see Table 21.3) Compared with SHAM, JIB resulted in a reduction in food intake and failure to gain weight. No overt signs of maldigestion or steatorrhea were present. The intestinal absorption of calcium and phosphorus was significantly decreased, while magnesium absorption was normal. The serum levels of 25-hydroxyvitamin D, 1,25-dihydroxyvitamin D as well as parathyroid hormone were significantly lower in JIB as compared with SHAM. Serum minerals were not changed from SHAM by JIB, and also unchanged was bone dry and ash weight as well as mineral content when factorized for the unit of bone volume. However, bone volume, dry and ash weight as well as specific density were elevated above SHAM when factorized for the body weight. Small intestinal oxalate absorption was unchanged, while colonic oxalate disappearance was increased (data not shown in Table 21.3). Urinary oxalate excretion in JIB rats was 30% lower than in SHAM, while the excretion of calcium was increased by three and a half times. No urinary calculi were found in the JIB group.

Table 21.3 Influence of jejunoileal bypass (JIB) or sham operation (SHAM) on bone and mineral metabolism of the rat, as assessed 7 weeks postoperatively.

	SHAM ($n = 10$)	JIB ($n = 13$)
Body weight; g		
Initial	283 ± 4	286 ± 3
Final	369 ± 6	275 ± 5*
Food intake per day/g	17.4 ± 1.5	11.8 ± 0.6*
Serum		
Total Ca†; mmol/l	2.8 ± 0.03	2.8 ± 0.03
PTH; ng/l	372 ± 32	275 ± 32*
25-Hydroxyvitamin D; nmol/l	205 ± 37	25 ± 12*
1,25-Dihydroxyvitamin D; pmol/l	216 ± 14	154 ± 22*
Oxalate; μmol/l	2.65 ± 0.2	3.26 ± 0.2*
Fractional absorption;		
Calcium	0.47 ± 0.1	0.33 ± 0.03*
Magnesium	0.38 ± 0.07	0.39 ± 0.05
Phosphorus	0.31 ± 0.08	0.18 ± 0.04*
Urinary excretion; μmol/100g body weight		
Creatinine	15 ± 1	13 ± 1**
Calcium	7 ± 2	26 ± 5**
Oxalate	1.7 ± 0.1	1.2 ± 0.1*
Variables of bone/100 g body weight		
Volume; μl	158 ± 3	185 ± 3*
Dry weight; mg	168 ± 4	198 ± 4*
Ash weight; mg	110 ± 4	126 ± 2*
Mean specific density;	1.06 ± 0.02	1.07 ± 0.01*
Variables per unit bone volume		
Ash weight; g/ml	0.68 ± 0.02	0.68 ± 0.01
Dry weight; g/ml	1.08 ± 0.02	1.07 ± 0.01
Calcium; mmol/ml	7.35 ± 0.43	8.05 ± 0.43

*: $p < 0.05$ vs SHAM
†: corrected for total serum protein

Discussion of the Results Our results indicate that JIB has proved to be a useful tool for reducing the normal growth of young male rats. However, our animal model has completely failed with respect to the above aims, namely to gain insight into the mechanisms of post-JIB osteopenia, as well as to obtain a rat CaOX urolithiasis model. Neither were any negative influences of JIB detectable with respect to bone mass, specific weight, ash weight, as well as calcium, magnesium and phosphorus content as compared with SHAM, nor was hyperoxaluria or urolithiasis present. On the contrary, urinary oxalate was lower than normal in the presence of elevated plasma oxalate levels, which indicates some JIB-induced changes in the renal handling of oxalate.

The intestinal fractional absorption of calcium and phosphorus was decreased in proportion to the extent to which the body weight was reduced. Signs of secondary hyperparathyroidism were absent, since the levels of parathyroid hormone, 25-hydroxyvitamin D as well as 1,25-dihydroxyvitamin D were reduced as compared to SHAM. The combination of normal serum calcium, low 1,25-dihydroxyvitamin D, low intestinal calcium absorption, low serum parathyroid hormone, unchanged urinary cyclic AMP, and an almost four-fold increase in urinary calcium means that calcium homeostasis has reached a new equilibrium after JIB, which is characterized by the suppression of parathyroid gland function and the conservation of bone mass in relation to the body weight. Apparently, JIB has altered satiety control of the growing rats, since signs of gross malabsorption or maldigestion were absent. There was neither steatorrhea nor diarrhea, but the food intake was decreased to a degree such as to freeze the body weight at pre-experimental levels. These findings differ substantially from those of Simmons *et al.*[150] who observed osteopenia in adult female rats after JIB. Apparently their rat model was more appropriate for the study of human post-JIB bone disease than ours using male rats. Unlike our studies on postgastrectomy bone disease (see the respective chapters of this article), our findings in the present context were unexpected and argue against the usefulness of our animal model, all the more so as it failed also as a hyperoxaluria model.

Our studies indicate that animal models, even if well planned and performed with painstaking care, may also be misleading in the study of pathophysiological situations observed in man. Species and sex differences must be taken into account.

4. Portacaval shunt (PCS)

4.1. Introduction

Splanchnic organs such as the stomach, intestine and liver may be involved in the control of calcium and vitamin D metabolism in general, and in particular the homeostasis of serum calcium. For example, native vitamin D is hydroxylated in the liver at the 25-position, and the hydroxylated metabolite serves as substrate for further hydroxylation by the kidneys. The small intestine is a target organ for dihydroxylated vitamin D, and the hypocalcemic action of the stomach has been addressed in preceding sections.

On the other hand, gastric acid is also influenced by the so-called intestinal phase of gastric secretion: the factors involved may be among those originating in the central nervous system, endocrine, paracrine and autocrine active cells of the intestine and stomach, and in intragastric events such as mechanical distension and short-loop reflexes, respectively.

The interaction of the various splanchnic organs long went unnoticed due mainly to the absence of suitable techniques permitting further insight(s) into the true role of the liver. Accordingly, the existence of a network – the nature of which remains unknown – capable of permitting splanchnic and extra-splanchnic inter-organ communication is suggested by the results of more recent research. The liver extracts and breaks down many nutrients and hormones, and thereby contributes to their concentration in peripheral blood. Hence, this function of the liver is critical not only for the performance of its own specific tasks, but possibly also for its influence on remote tissues like the brain, bone and kidney, as well as on tissue in the stomach and pancreas.

In this situation clinical research must concern itself with investigating 1) the contribution by the liver to disease states in small populations of patients with symptoms characteristic for the failure of one or the other hepatic functions, and 2) the response of the liver to loading tests, as reflected by laboratory parameters and their course over time. A particularly powerful tool has emerged with the advent of the surgical procedure PCS. Although highly invasive, PCS has become a life-saving operation for many patients with unstopable bleeding, for example from esophageal varices due to liver cirrhosis. Another indication for PCS may be the treatment of hypercholesterolemia.[152] It is ironic that PCS attracted the interest of researchers not only in physiology but also in clinical medicine, because of the fact that only after depriving the liver of portal-venous blood the true function of this organ becomes accessible to more in-depth investigation. In man, both the number of studies on PCS individuals and the spectrum of metabolic processes investigated in this state are limited. In contrast, once the performance of PCS in laboratory animals, especially the rat, was shown to be feasible, pertinent experimental work and scientific reports increased rapidly. There is no need to emphasize that in its basic form PCS or, for comparison purposes, its modification as spleno-caval shunt,[153] will help improve our understanding of the body's metabolic control mechanisms. In this article, we have elected to summarize data in the literature on PCS-induced changes of calcium metabolism and gastric secretion, two areas in which the PCS technique has permitted substantial progress to be made. In addition, data from pilot studies in our laboratory are presented.

4.2. Effects of PCS

PCS was carried out initially as a side-to-side anastomosis in 1877, by the Russian physician von Eck; the reverse Eck fistula means that after transection of the inferior vena cava its distal stump is connected to the portal vein.[154] Not until the nineteen-sixties did Warren and coworkers[153] introduce into clinical surgery the spleno-renal shunt, a more selective procedure in terms of the volume of diverted blood; its aim was to keep to a minimum the deleterious effects of complete shunting, including multiple disorders of nutrient metabolism, with development of encephalopathy,[155] loss of liver mass, appetite and body weight. Despite these adverse effects, PCS remains a valuable tool for highly motivated surgeons, prepared to provide appropriate postoperative patient management after elective PCS.

In the rat, the technique proposed by Lee and Fisher[156] has found wide acceptance, but in this species too models have been developed with only partial diversion.[157–159] Our own experience is limited to the Lee-Fisher technique of PCS which, in terms of the patency of the anastomosis, appears safe. On the other hand, ammonium urate urolithiasis is a regular event over the long term, being initiated by the excessive production of ammonia by tissues, leading (via ammonia trapping in acidic renal tubular fluid) to increased urinary ammonium.[160]

Radioimmunoassays of circulating parathyroid hormone with proven bioactivity at end organs were long plagued by a number of problems mainly because of the difficulty inherent in detecting the intact (full amino acid chain length) hormone. For this reason, studies on calcium metabolism in the rat involved surgical en bloc removal of parathyroids and thyroid with subsequent restoration of the euthyroid state by the addition of iodine hormone(s) to drinking liquid in many, but not all, investigations. Thus, this procedure does not compensate for the absence of calcitonin from thyroidal C-cells. Isolated parathyroidectomy would appear mandatory, and is frequently done by cauterizing the two proximal glands, which are responsible for maintaining normocalcemia. However, during cautery thyroidal calcitonin is released and, it is worth noting, it was this fact that led to the discovery of calcitonin in the first place.[161] Microsurgical parathyroidectomy leaving the thyroid intact is the most elegant tool that leaves the residual calcium metabolism uncompromised, as was shown in a number of studies in this laboratory; the same approach may be applied to thyroidectomy.[162]

4.2.1. Data from Clinical Studies

4.2.1.1. Influence of PCS on Calcium Metabolism. The concept of the existence of calcemic factors produced in abdominal organs was fostered by the fact that parathyroid hormone immunoreactivity was found to be elevated in the blood of hypercalcemic patients with extra-parathyroid malignant disease[163] and also in the tumor tissue.[164] Subsequently, a porta-systemic gradient of immunoreactive parathyroid hormone was found both in patients with abdominal carcinoma and in those suffering from benign abdominal organ disease.[165] However, the assay then in use did not sufficiently crossreact with the bioactive amino-terminal portion of the human molecule, and the porta-systemic gradient was small (approx. 1.1). Furthermore, the liver contributes to the metabolic clearance of many hormones, circulating parathyroid hormone included, via extraction and degradation;[166] these facts, taken together, suggest that there may be no room for an abdominal source of parathyroid and other calciotropic hormone(s), at least in healthy humans. To our knowledge, for rat and man, there is no information available in the literature about the porta-systemic gradient of the bioactive intact parathyroid hormone or its amino-terminal regions. This also applies to studies in patients with a portacaval shunt, which would be even more desirable. A new clinical syndrome comprising hyperparathyroidism, cirrhosis of the liver and portacaval shunt has been described.[167] The leading symptom was bone pain, explainable by radiological evidence of osteoporosis; peripheral blood parathyroid hormone concentration was elevated in all cases, but the data presented indicate that the immunoreactivity reflected biologically inactive portions of the hormone. Theoretically, removal by the liver of a stimulus to parathyroid glands could have been unmasked by the PCS, thereby generating parathyroid hyperplasia and adenoma within a period of several months to a few years. Considering that calcium and bone metabolism is frequently affected by hepatic diseases,[168] and that there is a frequent need for PCS in order to stop variceal bleedings, clinical research into this area to clarify the true situation is clearly indicated.[169]

4.2.1.2. Influence of PCS on Gastric Secretion and Hormones. Not long after the introduction of shunting procedures into clinical surgery by Whipple[170] and Blakemore[171] it was recognized that they were frequently associated with ulcers, a problem that increasingly attracted the interest of researchers. In several patients with PCS it was thought that ulcers had developed as a sequela of increased gastric acid.[172] In 1958 Clarke and coworkers described four patients out of sixty-two receiving PCS who developed ulcers within a year of the procedure after previously being ulcer-free. As the causal agent they postulated a humoral secretagogue of gastric acid originating in the abdominal viscera and normally inactivated by the liver. Thus the idea of a so-called intestinal phase hormone capable of stimulating gastric acid, was not new when Orloff *et al.*[173] initiated and pioneered research in this field. In an excellent 1979 review by the Orloff group,[174] most of the knowledge then available on the issue was appraised, and the authors concluded that the intestinal phase hormone existed not only in dogs, rats and pigs, but also in humans, was different from known gastric stimulatory hormones, might have a low molecular weight, and might arise from the jejunum. In a commentary on that publication[175] the possible role of intestinal gastrin release, a mediator role of the gastric antrum, and the contribution of the autonomic innervation to gastric acid hypersecretion was suggested, based on independent research. In the intact organism the portasystemic gastrin gradient is only slightly above 1.0,[176] indicating that gastrin extraction by the liver is small, and that the increase in blood gastrin after PCS may be small, too, so that oxyntic gland stimulation via gastrin should be less likely. On the other hand, when applied to the jejunum, balloon distension, a procedure activating neural reflexes, stimulated acid secretion in subjects with PCS, but not in normals,[177] which suggests neural rather than humoral factors. However, in an additional study, the same authors found that gastric acid secretion in cirrhotic patients with PCS was abnormally sensitive to both exogenously administered or endogenously released gastrin.[178] Similar findings were also reported for the PCS rat ([179]; see below). The existence of altered sensitivity toward gastrin was further corroborated by the finding of low serum gastrin, but unchanged gastric acid secretion in patients with cirrhosis, whether or not they had undergone PCS.[180] It appears that, at least in man, the postulated intestinal phase hormone may be called into question.

4.2.2. Data from Animal Experiments

4.2.2.1. PCS and Calcium Metabolism. All the work described below was carried out in the rat. In 1976, Kaplan *et al.*[181] reported on a serum-calcium-elevating polypeptide from the pancreas, which they believed to be contained in extracts of this tissue and which is indistinguishable from insulin on gel chromatography. Somewhat later these authors postulated that the new peptide had approximately the molecular size of insulin and acted directly on bone, i.e. independently of the parathyroid hormone.[182] In 1977, another group showed that PCS prevented the decrease in serum calcium that was expected to result from subsequent parathyroidectomy; when the procedures were performed in the reverse order, recovery from hypocalcemia was almost complete, provided dietary calcium was normal.[183] Again the same Chicago group[184] went on searching for the splanchnic calcium-elevating peptide whose existence they had already postulated.[181] The authors finally concluded that the substance is similar to but not identical with parathyroid hormone, that in the presence of PCS it develops a corrective effect on parathyroidectomy-caused hypocalcemia, and that a normal calcium diet was a prerequisite for this effect. Intestinal calcium absorption was unchanged by PCS alone and decreased by parathyroidectomy alone, but was restored with combined surgery.[185] This latter fact, together with the associated maintenance of normal bone calcium, magnesium, phosphate and hydroxyproline as found by the same authors,[186] indicates that the decrease in bone matrix formation following parathyroidectomy had been prevented by PCS. Among all these papers[181–186] only the last[186] considered a possible role of 1,25-dihydroxyvitamin D as a factor that might be involved in post-PCS serum calcium control, and none mentioned calcitonin, although this peptide responds to oral calcium.[187,188] Interestingly, the Chicago-Rochester cooperation resulted in the postulation of a hypocalcemic gastric factor, which may be unrelated to thyroidal calcitonin, but appears to be related to gastric acid secretion.[83]

4.2.2.2. Influence of PCS on Gastric secretion and Hormones. Compared with studies done in humans, many more have been carried out in animals, especially in rats and dogs. These latter extend the first observation on hypersecretion of acid, not pepsin, in the PCS dog with normal liver,[189] by the possible role of secretin, an inhibitor of acid in the rat[190] which may potentially be removed by PCS, by stimulants of acid appearing in greater quantity in the systemic blood during digestion after prior PCS,[191] and by the removal of nervous influences when using the Heidenhain pouch preparation in the dog.[192] Recently, evidence was provided for PCS-induced changes in the morphology of the oxyntic and antral gastric mucosa, changes in the responsiveness of oxyntic mucosa to endocrine stimuli, as well as altered dose-response curves in the interaction of the mucosa with pentagastrin, cholecystokinin-tetrapeptide, and the histamine-forming enzyme histidine decarboxylase.[193] In contrast to other authors[179] the basal acid output was found to be elevated by 50 % after PCS, while the pentagastrin-stimulated acid output was normal. Following PCS, serum gastrin was low, and antral gastrin content and G cell number were reduced,[80] indicating some G cell suppression. Nevertheless, some role of gastrin cannot be ruled out; for example, PCS might increase gastrin receptors on the histamine producing cells[194] which are the predominant endocrine cell population of the gastric oxyntic mucosa.

4.3. Data from Our Own Animal Studies

4.3.1. PCS and Calcium Metabolism

To approach the role of PCS in more detail, we carried out two kinds of pilot experiment, keeping in mind that the assays employed were not sufficiently suitable for measurement of rat serum bioactive parathyroid hormone, that bone resorption might be increased independently, that associated changes of serum total protein and albumin as well as of acid-base status can substantially alter serum total calcium, and that the feeding factor influencing the results may be identical with or related to dietary calcium (see above). Thus, in the first trial pair-feeding was instituted, animals were pre-labeled with 45-calcium as a bone marker; in one group ($n = 18$) the parathyroids were removed, while in the other ($n = 19$) the glands remained in situ; appropriate sham operations were also performed. Analyses included serum 25-hydroxy- and 1,25-dihydroxyvitamin D, calcitonin, alkaline phosphatase, 45-calcium, and urinary cyclic AMP. Established laboratory techniques were used throughout. Pair-feeding of sham-operated controls ruled out the possibility that unspecific changes of metabolism associated with food might mask the specific PCS effects; such changes may include the lower body weight seen after PCS.

The second trial – in which ad lib feeding was practiced - aimed at re-evaluating serum total protein, calcium, calcitonin and urinary cyclic AMP, and included an assessment of blood pH.

Results (Table 21.4) As expected, the body weight of pair-fed control rats (SHAM-SHAM) was of the order of magnitude shown by animals with and without PCS, and with parathyroidectomy, respectively. After PCS alone total calcium was low, possibly reflecting the low calcium intake associated with the reduced food intake in these rats, but possibly also low serum protein (and albumin; data not shown), both of which result in impaired calcium binding and hence a smaller concentration of total calcium. PCS was, however, able to ameliorate parathyroidectomy-induced hypocalcemia, and the combination of the two procedures was accompanied by even lower total protein; together with the raised 45-calcium of these PX-PCS rats the situation is indicative of excess calcium in the vascular space,

Table 21.4 Data on body weight and mineral metabolism of rats with sham-operation (SHAM), PCS alone, or PCS plus parathyroidectomy (PX). Mean values ± SEM.

	Parathyroids intact		*Parathyroidectomized*	
	SHAM^{+}-SHAM^{++}	SHAM^{+}-PCS	PX-SHAM^{++}	PX-PCS
Pair-fed animals	($n = 10–11$)	($n = 7–8$)	($n = 11$)	($n = 7$)
Body weight; g	246 ± 5	233 ± 5	248 ± 6	244 ± 8
Total protein; g/l	50 ± 1	45 ± 1[a]	51 ± 1	43 ± 1[b]
Total calcium; mg/dl	9.04 ± 0.16	8.44 ± 0.16[a]	5.04 ± 0.28	6.84 ± 0.40[b]
Corrected[1]; mean values	–	9.56	4.96	8.0
Calcium^{45}; (cpm/ml) × 10^{-3}	5.0 ± 0.2	4.1 ± 0.3[a]	2.9 ± 0.1	3.6 ± 0.4[b]
Phosphorus; mg/dl	6.3 ± 0.6	5.2 ± 0.5	11.2 ± 1.0	8.2 ± 0.7[b]
Akaline phosphatase; U/l	165 ± 10	244 ± 12[a]	152 ± 9	289 ± 32[b]
Calcitonin; pg-equiv/ml	45 ± 7	64 ± 10	62 ± 6	70 ± 13
1,25-Dihydroxyvitamin D; pg/ml	17 ± 2	22 ± 3	17 ± 2	13 ± 2
25-Hydroxyvitamin D; ng/ml	11 ± 1	15 ± 3	14 ± 2	16 ± 6
Urinary cyclic AMP; nmol/mg[2]	15 ± 1	13 ± 1	–	–
Ad lib fed animals	SHAM^{++}	PCS		PX-PCS
	($n = 7$)	($n = 9$)		($n = 6$)
Body weight; g	390 ± 8	238 ± 10[c]		212 ± 13[c]
Arterio-venous pH;	7.38 ± 8	7.42 ± 0.01[c]		7.37 ± 0.01
Total protein; g/l	51 ± 1	46 ± 1[c]		42 ± 1[c]
Total serum calcium; mg/dl	9.88 ± 0.08	9.04 ± 0.12[c]		5.76 ± 0.2[c]
Corrected[1]; mean values	–	10.04		7.0
Phosphorus; mg/dl	7.1 = 0.1	6.3 ± 0.5		11.4 ± 0.4[c]
Calcitonin; pg-equiv/ml	20 ± 7	42 ± 6[c]		106 ± 13[c]
Urinary cyclic AMP; nmol/mg[2]	18 ± 1	11 ± 1[c]		10 ± 2[c]

$^{+;++}$: incision in the anterior neck region, and or laparotomy, respectively;
[1]: for the mean of total protein in SHAM-SHAM; [2]: creatinine in the same urine; [a]: p = 0.05 or smaller vs SHAM-SHAM; [b]: p ± 0.05 or smaller vs PX-SHAM; [c]: p = 0.025 or smaller vs SHAM (with Bonferronis's protection).

most likely originating from bone. Support for this is provided by the PCS rats with increased alkaline phosphatase, lower-than-normal urinary cyclic AMP, and higher-than-normal calcitonin; these changes underscore the fact that PCS in this species can uncover a bone resorbing factor of non-parathyroid origin, that this factor is not sufficiently counterbalanced by the reactive (calcium-mediated) but only moderate release of the bone-protecting calcitonin, and that (speculatively) one of its characteristics is the modification of bone turnover. Vitamin D metabolites were unchanged. The state of bone minerals, and also of markers of bone turnover needs clarification by further work. Ad lib feeding corroborates data obtained in pair-feeding, in that PCS lowers protein and calcium, although the calcium-lowering effect of low protein was at least partially offset by the more alkaline blood pH (which leads to increased binding of calcium to protein). Also, total calcium in control rats (SHAM) is higher, due to adequate dietary calcium. However, despite this abundance of calcium serum total calcium in the PX-PCS rats remained low. In other words: under normal circumstances PCS was unable to elevate or even normalize calcemia after prior parathyroidectomy. Urinary cyclic AMP was lowered by PCS alone to a degree otherwise usually seen after parathyroidectomy. Therefore, from the combined data it may be deduced that some non-parathyroid calcium mobilizing factor does in fact exist, which in turn may be able to suppress the parathyroids if present (PCS), and to stimulate the release of calcitonin if parathyroids are absent (PX-PCS). As a result of this interaction, serum calcium concentration in the latter situation is determined by the equilibrium as dictated by parathyroidectomy. This sequence of events, characteristic of normal feeding conditions, fits well into the calcium-driven feed-back regulation of parathyroid hormone and calcitonin.

Conclusions

PCS raises questions as to the factors controlling serum calcium homeostasis. In the rat these can be answered adequately with the aid of suitable surgery, study design, and laboratory technology. While the clinical situation is still unclear, the data from our preliminary animal experiments support the presence of a splanchnic calcemic factor acting on bone that can be unmasked by proper evaluation of the traditional key actions known to govern the state of serum calcium. It is not yet known whether the same or a similar calcemic factor is present in the splanchnic area in man.

4.3.2. PCS, Gastric Secretion, Hormones

Taking together everything that is known about gastric secretion, ulcers, and hormones, from both clinical studies and animal experiments, one is left with a number of unanswered questions. Firstly, with regard to the reported increase in gastric acid output, all published figures are based on titratable acid determinations in gastric juice which, however, may not reflect hydrogen ion activity (pH) as obtained by glass electrodes; in other words: cotitration of free acids might have masked the fact that proton production remained unchanged. Secondly, the state of pepsin secretion did not seem to interest the investigators although pepsin has a documented role in gastric ulcer pathophysiology. Thirdly, ulcers developing spontaneously after PCS have not been graded in terms of ulcer frequency and severity (for details see ref.[195]). Finally, to our knowledge no systematic work-up of blood levels of known acid-stimulating and acid-inhibiting hormones has been reported in the literature, nor a synopsis of these hormones including VIP, which acts as a neurotransmitter in gastrointestinal and extra-gastrointestinal tissues. From these reasons we decided to carry out a pilot study on PCS in the rat, and to address several of these questions.

Study design Male Sprague Dawley rats (body weight 240–270 g) were starved over a period of twenty-four hours, then either sham-operated ($n = 10$) or provided with a portacaval shunt ($n = 10$). In all rats, an antrofundic gastric metal fistula was implanted. On the eleventh day after surgery basal gastric secretion was investigated in the conscious animals over a three-hour period and expressed on an hourly basis. Thereafter, under anesthesia, blood was drawn from the abdominal aorta; after exsanguination the stomach was removed, immediately cleaned and inspected for ulcers by microscopy, while an assisting investigator cut transverse slices from the antrum, fundus, duodenum, jejunum, ileum, and colon. The tissues were shock-frozen in liquid nitrogen and stored at -80°C prior to hormone extraction by the addition of acetic acid and subsequent boiling in water. With the exception of gastrin, other plasma hormones were extracted in alcohol. The radioimmunoassays were performed using established technologies. Gastric juice volume, pH, and titratable (up to pH 7.4) acid were assessed, and in pooled samples of control and PCS rats an enhanced bicarbonate loss was sought. Ulcerations were scored as outlined elsewhere;[195] a few ulcers were examined histologically by light microscopy.

Results (Table 21.5) and comments Acid output was approximately two and a half times as high in PCS rats, reflecting a combination of an increase in secretory volume and of titratable acid, although the latter was more marked. Pepsin secretion was unchanged, indicating that parietal chief cell activity was not altered. The striking finding was that the pH of the secretory fluid was not decreased as would have been expected from increased titratable acid. This means that the in vivo proton generation by oxyntic cells in fact remained stable after PCS, and that the titrated protons arose from dissociation of excess acid(s) in the gastric juice. A number of organic (citric, uric) and mineral (sulfuric, phosphoric) acids with pK-values between two and six were determined in these juices, but on a molar base the magnitude of the changes observed could not have accounted for the increase in titratable acid. To detect an increase in gastric bicarbonate secretion/or in duodeno-gastric bicarbonate reflux that, either alone or in combination, might have prevented the pH from falling, two pooled juice samples each from the SHAM and PCS groups were analyzed as described by Ohe *et al.*[196] This was done on the assumption that pH values above the pK_1 of carbonic acid, which is 6.1, would allow CO_2 to escape in the event that excess free acid had accumulated in the juice. However, the calculated bicarbonate was comparable in the samples from SHAM and PCS rats. Separate experiments revealed that ammonium was dramatically increased in juice from the latter, in accordance with the high levels of ammonium found in the urine of these rats.[160] But, in view of the pK of ammonia, which is approximately 9, any contribution to titratable acid of protons dissociated from ammonium may be small. Thus, the data obtained from our study still fails to identify the substance(s) responsible for PCS-increased gastric acid. In this situation a large number of sizeable ulcers develop in the gastric glandular area, which are similar histologically to the well documented stress ulcers; the major characteristic of these is the presence of focal areas of necrosis restricted to the epithelium, i.e. not penetrating the muscularis mucosae.

Possibly, a bacterial infection is responsible for this type of gastric ulcers, similar as for peptic ulcer disease in humans in which Helicobacter pylori is involved. We suppose an analogy, since the urease activity of Helicobacter pylori increases ammonium in gastric juice, as mentioned above for the rat. Furthermore, rodents spontaneously infected with Helicobacter muridarum, another member of the Helicobacter species, develop chronic gastritis.[197] On the other hand, Helicobacter pylori is suspected to cause hyperplasia of G cells and elevate serum gastrin levels via a long-term increase in gastric juice and tissue ammonium concentration,[198] whereas in our experiments hypergastrinemia was absent. For future work, it remains an open question whether changes of gastric mucosal blood circulation and a PCS-mediated increase of serum urea facilitate gastric infection by some Helicobacter species also in the rat; speculatively, a combination of such events can provoke gastric ulcers, but this hypothesis needs proof. If there is substantiation, PCS associated ulcers may constitute a suitable model for gastric ulcer research.

Compared with acid and mucosal morphology there are only moderate changes in hormones. In fact, gastrin is low in the blood and probably in the antrum as well, and among the hypothesized acid inhibitors (secretin, glucagon, GIP, VIP, somatostatin) only plasma glucagon was raised. Considering that (in terms of pH) proton generation was found to be unaltered, one is tempted to assume that the high acid output, the increase seen in ulcerations, and the state of the traditional gastrointestinal peptide hormones after PCS are not causally related. Future work on peptides in gastrointestinal tissue should assess their content separately in mucosa, especially

Table 21.5 Overview on gastric secretion, ulcers and gastrointestinal hormones in the rat 12 days after sham-operation (SHAM) or portacaval shunt (PCS).

		Secretion					Ulcers	
	n	*Volume secretion*[b] ml/kg/h	*pH*[b]	*Titratable acid*[b] mmol/l	*Acid*[b] *secretion* mmol/kg/h	*Pepsin*[c] *secretion* kU/kg/h	*Ulcer index*[c] AU[a]	*Ulcer severity*[c] mm
SHAM	10	19.0 ± 19	2.5 ± 0.1	73 ± 8	1.4 ± 0.2	7.9/0.4–34	0/0–6	0/0–2
PCS	10	28.9 ± 2.1•	2.5 ± 0.1	162 ± 41•	3.7 ± 0.8•	11.0/0.4–33	21/10–31•	6/1–10•
		Hormones						
		Peripheral blood[d]				*Gastric cross-sectioned tissue*[e]		
		Gastrin[c]	Secretin[c]	Glucagon[c]	GIP[c]	Gastrin[f;c]	Somatostatin[f;c]	VIP[g;c]
SHAM	10	18/9–40	0/0–19	36/0–67	0/0–114	0.4/0.1–100	49/5–687	360/33–978
PCS	10	11/6–43	9/0–20	313/185–1030•	46/0–91	1.6/0.1–100	51/15–3100	515/86–910

[a]: arbitrary units; [b]: mean ± 1 SEM; [c]: median/range of individual values; [d]: all data given in pg-equivalents of the standard preparation used; [e]: data given in ng per g wet tissue; [f]: antrum; [g]: fundus. •: $p = 0.05$ or smaller vs SHAM

in mucosal endocrine cells, and in the muscle layer, in particular in the regions with ganglia of the autonomous nervous systems.

Conclusions

From our preliminary results on the starved PCS rat there may be no necessity to postulate the existence of some intestinal phase hormone capable of stimulating or inhibiting basal gastric acid; however, it cannot be ruled out that our experimental conditions do not reflect those prevailing in the postprandial period, where food is present in the intestinal lumen, and when nutritive, endocrine and nervous factors may interact with acid at a different level. There would appear to be a need to identify the mechanism(s) underlying the development of ulcers, among which may be the state of the mucosal barrier, hydrogen ion back diffusion, microcirculation, etc. with special emphasis on the associated titratable acid in the luminal fluid.

References

1. Schusdziarra, V. (1985). *Gastrointestinale Hormone und Neuropeptide*. Stuttgart: Kohlhammer.
2. Stammers, F. A. R., Williams, J. A. (1963). *Partial Gastrectomy. Complications and Metabolic Consequences.*, London: Butterworths.
3. Clark, C. G. (1973). Nutritional and metabolic complications of partial gastrectomy. In *Vagotomy on trial*, edited by A. G. Cox and J. A. Williams, pp. 53–65. London: Heinemann.
4. Morgan, D. B., Paterson, C. R., Woods, C. G., Pulvertaft, C. N. and Fourman, P. (1965). Search for osteomalacia in 1228 patients after gastrectomy and other operations on the stomach. *Lancet*, II, 1085–1088.
5. Higgins, P. McR. and Pridie, R. B. (1966). Postgastrectomy osteomalacia: Incidence after the no-loop and other types of gastrectomy. *British Journal of Surgery*, **53**, 881–885.
6. Harvald, B., Krogsgaard, A. R. and Lous, P. (1962). Calcium deficiency following partial gastrectomy. *Acta Medica Scandinavica*, **172**, 497–503.
7. Jones, C. T., Williams, J. A. and Nicholson, G. (1963). Disturbances of bone metabolism after partial gastrectomy. In: *Partial Gastrectomy: Complications and Metabolic Consequences*, edited by F. A. R. Stammers and J. A. Williams, pp. 190–226. London: Butterworths.
8. Deller, D. J., Begley, M. D., Edwards, R. G. and Addison, M. (1964). Metabolic effects of partial gastrectomy with special reference to calcium and folic acid. Part I Changes in calcium metabolism and the bones. *Gut*, **5**, 218–225.
9. Clark, C. G., Crooks, J., Dawson, A. A. and Mitchell, P. E. G. (1964). Disordered calcium metabolism after Polya partial gastrectomy. *Lancet*, **1**, 734–738.
10. Garrick, R., Ireland, A. W. and Posen, S. (1971). Bone abnormalities after gastric surgery. *Annals of Internal Medicine*, **75**, 221–225.
11. Eddy, R. L. (1971). Metabolic bone disease after gastrectomy. *American Journal of Medicine*, **50**, 442–449.
12. Bussabarger, R. A., Freeman, S. and Ivy, A. C. (1938). Experimental production of severe homogenous osteoporosis by gastrectomy in puppies. *American Journal of Physiology*, **121**, 137–148.
13. Sarasin, C. (1941). Osteomalacie und hypochrome Anämie nach Magenresektion. *Gastroenterologia*, **66**, 182–197.
14. Pyrah, L. N. and Smith, I. B. (1956). Osteomalacia following gastrectomy. *Lancet*, **1**, 935–937.
15. Baird, I. McL. and Oleesky, S. (1957). Osteomalacia following gastric surgery. *Gastroenterology*, **33**, 284–292.
16. Melick, R. A. and Benson, J. A. (1959). Osteomalacia following partial gastrectomy. *New England Journal of Medicine*, **260**, 976–978.
17. Hall, G. H. and Neale, G. (1963). Bone rarefication after partial gastrectomy. *Annals of Internal Medicine*, **59**, 455–463.
18. Klein, K., Orwoll, E. S., Lieberman, D. A., Meier, D. E., McClung, M. R. and Parfitt, A. M. (1987). Metabolic bone disease in asymptomatic men after partial gastrectomy with Billroth II anastomosis. *Gastroenterology*, **92**, 608–616.
19. Ivy, A. C. (1940). The effects of gastrectomy in animals. *American Journal of Digestive Diseases*, **7**, 500–508.
20. Freeman, S., Hough, V., Wigodsky, H. and Ivy, A. C. (1943). The effect of gastrectomy on growing monkeys. *Gastroenterology*, **1**, 199–210.
21. Johnston, I. D. A., Nevin, H. O. and Welbourn, R. B. (1960). The syndrome of nutritional deficiency after gastrectomy in the rat. *British Journal of Surgery*, **48**, 81–86.
22. Kuhlencordt, F. and Jowsey, J. (1971). Knochenveränderungen nach subtotaler Gastrektomie und verschiedenen Dünndarmresektionen bei Hunden. *Zeitschrift für die gesamte experimentelle Medizin*, **156**, 87–103.
23. Barreras, R. F. (1973). Calcium and gastric secretion. *Gastroenterology*, **64**, 1168–1184.
24. Hellström, J. (1959). Hyperparathyroidism and gastroduodenal ulcer. *Acta Chirurgica Scandinavica*, **116**, 207–216.
25. Dent, R. I., James, J. H. and Wang, C. A., *et al.* (1972). Hyperparathyroidism: Gastric acid secretion and gastrin. *Annals of Surgery*, **176**, 360–368.
26. Reeder, D. D., Jackson, B. M., Ban, J., Clendinnen, B. G., Davidson, W. D. and Thompson, J. C. (1970). Influence of hypercalcemia on gastric secretion and serum gastrin concentrations in man. *Annals of Surgery*, **172**, 540–546.
27. Levant, J. A., Walsh, J. H. and Isenberg, J. I. (1973). Stimulation of gastric secretion and gastrin release by single oral doses of calcium carbonate in man. *New England Journal of Medicine*, **289**, 555–558.
28. Behar, J., Hitchings, M. and Smyth, R. D. (1977). Calcium stimulation of gastrin and gastric acid secretion:

Effect of small doses of calcium carbonate. *Gut*, **18**, 442–448.

29. Cooper, C. W., Bolman III, R. M., Linehan, W. M. and Wells, S. A. (1978). Interrelationships between calcium, calcemic hormones and gastrointestinal hormones. *Recent Progress in Hormone Research*, **34**, 259–278.
30. Selking, O., Gustavsson, S., Johansson, H. and Lundquist, G. (1981). Effect of parathyroid hormone on gastrin and somatostatin release from the gastric antrum. *Upsala Journal of Medical Science*, **86**, 259–267.
31. Zaniewski, M., Jordan, P. H., Yip, B., Thornby, J. I. and Mallette, L. E. (1986). Serum gastrin level is increased by chronic hypercalcemia of parathyroid or nonparathyroid origin. *Archives of Internal Medicine*, **146**, 478–482.
32. Vantini, J., Cominacini, L., Piubello, W. *et al.* (1981). Effect of exogenous gastrointestinal peptides containing the C-terminal tetrapeptide of gastrin on calcium, calcitonin and parathormone serum levels in man. *Hepato-Gastroenterology*, **28**, 43–48.
33. Windeck, R., Brown, E. M., Gardner, D. G. *et al.* (1978). Effect of gastrointestinal hormones on isolated bovine parathyroid cells. *Endocrinology*, **104**, 2020–2026.
34. Grimelius, L., Johansson, H., Lundquist, G., Olazabal, A., Polak, J. H. and Pearse, A. G. E. (1977). The parathyroid glands in experimentally induced hypergastrinemia in the rat. *Scandinavian Journal of Gastroenterology*, **12**, 739–744.
35. Care, A. D., Bruce, J. B., Boelkins, J., Kenny, A. D., Conaway, H. and Anast, C. S. (1971). Role of pancreozymin-cholecystokinin and structurally related compounds as calcitonin secretagogues. *Endocrinology*, **89**, 262–271.
36. Cooper, C. W., Schwesinger, W. H., Ontjes, D. A., Mahgoub, A. M. and Munson, P. L. (1972). Stimulation of secretion of pig thyroid calcitonin by gastrin and related hormonal peptides. *Endocrinology*, **91**, 1079–1089.
37. Hesch, R. D., Hüfner, M., Hasenhager, B. and Creutzfeld, W. (1971). Inhibition of gastric secretion by calcitonin in man. *Hormone and Metabolic Research*, **3**, 140.
38. Hotz, J., Widmaier, F., Minne, H. and Ziegler, R. (1971). Serum calcium and gastric acid secretion in chronic gastric fistula rat: influence of parathyroid hormone and calcitonin. *European Journal of Clinical Investigation*, **1**, 486–490.
39. Becker, H. D., Reeder, D. D., Scurry, M. T. and Thompson, J. C. (1974). Inhibition of gastrin release and gastric secretion by calcitonin in patients with duodenal ulcer. *American Journal of Surgery*, **127**, 71–75.
40. Ito, H., Sakurada, T., Orimo, H. (1977). Role of endogenous calcitonin in the secretion of gastrin and gastric acid in rats. *Hormone and Metabolic Research*, **9**, 89–92.
41. Morley, J. E., Levine, A. S. and Silvis, S. E. (1981). Intraventricular calcitonin inhibits gastric acid secretion. *Science*, **214**, 671–673.
42. Woloszczuk, W., Reich-Hilscher, B., Benke, A. and Dinstl, K. (1986). Effect of infusion of salmon calcitonin on the secretion of somatostatin and gastrin in man. *Hormone and Metabolic Research*, **18**, 197–200.
43. Filipponi, P., Gregorio, F., Cristallini, S., Mannarelli, C., Blass, A., Scarponi, A. M. and Vespasiani, G. (1990). Partial gastrectomy and mineral metabolism: Effects on gastrin-calcitonin release. *Bone and Mineral*, **11**, 199–208.
44. Cooper, C. W., Biggerstaff, C. R., Wiseman, C. W. and Carlone, M. F. (1972). Hypocalcemic effect of pentagastrin and related gastrointestinal hormonal peptides in the rat. *Endocrinology*, **91**, 1455–1450.
45. Care, A. D., Bruce, J. B., Boelkins, J., Kenny, A. D., Conaway, H. and Anast, C. S. (1971). Role of pancreozymin-cholecystokinin and structurally related compounds as calcitonin secretagogues. *Endocrinology*, **89**, 262–271.
46. Kaplan, E. L., North, P. T., Norberg, H. P., Schulak, J. A. and Hill, B. J. (1977). Evidence for a role of the stomach in serum calcium regulation. *Journal of Surgical Research*, **22**, 237–241.
47. Hakanson, R., Persson, P., Axelson, J., Johnell, O. and Sundler, F. (1990). Evidence that gastrin enhances 45Ca uptake into bone through release of a gastric hormone. *Regulatory Peptides*, **28**, 107–118.
48. Persson, P., Hakanson, R., Axelson, J. and Sundler, F. (1989). Gastrin releases a blood calcium-lowering peptide from the acid-producing part of the rat stomach. *Proceedings of the National Academy of Sciences USA*, **86**, 2834–2838.
49. Johnson, L. R., Aures, D. and Yuen, L. (1969). Pentagastrin induced stimulation of the *in vitro* incorporation of 14C-leucine into protein of the gastrointestinal tract. *American Journal of Physiology*, **217**, 251–254.
50. Dragstedt, L. R. and Owens, F. M. Jr. (1943). Supradiaphragmatic section of the vagus nerves in treatment of duodenal ulcer. *Proceedings of the Society of Experimental Biology and Medicine*, **53**, 152–154.
51. Weinberg, J. A., Stempien, S. J., Movius, H. J. and Dagradi, A. E. (1956). Vagotomy and pyloroplasty in the treatment of duodenal ulcer. *American Journal of Surgery*, **92**, 202–207.
52. Amdrup, E. and Jensen, H. E. (1970). Selective vagotomy on the parietal cell mass preserving innervation of the undrained antrum. *Gastroenterology*, **59**, 522–527.
53. Johnston, D. and Wilkinson, A. R. (1970). Highly selective vagotomy without a drainage procedure in the treatment of duodenal ulcer. *British Journal of Surgery*, **57**, 289–296.
54. Jordan, P. H. (1989). Operations for duodenal ulcer disease. *Annual Review of Medicine*, **40**, 1–15.
55. Lavigne, M. E., Healous, D. W., Martin, M., Way, L. W., Mayer, J. H., Sleisenger, M. H. and McGregor, J. L. (1979). Gastric, pancreatic and biliary secretion and the rate of gastric emptying after parietal cell vagotomy. *American Journal of Surgery*, **138**, 644–651.
56. Faxen, A., Rossander, L. and Kewenter, J. (1979). The effect of parietal cell vagotomy and selective vagotomy with pyloroplasty on body weight and dietary habits. *Scandinavian Journal of Gastroenterology*, **14**, 7–12.
57. Berstad, A., Roland, M., Petersen, H. and Liavag, J. (1976). Altered pancreatic function after proximal

gastric vagotomy in man. *Gastroenterology*, **71**, 958–963.

58. Mordes, J. P., Guillermo Herrera, M. and Silen, W. (1977). Decreased weight gain and food intake in vagotomized rats. *Proceedings of the Society of Experimental Biology and Medicine*, **156**, 257–260.
59. Hoffman, H. H. and Schnitzlein, H. N. (1969). The number of vagus nerves in man. *Anatomical Record*, **139**, 429–435.
60. Morgan, D. B. and Pulvertaft, C. N. (1969). Effects of vagotomy on bone metabolism. In: *After vagotomy*, edited by J. A. Williams and A. G. Cox, pp. 161–171. London: Butterworths.
61. Morgan, D. B., Pulvertaft, C. N. and Fourman, P. (1966). Effects of age on the loss of bone after gastric surgery. *Lancet*, **2**, 772–775.
62. Imamura, M., Yamauchi, H., Fukushima, K., Sasaki, I. and Ouchi, A. (1988). Bone metabolism following gastric surgery: Microdensitometry and single-photon absorptiometry. *Tohoku Journal of Experimental Medicine*, **156**, 237–249.
63. Scholz, D., Schwille, P. O., Engelhardt, W. and Morcinietz, C. (1979). Calcium metabolism following proximal gastric vagotomy in man. In: *Vitamin D. Basic Research and its Clinical Application*, edited by A. W. Norman, K. Schäfer, D. vHerrath, H. G. Grigoleit, J. W. Coburn, H. F. DeLuca, E. B. Mawer and T. Suda, pp. 1019–1022. Berlin: de Gruyter.
64. Schwille, P. O., Mühe, E., Gentsch, H., Schellerer, W. and Scholz, D. (1980). Gastrinemia, serum minerals and calciotropic hormones following highly selective vagotomy in duodenal ulcer patients. *Hepato-Gastroenterology*, **27**, 121–129.
65. Scholz, D., Schwille, P. O., Schley, H. W., Hanisch, E., Bieger, D., Zeuner, E. and Engelhardt, W. (1983). Mineral metabolism and vitamin D status before and up to five years following highly selective vagotomy in duodenal ulcer patients. Hepato-*Gastroenterology*, **30**, 102–106.
66. Swaminathan, R. (1973). The inhibitory effect of pentagastrin on intestinal calcium absorption. *Journal of Physiology*, **238**, p. 75.
67. Scholz, D. and Schwille, P. O. (1982). Effects of acutely infused pentagastrin and somatostatin on intestinal calcium absorption in humans. *Mineral and Electrolyte Metabolism*, **7**, 76–85.
68. Fujita, T. (1973). Effects of tetragastrin on calcium absorption from rat duodenal loop *in vivo*. *Endocrinologia Japonica*, **20**, 407–410.
69. Engelhardt, W., Grohmann, C., Schwille, P. O. and Geus, A. (1982). Calcium absorption in the rat as influenced by highly selective vagotomy with special regard to endogenous gastrin. *Research in Experimental Medicine*, **180**, 1–9.
70. Engelhardt, W., Rümenapf, G. and Schwille. P. O. (1984). Effects of highly selective vagotomy on small intestinal calcium transport in the rat. *Mineral and Electrolyte Metabolism*, **10**, 239–243.
71. Rümenapf, G., Beer, T., Meyer, E. and Schwille, P. O. (1990). Effects of gastrin and secretin on intestinal calcium absorption of the rat. *Clinical Physiology and Biochemistry*, **8**, 333–340.
72. Johnson, L. R. (1976). The trophic action of gastrointestinal hormones. *Gastroenterology*, **70**, 278-288.
73. Rümenapf, G., Schwille, P. O., Wagner, W., Tiecks, F. P., Fries, W. and Galewski, D. (1994). Highly selective vagotomy in the rat: effects on bone and mineral metabolism. *Scandinavian Journal of Gastroenterology*, **29**, 232–237.
74. Poitras, P., Modigliani, R. and Bernier, J. J. (1980). Effect of a combination of gastrin, secretin, cholecystokinin, glucagon and gastric inhibitory polypeptide on jejunal absorption in man. *Gut*, **21**, 299–304.
75. Bo-Linn, G. W., Davis, G. R., Buddrus, D. J., Morawski, S. G., Santa Ana, C. A. and Fordtran, J. S. (1984). An evaluation of gastric acid secretion in the absorption of dietary calcium. *Journal of Clinical Investigation*, **70**, 640–47.
76. Recker, R. R. (1985). Calcium absorption and achlorhydria. *New England Journal of Medicine*, **313**, 70–73.
77. Holle, F. and Heinrich, G. (1955). Subdiaphragmatic fundusectomy in gastric surgery. *Surgery Gynecology and Obstetrics*, **101**, 385–394.
78. Hollender, L. F., Meyer, C., Papastratis, G. and Seyer, B. (1984). Resecting procedures in ulcer surgery. In: *A Century of Ulcer Surgery*, edited by H. Bünte and P. Langhans, pp. 109–118. Munich: Urban & Schwarzenberg.
79. Gower, W. R., McSweeney, E. M. and Fabri, P. J. (1985). Hyperplastic G cell responsiveness *in vitro*. Surgery, **98**, 1045–1053.
80. Alumets, J., El Munshid, H. A., Hakanson, R., Hedenbro, J., Liedberg, G., Oscarson, J., Rehfeld, J. F., Sundler, F. and Vallgren, S. (1980). Gastrin cell proliferation after chronic stimulation: effect of vagal denervation or gastric surgery in the rat. *Journal of Physiology*, **298**, 557–569.
81. Fabian, M. A., Fabri, P. J., Knierim, T. H. and Gower, W. R. (1989). Fundusectomy-induced hypergastrinemia is not due to acid inhibition alone. *Surgery*, **106**, 1103–1107.
82. Schulak, J. A. and Kaplan, E. L. (1974). Gastrin-induced hypocalcemia in thyroparathyroidectomized rats. *Metabolism*, **23**, 1103–1106.
83. Klementschitsch, P., Kaplan, E. L., Heath III, H., North, P. and Lee, C. (1979). A gastric factor, calcitonin and the hypocalcemia induced by gastrointestinal hormones. *Endocrinology*, **105**, 1243–1247.
84. Rümenapf, G. and Schwille, P. O. (1995). Disturbances of bone and mineral metabolism following total gastrectomy, antrectomy, fundectomy and highly selective vagotomy in the rat - do functional endocrine interrelationships between the stomach, the thyroid, the parathyroids and the bone play a role? *Langenbecks Archiv für Chirurgie* (Supplement Forum 1995), 567–571.
85. Dembinski, A. B. and Johnson, L. R. (1979). Growth of pancreas and gastrointestinal mucosa in antrectomized and gastrin-treated rats. *Endocrinology*, **105**, 769–773.
86. French, J. M. and Crane, C. W. (1963). Undernutrition, malnutrition and malabsorption after gastrectomy. In: *Partial Gastrectomy*, edited by F. A. R. Stammers and J. A. Williams, pp. 227–262. London: Butterworths.

87. Bisballe, S., Eriksen, E. F., Melsen, F., Mosekilde, L., Sorensen, O. H. and Hessov, I. (1991). Osteopenia and osteomalacia after gastrectomy: interrelations between biochemical markers of bone remodelling, vitamin D metabolites and bone histomorphometry. *Gut*, **32**, 1303–1307.
88. Caniggia, A., Gennari, C. and Cesari, L. (1964). Intestinal absorption of 45Ca and dynamics of 45Ca in gastrectomy osteoporosis. *Acta Medica Scandinavica*, **176**, 599–605.
89. Arman, E., Nilsson, L. H. and Reizenstein, P. (1970). Studies in the dumping syndrome. VI.Calcium deficiency after partial gastrectomy. *American Journal of Digestive Diseases*, **15**, 455–462.
90. Päkkänen, M., Alhava, M., Karjalainen, P., Savolainen, K. and Syrjönen, K. (1984). Long-term follow-up after Billroth I and II partial gastrectomy. *Acta Chirurgica Scandinavica*, **150**, 485–488.
91. Fukuda, M., Shibata, H., Hatakeyama, Y., Soga, J., Koyama, S. and Muto, T. (1979). Differences in calcium metabolism following Billroth-I and Billroth-II procedures for gastric and duodenal ulcers. *Japanese Journal of Surgery*, **9**, 295–303.
92. Kocian, J. and Brodan, V. (1975). New observations on the absorption of 47Ca in patients with partial gastrectomy. *Digestion*, **12**, 193–200.
93. Mergenthaler, H. G., Fink, M., Sauer, H., Bartl, R. and Wilmanns, W. (1989). Multiple brown tumors in a patient with nutritional secondary hyperparathyroidism. *Klinische Wochenschrift*, **67**, 42–46.
94. Alhava, E. M., Aukee, S. and Karjalainen, P. (1974). Bone mineral after partial gastrectomy I. *Scandinavian Journal of Gastroenterology*, **9**, 463–466.
95. Nilas, L., Christiansen, C. and Christiansen, J. (1985). Regulation of vitamin D and calcium metabolism after gastrectomy. *Gut*, **26**, 252–257.
96. Imawari, M., Kozawa, K., Akanuma, Y., Koizumi, S., Itakura, H. and Kosaka, K. (1980). Serum 25-hydroxyvitamin D and vitamin D-binding protein levels and mineral metabolism after partial and total gastrectomy. *Gastroenterology*, **79**, 255–258.
97. Hoikka, V., Alhava, E. M., Savolainen, K., Karjalainen, P. and Parviainen, M. (1982). The effect of partial gastrectomy on bone mineral metabolism. *Scandinavian Journal of Gastroenterology*, **17**, 257–261.
98. Jones, C. T. and Williams, J. A. (1963). The effect of partial gastrectomy on the jejunal mucosa. In: *Partial gastrectomy*, edited by F. A. R. Stammers and J. A. Williams, pp. 141–147. London: Butterworths.
99. Fries, W., Rümenapf, G. and Schwille, P. O. (1992). Disturbances of mineral and bone metabolism following gastric antrectomy in the rat. *Bone and Mineral*, **19**, 245–256.
100. Wallach, S. (1990). Effects of magnesium on skeletal metabolism. *Magnesium and Trace Elements*, **9**, 1–14.
101. Abraham, G. E. and Grewal, H. (1990). A total dietary program emphasizing magnesium instead of calcium. Effect on the mineral density of calcaneous bone in postmenopausal women on hormonal therapy. *Journal of Reproductive Medicine*, **35**, 503–507.
102. Longmire, W. P. and Beal, J. M. (1952). Construction of a substitute gastric reservoir following total gastrectomy. *Annals of Surgery*, 135, 637–645.
103. State, D., Barcley, T. and Kelly, W.D. (1951). Total gastrectomy with utilization of a segment of transverse colon to replace the excised stomach. *Annals of Surgery*, **134**, 1035–1041.
104. Nicolaysen, R., Ragard, R. (1955). The efficiency of digestion in gastrectomized patients. *Scandinavian Journal of Clinical and Laboratory Investigation*, **7**, 271–273.
105. Nicolaysen, R. and Ragard, R. (1955). The calcium and phosphorus metabolism in gastrectomized patients. *Scandinavian Journal of Clinical and Laboratory Investigation*, **7**, 298–299.
106. Nishimura, O., Sakamoto, H., Furumoto, T., Nosaka, K., Kouno, K., Hisaki, T. and Koga, S. (1988). Bone disorder in long-term survival after gastrectomy with reference to bone mineral contents. *Journal of the Japanese Surgical Society*, **88**, 1684–1694.
107. Fischermann, K., Harly, S., Worning, H. and Zacho, A. (1967). Pancreatic function and the absorption of fat, iron, vitamin B_{12} and calcium after total gastrectomy for gastric cancer. *Gut*, **8**, 260–266.
108. Petri, S. and Norgaard, F. (1938). *American Journal of Medical Science*, **195**, 717.
109. Freeman, S., Hough, V., Wigodsky, H. and Ivy, A.C. (1943). The effect of gastrectomy upon growing monkeys. *Gastroenterology*, **1**, 199–210.
110. Welbourn, R. B. and Doggart, J. R. (1956). *British Journal of Surgery*, **44**, 320–322.
111. Rümenapf, G., Schwille, P. O., Steiner, G., Schreiber, M., Kißler, H. and Gall, F. P. (1994). The effect of duodenal passage and preservation of the vagus nerve after total gastrectomy on bone metabolism in the rat: studies on the pathogenesis of postgastrectomy bone disease. *Langenbecks Archiv für Chirurgie* (Supplement Forum 1994), 83–86.
112. Axelson, J., Persson, P., Gagnemo-Persson, R. and Hakanson, R. (1991). Importance of the stomach in maintaining calcium homeostasis in the rat. *Gut*, **32**, 1298–1302.
113. Persson, P., Gagnemo-Persson, R., Chen, D., Axelson, J., Nylander, A. G., Johnell, O. and Hakanson, R. (1993). Gastrectomy causes bone loss in the rat: is lack of gastric acid responsible? *Scandinavian Journal of Gastroenterology*, **28**, 301–306.
114. Kremen, A. J., Linner, I. H. and Nelson, C. H. (1954). An experimental evaluation of the nutritional importance of proximal and distal small intestine. *Annals of Surgery*, **140**, 439–448.
115. Payne, J. H., DeWind, L. T. and Commons, R. R. (1963). Metabolic observations in patients with jejunocolic shunts. *American Journal of Surgery*, **106**, 273–283.
116. Payne, J. H. and DeWind, L. T. (1969). Surgical treatment of obesity. *American Journal of Surgery*, **118**, 141–147.
117. The Danish Obesity Project. (1979). Randomized trial of jejunoileal bypass versus medical treatment in morbid obesity. *Lancet*, **2**, 1255–1258.

118. Buchwald, H., Varco, L., Matts, J. P. and Long, J. M. (1990). Effect of partial ileal bypass surgery on mortality and morbidity from coronary heart disease in patients with hypercholesterolemia. *New England Journal of Medicine*, **323**, 946–955.
119. Rannem, T., Hylander, E., Jarnum, S., Ladefoged, K., Schaadt, O., Stann, M. and Thale, M. (1990). Calcium absorption and bone mineral content in patients subjected to ileal bypass because of familial hypercholesterolemia. *Scandinavian Journal of Gastroenterology*, **25**, 897–905.
120. Wise, L. and Stein, T. (1976). The pathogenesis of diarrhea after bypass of the small intestine. *Surgery Gynecology and Obstetrics*, **142**, 686–688.
121. Manubio, A. T., Ruche, R. D. Jr., Schneider, P. D., Horstmann, J. P., Varco, R. L. and Buchwald, H. (1979). The liver in morbid obesity and following bypass surgery for obesity. *Surgical Clinics of North America*, **59**, 1079–1093.
122. Shagrin, J. W., Frame, B. and Duncan, H. (1971). Polyarthritis in obese patients with intestinal bypass. *Annals of Internal Medicine*, **75**, 377–380.
123. Hocking, M. P., Duerson, M. C., O'Leary, J. P. and Woodward, E. R. (1983). Jejunoileal bypass for morbid obesity. Late follow-up in 100 cases. *New England Journal of Medicine*, **308**, 995–999.
124. Juhl, E., Bruusgaard, A., Hippe, E., Korner, B., Quadde, F. and Baden, H. (1974). Vitamin B 12 depletion in obese patients treated with jejunoileal shunt. *Scandinavian Journal of Gastroenterology*, **9**, 543–547.
125. Campbell, J. M., Hunt, T. K., Karam, J. H., Forsham, P. H. (1977). Jejunoileal bypass as a treatment of morbid obesity. *Archives of Internal Medicine*, **137**, 602–610.
126. Teitelbaum, S. L., Halverson, J. D., Bates, M., Wise, L. and Haddad, J. G. (1977). Abnormalities of circulating 25-OH vitamin D after jejuno-ileal bypass for obesity. Evidence of an adaptive response. *Annals of Internal Medicine*, **86**, 289–293.
127. Parfitt, A. M., Miller, M. J., Frame, B., Villanueva, A. R., Rao, D. S., Oliver, I., and Thomson, D. L. (1978). Metabolic bone disease after intestinal bypass for treatment of obesity. *Annals of Internal Medicine*, **89**, 193–199.
128. Compston, J., Horton, L., Laker, M., Ayers, A., Woodhead, J., Bull, H., Gazet, J., and Pilkington, T. (1978). Bone disease after jejuno-ileal bypass for obesity. *Lancet*, **2**, 1–4.
129. Hey, H., Lund, B., Sorensen, O. H., Lund, B. J., and Christensen, M. S. (1979). Impairment of vitamin D and bone metabolism in patients with bypass for treatment of obesity. *Acta Medica Scandinavica*, **624** (Suppl.), 73–78
130. Mosekilde, L., Melsen, F., Hessov, I., Christensen, M., Lund, B., and Sprensens, O. (1980). Low serum levels of 1,25-dihydroxyvitamin D and histomorphometric evidence of osteomalacia after jejunoileal bypass for obesity. *Gut*, **21**, 624–631.
131. Kruse, H. P., Kuhlencordt, F., Montz, R., Ringe, J. D., and Koch, G. (1981). Calcium- und Knochenstoffwechsel nach Dünndarmbypass-Operation wegen alimentärer Adipositas. *Deutsche Medizinische Wochenschrift*, **106**, 1219–1223.
132. Taylor, H. C., Teitelbaum, S. L., and Lambert, P. W. (1983). Symptomatic osteomalacia after jejunoileal bypass surgery in a patient with primary hyperparathyroidism. A study of the change in bone morphology and vitamin D metabolites before and during treatment. *Gastroenterology*, **85**, 735-742.
133. Sellin, J. H., Meredith, S. C., Kelly, S., Schneir, H., and Rosenberg, I. H. (1984). Prospective evaluation of metabolic bone disease after jejunoileal bypass. *Gastroenterology*, **87**, 123–129.
134. Charles, P., Mosekilde, L., Sondergard, K., and Jensen, F. T. (1984). Treatment with high-dose oral vitamin D2 in patients with jejunoileal bypass for morbid obesity. *Scandinavian Journal of Gastroenterology*, **19**, 1031–1038.
135. Rickers, H., Christiansen, C., Balslev, I., and Rodbro, P. (1984). Impairment of vitamin D metabolism and bone mineral content after intestinal bypass for obesity. A longitudinal study. *Scandinavian Journal of Gastroenterology*, **19**, 184–189.
136. Dalen, N., and Hallberg, D. (1974). Bone mineral content in four obese subjects before and after intestinal shunt operation. *Acta Chirurgica Scandinavica*, **140**, 267–270.
137. Scholz, D., Schwille, P. O., Husemann, B., Herzog, T., Schley, H. W., Morcinietz, C., and Sigel, A. (1982). Mineral metabolism during prolonged calcium substitution after jejunoileal bypass for morbid obesity. *Klinische Wochenschrift*, **60**, 803–809.
138. Rickers, H., Christiansen, C., Balslev, I., Foltved, H., Rodbro, P., and Christensen, M.S. (1983). Vitamin D and bone mineral content after intestinal bypass operation for obesity. *Gut*, **24**, 67–72.
139. Dano, P., and Christiansen, C. (1978). Calcium absorption and absence of bone decalcination following intestinal shunt operation for obesity. *Scandinavian Journal of Gastroenterology*, **13**, 81–85.
140. Halverson, J., Teitelbaum, S., Haddad, J. and Murphy, W. (1979). Skeletal abnormalities after jejunoileal bypass. *Annals of Surgery*, **189**, 785–790.
141. Nunan, T. O., Compston, J. E. and Tonge, C. (1986). Intestinal calcium absorption in patients after jejunoileal bypass or small intestinal resection and the effect of vitamin D. *Digestion*, **34**, 9–14.
142. Schoen, M. S., Lindenbaum, J., Roginsky, M. S. and Holt, P. R. (1978). Significance of serum level of 25-hydroxycholecalciferol in gastrointestinal disease. *Digestive Diseases*, **23**, 137–142.
143. Tilson, D. M. and Wright, H. K. (1972). The effect of resection of the small intestine upon the fine structure of the intestinal epithelium. *Surgery Gynecology and Obstetrics*, **134**, 992–994.
144. Staun, M. and Jarnum, S. (1988). Measurement of the 10.000-molecular weight calcium-binding protein in small-intestinal biopsy specimens from patients with malabsorption syndromes. *Scandinavian Journal of Gastroenterology*, **23**, 827–832.
145. Melsen, F., Mosekilde, L., Hessov, I. and Christensen, M. S. (1979). A histomorphometric analysis of bone changes in patients after intestinal shunt for obesity. *Calcified Tissue International*, **27** (suppl), A28.

146. Atkinson, R. L., Dahms, W. T., Bray, G. A. and Schwartz, A. A. (1978). Parathyroid hormone levels in obesity: effects of intestinal bypass surgery. *Mineral and Electrolyte Metabolism*, **1**, 315–320.
147. Gregory, J. G., Park, K. Y. and Schoenberg, H. W. (1977). Oxalate stone disease after intestinal resection. *Journal of Urology*, **117**, 631–634.
148. Stauffer, J. Q. (1977). Hyperoxaluria and calcium oxalate nephrolithiasis after jejunoileal bypass. *American Journal of Clinical Nutrition*, **30**, 64–71.
149. Dobbins, J. W. (1985). Nephrolithiasis associated with intestinal disease. In *Urolithiasis and Related Clinical Research*, edited by P. O. Schwille, L. H. Smith, W. G. Robertson, W. Vahlensieck, pp. 119–126. New York: Plenum.
150. Simmons, D. J., Hyland, G., Lesker, P. A., Cohen, M., Stein, T. and Wise, L. (1975). The effect of small-bowel resection or bypass on the rat skeleton. *Surgery*, **78**, 460–471.
151. Blaurock, P., Schwille, P. O., Manoharan, M., Scheele, J., Fries, W. and Rümenapf, G. (1992). Effects of jejuno-ileal bypass on oxalate and mineral metabolism - A reappraisal in the rat. *European Journal of Surgery*, **158**, 595–602.
152. Starzl, T. E., Chase, H. P., Ahrens jr, E. H., McNamara, D. J., Bilheiner, D. W., Schaefer, E. J., Rey, J., Porter, K. A., Stein, E. and Francavilla, A. (1983). Portacaval shunt in patients with familial hypercholesterolemia. *Annals of Surgery*, **198**, 273–278.
153. Warren, W. D., Zeppa, R. and Fornon, Y. J. (1967). Selective trans-splenic decompression of gastroesophageal varices by distal splenorenal shunt. *Annals of Surgery*, **166**, 437–455.
154. Markowitz, J., Archibald, J. and Downie, H. G. (1959) *Experimental Surgery*, fourth edition. London: Bailliere, Tindall, Cox.
155. Jeppson, B., James, I. H., Edwards, L. L. and Fischer, J. E. (1985). Relationship of brain glutamine and brain neutral amino acid concentrations after portacaval anastomosis in rats. *European Journal of Clinical Investigation*, **15**, 179–187.
156. Lee, S. A., and Fisher, B. (1961). Portacaval shunt in the rat. *Surgery*, **50**, 668–572.
157. Wendling, P., Görninger, J. and Rudigier, J. (1980). Mesenterico-caval shunt in rats - Technique and results. *Research in Experimental Medicine*, **177**, 79–84.
158. Castaing, D., Beaulernard, C., Ariogul, O., Gigou, M., Franco, D. and Bismuth, H. (1982). Liver atrophy and encephalopathy after portacaval shunt in the rat. *European Surgical Research*, **14**, 192–202.
159. Schröder, R., Müller, O. and Bircher, J. (1985). The portacaval and splenorenal shunt in the normal rat. A morophometric and functional re-evaluation. *Journal of Hepatology*, **1**, 107–123.
160. Linnemann, U., Kuch, P. and Schwille, P. O. (1986). Ammonium urate urolithiasis in the rat with portocaval shunt – Some aspects of mineral metabolism and urine composition. *Urological Research*, **14**, 319–322.
161. Copp, D. H., Cameron, F. C., Cheney, B. A., Davidson, A. G. F. and Henze, K. G. (1962). Evidence for calcitonin - a new hormone from the parathyroid that lowers blood calcium. *Endocrinology*, **72**, 638–649.
162. Schwille, P. O. and Steiner, H. (1975). Mikrochirurgische Thyroidektomie unter Schonung der Nebenschilddrüsen. *Zeitschrift für Versuchstierkunde*, **17**, 304–307.
163. Benson, R. C. Jr., Riggs, L. and Pickard, B. M. (1974). Radioimmunoassay of parathyroid hormone in hypercalcemic patients with malignant disease. *American Journal of Medicine*, **56**, 821–826.
164. Runson, P. L., Tashjian, A. H. Jr. and Levine, L. (1965). Evidence for parathyroid hormone in nonparathyroid tumors associated with hypercalcemia. *Cancer Research*, **25**, 1062.
165. Moossa, A. R., Hall, J. T., Hughes, R. G., Wood, R. A. B., Mackie, C. R. and Hale, A. W. (1978). Portosystemic gradient of immunoreactive parathyroid hormone in man. *Journal of Surgical Research*, **25**, 445–448.
166. Martin, K., Hruska, K., Greenwalt, A., Klahr, S. and Slatopolski, E. (1976). Selective uptake of intact parathyroid hormone by the liver. Differences between hepatic and renal uptake. *Journal of Clinical Investigation*, **58**, 781–788.
167. Orloff, M. J. (1988). Hyperparathyroidism, cirrhosis and portacaval shunt. *American Journal of Surgery*, **155**, 76–81.
168. Long, R. G., Minshard, E., Skinner, R. K., Varghese, Z., Wills, M. R. and Sherlock, S. (1978). Clinical, biochemical and histological studies of osteomalacia, osteoporosis and parathyroid function in chronic liver disease. *Gut*, **19**, 85–90.
169. Urban, E., Dinzans, V. J. and van Thiel, D. H. (1989). Hyperparathyroidism and portacaval shunt: a syndrome or a coincidence? *Hepatology*, **9**, 506–507.
170. Whipple, A. O. (1945). The problem of portal hypertension in relation to the hepatosplenopathies. *Annals of Surgery*, **122**, 449–454.
171. Blakemore, A. H. (1952). Portacaval shunting for portal hypertension. *Surgery Gynecology and Obstetrics*, **94**, 443–449.
172. Clarke, J. S., Ozeran, R. S., Hart, J. C., Cruze, K. and Crevling, V. (1958). Peptic ulcer following portacaval shunt. *Annals of Surgery*, **148**, 551-566.
173. Orloff, M. J., Chandler, J. G., Alderman, S. J., Keiter, J. E. and Rosen, H. (1969). Gastric secretion and peptic ulcer following portacaval shunt in man. *Annals of Surgery*, **170**, 515–524.
174. Orloff, M. J., Hyde, P. V. B., Kosta, L. D., Guillemin, R. C. L. and Bell, jr., R. H. (1979). The intestinal phase hormone. *World Journal of Surgery*, **3**, 523–537.
175. Konturek, S. J. (1979). Invited Commentary. *World Journal of Surgery*, **3**, 537–538.
176. Walsh, J. H. and Grossman, M. I. (1975). Gastrin. *New England Journal of Medicine*, **292**, 1324–1334.
177. Lenz, H. J., Hogan, D. L. and Isenberg, J. I. (1985). Intestinal phase of gastric acid secretion in humans with and without portacaval shunt. *Gastroenterology*, **89**, 791–796.
178. Lenz, H. J., Struck, T., Greten, H., Koss, M. A., Eysselein, V. E., Walsh, J. H. and Isenberg, J. I. (1987). Increased sensitivity of gastric acid secretion to gastrin in cirrhotic patients with portacaval shunt. *Journal of Clinical Investigation*, **79**, 1120–1124.

179. Kutz, K., Herz, R., Halter, R., Leuenberger, J. and Bircher, J. (1974). Increased responsiveness of gastric mucosa to gastrin stimulation in the Eck fistula. *Gastroenterology*, **66**, 73–78.
180. Albillos, A., Abreu, L., Alvarez-Mon, M., Gea, F., Fonzalo, M. A., Rossi, J., Garrido, A., Barrios, C. and Escartis, P. (1988). Study of the secretion of pepsinogen I in cirrhotic humans with and without portacaval shunt. *Americal Journal of Gastroenterology*, **83**, 37–41.
181. Kaplan, E. L., Hill, B. J., Marhese, J. and Tager, H. S. (1976). A calcium-elevating polypeptide from pancreas. *Gastroenterology*, **70**, 971 (abstract).
182. Kaplan, E. L., Tager, H. S., Klementschitsch, P. and Lagocki, R. (1978). The calcium-elevating peptide from the pancreas. *Scandinavian Journal of Gastroenterology*, 13 (Suppl. 49), 96 (abstract).
183. Pfeffermann, R., Sakai, A., Kashiwabara, H., Fisch, H., Taha, M., Goldson, H. and Kountz, S. L. (1977). The prevention and correction of hypocalcemia in the parathyroidectomized rat by portacaval shunt. *Surgery*, **82**, 266–270.
184. Lee, C. S., Kaplan, E. L., Sugimoto, J. and Heath III, H. (1980). A portal factor influences serum calcium hormeostasis. *Annals of Surgery*, **192**, 459–464.
185. Al-Jurf, A., Gishan, F. and Younoszai, K. (1979). Effects of portacaval shunt on calcium homeostasis. *Surgical Forum*, **30**, 105–107.
186. Al-Jurf, A. S. and Smith, G. L. (1981). Changes in bone and serum minerals after parathyroidectomy and their prevention by portacaval shunt. *Metabolism*, **30**, 544–548.
187. Koroma, S. J. (1992). Fundektomie bei der Ratte – Wirkungen auf Gastrinämie, calciotrope Hormone und Calciumtoleranz, unter besonderer Beachtung von Mineralien in Knochen, Niere, Urin. *MVD Thesis*, University of Munich, Germany.
188. Peng, T. C. and Garner, S. G. (1980). Sex difference in serum calcitonin level in rats as related to feeding, fasting and age. *Endocrinology*, **107**, 289–293.
189. Lebedinskaya, S. I. (1932). Über die Magensaftsekretion bei Eckschen Fistelhunden. *Zeitschrift für die gesamte experimentelle Medizin*, **88**, 264–270.
190. Stollmaier, W. and Schwille, P. O. (1992). Endogenous secretin in the rat – evidence for a role as enterogastrone but failure to influence serum calcium homeostasis. *Clinical and Experimental Endocrinology*, **99**, 169–174.
191. Gerez, L. and Weiss, A. (1936). Über die Magensaftsekretion bei Eckscher Fistel. *Zeitschrift für die gesamte experimentelle Medizin*, **100**, 281–288.
192. Gregory, R. A. (1958). Gastric secretory responses after portal venous ligation. *Journal of Physiology*, **144**, 123–137.
193. Ekelund, M., Hakanson, R., Holumi, T., Oscarson, J., Rehfeld, J., Sundler, F. and Westtrin, P. (1985). Effects of portacaval shunt on the rat stomach. *Acta Physiologica Scandinavica*, **1244**, 437–444.
194. Axelson, J., Ekelund, M., Sundler, R. and Hakanson, R. (1990). Enhanced hyperplasia of gastric enterochromaffine cells in response to omeprazole-evoked hypergastrinemia in rats with portacaval shunts. An immunocytochemical and chemical study. *Gastroenterology*, **99**, 635–640.
195. Lotter, C. (1990). Der portakavale Shunt bei der Ratte – eine kritische Bewertung von Magensekretion und Peptidhormonen im gastrointestinalen Gewebe und peripheren Blut. *MD Thesis*, Universität Erlangen, Germany.
196. Ohe, K., Mura, Y., Taoha, Y., Okada, Y. and Miyoshi, A. (1988). Cysteamine-induced inhibition of mucosal and pancreatic alkaline secretion in rat duodenum. *Digestive Diseases and Sciences*, **33**, 330–337.
197. Lee, A., Chen, M., Coltro, N., O'Rourke, J., Hazell, S., Hu, P. and Li, Y. (1993). Long term infection of the gastric mucosa with Helicobacter species does induce atrophic gastritis in an animal model of Helicobacter pylori infection. *Zentralblatt Bakteriologie*, **280**, 38–50.
198. Lichtenberger, L. M., Dial, E. J., Romero, J. J., Lechago, J., Jarboe, L. A. and Wolfe, M. M. (1995). Role of luminal ammonia in the development of gastropathy and hypergastrinemia in the rat. *Gastroenterology*, **108**, 320–329.

PART VII

22. Experimental Islet Transplantation

Bo Ahrén[1] and Aamer Ar 'Rajab[2]

Department of [1]Medicine, Malmö and [2]Surgery Lund, Lund University, Swedan

Introduction

In the 1960s, it became possible to isolate the islets of Langerhans from the surrounding exocrine pancreatic tissue by the use of the collagenase digestion technique,[1,2] and in 1972, it was shown that transplantation of such islets to diabetic rats reversed the diabetic condition.[3] These findings initiated an era of experimental research with the ultimate goal to offer islet transplantation to patients with insulin-dependent diabetes mellitus. This research has dealt with experimental diabetes models, islet isolation and purification techniques, the effects of transplanted islets on the metabolism of diabetes, immunogenicity of rejection, as well as technical aspects of the transplantation procedure. This review summarizes this development and describes the position of the field in the 1990s.

Experimental Diabetes

Islet transplantation research needs well defined and reliable diabetes models. The ideal model should be stable and permanent and should show clinical and immunologic resemblences to insulin-dependent (type 1) diabetes mellitus in man. Such diabetes models exist mainly in rodents and in the dog, and they may be divided into spontaneous diabetes, surgically-induced diabetes and chemically-induced diabetes.

Spontaneous Experimental Diabetes

Diabetes develops rarely in experimental animals. One exception is the BB (BioBreeding) Wistar rat. In this strain of rat, a type of diabetes resembling human type 1 diabetes spontaneously develops in approximately 30–50% of the rats, usually between the age of 60 and 120 days.[4] The pathogenesis of this diabetes is supposed to be an autoimmune process mediated by T-lymphocytes.[4–6] This autoimmune process causes insulitis which is followed by a short period with glucose intolerance and impaired insulin secretion, whereafter overt diabetes develops. By that time, the pancreatic insulin content is reduced, insulin secretion is markedly impaired and hepatic and peripheral insulin insensitivities exist.[7] However, the pancreatic insulin content has been found to be reduced prior to the development of insulitis, which suggests that a primary immunoreaction against the islet B-cells is an important pathogenetic feature.[8] In fact, antibodies to islet cell surface and an islet protein with a molecular weight of 64 kd have been detected in the BB rat already before diabetes has developed.[9,10] This 64 kd protein has been shown to be identical with glutamic acid decarboxylase (GAD).[11] GAD has been shown to be a major autoantigen for both humoral and cellular autoimmunity in type 1 diabetes.[12] The early occurrence of this and other antibodies suggests that they initiate the destruction of the B-cells. These pathogenetic processes are very similar to those of insulin-dependent diabetes mellitus in man, and, therefore, the diabetes in the BB rat may be used in studies on this type of diabetes. The BB rat has also been frequently used in studies on experimental islet transplantation[13] However, the BB rat is not the only experimental animal that spontaneously develops diabetes. For example, the so called NOD (non-obese diabetic) mouse also develops diabetes spontaneously, usually at the age of approximately 120 days.[14] Approximately 80% of the females and 30% of the males belonging to this strain of mouse develop a type of diabetes which shows immunologic and metabolic resemblances to human type 1 diabetes.[15]

An experimental model of diabetes in the rat resembling non-insulin-dependent (type 2) diabetes has been developed; the so called GK rat (Goto-Kakizaki rat).[16] These rats were developed by selective inbreeding of Wistar rats with glucose intolerance, which after ten generations resulted in a stable diabetes condition. However, although this model has been used for studies on pathogenesis of type 2 diabetes,[17] it has not been used in islet transplantation studies.

Surgically Induced Diabetes

As originally demonstrated by Minkowski, total pancreatectomy in dogs results in overt diabetes.[18] Pancreatectomy-induced diabetes has therefore been used as a diabetes model for studies on islet transplantation in large animals.[19] Usually, the islet transplantation is performed as an autotransplantation on the same day as the pancreatectomy. In small animals, however, total pancreatectomy is a difficult procedure and is also associated with mortality. Therefore, pancreatectomy-induced diabetes is rarely used for islet transplantation studies in rats or mice. It is also possible to perform subtotal or partial pancreatectomy for inducing diabetes. However, it is necessary to remove more than 90% of the pancreas for inducing diabetes and even after such a major pancreatectomy, significant regeneration of the pancreas occurs.[20] Therefore, diabetes evolving after subtotal pancreatectomy is not stable and thus not suitable for studies on islet transplantation.

Chemically Induced Diabetes

The most commonly used models for experimental diabetes in islet transplantation research are those induced by various drugs. A variety of different chemicals have been used to destroy the islet B-cells and thereby cause diabetes, of which the most frequently used chemicals are alloxan and streptozotocin.[21] Alloxan is a pyrimidine derivative shown in 1943 by Dunn *et al.* to induce a selective pancreatic islet necrosis and diabetes in rabbits.[22] It was later demonstrated that the drug causes diabetes by selectively destroying the islet B-cells by a direct action.[23] The exact mechanism underlying the B-cell cytotoxicity and diabetogenic effect of alloxan is, however, not yet established and several hypotheses have been put forward.[21,24] One hypothesis is that the drug disrupts the DNA chain in the B-cells,[25] whereas another hypothesis suggests that the drug initiates production and accumulation of free oxygen radicals. A recent hypothesis is that the drug acts by oxidizing two adjacent SH groups in glucokinase, thereby inactivating the enzyme.[26] Since glucokinase is thought to be of importance as a glucose recognition enzyme in the B-cells,[27] its destruction would deleteriously affect the insulin secretion process. However, the exact mechanism underlying the diabetogenic action of alloxan remains to be established.

It has been difficult to establish a relationship between the dose of alloxan and the resulting diabetes under *in vivo* conditions, due to both instability of the drug and a variety of animal-related factors, such as diet, age and species, as has been reviewed by Lenzen and Panten.[26] Due to its short half life, less than one minute,[21] it is necessary to inject alloxan intravenously. By doing that in mice, a characteristic triphasic glucose response is seen.[28] During the first four hours after injection, a marked and transient hyperglycemia is seen, most likely due to a rapid mobilization of liver glycogen. Thereafter, hypoglycemia develops which lasts for forty eight hours. The hypoglycemia is due to hyperinsulinemia caused by release of insulin from the damaged B-cells. Finally, chronic hyperglycemia develops due to permanent destruction of the islet B-cells. This period starts to develop at two days after injection of alloxan and is usually, but not always, life-long.[21] It is during this period that the animals are suitable for experimental islet transplantation. Usually, the islet transplantation is performed at six to eight days after administration of alloxan, when, for example, the serum glucose levels after injection of alloxan at 75 mg/kg are approximately 40 mM.[29] Alloxan injected at lower dose levels (40–50 mg/kg) results in a transient form of diabetes.[30] It has been demonstrated that preserved cholinergic influences are required for the transient nature of the diabetes.[31] This transient form of alloxan-induced diabetes is, however, not suitable for islet transplantation studies.

The other commonly used diabetogenic substance is streptozotocin. This drug is a broad-spectrum antibiotic extracted from *streptomyces acromogenes* and contains a nitrosmethylamide residue coupled to glucose.[32] In 1963 its diabetogenic action was demonstrated and the drug has been shown to induce B-cell necrosis and experimental diabetes in a variety of species, such as rats, mice, guinea pigs, hamsters and dogs.[21,33] The diabetogenic action of streptozotocin is thought to be due to breaking the DNA molecule which, due to stimulation of nuclear poly (ADP-ribose) synthetase, is followed by depletion of B-cell NAD.[25] Streptozotocin has been used extensively as an experimental diabetes model in several species. In rats, streptozotocin induces a stable, long-term, diabetes when injected intravenously at dose levels above 50 mg/kg, whereas at lower dose levels, the diabetes is reversible.[34,35] However, spontaneous reversal of the diabetes has also been demonstrated after streptozotocin at higher dose levels (55–65 mg/kg),[36] which suggests that different sensitivity to streptozotocin in different strains or even in individual animals might exist. Therefore, it is important to establish the diabetogenic dose of streptozotocin in the actual strain of rats used for a given study. For example, if the dose of streptozotocin is too low, spontaneous recovery of the diabetes will occur which will influence the interpretation of an islet transplantation study. It has been demonstrated that a dose of 70 mg/kg induces a long-term and stable diabetes in the rat.[35] As in the case of alloxan-induced diabetes, streptozotocin induces a triphasic glycemic response.[37] However, the development of stable diabetes is more rapid after streptozotocin than after alloxan. Because of this rapidity and also because of the reliable induction and the stability of the diabetes, streptozotocin-induced diabetes has become the most commonly used diabetes model in islet transplantation studies. Furthermore, since the induced diabetes is stable, provided the dose of streptozotocin is sufficiently high, it is possible to perform islet transplantation already the day after administration of the drug.

Other types of chemically-induced diabetes models have also been described. For example, a multi-insult diabetes model has been developed in the rat.[38] The idea underlying such a model is that type 1 diabetes in humans is caused by the cooperation of several factors, such as, for example, virus infection in conjunction with immune reaction. Therefore, a diabetes model could include several different initiators. In the multi-insult diabetes model, streptozotocin and complete Freund's adjuvant are administered together to rats at low dose levels. At these dose levels, neither of the substances is diabetogenic when given alone, yet a long-standing diabetes is developed when they are given together.[38,39] Models resembling non-insulin-dependent diabetes mellitus have also been developed, both in rats[40,41] and mice.[42] These models rely on streptozotocin given neonatally during the first week of life. At two to six weeks after the injection, a type of diabetes, resembling type 2 diabetes, develops. However, these diabetes models have not been used in islet transplantation research.

Thus, several different types of experimental diabetes exist with different characteristics. In islet transplantation research, however, the vast majority of studies have used diabetes induced by total pancreatectomy in large animals and diabetes induced by high dose levels of either alloxan or streptozotocin in small animals.

Islet Isolation, Preservation and Storage

A most important topic in islet transplantation research is the development of techniques allowing the isolation of a sufficiently large number of pure islets. This topic includes the development of isolation techniques and of techniques for the control of islet quality and for storage of islets.

Isolation Techniques

Islet isolation techniques rely on either microdissection or collagenase digestion of the pancreas. Islets isolated by microdissection have been demonstrated to exhibit a normal function.[43] However, this technique is too slow for allowing isolation of islets on such a large scale basis which is required for transplantation. Therefore, isolation of islets by the collagenase digestion technique, which was originally described in the 1960s,[1,2] has instead become the standard method. Collagenase is produced by *clostridium histolyticum* and is commercially available. The preparations available on the market are mixtures of several different enzymes and the quality of different batches of collagenase differs.[44] Therefore, it is important to characterize and test each batch of collagenase. The optimal collagenase preparation for islet isolation has not yet been established. With the collagenase isolation technique, the enzymes in the preparation attack the islet-acinar interface. Thereafter, a dissociation of the islets from the surrounding exocrine pancreatic tissue is performed. The technique is tricky and may be hazardous. Thus, if the attack of collagenase is too strong, by the use of too much collagenase or by allowing the digestion to last too long, the islets will be damaged. Conversely, if the attack is too weak, the dissociation of the islets from the exocrine tissue becomes difficult.

For an optimal result, the collagenase digestion technique includes several steps. The initial collagenase digestion of the tissue is performed to a moderate degree and is followed by gentle dissociation of the islets from the surrounding tissue. Thereafter, a further purification step, to ensure a high purity yield, is often performed. Usually, the pancreas is initially distended by injection of saline, Hank's balanced salts solution or a collagenase-containing solution into the pancreatic duct or into the pancreatic parenchyma. This procedure optimizes the possibility of the enzyme reaching the islet-acinar space and has been shown to be a crucial step.[45] Initial chopping of the tissue in the collagenase-containing solution may also be performed. In any case, after removal of the organ, the tissue is chopped and shaken in a collagenase-containing solution for a standardized period of time, usually between five and twenty minutes. Finally, the islets are separated from the exocrine tissue, either individually by hand picking or by a density gradient centrifugation technique. The standard isolation technique with hand picking of the islets under a stereomicroscope results in isolation of a comparatively low number of islets, approximately 200–250 per rat pancreas.[46] The advantage of this technique is that the isolated islets are essentially pure, which is of importance for a good transplantation result.[47] In contrast, the separation of islets from the exocrine tissue by gradient centrifugation results in the isolation of a higher number of islets. For example, by separation through centrifugation in Ficoll gel, approximately 350–400 islets per rat pancreas are usually isolated, and this number was in one study increased to approximately 700 islets per rat pancreas by the use of a gradient of bovine serum albumin.[48] The use of gradient centrifugation techniques is also easier than hand picking and although the purity of the resulting islet yield is not as good as with hand picking, several studies have shown that islets isolated by this technique show normal morphological and functional characteristics and may be used with good result in transplantation.[48–52] Therefore, gradient centrifugation techniques are the most commonly used isolation and purification techniques today.

Other techniques have also been developed for separation of the islets from the surrounding exocrine tissue after collagenase digestion. These methods include magnetic separation of immunomicrospheric-attached exocrine tissue, anti-acinar cell monoclonal antibodies or fluorescence-activated sorting.[53–57] The experience with these techniques is still, however, limited.

In large animals, a huge number of islets is necessary for transplantation. Furthermore, it has been found that each species is encumbered with separate and specific problems connected with islet isolation. Therefore, different alterations have been introduced in the technique of isolating islets in these species. In the *pig pancreas*, the experience is that the islets are fragile and that they rapidly dissociate into single cells during the isolation procedure.[51] A special pig islet isolation technique has therefore been developed. It uses a mechanical method for disrupting the pancreas. Collagenase is initially injected into the pancreatic duct and the tissue is thereafter mechanically macerated with the use of a special motor-driven device in a standardized manner. Thereafter, separation is performed with the Ficoll density gradient centrifugation technique. With this technique, it was shown that a remarkably high number of islets (approximately 79,000 per pig pancreas) was isolated and the purity was good (approximately 90-95%). Furthermore, the islets isolated by this technique have been shown to be morphologically and functionally normal.[51] Some authors have used filtration of collagenase digested pancreas through filters before applying to the Ficoll density gradient centrifugation when isolating pig islets.[58] Another technique for isolation of pig islets uses three different enzymes in combination (collagenase, elastase and deoxyribonuclease) in a solution that is continuously pumped through the pancreatic duct.[59] The solution is seeped through the organ, thereby attacking the tissue. Following a forty minute digestion in this manner, a Ficoll density gradient centrifugation is performed. The islet yield with the use of this technique has been shown to be good (approximately 11,000 per g pancreas), and the islets have been shown to exhibit normal function upon subsequent *in vitro* studies.[59]

In the *dog pancreas*, a well controlled study on three different isolation techniques has been performed.[19] The first technique involves digestion of saline-distended pancreas with collagenase for twenty minutes with a separation based on washing and centrifugation.[60] In the second technique, intraductal infusion of collagenase for twenty three to forty eight minutes is performed, followed by sieving of the tissue through a 400 μm mesh filter and a second digestion of undigested remnant tissue with trypsin.[61] Finally, the third isolation technique uses a tissue fastener (''velcro'') to trap fibrous tissue in the pancreas allowing ''release'' of islets during collagenase digestion.[62] By comparing these three techniques, Griffin *et al.* showed the second technique to be superior in isolating islets for a subsequent transplantation.[19] However, other techniques have also been efficient. Thus, Warnock *et al.* isolated $> 100,000$ pure islets per dog pancreas by perfusing the pancreas with collagenase followed by gentle dissociation and a subsequent Ficoll gradient centrifugation.[63]

Isolation of islets from the *human pancreas* has been found to be very difficult, largely because the human pancreas contains a compact dense connective tissue stroma uniting the islets with the acini.[64] However, injection of collagenase in the pancreatic duct, followed by incubation and separation on a Ficoll gradient centrifugation technique, has been found to result in a good number of isolated human islets.[65] Warnock *et al.* have found that the initial delivery of collagenase to the islet-acinar interface is a critical step.[52] They therefore exercise care to avoid disruption of the pancreatic capsule during excision to allow maximal distention with collagenase and they insert several cannulas for administration of collagenase by performing several cutdowns on the duct. The technique has been further improved by performing whole pancreas perfusion with collagenase solution for thirty five minutes in combination with a slow warming before dissociation and Ficoll purification.[66,67] Thus, initial intraductal collagenase administration followed by gentle dissociation and purification with Ficoll density gradient centrifugation seems to be the optimal technique for islet isolation from the human pancreas. The dissociation of the collagenase-digested tissue may be achieved by teasing, trituration or shaking. Which technique to use seems less important, as long as the digest is regularly checked under the microscope during the procedure.[52] Approximately 250,000 to 350,000 islets have been isolated per human pancreas by the use of this approach, and the isolated islets have been shown to be of a high purity, to exhibit biphasic glucose-stimulated insulin secretion when studied *in vitro* and to reverse streptozotocin-induced diabetes when transplanted in the rat.[50,52]

The interest in fetal islets for clinical transplantation has stimulated studies on isolation also of such islets. A problem has thereby been the low number of islets obtained from a fetal pancreas. However, it has been demonstrated that fetal islets are enriched in culture. Thus, in one study, the digest obtained after dispersion of fetal pancreatic tissue with collagenase was incubated in a culture medium, and it was demonstrated that after forty eight hours, the resulting tissue consisted of almost pure islets.[68]

To conclude, modifications of the collagenase digestion technique and various separation techniques are the basis of the studies undertaken so far to increase the yield of highly purified islets from the pancreas of large animals, including man. This development is of importance, since isolating a sufficient number of islets remains one of the obstacles for successful clinical transplantation and this development thereby represents a great challenge for the future.[69]

Quality Control

Although the number of islets is of great importance in isolation techniques, good quality of the isolated islets is another extremely important goal. Therefore, it is necessary to develop reliable and reproducible means for assessing the islet quality. The optimal quality control is to verify the function of the isolated islets *in vivo* following transplantation.[49] This technique is, however, complicated and cannot be used in each case after isolation. Therefore, the quality control of an isolation procedure relies mainly on different *in vitro* techniques.[70,71] Following isolation, the *number* of islets and the purity should be assessed. This may be performed by means of histology or by measuring the tissue content of amylase, preferable in relation to the tissue content of insulin. Morphologically, dithizone staining of the islets[72] or immunocytochemistry may be used for islet identification, whereas electron microscopy has been shown to be difficult in this regard.[71] For controlling the *viability* and function of the isolated islets, neutral red inclusion or trypan blue exclusion may be used but these techniques have been stated to be of limited value.[71] Instead, the combination of inclusion and exclusion dyes (such as propidium iodide with either acridine orange or fluorescein diacetate) has been suggested to be of interest for testing islet tissue viability.[73,74] Functional *in vitro* tests for isolated islets aimed at transplantation include insulin secretion, proinsulin synthesis, glucose utilization and oxygen consumption.[71] Such tests should be used for the characterization of a specific isolation technique. Also within this field, new techniques are continuously developed, like ^{31}P NMR spectroscopy of isolated islets, which has been found to distinguish normal from damaged islets.[75] However, it has been stated that these tests "do not necessarily predict islet performance after implantation",[71] which, again, stresses that the main test on viablity is the function after transplantation.

Storage

The difficulties in obtaining a sufficiently large number of islets from each donor remains one of the major challenges for successful clinical transplantation.[69] Therefore, different modes of storage of islets have been developed. If such storage proves successful, islet isolation for one transplantation can be prolonged over time and thereby include several donors. One mode of storage is simply culturing the islets, which has been demonstrated to be successful, also for human islets, for several months with retained islet viability and function.[76] Another approach is storing the islets at low temperature.[77] It has been demonstrated that cold storage in University of Wisconsin (UW) solution is possible with preserved function for up to 3–5 days, whereas culturing is required for storage of islets for 10-14 days.[78] The most interesting recent development is banking of cryopreserved islets.[78,79] For example, it has been shown that isolated human islets are morphologically and functionally intact after cryopreservation at $-196°C$.[80] In addition, cryopreserved islets have also been suggested to express a weaker immunogenicity than normal islets.[81] Also isolated pancreatic endocrine cells are possible to cryopreserve and they work properly when transplanted to diabetic rats.[82] However, a problem associated with cryopreservation is loss of viable cells during the process, particularly during the thawing phase.[78,83] Nevertheless, a combination of improvements of the islet isolation and islet banking techniques might in the future allow availability of a sufficiently large number of islets also for clinical transplantation.

Transplantation Site, Islet Number, and Timing

Important topics in islet transplantation research are to define the optimal site for placing the graft and to establish the lowest number of islets that is sufficient for correcting the diabetes. Finally, for experimental islet transplantation research, it is also of importance to define when the transplantation may be performed in relation to the induction of the experimental diabetes.

Transplantation Site

The optimal site for islet transplantation should offer an optimal engraftment with long-term graft function. It should also be practically easy to access with minor surgical complications, and it should be immunologically privileged. However, no site is completely privileged, i.e. protects the islets from rejection, and surgical complications may be seen following transplantation to all possible sites. In the rat, it has been demonstrated that intramuscular or subcutaneous placing of the graft is surgically easy but it does not offer reliable long-term results.[52] Furthermore, placing the islets in the portal vein has been shown to be efficient and superior to intraperitoneal administration.[84] On the other hand, intraportal graft placing has been accompanied by serious complications, such as portal hypertension,[52,85] although recent technical development has reduced the risk of this complication.[66] Compared to these sites, improvement of the transplantation result has been demonstrated by implanting the islet graft in well-vascularized organs like the spleen[86] or the liver.[87] Technically, it is easy to implant islets intraperitoneally or intraportally, into which approximately 1, 000 islets suspended in 0.1-0.3 ml of medium may be injected under direct vision.[88] When implanting in the liver, the suspended islets may be injected directly into the right liver lobe parenchyma after clamping of the portal branches to the median, left and caudate lobes.[89] Alternatively, a mesenteric ileal vein[82] or the umbilical vein[90] may be used for catheterization and administration of suspended islets. When transplanting to the spleen, a technique involving isolation of three to five tributaries of the splenic vein has been described.[91] These tributaries are catheterized individually and islets injected, whereafter the veins are ligated to avoid the transplanted islets leaving the spleen. It should be emphasized here that it is important not to place the graft in the splenic pulp, in which engraftment is delayed.[52] By directly comparing the hepatic and the splenic sites for autotransplantation of islets in dogs, the splenic site was found to be superior.[92] This might be due to exhaustion of insulin secretion or graft fibrosis after placing the graft in the liver, or, alternatively, to local trophic substances stimulating the engraftment in the spleen.

Immunologically privileged sites are the testis,[93] the cerebral ventricles[94] and the thymus,[95] although the experience of islet transplantation to these areas is still limited. Several studies have shown that also the kidney subcapsule is immunologically privileged,[52,96] although this is at present not established since other studies have not supported this conclusion.[49,90] Nevertheless, islet transplantation to the kidney subcapsule has been used by several authors with good result,[29,46,97,98] and, in fact, been demonstrated to be superior to transplantation to the portal vein.[99] Technically, a laparatomy is performed and the left kidney capsule is isolated. Several incisions are made in the capsule into which the isolated islets are implanted, usually after suspension in blood. To conclude, transplantation of islets seems to be possible to perform to the spleen, kidney subcapsule, liver, intraportally and intraperitoneally as well as to more immunologically privileged sites. Varying degrees of success have been reported with the use of all these transplant sites. The optimal transplant site remains, however, to be finally established.

Number of Islets

It is important to define the number of islets needed for a successful transplantation. Experimentally, it has been demonstrated that the required number of islets depends on the pretransplant status of the islets, the volume and the place of grafting the islets, the immunologic barrier between donor and recipient, and the immune treatment of the recipient. In streptozotocin diabetic rats, approximately 1,000 islets seem sufficient for correcting the hyperglycemia, as has been demonstrated after transplantation intraportally,[88] in the spleen[86] and in the kidney subcapsule.[100] However, it has also been shown that for complete restoration of glucose metabolism, a larger number of islets is required. Thus, transplanting a higher number of islets (3,000) was in one study shown to normalize not only the glycemia but also the glucose tolerance.[101] In contrast, the use of a low number of islets, 400, has been shown to be insufficient.[100] For islet transplantation in large animals, a huge number of islets seems necessary. Thus, in dogs, it has been demonstrated that autotransplantation of 3,000 to 5,000 islets/kg body mass results in normoglycemia in approximately 60% of the cases, whereas more than 5,000 islets/kg are needed for normoglycemia in all cases.[66] Islet volume is probably a more accurate determination of the islet mass than the number of islets and it has also been calculated that the critical islet mass is, in volume, 4 μl/kg body mass for correction of the diabetic metabolism.[102]

In man, it is known that more than 90% of the pancreas may be removed without the development of diabetes.[103] However, probably much more than 10% of the islet mass is required for the correction of diabetes after islet transplantation since the islet graft probably does not function as properly as endogenous islets. This impaired function is most likely due to poorer viability, altered patterns of vascularization and innervation, and altered site of the islets. It has been calculated that approximately 250,000 islets are needed for the correction of the diabetic state in humans.[52,69] This calculation is based on the clinical experience that when performing pancreas transplantation, approximately 50% of the gland can normalize the hyperglycemia[104] and this fraction of a human pancreas contains approximately 250,000 islets. Furthermore, a critical islet mass of 4 μl/kg body weight corresponds to approximately 200,000 islets.[102] However, no direct study on the needed number of islets in islet transplantation has been performed in man.

Timing of Experimental Islet Transplantation

When performing islet transplantation in pancreatectomized animals, transplantation is usually performed in close connection to the pancreatectomy.[91] In contrast, in the more commonly used chemically-induced diabetes models in small animals, it has been considered of importance not to transplant the islets until a stable diabetes is evident in the individual animal. If the diabetes is not stable, regeneration of the endogenous B-cells remains a possibility, and a good long-term result after islet transplantation might then be due simply to regeneration of the pancreas and not to graft function. Therefore, in alloxan diabetic mice, transplantation is usually performed at six to eight days after administration of the drug,[29] and in streptozotocin-diabetic rats, the transplantation is performed after two to three weeks.[87,88] However, by using streptozotocin at a high dose level ($\geqslant$60 mg/kg), the diabetes induced has been demonstrated to be permanent.[35] Therefore, after administration of streptozotocin at such high dose levels, islet transplantation also during the first few days after diabetes induction is safe, as has been shown in several reports.[3,46,100,105] Similarly, alloxan injected at $\geqslant$70 mg/kg creates permanent diabetes in the mouse.[30] Nevertheless, graftectomy should be undertaken in all models used to verify that it is the islet graft, and not any regenerated B-cells, that correct the diabetic metabolism following transplantation. When normoglycemia depends on the graft, hyperglycemia usually returns within one to two days after graftectomy.[106,107]

Immunology

Besides obtaining a sufficient number of islets for transplantation, a main challenge for islet transplantation research is to prevent the graft rejection. The transplanted islets may undergo rejection because antigens present in the graft initiate an immunologic response. Transplanted islets may also undergo rejection if the graft is attacked by the immune process that afflicts the patient due to its involvement in the pathogenesis of insulin-dependent diabetes. To study this latter immune response, it is necessary to use diabetes models that resemble human insulin-dependent diabetes mellitus, like the BB rat or the NOD mouse. Since most studies so far have been undertaken in chemically-induced diabetes, most knowledge has so far been accumulated on the rejection due to antigens present in the graft.

Rejection

It has been demonstrated that it is mainly passenger lymphoid cells within the islet graft that are responsible for the rejection.[108] These cells, in contrast to the endocrine cells, express under normal conditions the major histocompatibility (MHC) antigen class II.[109] If the recipient is immunized against these antigens, rejection will evolve. Such immunization includes activation of cytotoxic T cells as a response to the MHC class II antigens.[110] When this activation is initiated, lymphokines, like gamma-interferon, interleukins and tumor necrosis factor, are released and these will enhance the expression of the antigens and mediate B-cell cytotoxicity,[111,112] which amplifies the rejection. Means to inhibit the rejection may include the selection of an immunologically privileged site for the transplantation, like the thymus,[95] or the use of several different donors,[113] which both might avoid any rejection being initiated despite the existence of foreign antigens in the graft. The idea behind the first approach is that in a completely immunologically privileged site, foreign antigens will not be recognized as foreign.[95] The idea behind the second approach is that grafts consisting of islets from several donors initiate several different "minor" rejection reactions, but none of these reach the "threshold" for a clinically relevant rejection.[113] However, this approach has not gained general acceptance. Thus, the hypothesis was supported by the result in one study,[113] whereas it was not supported by another study.[114].

Two other means of reducing rejection are treating the graft and treating the recipient. Reduction of expression of MHC class II antigens in the transplanted islets or removal of the antigen-expressing cells, will reduce the initiation of the immune response in the recipient after transplantation. If this succeeds, the need for immunosuppression of the recipient would be diminished or eliminated, which in turn will be of great benefit for clinical islet transplantation. It is performed by pretreating the graft *in vitro* prior to transplantation. Finally, if needed, immunosuppression of the recipient will inhibit the immune-response, thereby reducing the risk of rejection.

Islet Pretreatment

Since islet MHC class II antigens are responsible for the immune reaction, islet pretreatment should aim at reducing the expression of these antigens. Furthermore, since these antigens are mainly expressed on the passenger lymphoid cells and not on the endocrine cells,[115] islet pretreatment could also aim at removing such cells from the islet graft. Non-specific modes to reduce the immunogenicity of the transplanted islets are gamma irradiation[115] and ultraviolet light irradiation.[104,116] By irradiation, the expression of the graft MHC class II antigens is reduced. Gamma irradiation is usually performed at a low dose irradiation (250 rads) and is optimal in combination with immunosuppressive treatment of the recipient.[110] Similarly, ultraviolet light irradiation, usually performed at the dose level of 900 J/m^2, is optimized by concomitant immunosuppressive treatment of the recipient.[114,116]

An alternative islet pretreatment modality is immunoisolation.[117–121] With this technique, the islet graft is protected from entry by host antibodies and lymphoid cells and rejection is thus prevented. Immunoisolation of the islets can be performed either by diffusion chambers or by microencapsulation. In the diffusion chambers, the islet tissue is enclosed between membranes which are attached to a plastic support ring with glue.[118] Two different types of diffusion chamber exist. In the Millipore[R] chambers, the membranes are placed outside the rings.[118] This creates a large volume for the graft. However, the experience with this type of chamber has shown that fibrous overgrowth of the graft evolves, and, furthermore, there is a considerable diffusion time within the chamber. Therefore, the chambers were modified by Algire *et al.* who placed the membranes close together.[118,122] It was demonstrated that the immune response after transplantation of islets within these chambers was delayed due to prevention of direct contact between recipient cells and the graft.[121] This technique was further improved by using intravascularly placed devices in which blood was made to flow through the chamber with the graft enclosed behind a membrane.[118,123] Nevertheless, with diffusion chambers, the transplant results have been poor[119], and the result seems to be superior with the use of microencapsulated islets.[117–120]

The most commonly used microencapsulation technique uses a semi-permeable sodium alginate-polylysine-sodium alginate membrane, in which the islets are encapsulated before transplantation.[117,118] With this technique, islets are suspended in sodium alginate and the solution is sprayed into $CaCl_2$ to form microspheres. The spheres are then soaked first in polylysine and then, after washing, in sodium alginate. This forms sodium alginate-polylysine-sodium alginate microencapsulated islets. Alginate within the capsules is removed by dissolving in sodium citrate and then, after culturing, the capsules are transplanted. Microencapsules may contain several islets each or they may contain only one islet per capsule. The capsules allow passage of glucose and insulin, whereas larger molecules like antibodies and also lymphoid cells cannot pass. Thus, the islets within a microcapsule are protected from rejection but have the opportunity of functioning normally. The advantage of microcapsules over diffusion chambers involves the possibility of performing transplantation to multiple sites and the ease with which the transplantation is performed. Furthermore, small microcapsules have a small diffusion lag. By intraperitoneal transplantation of microencapsulated islets, some encouraging results have been reported.[118–121] Recently it was also demonstrated that xenotransplantation of porcine and bovine islets to non-immunosuppressed streptozotocin-diabetic mice was possible with the use of islets placed in alginate microspheres.[124] However, improvements in this technique have still to be performed, particularly since the capsules are unstable over a long-term period of time[120] and since the alginate-polylysine-alginate capsule by itself seems to inhibit glucose-stimulated insulin secretion.[126]

Islet immunoisolation in diffusion chambers or in microcapsules represents non-specific immunomodulating treatment. Recently, more specific immunomodulating treatment modes for islets before transplantation have been developed. These include incubation of the islets with antibodies to MHC class II antigens,[127] treatment with antibodies to dendritic cells,[128] and pretreatment with transforming growth factor β.[129] Though all these treatment modalities have given encouraging results, this interesting field is still only in its infancy. Most likely, when knowledge of the exact immune mechanism underlying the rejection is increased, more specific and efficient immune-modulating treatment modes will be developed. When optimized, these procedures will make islet transplantation possible without any immunosuppression of the recipient.

Immunosuppressive Treatment of Recipient

Until the risk of rejection is completely eliminated by graft treatment, it is necessary to use immunosuppressive treatment in the recipient. This is a drawback for the clinical application of islet transplantation since, even if successful, the transplantation will turn the patient from being insulin dependent to being dependent on immunosuppressants. Furthermore, most immunosuppressive agents have deleterious effects on islet function. The most commonly used immunosuppressive agent is cyclosporine A, which is a fungus metabolite with antilymphocytic activitiy.[130,131] The drug reduces the activation of cytotoxic T cells and their production and release of lymphokines.[132]. Cyclosporine A has been found to prolong graft survival after allogenic islet transplantation in rats.[133]

However, cyclosporine A has serious drawbacks when used in the recipient after islet transplantation. Thus, the drug has been shown to inhibit insulin secretion both *in vivo*, as reported for example in the dog,[134] and *in vitro* in isolated islets.[135] Furthermore, cyclosporine A has been demonstrated to inhibit islet cell proliferation and differentiation[136] and islet vascularization,[137] effects which are deleterious after transplantation. Hence, the use of cyclosporine A in immunosuppression of the recipient after islet transplantation seems to have serious negative effects. However, other immunosuppressive agents also inhibit insulin secretion, at least at high dose levels, like azathioprine[138] and the newly developed FK 506.[139,140] Finally, corticosteroids, which have also been used as immunosuppressive agents, are known to induce insulin resistance. Other modes to treat or prevent the rejection have also been developed, like the use of a low dose of cyclosporine in combination with antilymphocyte globulin[141] and the use of a specific interleukin 2-receptor antagonist.[142] However, the development of new, islet inert, substances as immunosuppressants is of importance. One such recent development is a chimeric interleukin 2 diphteria toxin fusion protein, which specifially affects *T* cells, and which has been shown to prolong islet graft survival in mice when given once weekly to the recipient.[143]

Effects of Islet Transplantation

Effects on Metabolism

Numerous studies have been performed demonstrating that islet transplantation improves and normalizes the metabolic disturbances seen in diabetes, also on a long-term basis. For example, several studies have shown that diabetic rats and dogs become normoglycemic after islet transplantation[87,91,144,145] and that insulin and glucagon secretion are almost completely normal following islet transplantation, also after several months.[144,146,147] The glucose elimination rate after intravenous glucose challenge has also been shown to be normal or only slightly reduced in islet transplanted animals[52,148–150]. Furthermore, hemoglobin A_{1c} levels have became normal after islet transplantation in diabetic rats,[151] and clinical parameters like body weight, urinary volume and water intake have also normalized.[52,144]. Hence, several studies have shown that islet transplantation corrects the metabolic disturbances in diabetes on a long-term basis. In contrast, a series of reports suggests that this normalization is not lifelong, since after transplantation of 1,500 islets intraportally in diabetic rats, hyperglycemia returned after an initial success.[152,153] However, as discussed in detail previously,[52] this long-term failure might be due to the transplantation of a suboptimal mass of islets to this site. Therefore, most evidence suggest an efficient long-term metabolic result after islet transplantation.

Islet transplantation has also been shown to normalize the abnormalities seen in the liver and the pancreas in diabetic rats.[89,154] Thus, the reduced plasma levels of amylase and the alterations in plasma levels of liver enzymes as well as the morphological changes in the exocrine pancreatic and liver parenchyma accompanying streptozotocin diabetes in the rat have been demonstrated to be normalized by intraportal transplantation of 1,000 islets.[89,154] Islet transplantation in the liver has also been shown to normalize the reduced activity of the liver enzyme, glycogen synthase phosphatase, in streptozotocin diabetes rats.[155] Hence, with few exceptions, islet transplantation seems to normalize all metabolic disturbances that accompany experimental diabetes completely, provided that a sufficient mass of islets is transplanted.

Secondary Complications

An important aspect of islet transplantation is whether secondary complications to diabetes are prevented or reversed. Several experimental studies have shown this to be the case. Thus, both morphological and functional evidence of diabetic nephropathy, retinopathy and neuropathy have been shown to be prevented by islet transplantation in the rat.[156–161] Similar studies in larger animals are still lacking, mainly due to the long period of time that is needed for allowing reliable conclusions. Besides this also carefully undertaken studies on the temporal relation between the transplantation and the reversal of complications as well as on the precise influence of islet transplantation on the pathophysiologic disturbances underlying the secondary complications are required. In fact, recent studies have shown that in the case of established advanced alterations in kidney function in diabetic rats, islet transplantation had only a poor effect.[162] To conclude, several studies in different species and with the use of different diabetes models have shown that islet transplantation improves or prevents the metabolic, morphological and clinical consequences to diabetes, both on a short-term and on a long-term basis, although more studies are required on the long-term effects on complications.

Nevertheless, based on experimental studies, islet transplantation seems to offer an excellent approach to treating diabetes.

Graft Function

A question that has been discussed in some detail is whether the transplanted islets function as normal islets or whether the diabetic and engraftment environment alter their function. It has then to be emphasized that the microenvironment of the islets is not normal after transplantation. This may be of special importance with regard to the microvasculature and the innervation pattern. In normal islets of most species, an afferent vessel reaches the center of an islet, whereas the efferents leave the islets from the periphery,[163] although the microvasculature architecture may differ in different species.[164] Though not established, this pattern may be altered in transplanted islets. Thus, such islets may be nourished by surrounding vessels initially reaching the islet periphery. Furthermore, it is known that normal islets are innervated by adrenergic, cholinergic as well as peptidergic nerves and that these nerves are of great importance for the precise regulation of islet function.[165] Whether transplanted islets become reinnerrated have been the subject of several studies.[166] One study has shown that islets transplanted to the liver do indeed become innervated but careful analysis of the nerves has shown that they primarily innervate the islet vessels and not the islet endocrine cells.[167] Furthermore, in the mouse, nerve fibers have been demonstrated to exist in islets transplanted to the kidney capsule.[168,169] The nerve fibers were visualized at twelve weeks after transplantation in normoglycemic animals and at twenty weeks after transplantation in hyperglycemic animals.[168,169] It has also been shown that the reinnervation of transplanted islets results in a neural distribution within the islets which is different from that in normal islets.[170] However, more important functional studies of examining the precise physiology of transplanted islets are few. Perfused livers containing islet grafts have been found to exhibit normal biphasic insulin secretion after stimulation with glucose[87] and almost normal insulin and glucagon secretory responses to arginine,[171] indicating normal function of the transplanted islets. However, a careful *in vivo* study in the rat has shown that the first phase ("cephalic phase") of insulin secretion after oral intake of glucose is defective after islet transplantation into the portal system in diabetic rats, which suggests a functional (parasympathetic) denervation of the islets.[172] In contrast, in diabetic dogs transplanted with islets into the spleen, an early insulin response into the splenic vein after the intake of a meal was reported,[149] suggesting a normal parasympathetic innervation. The reinnervation of the graft might therefore be dependent on the transplant site or the species. In streptozotocin-diabetic rats transplanted with islets to the kidney subcapsule, the α_2-adrenoceptor antagonist yohimbine was found not to increase plasma insulin levels, as it does in normal rats, which suggests a state of functional (sympathetic) denervation in the transplanted islets.[46] In contrast, a normal islet adaptation to peripheral insulin resistance has been demonstrated in rats after islet transplantation.[46] Finally, the islet response to exercise seems well preserved after islet transplantation in dogs.[173] Hence, whether or not transplanted islets function completely normally is not yet finally established and more studies are required, particularly with regard to the possible reinnervation of the islets after transplantation and its functional consequences.

Whether the metabolic disturbances seen in diabetes adversely affect the transplanted islets has been the matter of a few studies, the results of which seem to suggest this to be the case. Thus, hyperglycemia has been demonstrated to impair the function in transplanted islets,[105,174,175] and the diabetic state seems to impair the neovascularization of the islet graft.[176] This also means that, conversely, by carefully controlling the diabetes in the recipient, the long-term effect of the islet transplantation might be improved. This finding has clinical implications, since if the metabolic treatment of the diabetes is strictly controlled during the peritransplantation period, the number of islets needed for a successful transplantation may be reduced. This has been established and verified experimentally, since the prevention of hyperglycemia in the peri-transplantation period was found to improve the long-term islet function.[105,174,175]

Clinical Islet Transplantation

Since it has been calculated that approximately 250,000 islets would be needed for a successful islet transplantation into a diabetic patient,[69] it is not strange that clinical islet transplantation is still in its infancy. Several reports on the progress of clinical islet transplantations have been published.[177–180] In 1993, it was thereby reported that 167 adult islet transplants had been performed in 25 institutions worldwide.[179] A few recent reports may be mentioned. Sharp *et al.* reported one case with insulin-dependent diabetes transplanted with 800,000 islets into the portal vein who become insulin-independent for twelve days.[181] Furthermore, Warnock *et al.* reported one case transplanted into the liver with 250,000 islets who, however, never became insulin independent.[182] To use a higher number of islets, Warnock *et al.* then proceeded and transplanted a total of 611,000 islets to a type 1 diabetic patient with severe late complications.[90]. The islet transplantation was performed in combination with kidney transplantation. Of the transplanted islets, 243,000 were freshly isolated from the kidney donor, whereas 368,000 had been banked by cryopreservation after isolation from four other donors. The transplantation was performed via the umbilical vein into the liver, and, postoperatively, immunosuppression with antilymphocyte globulin, cyclosporine and azathioprine was performed. The patient became insulin independent, and was reported as such after five months. Finally, in 1992 a report on 5 patients transplanted with islets placed in the liver was published.[183] The patients were transplanted with 110,000 to 412,000 islets and they remained insulin independent for 1 to 7.5 years after transplantation. Insulin secretion was found to be preserved in the islets, and the conclusion was that 265,000 islets seem sufficient or long-term appropriate islet function.[183] Thus, it seems as if islet transplantation might now become a clinical reality, and the next few years will most likely be very productive within this field.

Concluding Remarks and Future Aspects

Experimental islet transplantation has been a large and productive area of research since the 1960s. During this period, a tremendous amount of knowledge on islet isolation and purification techniques, immunology of transplantation, carbohydrate metabolism, islet physiology and diabetes has been generated. In the 1990s, this research area started to pay off with clinical results. It might be recalled that it was experimental islet transplantation that created the basic foundation for the clinical trials.[184] The initial description of collagenase isolation of islets by Moskalewski in 1965[1] and by Lacy and Kostianovsky in 1967[2] represents the historical landmark since this technique allows rapid isolation of islets. The development of automated large-scale isolation techniques and purification steps has further improved the isolation, making it possible to isolate several hundred thousands of islets within one day. Furthermore, the addition of banking techniques, particularly cryopreservation, enables isolation from several donors and therefore transplantation of sufficient amount of islets into diabetic patients.

Improvements, however, are still required, since the used techniques are time-consuming and expensive and cannot be applied to more than extraordinary cases. Furthermore, the immunological processes underlying islet rejection have initiated important studies on possibilities for immune treatment of the islets prior to transplantation and for using an immunosuppressive regimen in the recipient that does not adversely affect islet function. Studies on this topic need an experimental approach with reliable diabetes models, islet isolation techniques and defined transplantation procedures. In addition, the clinically oriented researcher also needs to define the patient to be treated with islet transplantation and whether islet transplantation should be combined with kidney transplantation. In addition, the optimal source of islets for clinical transplantation remains to be studied in more detail, since islets for clinical use may be obtained from living donors, from cadavers, from fetuses and from animals (xenotransplantation). However, the main issues for islet transplantation research in the near future will be to improve and optimize the islet yield after isolation further and to establish means to avoid islet rejection after transplantation.

Acknowledgements

The studies by the authors have been supported by the Swedish Medical Research Council (Grant No 14X-6834), Novo Nordic, Albert Pahlsson, Ernhold Lundström and Crafoord Foundations, and the Faculty of Medicine, Lund University.

References

1. Moskalewski, S. (1965). Isolation and culture of the islets of Langerhans of the guinea pig. *Gen. Comp. Endocrinol.*, **5**, 342–353.
2. Lacy, P. E. and Kostianovsky, M. (1967). Method for the isolation of intact islets of Langerhans from the rat pancreas. *Diabetes*, **16**, 35–39.
3. Ballinger, W. F. and Lacy, P. E. (1972). Transplantation of intact pancreatic islets in rats. *Surgery*, **72**, 175–186.
4. Marliss, E. B., Nakhooda, A. F., Poussier, P. and Sima, A. A. F. (1982). The diabetic syndrome of the "BB" Wistar rat: possible relevance to type 1 (insulin-dependent) diabetes in man. *Diabetologia*, **22**, 225–232.
5. Seemayer, T. A., Tannenbaum, G. S., Goldman, H. Y. and Colle, E. (1982). Dynamic time course studies of the spontaneously diabetic BB Wistar rat. III. Light-microscopic and ultrastructural observations of the pancreatic islets of Langerhans. *Am. J. Pathol.*, **106**, 237–249.
6. Prud'homme, G. J., Fuks, A., Colle, E. and Guttmann, R. D. (1984). Isolation of T-lymphocyte lines with specificity for islet cell antigens from spontaneously diabetic (insulin-dependent) rats. *Diabetes*, **33**, 801–803.
7. Baudon, M. A., Ferré, P., Pénicaud, L., Maulard, P., Ktorza, A., Castano, L. and Girard, J. (1989). Normal insulin sensitivity during the phase of glucose intolerance but insulin resistance at the onset of diabetes in the spontaneously diabetic BB rat. *Diabetologia*, **32**, 839–844.
8. Svenningsen, A., Dyrberg, T., Markholst, H., Binder, C. and Lernmark, Å. (1986). Insulin release and pancreatic insulin is reduced in young prediabetic BB rats. *Acta. Endocrinol.*, **112**, 367–371.

9. Baekkeskov, S., Dyrberg, T. and Lernmark, Å. (1984). Autoantibodies to a 64-kilodalton islet cell protein precede the onset of spontaneous diabetes in the BB rat. *Science*, **224**, 1348–1350.
10. Dyrberg, T., Poussier, P., Nakhoda, F., Marliss, E.B. and Lernmark, Å. (1984). Islet cell surface and lymphocytic antibodies often precede the spontaneous diabetes in the BB rat. *Diabetologia*, **26**, 159–165.
11. Baekkeskov, S., Aanstoot, H.J., Christgau, S., Reetz, A., Solimena, M., Cascalho, M., Folli, F., Richter-Olesen, H. and Camilli, P. D. (1990). Identification of the 64K autoantigen in insulin-dependent diabetes as the GABA-synthesizing enzyme glutamic acid decarboxylase. *Nature*, **347**, 151–156.
12. Clare-Salzler, M. J., Tobin, A. J. and Kaufman, D. L. (1992). Glutamic decarboxylase: an autoantigen in IDDM. *Diabetes Care*, **15**, 132–135.
13. Hegre, O. D., Enriquez, A. J., Kethcum, R. J., Weinhaus, A. J. and Serie, J. R. (1989). Islet transplantation in spontaneously diabetic BB(Wor) rats. *Diabetes*, **38**, 1148–1154.
14. Makino, S., Kunimoto, K., Muraoka, Y., Mizushima, Y., Katagiri, K. and Tochino, Y. (1980). Breeding of a non-obese, diabetic strain of mice. *Exp. Anim.*, **29**, 1–13.
15. Kano, Y., Kanatsuna, T., Nakamura, N., Kitagawa, Y., Mori, H., Kajiyama, S., Nakano, K. and Kondo, M. (1986). Defect of the first-phase insulin secretion to glucose stimulation in the perfused pancreas of the nonobese diabetic (NOD) mouse. *Diabetes*, **35**, 486–490.
16. Goto, Y. and Kakizaki, M. (1981). The spontaneous-diabetes rat: a model of noninsulin dependent mellitus. *Proc. Jpn. Acad.*, **57**, 381–384.
17. Östenson, C. G., Khan, A., Abdel-Halim, S. M., Guenifi, A., Suzuku, K., Goto, Y. and Efendic, S. (1993). Abnormal insulin secretion and glucose metabolism in pancreatic islets from the spontaneously diabetic GK rat. *Diabetologia*, **36**, 3–8.
18. Minkowski, O. (1989). Historical development of the theory of pancreatic diabetes. *Diabetes*, **38**, 1–6.
19. Griffin, S. M., Alderson, D., Farndon, J. R. (1986). Comparison of harvesting methods for islet transplantation. *Br. J. Surg.*, **73**, 712–715.
20. Leahy, J. L., Bonner-Weir, S. and Weir, G. C. (1984). Abnormal glucose regulation of insulin secretion in models of reduced B-cell mass. *Diabetes*, **33**, 667–674.
21. Rerup, C. C. (1970). Drugs producing diabetes through damage of the insulin secreting cells. *Pharmacol. Rev.*, **22**, 485–518.
22. Dunn, J. S., Sheehan, H. L. and McLetchie, N. G. B. (1943). Necrosis of islets of Langerhans produced experimentally. *Lancet*, **1**, 484–487.
23. Hellman, B. and Diderholm, H. (1955). The diabetogenic effect of alloxan after elimination of extrapancreatic factors. *Acta. Endocrinol.*, **20**, 81–87.
24. Malaisse, W. J. (1982). Alloxan toxicity to the pancreatic B-cell. *Biochem. Pharmacol.*, **31**, 3527–3534.
25. Yamamoto, H., Uchigata, Y. and Okamoto, H. (1981). Streptozotocin and alloxan induce DNA strand breaks and poly (ADP-ribose) synthetase in pancreatic islets. *Nature*, **294**, 284–285.
26. Lenzen, S. and Panten, U. (1988). Alloxan: history and mechanism of action. *Diabetologia*, **31**, 337–342.
27. Lenzen, S. and Panten, U. (1988). Signal recognition by pancreatic B-cells. *Biochem. Pharmacol.*, **37**, 371–378.
28. Lundquist, I. and Rerup, C. (1967). On the development of alloxan diabetes in mice. *Eur. J. Pharmacol.*, **2**, 35–41.
29. Korsgren, O., Jansson, L., Sandler, S. and Andersson, A. (1990). Hyperglycemia-induced B cell toxicity. The fate of pancreatic islets transplanted into diabetic mice is dependent on their genetic background. *J. Clin. Invest.*, **86**, 2161–2168.
30. Ahrén, B. and Sundkvist, G. (1995). Long-term effects of alloxan in mice. *Int. J. Pancreatol.*, **17**, 197–202.
31. Ahrén, B., Sundkvist, G., Mulder H. and Sundler, F. (1996). Blockade of muscarinic transmission increases the frequency of diabetes after low-dose alloxan challenge in the mouse. *Diabetologia*, **39**, 383–390.
32. Evans, J.S., Gerritsen, G.C., Mann, K.M. and Owen, S. P. (1965). Antitumor and hyperglycemic activity of streptozotocin (NSC-37917) and its cofactor, U-15,774. *Canc. Chemother. Rep.*, **48**, 1–6.
33. Rakieten, N., Rakieten, M.L. and Nadkami, V. (1963). Studies on the diabetogenic action of streptozotocin (NSC-37919) *Canc. Chemother. Rep.*, **29**, 91–98.
34. Junod, A., Lambert, A. E., Strauffacher, W. and Renold, A. E. (1969). Diabetogenic action of streptozotocin: Relationship of dose to metabolic response. *J. Clin. Invest.*, **48**, 2129–2139.
35. Ar'Rajab, A. and Ahrén, B. (1993). Long-term diabetogenic effect of streptozotocin in rats. *Pancreas*, **8**, 50–57.
36. Reckard, C. R., Ziegler, M. M. and Barker, C. F. (1973). Physiological and immunological consequences of transplanting isolated pancreatic islets. *Surgery*, **74**, 91–99.
37. Junod, A., Lambert, A. E., Orci, L., Pictet, R., Gonet, A. E. and Renold, A. E. (1967). Studies on the diabetogenic action of streptozotocin. *Proc. Soc. Biol. Exptl. Biol. Med.*, **126**, 201–205.
38. Ziegler, M., Ziegler, B., Hehmke, B., Dietz, H., Hildmann, W. and Kauert, C. (1984). Autoimmune response directed to pancreatic beta cells in rats induced by combined treatment with low doses of streptozotocin and complete Freund's adjuvant. *Biomed. Biochim. Acta.*, **43**, 675–681.
39. Kohnert, K. D., Ziegler, B., Fält, K., Odselius, R., Ziegler, M. and Falkmer, S. (1987). Biochemical, immunohistochemical and ultrastructural studies of the alteration of pancreatic beta cells resulting from the combined effects of complete Freund's adjuvant and non-diabetogenic doses of streptozotocin. *Exp. Clin. Endocrinol.*, **89**, 259–268.
40. Weir, G. C., Clore, E. T., Zmachinski, C. J. and Bonner-Weir, S. (1984). Islet secretion in a new experimental model for non-insulin-dependent diabetes. *Diabetes*, **33**, 667–673.
41. Giroix, M. H., Portha, B., Kergoat, M., Bailbe, D. and Picon, L. (1983). Glucose insensitivity and amino acid hypersensitivity of insulin release in rats with non-

insulin-dependent diabetes: a study with the perfused pancreas. *Diabetes*, **32**, 445–451.
42. Ahrén, B. and Skoglund, G. (1989). Insulin secretion in neonatally streptozotocin-injected mice. *Diabet. Res.*, **11**, 185–190.
43. Hellerström, C. (1964). A method for the microdissection of intact pancreatic islets of mammals. *Acta. Endocrinol.*, **45**, 122–131.
44. Wolters, G. H. J., van Suylichen, P. T. R., van Deijnen, J. H. M. and van Schilfgaarde, R. (1990). Factors influencing the isolation process of islets of Langerhans. *Horm. Metab. Res.*, suppl., **25**, 20–26.
45. Gotoh, M., Ohzato, H., Porter, J., Maki, T. and Monaco, A.P. (1990). Crucial role of pancreatic ductal collagenase injection for isolation of pancreatic islets. Horm. Metab. *Res.*, suppl., **25**,10–16.
46. Ar'Rajab, A. and Ahrén, B. (1991). Effects of yohimbine and nicotinic acid on insulin secretion in islet transplanted streptozotocin-diabetic rats. *Diab. Res. Clin. Pract.*, **11**, 81–88.
47. Gray, D. W. R., Sutton, R., McShane, P., Peters, M. and Morris, P. J. (1986). Exocrine contamination adversely affects the implantation of rat islets transplanted to the kidney capsule site. *Transpl. Proc.*, **18**, 1823–1824.
48. Lake, S. P., James, R. F. L., Anderson, J. L., Chamberlain, J., Gardner, S. and Bell, P. R. F. (1986). An improved isolation method for rat pancreatic islets. *Transpl. Proc.*, **6**, 1817–1818.
49. Lake, S. P., Chamberlain, J., Husken, P., Bell, P. R. F. and James, R. F. L. (1988). *In vivo* assessment of isolated pancreatic islet viability using the streptozotocin-induced diabetic nude rat. *Diabetologia*, **31**, 390–394.
50. Warnock, G. L., Ellis, D., Rajotte, R. V., Dawidson, I., Baekkeskov, S. and Egebjerg, J. (1988). Studies of the isolation and viability of human islets of Langerhans. *Transplantation*, **45**, 957–963.
51. Ricordi, C., Finke, E. H. and Lacy, P. E. (1986). A method for the mass isolation of islets from the adult pig pancreas. *Diabetes*, **35**, 649–653.
52. Warnock, G. L., Kneteman, N. M. and Rajotte, R. V. (1991). Islet transplantation. Experimental results and evolution of clinical trials. In *The Endocrine Pancreas*, edited by E. Samols, pp. 487–517. New York: Raven Press.
53. Winoto-Morbach, S., Ulrichs, K., Layhausen, G. and Muller-Rucholtz, W. (1989). New principle for large-scale preparation of purified human pancreas islets. *Diabetes*, **38** (suppl., 1), 146–149.
54. Gray, D. W., Gohde, W., Carter, N., Heiden, T. and Morris, P. J. (1989). Separation of pancreatic islets by fluorescence-activated sorting. *Diabetes*, **38** (suppl., 1), 133–135.
55. Soon-Shiong, P., Heintz, R., Fujioka, T., Terasaki, P., Merideth, N. and Lanza, R. P. (1990). Utilization of anti-acinar cell monoclonal antibodies in the purification of rat and canine islets. *Horm. Metab. Res.*, suppl., **25**, 45–50.
56. Soon-Shiong, P., Fujioka, T., Terasaki, P., Heintz, R. and Lanza, R. P. (1990). Islet purification by a novel immunomicrosphere cell depletion technique. *Transpl. Proc.*, **22**, 780–781.
57. v Specht, B. U., Finke, M., Eckhardt, A. and Permanetter, W. (1990). Large-scale purification of pig islets of Langerhans by fluorescence activated cell sorting. *Horm. Metab. Res.*, suppl., **25**, 54–57.
58. Crowther, N. J., Gottfredsen, C. F., Moody, A. J. and Green, I. C. (1989). Porcine islet isolation, cellular composition and secretory response. *Horm. Metab. Res.*, **21**, 590–595.
59. Calafiore, R., Calcinaro, F., Basta, G., Pietropaolo, M., Falorni, A., Piermattei, M. and Brunetti, P. (1990). A method for the massive separation of highly purified, adult porcine islets of Langerhans. *Metabolism*, **39**, 175–181.
60. Alderson, D., Walsh, T. N. and Farndon, J. R. (1984). Islet cell transplantation in diabetic dogs: studies of graft function and storage. *Br. J. Surg.*, **71**, 755–760.
61. Horaguchi, A. and Merrell, R. C. (1981). Preparation of viable islet cells from dogs by a new method. *Diabetes*, **30**, 455–457.
62. Lacy, P. E., Lacy, E. T., Finke, E. H. and Yasunami, Y. (1982). An improved method for the isolation of islets from the beef pancreas. *Diabetes*, **31** (suppl., 4), 109–111.
63. Warnock, G. L., Cattral, M. S., Evans, M. G., Kneteman, N. M. and Rajotte, R. V. (1989). Mass isolation of pure canine islets. *Transpl. Proc.*, **21**, 3371–3372.
64. Sharp, D. W., Downing, R., Merrell, R. C. and Greider, M. (1980). Isolating the elusive islet. *Diabetes*, **29** (suppl. 1), 19–30.
65. Gray, D. W. R., McShane, P., Grant, A. and Morris, P. J. (1984). A method for isolation of islets of Langerhans from the human pancreas. *Diabetes*, **33**, 1055–1061.
66. Warnock, G. L., Rajotte, R. V. and Procyshyn, A. W. (1983). Normoglycemia after reflux of islet-containing pancreatic fragments into the splenic vascular bed in dogs. *Diabetes*, **32**, 452–459.
67. Warnock, G. L., Kneteman, N. M., Evans, M. G. and Rajotte, R. V. (1990). Isolation of purified large mammal and human islets of Langerhans. *Horm. Metab. Res.*, suppl., **25**, 37–44.
68. Farkas, G., Lazar, G. and Herczegh, J. (1990). The long-term cultivation and cryopreservation of human fetal pancreatic tissue. *Horm. Metab. Res.*, suppl., **25**, 64–68.
69. Weir, G. C., Bonner-Weir, S. and Leahy, J. L. (1990). Islet mass and function in diabetes and transplantation. *Diabetes*, **39**, 401–405.
70. Sharp, D. W., Alderson, D. and Kneteman, N. M. (1986). Optimization of islet preparations for clinical transplantation. *Transpl. Proc.*, **18**, 1814–1816.
71. Ricordi, C., Gray, D. W. R., Hering, B. J., Kaufman, D. B., Warnock, G. L., Kneteman, M. N., Lake, S. P., London, N. J. and Alejandro, R. (1990). Islet isolation assessment in man and large animals. *Acta. Diabetol. Lat.*, **27**, 185–195.
72. Latif, Z. A., Noel, J. and Alejandro, R. (1988). A simple method of staining fresh and cultured islets. *Transplantation*, **45**, 827–830.
73. Bank, H. L. (1988). Assessment of islet cell viability with acridine orange and propidium iodide. *In Vitro*, **24**, 266–272.

74. London, N. J. M., Contractor, H., Lake, S. P. Aucott, G. C., Bell, R. P. F. and James, R. F. L. (1990). A fluorometric viability assay for single human and rat islets. *Horm. Metab. Res.*, suppl., **25**, 82–87.
75. Danis, J., Tunggal, B., Weyer, J., Meyer, G., Saad, S., Isselhard, W., Stoffel, W. and Hesse, U. J. (1990). ^{31}P NMR spectroscopy for *in vitro* viability testing of porcine pancreatic islets. *J. Surg. Res.*, **49**, 534–538.
76. Nielsen, J. H., Burnstedt, J., Andersson, A. and Frimodt-Møller, C. (1979). Preservation of beta cell function in adult human pancreatic islets for several months *in vitro*. *Diabetologia*, **16**, 97–100.
77. Frankel, B. J., Gylfe, E. and Hellman, B. (1976). Maintenance of insulin release from pancreatic islets stored in the cold for up to 5 weeks. *J. Clin. Invest.*, **57**, 47–52.
78. Jindal, R. and Gray, D. (1994). Preservation and storage of pancreatic islets. *Transplantation*, **57**, 317–321.
79. Rajotte, R. V., Evans, M. G., Warnock, G. L. and Kneteman, M. N. (1990). Islet cryopreservation. *Horm. Metab. Res.*, **25**, 72–81.
80. Kneteman, M. N., Alderson, D., Sharp, D. W. and Lacy, P. E. (1989). Long-term cryogenic storage of purified adult human islets of Langerhans. *Diabetes*, **38**, 386–396.
81. Bank, H. L. (1983). Cryobiology of isolated islets of Langerhans circa 1982. *Cryobiology*, **20**, 119–128.
82. Tze, W. J. and Tai, J. (1990). Successful banking of pancreatic endocrine cells for transplantation. *Metabolism*, **39**, 719–723.
83. Rajotte, R. V., Warnock, G. L., Bruch, L. C. and Procyshyn, A. W. (1983). Transplantation of cryopreserved and fresh rat islets and canine pancreatic fragments: comparison of cryopreservation protocols. *Cryobiology*, **20**, 169–184.
84. Kemp, C. B., Knight, M. J., Sharp, D. W., Ballinger, W. F. and Lacy, P. E. (1973). The effect of transplantation site on the results of pancreatic islet isographs in diabetic rats. *Diabetologia*, **9**, 486–491.
85. Mehigan, D. G., Bell, D. W. R., Zuidema, G. D., Eggleston, J. C. and Cameron, J. L. (1980). Disseminated intravascular coagulation and portal hypertension following pancreatic islet autotransplantation. *Ann. Surg.*, **191**, 287–293.
86. Wohlrab, F., Wilke, B., Schmidt, S. and Cossel L. (1988). Syngeneic transplantation of rat pancreatic islets into the spleen. Light and electron microscopical findings. *Int. J. Pancreatol.*, **3**, 59–72.
87. Brömme, H. J., Hahn, H. J. and Blech, W. (1988). Biphasic release of insulin from islets of Langerhans after their transplantation into the liver of rats. *Horm. Metab. Res.*, **20**, 138–140.
88. Trimble, E. R., Karakash, C., Malaisse-Lagae, F., Vassutine, I., Orci, L. and Renold, A. E. (1980). Effects of intraportal islet transplantation on the transplanted tissue and the recipient pancreas. I. Functional studies. *Diabetes*, **29**, 341–347.
89. Barneo, L., Esteban, M. M., Garcia-Pravia, C., Diaz, F. and Mavin, B. (1990). Islet transplantation restores normal serum amylase levels in diabetic rats. *Eur. Surg. Res.*, **22**, 143–150.
90. Warnock, G. L., Kneteman, N. M., Ryan, E., Seelis, R. E. A., Rabinovitch, A. and Rajotte, R. V. (1991). Normoglycaemia after transplantation of freshly isolated and cryopreserved pancreatic islets in type 1 (insulin-dependent) diabetes mellitus. *Diabetologia*, **34**, 55–58.
91. Merrell, R. C., Marincola, F., Maeda, M., Cobb, L. and Basadonna, G. (1985). The metabolic response of intrasplenic islet autografts. *Surg. Gyn. Obstet.*, **160**, 552–556.
92. Warnock, G. L., DeGroot, T., Untch, D., Ellis, D. K. and Rajotte, R.V. (1989). The natural history of pure canine islet autografts in hepatic or splenic sites. *Transpl. Proc.*, **21**, 2617–2618.
93. Selawry, H. P. and Whittington, K. (1984). Extended allograft survival of islets grafted into intra-abdominally placed testes. *Diabetes*, **33**, 405–406.
94. Tze, W. J. and Tai, J. (1984). Successful intracerebral allotransplantation of pancreatic endocrine cells in spontaneously diabetic BB rats without immunosuppression. *Metabolism*, **33**, 785–789.
95. Posselt, A. M., Barker, C. F., Tomaszewski, J. E., Markmann, J. F., Choti, M. A. and Naji, A. (1990). Induction of donor-specific unresponsiveness by intrathymic islet transplantation. *Science*, **249**, 1293–1295.
96. Reece-Smith, H., DuToit, D. F., McShane, P. and Morris, P. J. (1981). Prolonged survival of pancreatic islet allografts transplanted beneath the renal capsule. *Transplantation*, **31**, 305–306.
97. Evans, M. G., Warnock, G. L. and Rajotte, R. V. (1989). Comparison of sites for transplantation of canine pancreatic microfragments. *Diabetes Res.*, **10**, 35–41.
98. Woehrle, M. F., Markmann, J. F., Armstrong, J. and Naji, A. (1987). Effect of transplant site on islet allograft survival in BB rats. *Transpl. Proc.*, **19**, 925–927.
99. Woehrle, M., Markman, J. F., Beyer, K., Naji, A., Bretzel, R. G. and Federlin, K. (1990). The influence of the implantation site (kidney capsule vs. portal vein). on islet survival. *Horm. Metab. Res.*, suppl., **25**, 163–165.
100. Ar'Rajab, A., Ahrén, B. and Bengmark, S. (1989). Islet transplantation to the renal subcapsule space in streptozotocin-diabetic rats: short-term effects on insulin secretion. *Diab. Res. Clin. Pract.*, **7**, 197–204.
101. Hayek, A., Lopez, A. D. and Beattie, G. M. (1990). Factors influencing islet transplantation. Number, location, and metabolic control. *Transplantation*, **49**, 224–225.
102. Warnock, G. L., Dabbs, K. D., Evans, M. G., Cattral, M. S., Kneteman, M. N. and Rajotte, R. V. (1990). Critical mass of islets that function after implantation in a large mammalian. *Horm. Metab. Res.*, suppl., **25**, 156–162.
103. Child, C. G., Frey, C. J. and Fry, W. J. (1969). A reappraisal of removal of ninety-five percent of the distal portion of the pancreas. *Surg. Gynecol. Obstet.*, **129**, 49–56.
104. Sutherland, D. E. R., Goetz, F. C. and Najarian, J. S. (1984). Pancreas transplants from related donors. *Transplantation*, **38**, 625–633.
105. Ar'Rajab, A. and Ahrén, B. (1992). Prevention of hyperglycemia improves the long-term results of

islet transplantation in streptozotocin-diabetic rats. *Pancreas*, **7**, 435–442.

106. Finch, D. R. A., Wise, P. H. and Morris, P. J. (1977). Successful intra-splenic transplantation of syngenic and allogenic isolated pancreatic islets. *Diabetologia*, **13**, 195–199.
107. Woehrle, M., Beyer, K., Bretzel, R. G. and Federlin, K. (1989). Prevention of islet allograft rejection in experimental diabetes of the rat without immunoressión of the host. *Transpl. Proc.*, **21**, 2705–2706.
108. Lacy, P. E. and Sharp, D. W. (1986). Islet transplantation in treating diabetes. *Ann. Rev. Med.*, **37**, 33–40.
109. Pujol-Borrell, R., Todd, I., Doshi, M., Bottazzo, G. F., Sutton, R., Gray, F., Adolf, G. R. and Feldmann, M. (1987). HLA class II induction in human islet cells by interferon-gamma plus tumour necrosis factor or lymphotoxin. *Nature*, **326**, 304–306.
110. Wang, Y., Pontesilli, O., Gill, R. D., La Rosa, F. G. and Lafferty, K. J. (1991). The role of $CD4^+$ and $CD8^+$ T cells in the destruction of islet grafts of spontaneously diabetic mice. *Proc. Natl. Acad. Sci.* USA, **88**, 527–531.
111. Wright, J. R., Lacy, P. E., Unanue, E. R., Muszynski, C. and Hauptfeld V. (1986). Interferon-mediated induction of Ia antigen expression on isolated murine whole islets and dispersed islet cells. *Diabetes*, **35**, 1174–1177.
112. Hao, L., Wang, Y., Gill, R. G., La Rosa, F. G., Talmage, D. W. and Lafferty, K. J. (1990). Role of lymphokines in islet allograft rejection. *Transplantation*, **49**, 609–614.
113. Gotoh, M., Porter, J., Kanai, T., Monaco, A. P. and Maki, T. (1988). Multiple donor allotransplantation. A new approach to pancreatic islet transplantation. *Transplantation*, **45**, 1008–1013.
114. Ar'Rajab, A., Bengmark, S. and Ahrén, B. (1991). Islet allotransplantation to streptozotocin-diabetic rats. Effects of multistrain donation and ultraviolet light irradiation. *Eur. Surg. Res.*, **23**, 302–308.
115. Jamdes, R. F. L., Lake, S. R., Chamberlain, J., Thirdborough, S., Bassett, P. D., Mistry, N. and Bell, P. R. F. (1989). Gamma irradiation of isolated rat islets pretransplantation produce indefinite allograft survival in cyclosporine-treated recipients. *Transplantation*, **47**, 929–933.
116. Lau, H., Reemtsma, K. and Hardy, M. A. (1984). Prolongation of rat islet allograft survival by direct ultraviolet irradiation of the graft. *Science*, **223**, 607–609.
117. Lim, F., Sun, A. M. (1980). Microencapsulated islets as bioartificial endocrine pancreas. *Science*, **210**, 908–910.
118. Clayton, H. A., James, R. F. L. and London, N. J. M. (1993). Islet microencapsulation: a review. *Acta. Diabetol.*, **30**, 181–189.
119. Sharp, D. W., Mason, N. S. and Sparks, R. E. (1984). Islet immuno-isolation: the use of hybrid artificial organs to prevent islet tissue rejection. *World. J. Surg.*, **8**, 221–229.
120. Ar'Rajab, A., Bengmark, S. and Ahrén, B. (1991). Insulin secretion in streptozotocin-diabetic rats transplanted with immunoisolated islets. *Transplantation*, **51**, 570–574.
121. Wu, Z., Lu, Z. and Sun, A. M. (1988). Preliminary report on microencapsulated islet transplantation in experimental diabetes mellitus in China. *Int. J. Pancreatol.*, **3**, 91–100.
122. Algire, G. H., Weaver, J. M. and Prehn, R. T. (1954). The growth of cells *in vivo* in diffusion chambers. II. The role of cells in the destruction of homografts in mice. *J. Natl. Cancer Inst.*, **15**, 1757–1767.
123. Reach, G., Poussier, P., Sausse, A., Assas, R., Itoh, M. and Gerich, J. E. (1981). Functional evaluation of a bioartificial pancreas using isolated islet perfused with blood ultrafiltrate. *Diabetes*, **30**, 296–301.
124. Lanza, R. P., Kühtreiber, W. M., Ecker, D., Staruk, J. E. and Chick, W. L. (1995). Xenotransplantation of porcine and bovine islets without immunoression using uncoated alginate microspheres. *Transplantation*, **59**,1377–1384.
125. O'Shea, G. M. and Sun, A. M. (1986). Encapsulation of rat islets of Langerhans prolongs xenograft survival in diabetic mice. *Diabetes*, **35**, 943–946.
126. Fritschy, W. M., Wolters, G. H. J. and v Schilgaarde, R. (1991). Effect of alginate-polylysine-alginate microencapsulation on *in vitro* insulin release from rat pancreatic islets. *Diabetes*, **40**, 37–43.
127. Faustmann, D., Hauptfield, V., Lacy, P. E. and Davie, J. (1981). Prolongation of murine islet allograft survival by pretreatment of islets with antibody directed to Ia determinants. *Proc. Natl. Acad. Sci.* USA, **78**, 5156–5159.
128. Faustman, D. L., Steinman, R. M., Gebel, H. M., Hauptfeld, V., Davie, J. M. and Lacy, P. E. (1984). Prevention of rejection of murine islet allografts by pretreatment with anti-dendritic cell antibody. *Proc. Natl. Acad. Sci.* USA, **81**, 3864–3868.
129. Carel, J. C., Schreiber, R. D., Falqui, L. and Lacy, P. E. (1990). Transforming growth factor β decreases the immunogenicity of rat islet xenografts (rat to mouse). and prevents rejection in association with treatment of the recipient with a monoclonal antibody to interferon gamma. *Proc. Natl. Acad. Sci.* USA, **87**, 1591–1595.
130. Borel, J. F., Feurer, C., Magnée, C. and Stählin, H. (1977). Effects of new anti-lymphocytic peptide cyclosporine A in animals. *Immunology*, **32**, 1017–1025.
131. Gordon, M. Y. and Singer, J. W. (1979). Selective effects of cyclosporine A on colony-forming lymphoid and myeloid cells in man. *Nature*, **279**, 433–434.
132. Paetkau, V., Havele, C. and Shaw, J. (1988). Direct and indirect modes of action of cyclosporine on cytotoxic T lymphocytes. *Ann. N. Y. Acad. Sci.*, **532**, 405–412.
133. Dibelius, A., Konigsberger, H., Walter, P., Permanetter, W., Brendel, W. and von Specht, B. U. (1986). Prolonged reversal of diabetes in the rat by transplantation of allogeneic islets from a single donor and cyclosporine treatment. *Transplantation*, **41**, 426–431.
134. Wahlström, H. E., Lavelle-Jones, M., Endres, D., Akimoto, R., Kolterman, O. and Moosa, A.R. (1990). Inhibition of insulin release by cyclosporine and production of peripheral insulin resistance in the dog. *Transplantation*, **49**, 600–604.

135. Martin, F. and Bedoya, F. J. (1990). Short-term effects of cyclosporine on secretagogue-induced insulin release by isolated islets. *Transplantation*, **50**, 551–553.
136. Healey, M. A., Clas, D., Duguid, W. P. and Rosenberg, L. (1990). Adverse effects of cyclosporine on islet growth and differentiation. *J. Surg. Res.*, **48**, 557–561.
137. Rooth, P., Dawidson, I., Lafferty, K., Diller, K., Armstrong, J., Pratt, P., Simonsen, R. and Täljedal, I. B. (1989). Prevention of detrimental effect of cyclosporin A on vascular ingrowth of transplanted pancreatic islets with verapamil. *Diabetes*, **38** (suppl., 1), 202–205.
138. Viviani, G. L., Borgoglio, M. G., Fontana, I., Borgoglio, A., Nocera, A., Leprini, A., Comaschi, M., Valente, U. and Adexati, L. (1989). Azathioprine decreases insulin secretion in human islets. *Transpl. Proc.*, **21**, 2714–2716.
139. Carroll, P. B., Goncalves, A. A., Boschero, A. G., Tzakis, A. G., Starzl, T. E. and Atwater, I. (1991). Effect of the immunoressant FK 506 on insulin release from adult rat islets of Langerhans. *Transpl. Proc.*, **23**, 337–339.
140. Strasser, S., Alejandro, R., Ricordi, C., Todo, S. and Mintz, D. H. (1991). The effect of FK 506 on canine pancreatic islet cell function in beagle dogs. *Transpl. Proc.*, **23**, 756–757.
141. Kneteman, N. M., Cattral, M. C., Warnock, G. L., Scharp, D. W., Jaspan, J. L. and Rajotte, R. V. (1989). Immunoression and pancreatic islet transplantation. *Horm. Metab. Res.*, **25**, suppl., 193–199.
142. Hahn, H. J., Kuttler, B., Volk, H. D., Lucke, S. and Diamantstein, T. (1990). Interleukin 2 receptor-targeted therapy in islet transplantation research. *Horm. Metab. Res.*, **25**, suppl., 177–180.
143. Pankewycz, O., Mackie, J., Hassarjian, R., Murphy, J. R., Ström, T. B. and Kelley, V. E. (1989). Interleukin-2-diphteria toxin fusion protein prolongs murine islet cell engraftment. *Transplantation*, **47**, 318–322.
144. Ar'Rajab, A., Ahrén, B., Alumets, J. and Bengmark, S. (1989). Islet transplantation to the renal subcapsular space in streptozotocin-diabetic rats. Long-term effects on insulin and glucagon secretion. *Acta. Chir. Scand.*, **155**, 503–508.
145. Coulumbe, M. G., Warnock, G. L. and Rajotte, R. V. (1988). Reversal of diabetes by transplantation of cryopreserved rat islets of Langerhans to the renal subcapsular site. *Diabetes Res.*, **3**, 9–15.
146. Gray, B. N., Caruso, G., Alford, F. and Chisholm, D. (1979). Insulin and glucagon responses of transplanted intrasplenic pancreatic islets. *Arch. Surg.*, **114**, 96–99.
147. Rajotte, R. V., Warnock, G. L., Procyshyn, A. W. and Wierczorek, K. (1984). Intrasplenic isografts of canine pancreatic islets: metabolic study. *Transpl. Proc.*, **16**, 834–837.
148. Ar'Rajab, A., Ahrén, B. and Bengmark, S. (1989). Insulin and glucagon secretion in streptozotocin-diabetic rats: influences of islets transplanted to the renal subcapsular space. *Diabetes Res.*, **12**, 37–41.
149. Kneteman, N. M., Evans N. G., Nason, R. W., and Rajotte, R. V. (1989). Prolonged function of intrasplenic canine pancreatic fragment autografts. *Transpl. Proc.*, **21**, 2619–2620.
150. Warnock, G. L., Rajotte, R. V., Kneteman, N. M., Ellis, D. and Toth, K. (1987). Oral glucose tolerance and response to mixed meals in dogs bearing long-term transplanted islets of Langerhans. *Transpl. Proc.*, **19**, 969–973.
151. Lake, S. P., Chamberlain, J., Swift, S., Wood, J. K., Bell, P. R. and James, R. F. (1988). Glycosylated hemoglobin: an index of glucose control in pancreatic islet transplanted rats. *Diabetes Res.*, **7**, 75–78.
152. Orloff, M. J., Macedo, C., Macedo, A. and Greenleaf, G. E. (1987). Comparison of whole pancreas and pancreatic islet transplantation in controlling nephropathy and metabolic disorders of diabetes. *Ann. Surg.*, **206**, 324–332.
153. Orloff, M. J., Macedo, A., Grennleaf, G. E. and Girard, B. (1988). Comparison of the metabolic control of diabetes achieved by whole pancreas transplantation and pancreatic islet transplantation in rats. *Transplantation*, **45**, 307–312.
154. Barneo, L., Esteban, M. M., Garcia-Pravia, C., Diaz, F. and Marin, B. (1990). Normalization of the altered liver function tests after islet transplantation in diabetic rats. *Diabet. Metab.*, **16**, 284–289.
155. Margolis, R. N., Selawry, H. P. and Curnow, R. T. (1985). Regulation of hepatic glycogen metabolism: effects of diabetes, insulin infusion and pancreatic islet transplantation. *Metabolism*, **34**, 62–68.
156. Ar'Rajab, A., Ahrén, B., Alumets, J., Lögdberg, L. and Bengmark, S. (1990). Islet transplantation to the renal subcapsular space improves late complications in streptozotocin-diabetic rats. *Eur. Surg. Res.*, **22**, 270–278.
157. Federlin, K., Bretzel, R. G. and Schmidtchen, U. (1976). Islet transplantation in experimental diabetes of the rat, V. Regression of glomerular lesions in diabetic rats after intraportal transplantation of isogeneic islets. *Horm. Metab. Res.*, **8**, 404–406.
158. Hu, Y. F., Cheng, R. L., Shao, A. H., Ye, R. S., Gu, Z. F., Zhang, H. D., Zhang, Z. G., Hen, L. R., Bi, H. F. and Sho, G. F. (1989). The influences of islet transplantation on metabolic abnormalities and diabetic complications. *Horm. Metab. Res.*, **21**, 198–202.
159. Buscher, C., Weis, A., Wöhrle, M., Bretzel, R. G., Cohen, A. M. and Federlin, K. (1989). Islet transplantation in experimental diabetes of the rat. XII. Effect on diabetic retinopathy, morphological findings and morphometrical evaluation. *Horm. Metab. Res.*, **21**, 227–231.
160. Federlin, K. F. and Bretzel, R. G. (1984). The effect of islet transplantation on complications in experimental diabetes of the rat. *World J. Surg.*, **8**, 169–178.
161. Schmidt, R. E., Plurad, S. B., Olack, B. J. and Scharp, D. W. (1983). The effect of pancreatic islet transplantation and insulin therapy on experimental diabetic autonomic neuropathy. *Diabetes*, **32**, 532–540.
162. Pugliese, G., Tilton, R. G., Chang, K., Speedy, A., Province, M., Eades, D. M., Lacy, P. E., Kilo, C. and Williamson, J. R. (1990). Effects of islet isografts on hemodynamic and vascular filtration changes in diabetic rats. *Diabetes*, **39**, 323–332.

163. Bonner-Weir, S. and Orci, L. (1982). New perspectives on the microvasculature of the islets of Langerhans in the rat. *Diabetes*, **31**, 883–889.
164. Brunicardi, F. C., Stagner, J., Bonner-Weir, S., Wayland, H., Kleinman, R., Livingston, E., Guth, P., Menger, M., McCuskey, R., Intaglietta, M., Charles, A., Ashley, S., Cheung, A., Ipp, E., Gilman, S., Howard, T. and Passaro Jr, E. (1996). Microcirculation of the islets of Langerhans. *Diabetes*, **45**, 385–392.
165. Ahrén, B., Taborsky Jr, G. J. and Porte Jr, D. (1986). Neuropeptidergic versus cholinergic and adrenergic regulation of islet hormone secretion. *Diabetologia*, **29**, 827–836.
166. Adeghare, E. (1995). Reinnervation of pancreatic tissue and islet grafts by aminergic, cholinergic and peptidergic nerves. A review. *Biogenic Amines*, **11**, 147–164.
167. Griffith, R. C., Scharp, D. W., Hartman. B. K., Ballinger, W. F. and Lacy, P. E. (1977). A morphologic study of intrahepatic-portal-vein islet isografts. *Diabetes*, **26**, 201–214.
168. Korsgren, O., Andersson, A., Jansson, L. and Sundler, F. (1991). Reinnervation of transplanted mouse pancreatic islets. *Transpl. Proc.*, **23**, 751–751.
169. Korsgren, O., Andersson, A., Jansson, L. and Sundler, F. (1992). Reinnervation of syngeneic mouse pancreatic islets transplanted into renal subcapsular space. *Diabetes*, **41**, 130–135.
170. Portis, A. J., Rajotte, R. V. and Krukoff, T. L. (1994). Reinnervation of isolated islets of Langerhans transplanted beneath the kidney capsule in the rat. *Cell Transplantation*, **3**, 163–170.
171. Trimble, E. R., Karakash, C., Malaisse-Lagae, F., Vassutine, I., Orci, L. and Renold, A. E. (1980). Effects of intraportal islet transplantation on the transplanted tissue and the recipient pancreas. I. Functional studies. *Diabetes*, **29**, 341–347.
172. Louis-Sylvestre, J. (1978). Feeding and metabolic pattern in rats with truncal vagotomy or with transplanted B-cells. *Am. J. Physiol.*, **235**, E119–E125.
173. Portis, A. J., Warnock, G. L., Belcastro, A. N. and Rajotte, R. V. (1987). Glycemic control during submaximal exercise in long-term islet cell autografted dogs. *Diabetes*, **36** (suppl., 1), 837
174. Mullen, Y. S., Clark, W. R., Molnar, I. G. and Brown, J. (1977). Complete reversal of experimental diabetes mellitus in rats by single fetal pancreas. *Science*, **195**, 68–70.
175. Hayek, A., Lopez, A. D. and Beattie, G. M. (1988). Decrease in the number of neonatal islets required for successful transplantation by strict metabolic control of diabetic rats. *Transplantation*, **45**, 940–942.
176. Sandler, S. and Jansson, L. (1987). Blood flow measurements in autotransplanted pancreatic islets of the rat. Impairment of the blood perfusion of the graft during hyperglycemia. *J. Clin. Invest.*, **80**, 17–21.
177. Sutherland, D. E. R. and Kendall, D. (1985). Clinical pancreas and islet transplantation registry report. *Trabspl. Proc.*, **17**, 307–311.
178. Federlin, K., Bretzel, R. G. and Hering, B. (1991). Pancreatic islet transplantation in diabetes. *Int. J. Art. Org.*, **14**, 74–77.
179. Hering, B. J., Browatzki, C. C., Schultz, A., Bretzel, R. G. and Federlin, K. F. (1993). Clinical islet transplantation - registry report, accomplishments in the past and future research needs. *Cell Transplantation*, **2**, 269–282.
180. Alejandro, R. (1995). Transplantation of islets of Langerhgans in patients with insulin-dependent diabetes mellitus. *Scand. J. Gastroenterol.*, **30**, suppl., 208, 125–128.
181. Scharp, D. W., Lacy, P. E., Santiago, J. V., McGullough, C. S., Weide, L. G., Falqui, L., Marchetti, P., Gingerich, R. L., Jaffe, A. S., Cryer, P. E., Anderson, C. B. and Flye M. W.(1990). Insulin independence after islet transplantation into type 1 diabetic patient. *Diabetes*, **39**, 515–518.
182. Warnock, G. L., Knetemamn, N. M., Ryan, E. A., Evans, M. G., Seelis, R. E. A., Halloran, P. F., Rabinovitch, A. and Rajotte, R. V. (1989). Continued function of pancreatic islets after transplantation in type I diabetes. *Lancet*, **2**, 570–572.
183. Pyzdrowski, K. L., Kendall, D. M., Halter, J. B., Nakhleh, R. E., Sutherland, D. E. R. and Robertson, R. P. (1992). Preserved insulin secretion and insulin independence in recipients of islet autografts. *N. Engl. J. Med.*, **327**, 220–226.
184. Federlin, K. F. (1993). Islet transplantation. The connection of experiment and clinic exemplified by the transplantation of islets of Langerhans. *Exp. Clin. Endocrinol.*, **101**, 334–345.

PART VIII

UROLOGY

PART VIII

23. Experimental Upper Urinary Tract Obstruction

Jens Christian Djurhuus

Institute of Experimental Clinical Research, Århis University and Aarhus University Hospital DK-8200 Aarhus N, Denmark

Introduction

Experimental obstruction of the urinary tract is induced in order to evaluate changes in transport function and changes in kidney function. Over the last decades, such investigations have gained increasing attention due to the fact that patients subjected to evaluation for treatment of transport dysfunctions have changed category. Most of the patients are now either patients with post-obstructional changes or patients where the dialtation has been diagnosed either pre-natally or as a coincidental finding during general evaluation.

The patients, therefore, are not subjected to evaluation for treatment due to pain or other complications, but merely because they show dilatation and perhaps some changes in kidney function which may be stationary, slowly progressing, fast progressing or may even be spontaneously resolving as is frequently seen in infancy. Consequently, studies over the last decades have concentrated on evaluation of the transport function and evaluation of kidney function with special reference to transport function dynamics, kidney function plasticity and possibilities of regaining functions secondary to obstruction.

Experimental obstruction designed to mimic clinical partial obstruction is difficult to achieve, but necessary if the renal and pelvic changes seen clinically are to be reproduced.

Galen[39] was probably the first to induce and study the effects of ureteric obstruction. He induced total obstruction in order to demonstrate that the ureters were the agents of transport of urine from the kidney to the bladder. More systematic studies of the effects of flow hindrance on renal and pelvi-ureteric function were first instituted in this century.

Methods of Inducement of Obstruction

Many factors determine the effect of obstruction on renal and pelvi-ureteric function: whether the obstruction is partial or total; the site of obstruction; and whether condition is apparent in the growing, or full-grown animal. High obstruction with no proximal peristaltically functional ureter segment, affects renal and pelvic function immediately. When there is a functional ureteric segment above the obstruction, the hydrodynamics of the kidney and pelvis first manifest when the ureteric peristaltic drive can no longer overcome the obstruction.

Many methods for induction of obstruction have been proved. The ureter has been obstructed by *intraluminal catheters*, and by catheters laid *extraluminally, parallel with, and externally ligated to, the ureter.*[1–5] These methods induce a rigid and progressive obstruction – progressive because of the local foreign body reactions.

Relatively inert material – which incites a foreign body reaction – has been used to *ensheathe the ureter.*[5–7] Results have not been uniform, and the obstruction is again progressive, though Tanagho[8] achieved the desired partial obstruction in two of eight experiments using this method.

Induction of *an inflammatory reaction in the ureter wall*, without stricture, has been demonstrated using cellophane investment of the ureter.[2,9] This method gave the desired partial obstruction in about 60% of the experiments.

Localized *crushing of the ureter* is also used.[10–12] The degree of obstruction depends on the duration of the crushing. After six hours crushing, rabbits exhibited progressive hydronephrosis,[13] but crushing at the pelvic-ureteric junction in the pig of up to one hour's duration (Kelly-clamp) was without effect.

Simultaneous *crushing and cauterisation* of the ureter accentuates the damage otherwise seen after crushing alone. The rapidly progressive change involves stenosing fibrosis and interruption of muscle continuity.[7]

External compression by implantation of a ureteric segment within the psoas muscle, induces a mild partial obstruction.[7,14,15] Changes are uniform and reproducible, which enables prolonged study of similar, induced disease, states. Other transpositions of the ureter;behind the iliac artery, the inferior mesenteric artery, or the aorta[16] only occasion obstruction in the latter instance.

Common to most of the proven obstruction methods, is the induction of severe changes which affect the ureteric musculature and adventitia, causing irregular and often progressive obstruction. Thus, comparison between experimental and clinical observations is seldom valid, nor are comparisons between different experimental series.

Intrapelvic Pressure and Obstruction

Acute obstruction is followed immediately by an increase in the intrapelvic pressure. The rate of increase and the maximum pressure depend upon the degree of the obstruction, the level of the obstruction, and the actual diuresis. High total obstruction causes an abrupt pelvic pressure increase, while the lower the obstruction, the less abrupt the pressure increase. The period of pressure increase is always only of hours duration, reaching maximums of 60–90 mm Hg.[17] Thereafter, pressure falls. After eight hours, maximums are about 40 mm Hg, and after twenty four hours pressures are only slightly above normal. After one month pelvic pressures are seldom above normal.[18] In another study of total obstruction after four to six weeks,[7] large variations in intrapelvic pressures were found. These may have been actual, but it is rather more likely that the variations reflect technical difficulty in measurements by direct puncture of the tense pelvis without leakage. Some part of the variations may depend upon the different time intervals between induction of the obstruction and its development.

Partial obstructions cause changes similar to total obstruction (Schweitzer), but pressure changes are smaller and slower, and the period of elevated basal pressure tends to be longer,[19] though often normal after five weeks (Schweitzer). In states of mild obstruction when kidneys are functioning well, a non significant increase in basal pressure was observed after five weeks (7.9 mm Hg in pigs with mild obstruction when urine flow averaged 2.6 cc / ureter / min, compared with 4.7 mm Hg and 1.4 cc / ureter / min in normal pigs). This difference might be casual; it might depend on the difference in the diuresis, or it might reflect the presence of the obstruction.

Peristaltic Activity and Obstruction

The immediate effects of obstruction on ureteric peristalsis are well documented.[20,21] An Increase in the basal pelvic pressure is followed by an increase in the frequency of peristalsis. The pressure waves become broad, and wave amplitude reduces to the point where only small undulations are seen.[18] The same changes occur in the pelvis, the degree depending on the volume diuresis and the level of obstruction. Thus, peristalsis changes are in principle the same as seen after a rapid increase in diuresis, though more marked.

Pressure waves. Under conditions of established mild obstruction, pressure waves during anti- diuresis are broader and of higher amplitude than normal, but of uniform appearance. These changes cannot be taken as certain signs of a change in peristalsis, but isolated waves are of even higher amplitude and of longer duration than their fellows, and the collected observations suggest that there is a significant change in peristaltic activity following mild obstruction of longer duration.

During total obstruction the pressure waves are not uniform, varying markedly in amplitude, configuration and duration. Total obstruction for five weeks in three pigs showed pressure waves with marked variations in amplitude and duration.[7] After six weeks peristaltic activity had decreased showing decreasing amplitude and frequency of pressure waves. This is an observation which is in accordance with the findings of Rose and Gillenwater. After total obstruction of long duration, peristaltic activity in the dog pelvis cannot be registered.[12]

Electromyographic Studies of the Pelvis in the Presence of Urinary Tract Obstruction

Electromyographic studies of the upper urinary tract under conditions of experimental obstruction are few, but the method is good. It allows not only registration of muscle activity, which cannot be demonstrated by other means (pressure curve studies), but also, using a multi-lead system, spread and coordination of muscle activity can be recorded and studied.

When the obstruction is partial and mild, activity begins proximally, as in the normal pelvis, and impulses spread at an even rate over the pelvis. But exceptions occur: there may be a retrograde spread of activity, also at an even rate, and this observation is paralleled by a pressure wave complex which is higher and broader than its fellows. This retrograde activity is probably initiated from the obstruction site, as has been clinically observed when there is a more distal obstruction of the ureter.[21–24]

Variations in activity conduction rates are also seen. Typically, the first peristaltic wave after an activity pause, spreads over the pelvis and down the ureter at a greater rate than the succeeding waves. *In vivo* observations in the pig are consistent with this finding: when ureteric activity is of low frequency, activity transmission rates are higher than when ureteric activity is of high frequency.[25] *In vitro* studies of the ureter show, likewise, that transmission rates are greater directly after an activity pause.[26] The explanation of this phenomenon might be that, when activity is of low frequency, the successive impulses do not occur during the relative refractory period,[26,27] while normal frequency impulses occur within the long relative refractory period after the proceeding impulse.

In the case of total obstruction, even after five weeks, a chronological relationship exists between the activities registered at different pelvic sites, but the incidence of retrograde and anterograde spread of activity is almost the same. After six weeks of total obstruction only isolated islands of muscle tissue survive amid proliferating fibrous connective tissue.[28] There is no longer any coordination of activity and sometimes no activity can be recorded. The net effect is, thus, a rapidly progressive disruption of activity coordination, and ultimate cessation of activity.

Activity Frequency under Conditions of Established Obstruction

The normal pelvis, as previously described, has a monomodal peristaltic activity. Frequency of this activity varies between individuals, but is characteristically constant within the individual. Monomodal activity persists in the pig pelvis despite mild partial obstruction, but it is not so well defined. Activity pauses of over a minute are followed by "salvoes" of activity, and this can be reflected by histogram representation of the time intervals between peristaltic episodes, which are less apparent when obstruction is present. That this can be an actual effect of the obstruction, is illustrated by a better collected histogram after the relief of the obstruction.[7] By contrast the mean frequencies of peristaltic activity in the non obstructed system, the partially obstructed system, and the latter after the relief of obstruction, cannot be diffentiated with certainty.[7] When obstruction is more severe, total peristaltic frequency falls. This may reflect that the muscle stimulation threshold is increased in the presence of an obstruction.[29] The monomodal activity distribution is, likewise, no longer seen. Obstructions of five weeks duration still contain several peristaltic events within a minute, whereas six weeks of obstruction only show occasional activity.[7]

The Relationship between Pelvic and Ureteric Activity in the Presence of Obstruction

The relationship between activity above and below a high ureteric or pelvi-ureteric junction obstruction may be preserved, partly preserved, or totally lost.

When the relationship is preserved, constant transmission times during obstruction, are recorded. The pelvic function is normal, despite dilatation, but, as previously mentioned, the pressure waves are of higher amplitude and broader. This preservation of normal transmission might depend on the increased degree of filling of the pelvis and increased basal pressure, but, as yet, such a pressure increase has not been demonstrated with certainty. Another plausible explanation, also compatible with the above, and with the observation of broader pressure waves of higher amplitude, might be that the contraction power of the muscle proximal to an obstruction is increased (Weiss, R.M., personal communication). Increased loading of the upper urinary tract is accompanied by increase in compliance,[30] and this might entail a more optimal resting lenght of the muscle fibres between contractions. In early and/or mild partial obstruction muscle hypertrophy and hyperplasia have been histologically seen.[31]

When the realtionship is only partly preserved, as after cauterization of the pelvi-ureteric junction,[7] or transection and neo-anastomosis of the ureter,[26] variable transmission times over the obstruction are recorded. This probably depends more on stricture formation with the interruption of muscle continuity, because activity is normal over a technically perfect ureter neo-anastomosis.[7,26]

Total loss of the realtionship between pelvic and ureteric activity can be seen when stricture formation is sufficiently severe. The pelvis cannot force a bolus through the stricture so that the ureter is not distended, but functions only as a passive drain. The ureter reacts only if there is adequate muscle stretch, and quite independent of pelvic activity.

When obstruction is total, despite cessation of urine production, some relationship between activity above and below the obstruction may yet be seen. Retrograde activity is seen as often as anterograde. It is unclear whether the activity originating from the ureter is initiated by propagation of a pelvic impulse, or whether there is stretch over the obstruction. Alternatively, ureteric activity retrogressively transmitted to the pelvis, together with the loss of orientation, could be interpreted as supportive of postulates made of subsidiary pacemakers in the ureter.[32,33] Finally, the ureteric activity might reflect only a recognized capacity of smooth muscle to depolarize spontaneously, as seen in isolated preparations in the form of spontaneous contractions[26,34–36]

After prolonged obstruction, there is no longer any coordinated activity within the pelvis itself, surely a reflection of progressive muscle degeneration above the obstruction. Sparse ureteric activity of the same frequency, as seen in isolated preparations, is reported,[26] but this is not genuine spontaneous activity. Methods hitherto used for registration of ureteric activity, all involve instrumentation of the ureter, and thus, mechanical stimulation of activity cannot be excluded.

Pelvi-Ureteric Adaptation to Change in Diuresis in the Presence of Obstruction

Pelvi-ureteric function depends on the capacity of the kidney to produce urine, on the degree of obstruction, on volume adjustment capacity, and on the degree of preservation of peristaltic function and coordination. Significant obstruction slowly impairs renal function.[1,7] Thus, any estimation of pelvi-ureteric function must take account of kidney function, and ureteric urine flow should be measured.

Obstruction, dependent on degree, induces changes in the pelvi-ureteric structure: muscle cell hypertrophy and muscle cell proliferation;[6,31,37] connective tissue formation between the muscle cells,[38] and after five to six weeks, if obstruction is total or nearly so, the majority of muscle cells are replaced by connective tissue.[28] The compliance of the system is thus determined by this connective tissue.

Five weeks after induction of partial obstruction in pigs, furosemide administration caused a massive increase in diuresis and a consequent increase in pelvic pressure; but this increase was less than that seen in normal pigs during the same volume diuresis. The mildly obstructed system seems thus, to have the capacity of increasing stroke volume more than the normal system. This capacity is retained despite the relief of the obstruction as evidenced by studies made five weeks after removal of the obstruction.[7] Otherwise, mild obstruction does not affect the response to increase in diuresis: the 1:1 transmission of pelvic activity down to the ureter, pressure increase, volume increase, and stroke volume increase, are as in a normal unobstructed system – even the activity increase during abrupt pressure increase is the same.

After five weeks of total obstruction and the consequent severe affection of renal function no pressure increase accompanies attempted induction of a diuresis,[1,7] nor is there any change in residual peristaltic activity. Reconstruction relief of the obstruction is occasionally followed by some minor restoration of both urine flow and peristaltic activity, but in general the changes induced experimentally are irreversible in contrast to clinical congenital disorders.

References

1. Boyarsky, S. and Martinez, J. (1964). Pathophysiology of the ureter. Patial ligation of the ureter in dogs. *Investigative Urology*, **2**, 173.
2. De Luca, F. G., Swenson, O. and Smyth, B. (1961). The effect of chronic mechanical obstruction on ureteral peristalsis. *Journal of Urology*, **85**, 497.
3. Radwin, H. M., O'Dell, R. M. and Schlegel, J. V. (1963). The renal response to acute partial obstruction. *Journal of Urology*, **80**, 243.
4. Bäcklund, L. and Reuterskjöld, A. G. (1969). Activity in the dilated ureter. *Scandinavian Journal of Urology and Nephrology*, **3**, 99.
5. Schreiter, R., Planz, K., Laese, W., Kunke, S. and Karus, F. (1973). Elektro-myographische Untersuchungen am normalen und partiel gestauten Harnleiter des Kaninschens. *Urologia Internationalis*, **28**, 268.
6. Gee, W. F. and Kiviat, M. D. (1975). Ureteral response to partial obstruction. Smooth muscle hyperplasia and connective tissue proliferation. *Investigative Urology*, **12**, 309.
7. Djurhuus, J. C., Nerstrφm, B., Gyrd-Hansen, N. and Rask-Andersen, H. (1976). Experimental hydronephrosis. An electrophysiologic investigation before and after release of obstruction. *Acta Chirugica Scandinavica suppl.*, **472**, 17.
8. Tanagho, E. A. (1972). Surgically induced partial urinary obstruction in the fetal lamb III. Ureteral obstruction. *Investigative Urology*, **10**, 35.
9. Finkle, A. L., Karg, S. J. and Smith, D. R. (1962). Ureteral atony and hydronephrosis following periureteral fibrosis in dogs. *Journal of Urology*, **87**, 535.
10. Sheehan, H. L. and Davis, J. C. (1960). Experimental hydronephrosis without mechanical obstruction of the ureter. *British Journal of Urology*, **32**, 53.
11. Mannes, H., Zimskind, P. D., Subbarao, Y. and Lewis, P. L. (1972). Crush injury of the lower ureter:An experimental study. *Journal of Urology*, **108**, 548.
12. Boyarsky, S. and Labay, P. C. (1972). In Ureteral Dynamics. P. 347. Baltimore, Willi ams & Wilkins Company.
13. Aung-Khin, M. (1972). The innervation of the ureter in experimental open hydronephrosis. *Investigative Urology* **10**, 379.
14. Ulm, A. H. and Miller, F. (1962). An operation to produce experimental reversible hydronephrosis. *Journal of Urology*, **88**, 337.
15. Olesen, S. and Madsen, P. O. (1968). Renal function during experimental hydronephrosis. Function during partial obstruction following contralateral nephrectomy in the dog. *Journal of Urology*. **99**, 692.
16. Cockett, A. T. K. and Fitzgerald, J. B. (1963). Experimetnal hydronephrosis: Retroaortic transposition of left ureter. *Journal of Urology*, **89**, 656.

17. Zimskind, P. D., Davis, D. M., Lewis, P. L., DeCaestecker, J. and Berwind, R. T. (1968). Complete obstruction of the ureter: Immediate and longterm effects on the renal pelvis pressure in dogs. *Surgical Forum*, **19**, 538.
18. Rose, J. G., Gillenwater, J. Y., Attinger, R., Sim, P. and Wyker, A. T. (1973). Ureteral wall tension: A new means of evaluating ureteral function. *Investigative Urology*, **10**, 480.
19. Wexler, N. S., McClinton, M. A., Clayton, J. D. and Roberts, J. A. (1972). Partial ureteral obstruction in the monkey. A comparison of hydrated versus dehydrated scintiphoto studies for evaluation. *Investigative Urology*, **10**, 473.
20. Kiil, F. (1957). The function of the ureter and renal pelvis. Oslo University Press.
21. Enhörning, G. and Weaver, R. G. (1964). Intraureteric pressure above and below an area of compression. *American Journal of Obstetrics and Gynecology*, **90**, 1332.
22. Ulmsten, U. (1974). Studies of ureteral function in women. Dissertation. Malmö, Sweden.
23. Noix, M. (1967). Cinéradiologie en urologie. In *Progress en Urologie*. Cit. Vereecken (1975).
24. Mortensen, S. O. and Djurhuus, J. C. (1976). Right ovarian vein syndrome. A case with pre- and peroperative electromyographic registration of ureteral activity. *Acta Chirugica Scandinavica*, suppl. **472**, 91.
25. Tscholl, R., Keller, U. and Spring, P. (1976). Investigation of urine transport in the ureter of the pig by x-ray video-densitometry I. The effect of increasing diuresis on velocity and rhytm of concentrations: Relationship between pressure wave and urine bolus. *Investigative Urology*, **13**, 404.
26. Vereecken, R. L. (1975). La physiologie et la physiopathologie de lúretere. *Acta Urologica Belgica*, **43**, 114.
27. Butcher, H. R. T. and Sleator, W. Jr. (1955). Action potentials and pressure changes in ureteral peristaltic waves. *American Journal of Physiology*, **180**, 261.
28. Ladefoged, O. and Djurhuus, J. C. (1976). Morphology of the upper urinary tract in experimental hydronephrosis in pigs. *Acta Chirugica Scandinavica*, suppl. **472**, 29–35
29. Weiss, R. M., Holcomb, W. and Bassett, A. L. (1972).b Excitability of *in vitro* normal and dilated human ureteral segments. *Investigative Urology*, **5**, 42.
30. Schwartzberg, J. E., Heilbron, D. and Hinman, F. Jr. (1971). Disuse and increased function of the dog ureter. *Urologia Internationalis*, **26**, 51.
31. Cussen, L. J. and Tymms, A. (1972). Hyperplasia of ureteral muscle in response to acute obstruction of the ureter. *Investigative Urology*, **9**, 504.
32. Weiss, R. M., Wagner, M. L. and Hoffman, B. F. (1967). Localization of the pacemaker for peristalsis in the intact canine ureter. *Investigative Urology*, **5**, 42.
33. Notley, R. G. (1968). Electron microscopy of the upper ureter and the pelvi-ureteric junction. *British Journal of Urology*, **40**, 37.
34. Bozler, E. (1942). The activity of the pacemaker previous to the discharge of a muscular impluse. *American Journal of Physiology*, **136**, 543.
35. Gosling, J. A. and Constaninou, C. E. (1976). The origin and propagation of upper urinary tract contraction waves. A new *in vitro* methodology. *Experientia*, **32**, 266
36 Longrigg, N. (1975). *In vitro* studies of the human renal calyces. *Journal of Urology*, **114**, 325.
37. Cussen, L. J. (1972). The effect of incomplete chronic obstruction on the ureteric muscle of the dog. *Investigative Urology*, **10**, 208.
38. Ladanyi, P. and Lelkes, G. (1966). Study of experimental fibrillogenesis in tunica muscularis of the rat ureter. *Acta Morphologica and Academic Science*, Hungaria **16**, 147
39. Galen (1916). On the Natural Faculties. Harvard University Press.

PART IX

CENTRAL NERVOUS SYSTEM

PART IX

24. The Vascularly Isolated Brain in Animals

Hans Stødkilde-Jørgensen

The MR Research Center, Skejby Hospital, Skejbygaardsvej, Aarhus, Denmark

Introduction

The isolated perfused brain preparation enables pharmacological and electrostimulatory studies of cerebral functions without the influence of secondary peripheral actions such as respiratory and cardiovascular effects. Furthermore, as elimination through the liver and kidney is abolished, a greater control of the experimental conditions in respect to composition and concentration of centrally active drugs in the perfusate is offered. Experimental models for perfusion of isolated brain preparation have been developed in different species of animals. The early studies were done in the cat,[33,34] in the calf,[79] in the rat,[105,2] in the monkey[109,110] and in the dog.[39–41] Later on especially the rat and the dog have been widely used in different isolated brain perfusion models.

This review resumes the most important papers on the establishment of isolated perfused brain models in different animal species. Aspects as anesthesia, surgical procedures, perfusion techniques and perfusion fluids are discussed. Furthermore, the results from both neurophysiological and neuroendocrinological studies in which the isolated brain models have been applied are included.

Anesthesia

Different types of anesthesia have been used in animals with vascularly isolated brains. In the small species like the rat, mouse and guinea pig, urethan 1.2 g/kg i.p., thiopentone-sodium 30–35 mg/kg i.p. or pentobarbital-sodium 50–60 mg/kg i.p. have been used. This may be supplemented with ketamine 10 mg/kg and atropine 20 μg/kg. The larger animals like the dog and the monkey have received pentobarbital 25 mg/kg. However, in the dog 1–2% halothane is the most frequently used anesthetic. Phenoxybenzamine hydrochloride (Dibenzyline) 1 mg/kg is often added to prevent surgical shock. Chloralose has furthermore been used for anesthesia in the dog. The dose for induction is 100 mg/kg i.v. and for maintenance 10 mg/kg administered in intervals of thirty minutes. In the preparation of the isolated pig brain, 0.5–1.0% halothane supplemented with 50% N_2O in the ventilatory gas has been applied. The pig is especially susceptible to cardiovascular collapse due to initial anxious episodes during induction of the anesthesia. Therefore, premedication should be applied before halothane, e.g. azaperone (Sedaperone) 0.1 7 ml/kg. Another combination of anesthesia for the pig solely consisting of intravenous drugs is: midazolam 0.5 mg/kg, i.m. for premedication and twenty minutes later ketamine 8mg/kg, i.m. and after the sleep induction of abolus of midazolam 0.25 mg/kg, ketamine 3mg/kg and pancuron 0.1mg/kg through an ear vein followed by intubation. The anesthesia is maintained with a mixture of ketamine 16 mg/kg/h, midazolam 0.8 mg/kg/h and pancuron 0.2 mg/kg/h continuously administered intravenously

The use of chemical anesthetic agents is undesirable from a functional point of view as most anesthetic drugs exhibit substantial effects on both electrophysiological and metabolic functions in the brain.[98,19,50] Even short-acting anesthetic drugs may not be metabolized in the isolated brain circulation without liver and kidneys, and the effects of the drugs may persist for a considerable period of time. But if a volatile anesthetic such as halothane is used during the surgical preparation, this can be removed through the ventilation of the extracorporeal blood circulation for e.g. forty five minutes before the actual measurements begin.[63] However, the normal pain reactions looked for by the anesthetist may not be present in an isolated brain preparation. Therefore, the concentration of analgetics should, at least in larger animals, be maintained on a level corresponding to that used in the intact animal. In the smaller animal species, the use of anesthesia during the surgical preparation can be avoided by inducing hypothermia. A rectal temperature of 16°C may in the rat be low enough to obtain surgical anesthesia.[2] Furthermore, hypothermia diminishes the tendency of bleeding from the wound surfaces, and as the metabolic activity in the brain continues on a lower level, the risk of brain cell damage due to transient anoxia is also diminished.

Surgical Preparation

The surgical methods necessary for the establishment of an isolated brain perfusion model depend on the species of animals used. The differences in size, in vessel topography and in flow distribution between the main vessels to the brain imply that the surgical procedures optimal for one species cannot automatically be transferred to another. Besides this, the most optimal way of cannulating the vessels to the brain is also dependent on the conditions for blood supply to the brain during the cannulation procedure.

If the vessel to the brain are not cannulated close to the brain, e.g. in the internal carotids, a problem may arise from the heterogeneity of the structures then perfused, e.g. extracranial tissues and muscles on the neck and upper limbs. These structures could imply a substantial contribution to the metabolic parameters measured, e.g. in the form of waste products from anerobic processes. In smaller animals, this problem may be overcome by removing all extracranial soft tissue from the face together with the lower jaw. But when used in a larger animal such as the dog this demands tremendous surgical effort.

The Rat, the Mouse and the Guinea pig

The preparation of the experimental model for isolated brain perfusion is identical for the rat, the mouse and the guinea pig.[106,89] The procedures for isolation of the rat brain initially proposed by Thomson and later modified by Andjus and Krieglstein[2,66,67,70] are followed. The methodical development has been reviewed by Woods.[118] The anatomy of the neck of the rat is so delicate that brain perfusion in this animal was initially performed by catheterization of the aorta in the thorax.[105] The venous blood is drained through the left and right superior caval vein by a catheter inserted in the inferior caval vein. To avoid venous return from another part of the body than the brain, the azygos vein is ligated, and to prevent blood from escaping into the right ventricle of the heart, a cardiac ligature is applied. Others have cannulated the common carotids directly after having ligated the external carotids and the pterygopalatine arteries.[18,123] By this method, the carotids are cannulated in turn. When the first artery has been cannulated, a slow perfusion is begun to ensure that there is no interruption of the blood flow to the brain. Furthermore, during the cannulation procedure, the brain can be blood supplied via the vertebral arteries. After cannulation the head is severed from the body and connected to the extracorporeal circulation. The benefit of the method of direct intubation of the internal carotids is that less of the perfusate enters the cranial bone and extracranial structures such as the muscles of the face. However, this has been questioned by others.[3]

A very simple model has been developed in the guinea-pig[84] and the ground squirrel.[86]

The isolated perfused brain technique has resulted in the development of special purpose models with isolation of selected areas of the CNS. An interesting model for perfusion of only the right cerebral hemisphere *in situ* has been developed in the rat.[102] Besides the possibility of comparing measurements with the left hemisphere, the method has the advantage that all infused blood reaches the brain without escape to bones or soft tissue of the face or mixing with blood from other places at the Circle of Willis. In other special perfusion models in adult rats, the cervical spinal cord was perfused for up to five hours with preserved conductivity and no structural changes apparent in electron micrographs.[17] In the guinea pig, the brain stem and the cerebellum have been used for studying neuronal circuit activity.[73]

The Dog

Isolated perfused brains in dogs primarily build on the work of David Gilboe.[39,40,43] Most of the soft tissue structures on the head and neck are cut and removed and furthermore, the mandible and the nose, maxilla and sinuses are removed. "T" tubes are placed in the common carotids for arterial inlet. This facilitates the shift to the extracorporeal circulation by preventing longer periods of interrupted blood supply to the brain. The venous blood is drained directly from the dural sinuses.

A special model developed in the dog allows the body to survive during the extracorporeal perfusion of the head.[30] The blood coming up through the common carotids from the heart is trapped and diverted into an extracorporeal circulation where it might have the oxygen or carbondioxide tension and other constituents changed before it is led into the common carotids again. The vertebral arteries are ligated.

The Pig

The pig has only seldom been used in isolated brain perfusion models. However, one model which allows the body to survive has been established.[100,101] The vessels between head and body are occluded, but the spinal cord and the sympathic and vagal trunks are left intact. While the brain is blood supplied in an extracorporeal circulation, the body of the animal survives by the function of its own heart and lungs. Besides internal function in the brain, the model provides the basis for differentiation of the ortotropic drug action on organs from the indirect effect mediated via the central nervous system.

As the body circulation is maintained, the surgical approach for isolation of the cerebral circulation differs from those experiments, where only the brain is kept alive such as in the dog and in the rat. The blood overflow between the two circulatory compartments has to be reduced as much as possible, and therefore both the internal and external vertebral venous plexus surrounding the spinal cord and the vertebral spine have to be occluded. In order to facilitate the shift to the extracorporeal perfusion of the brain in terms of avoiding transient hypoxia, the subclavian artery and vein on both sides are tubulated and connected to the extracorporeal circulation. At the moment of shifting to the extracorporeal circulation, both the superior caval vein, the left subclavian artery and the brachiocephalic trunk from which both common carotids arise are ligated close to the heart. As a consequence of this, the blood infused through the subclavian arteries will be forced up into the carotids and the vertebral arteries. Furthermore, the venous return from the internal and external jugular vein cannot be drained through the superior caval vein, but is forced out into the subclavian veins.

The Monkey

R. J. White introduced the monkey for studies of vascularly isolated brains.[109,110,112] This was followed by detailed descriptions of the preparation technique for connecting the isolated brain to either an extracorporeal circulation primed with donor blood or directly to the blood circulation of another animal.[111,113,114] All cephalic-cervical soft tissue is completely removed including exenteration of the orbits leaving only the skull and the cervical spinal column devoid of musculature. During the preparation process, both common carotids are cannulated with a "T" cannula followed by a later ligation of the vertebral arteries and external carotids. Before the shift to the extracorporeal circulation, the spinal cord is ligated in the height of C1–C2. The venous blood returning from the brain is collected in a reservoir.

Even though the surgical procedures for vascular isolation of the monkey brain are carefully described in the literature and the model has proved viability for up to two hours of perfusion, this has not been followed by any studies of brain functionality with use of the isolated monkey brain. However, a special model for isolated perfusion of the hypothalamus-thalamus area in the rhesus monkey has been developed for studies of the metabolism of sexual hormones.[13]

The Extracorporeal Circulation

Oxygenator and Pump

The extracorporeal circulation used for perfusion of isolated animal brains consists principally of a unit for blood oxygenation, an arterial infusion pump and sometimes also a venous pump.[123] The arterial pump is normally an ordinary roller pump capable of delivering either non-pulsatile or pulsatile flow.

The oxygenator used in the smaller animals is most often the bubble type oxygenator built in glass consisting of a reservoir for temperature stabilization and a unit for exchange of oxygen and carbon dioxide with a gas inlet. Normally a gas mixture of 94.6% oxygen and 5.6% carbon dioxide is used. The carbon dioxide is added to the ventilatoric gas in order to avoid vasoconstruction due to hypocapnia. These oxygenation units are without defoamer, but equipped with a mesh filter (e.g. 20–40 μm) of polyester, Dacron-Wool or of sintered glass for removal of platelet aggregates, fibrin particles or cell debris and other impurities which may be captured if the venous blood is not collected through tubes but simply collected in a bath below the brain preparation. In the larger animals such as the dog and the pig, the bubble type oxygenators known from human cardiopulmonary bypass operations are used either in their original form or modified, e.g. by removing the heat exchanger resulting in a lower priming volume or by reducing dimensions to suit the lower blood flow needed for ventilation of the brain compared to the whole body[100]. Furthermore, microdisc oxygenators,[25,41] and membrane oxygenators[30,63,121] have been used.

The tubes used to connect the infusion pump and the oxygenator equipment with the tubes inserted into the vessels are normally made of PVC, rubber or silicone rubber. If pulsatile flow is used, the arterial tubes must be made of rigid material in order to preserve the energy in the pulsations delivered by the pump. It should be taken into consideration that most soft plastic materials do uptake lipophilic drugs and this may result in experimental errors.[77] Furthermore, if whole blood is used as perfusate, platelets may adhere to the silicone tubes and the walls of the oxygenator.[88]

Specialized perfusion equipment with the possibility of shifting rapidly between two volumes of perfusate with different composition has been developed for perfusion of the rat brain[28]. In the larger animals, the same possibility for a rapid shift between perfusate has been done simply by just using two oxygenators.

Perfusion Fluid

In a small animal such as the rat, the fluid used for cerebral perfusion varies from allo or autotransfusion of whole blood, various mixtures of albumin, glucose, salts and washed erythrocytes to HCO_3-buffered saline solutions to which were added salts and perhaps H_2O_2.[73] For perfusion of the small animals like the rat, suspensions of erythrocytes may be obtained by collecting bovine blood and storing it in an acid-citrate-buffer or directly from heparinized donor rats. The washed erythrocytes should be prepared fresh for each day's experiments by centrifugation followed by suspension in Krebs-Henseleit buffer to which are added 2g% (Wt/vol) bovine serum albumin to prevent brain edema. The brain is very sensitive to changes in the osmotic pressure of the perfusate, and swelling is seen even after small falls in the protein concentration. The hematocrite should preferably be 30%.[2,65–69] The Krebs-Henseleit buffer has the following composition (in mM): NaCl 118.4, KCl 4.7, $CaCl_2$ 2.4, $MgSO_4$ 1.2, $NaHCO_3$ 15.0 and glucose 11.2. The pH will be between 7.35 and 7.40. If no erythrocytes are used in the perfusate, the Krebs-Henseleit could be solely or alternatively an HCO_3-buffered physiological saline solution consisting of (in mM): NaCl 142, $NaHCO_3$ 28, KH_2PO_4 4.2, $CaSO_4$ 1.7, $MgSO_4$ 1.0 and dextrose or glucose.[99,102] In this type of perfusate, the oxygen requirements of the brain are provided from oxygen dissolved directly in the water of the perfusion medium. In larger animals like the dog or the pig, diluted whole blood appears to be the perfusate of choice. The blood is taken from a donor animal, and blood type incompatibilities are normally not taken into consideration if the animals are from the same herd. A comparison between whole blood and 1) a suspension of washed erythrocytes suspended in plasma containing electrolytes, amino acids, and albumin and 2) erythrocytes mixed with plasma of the same composition as cerebrospinal fluid, showed an energy failure in the two latter mixtures, probably due to the absence of essential nutrients.[63]

In later years, artificial blood substitutes have gained more widespread use as perfusate for organ perfusion. In particular the inert perfluorochemicals (PFCs) such as fluorocarbons are known to dissolve more than 50 volumes per cent of oxygen and more than 400 volumes per cent of carbon dioxide.[35,36] Solutions in emulsified forms are commercially available from Green Cross Corporation of Osaka, Japan, either as Fluorosol-DA 20% fluorocarbon emulsion (0.1 μm) solution (FDA) or as FC-43, which is the original PFC used in animal experiments. A suitable mixture in the rat and other animals of the same size could be 20% (vol/vol) FC43 and 4.8% (wt/vol) of the nonionic surfactant, Pluronic F 68, in a Krebs-Henseleit buffer[18,65] or in a Krebs-bicarbonate buffer to which were added lactate 3.0 mM/l and pyruvate 0.3 mM/l and glucose 7–10 mM/l.[17,55,56] Even though FDA only contains approximately 1.2 ml oxygen per 100 ml of emulsion at a pO_2 of 16 kPa compared to approximately 20 ml in blood at the same tension, FDA delivers only a small amount less than the about 5 ml oxygen per 100 ml delivered by blood when the oxygen tension falls from 70 to 6 kPa in the tissue.[23] This is due to the S-formed shape of the relation between oxygen content and oxygen tension for hemoglobin. The anaphylactic reactions seen at random when blood and FDA were mixed were due to the surfactant, Pluronic F-68, used to prepare FDA. These effects have been eliminated when egg yolk phospholipids are used as surfactants. FDA is supplied frozen due to poor room-temperature emulsion stability. Even though the fluorocarbon emulsion are still carrying some drawbacks, further development is carried out.[14]

Fluorocarbon erythrocyte substitution in brain perfusion has especially been used in the rat[18,71,89,95] and has enabled the measurements of the metabolic turnover of sugar and amino acids which are not possible with blood as perfusate.[73,76,83] Even though perfluorocarbon emulsions have not been used for brain perfusion in larger animals, it has been used for perfusion of other organs in both the pig and dog.[24] In brains perfused with fluorocarbon substitutes, the cerebral blood flow resistance is significantly decreased due to the lower viscosity.[72] Furthermore, the synthetic blood substitutes do not have the rheological properties of whole blood, and though glucose and inorganic ions are added, it might still lack constituents seen in whole blood with importance for the metabolic processes in the brain.

Extracorporeal perfusion of an isolated brain preparation may suffer from the missing breakdown and excretion of waste products through the liver and the kidneys. With the use of whole blood perfusate, the lactate concentration increases in the same way as known from cardiac by-pass operations.[31,32] However, blood isolated in an extracorporeal well-oxygenated circulation is known *per se* to be responsible for a significant lactate formation.[85,112]. Emphasis has been put on the role of lactic acid in the survival of the brain nerve cells during extracorporeal perfusion since the accumulation during ischaemia of this product together with the falling pH may cause several structural changes.[54] Therefore, the lactate increase could indicate the need for dialysis of the perfusate to remove lactate.[89] Other waste products such as glutamine resulting from the ammonia detoxication[5,29] may also be removed. This is furthermore supported by the observation that an intact EEG is maintained for significantly longer periods when a non-recycling perfusion is adopted.

Anticoagulation

In perfusion of the brain of larger animals, anticoagulated whole blood from a donor animal is normally used. However, increasing bleeding from wound surfaces during perfusion is an often seen phenomenon and if the rebleeding is great enough it may empty the extracorporeal circulation and thus shorten the available perfusion time. The rebleeding, also seen in cardiopulmonary by-pass operations, has been attributed to a number of different factors such as surgical trauma, hypothrombinemia, thrombocytopenia, hyperfibrinolysis and other hemostatic defects.[48] It has earlier been demonstrated that mechanical trauma from an oxygenator and rotor pump does not really affect the concentration of coagulation factors, whereas platelets are markedly reduced, perhaps because of the adhering of platelets to the silicone tubes and to the oxygenator.[88] The rebleeding may to some extent be explained by the type of heparin used. This is not a homogeneous biopolymer, but rather a polydisperse with a molecular weight ranging from 3, 000 to 60, 000. The thrombolytic effect of heparin is most pronounced for molecular weights around 20, 000 whereas the inhibition of the clot formation is related to smaller heparin molecules.[4] Rebleeding can perhaps be reduced by using a type of heparin with a relatively higher content of the low molecules form of heparin. Furthermore, epsilon aminocaproic acid (EACA) can be added to prevent rebleeding. EACA is known to inhibit fibrinolysis by blocking the formation of plasmin[87] and thus minimizing the disintegration of already formed clots.[46]

Perfusion Pressure and Flow

The perfusion modus used in isolated brain perfusion can be based either on constant flow or on constant pressure. But constant perfusion, pressure is most often chosen, and in this situation, the flow varies according to the vasoresistence in the cerebral vessels. However, even though the nerve cells may survive on a lower metabolic rate, there is a minimal cerebral flow rate below which the supply of substrates and oxygen is insufficient to support even a basal metabolism. Furthermore, earlier observations of varying glucose uptake in the perfused brain and the relationship between perfusion rate and glucose uptake do indicate that insufficient regional cerebral blood supply can lead to erroneous interpretations of metabolic data.[2,123]

If the blood is diluted or an artificial blood substitute like the fluorocarbon perfusion medium with lower oxygen transport capacity is used, the infusion flow should be increased to secure sufficient oxygenation. This will not normally result in significant increase in infusion pressure because of the lower viscosity of these perfusates. If the perfusion of the isolated brain is continued with whole blood for more than one - two hours with the constant infusion flow, the resulting pressure increases to some extent. This phenomenon is also known from perfusion of other organs and can probably be related to thrombosis in the brain capillaries or an increasing amount of perivascular inflammation due to the appearance of denaturated proteins in the perfusate. If the perfusate is not filtered at all, this will inevitably result in a marked increase in the perfusion pressure.[2,28,118,123] However, others have demonstrated that the increase in cerebral vascular resistance in the isolated canine brain could be related to the appearance of vasoconstrictors in the perfusate such as leukotrienes from the leukocytes.[64]

Non-pulsatile flow is the perfusion method normally used in experiments with artificial perfusion of isolated brain preparations. However, even though the question of the proper execution of perfusion of the brain in terms of pulsation or non-pulsation remains controversial,[53] several studies indicate the benefit of using a pulsatile infusion flow, which carries relatively more energy at a comparable volume and mean pressure.[93,115] Pulsatile infusion flow is considered important for the transport capability of oxygen/carbon dioxide and nutritients in and out of the nerve cells. The ratio between regional blood flow and glucose uptake in the brain has been shown to be of the same size in pigs subjected to three hours of pulsatile cardiopulmonary bypass at normothermia as found in a control group.[1] Furthermore, it has been reported that pulsatile perfusion in rats seems to be essential for long-term viability of the brain[70] and that the diffuse nerve cell changes in the dog immediately after two - three hours of perfusion are more pronounced in non-pulsatile perfusion compared to pulsatile perfusion.[91] However, the differences were less pronounced twelve hours postoperatively.[120]

The extracorporeal perfusion of the vascularly isolated brain implies changes of the regional cerebral blood flow pattern in respect to infusion pressure and pulsatile or non pulsatile flow. The autoregulation is of major importance for the cerebral blood flow and thus for the amount of oxygen presented to the nerve cells.[51] If the infusion pressure is kept within normal ranges for body pressure (roughly between 60 and 150 mmHg), the autoregulation of the brain is known to be preserved during extracorporeal perfusion of the isolated brain.[103] Even during hypothermic (20°C) non pulsatile infusion in the canine brain an autoregulatory response exists and the oxygen uptake is preserved with an infusion pressure down to 30 mmHg.

The Perfused Isolated Brain in Research

Many aspect of cerebral function have been evaluated by means of the isolated perfused brain model. The metabolic rate of oxygen and glucose as well as aspects of the transport across the blood brain barrier have in particular been studied. Furthermore, besides the influence of different drugs on brain function *per se*, the survival of the brain cells in the isolated perfusion model has attracted special interest. The viability of the neurons is fundamental in order to validate the results of pharmacological or physiological studies of the brain function in vascularly isolated brain preparations. The establishment of extracorporeally perfused brain preparations involves extensive surgery and this, together with the perfusion technique and type of perfusate and anesthesia used, will be of importance for neurophysiological and biochemical mechanisms in the nerve cells and thus for the brain viability.[27] Evaluation of brain viability in the intact animal often relies on the brain's functional integrity with body visceral entities such as the cardiovascular system. Because of this, quantification of the viability of separate functional entities in the brain, as well as of the brain as a whole, is difficult in isolated brain preparations. Nevertheless, to evaluate cerebral viability in an experimental set-up, one has to look for different neurophysiological and metabolic parameters correlating to cerebral function, including the persistence of the blood brain barrier.

Brain viability has been correlated to different measurements such as gross and microscopic structure, level of high energy phosphates, cerebral blood flow, net flux of oxygen and carbon dioxide, metabolic rate for glucose, lactate accumulation, EEG and persistence of somatosensory evoked reflexes.

Energy Metabolism

The metabolic studies using the isolated brain model primarily concern the turnover of oxygen and glucose in the brain.

Glucose

The high energy requirements of the brain have made the brain vulnerable to failing supply of oxygen and nutritients such as glucose. With a normal blood concentration of glucose, 80–90% of the oxygen taken up by the brain in the dog is used for the metabolism of glucose.[25] Others have found that cerebral O_2 consumption was constant and unaffected by the rate of glucose uptake.[44] Other metabolic substrates such as ketone bodies, mannose and amino acids may also contribute to the energy production.[37] This is confirmed by data in the rat brain, which in the absence of glucose seems to be able to provide energy from amino acids such as glutamic acid.[80, 97] In the *in situ* brain of the rat, the uncoupling effect of 2, 4-Dinitrophenol on electron transfer reactions has by optical techniques and [^{31}P]-NMR spectroscopy been shown to result in a reduction of intramitochondrial cytochromes, but this has not been confirmed by measurements *in vitro*.[60] In the canine brain, reduced blood oxygen pressure at a 50% saturation by treating with carbamylation limits O_2 diffusion to the brain with a resulting reduced oxygen consumption and deterioration of the EEG quality.[119] In dogs, energy metabolism was sufficient to maintain substantial levles of ATP in the nerve cells during thirty minutes of hypoxia ($PaO_2 = 20$ mmHg).[62] Furthermore, normal ion gradients were sustained, the formation of edema was prevented, and after reperfusion with normal oxygenized blood, recovery of the parameters was total. However, after anoxia ($PaO_2 < 10$ mmHg) also during thirty minutes, the recovery was not total, and after total ischemia, parameters were only poorly restored. After only six minutes of perfusion with anoxic blood ($pCO_2 < 10$ mmHg) in the isolated dog brain, the glucose uptake and brain lactic acid increased considerably.[19] An activity increase was also observed for the regulatory enzymes hexokinase, phosphofructokinase and glycogen phosphorylase. In the rat the magnitude of ischemic changes such as increased ADP, AMP and fructose 1, 6-diphosphate varied with the severity of the ischemia.[65] The results were confirmed by the use of nuclear magnetic resonance analysis.

The ability of the brain to metabolize ketone bodies was preserved in the isolated rat brain, but the energy contribution from D-3-hydroxybutyrate was small compared to that of glucose.[122] The metabolism shifted towards ketone bodies during starvation.

The energy metabolism in the isolated prefused rat brain during addition of neuroleptics and antidepressants showed a significant decrease in glucose-6-P alone.[68] Toxic doses of these classes of drugs resulted in accumulation of P-creatine and glucose and in a decrease of glycogen, glucose-6-P and ammonia.

The mitrochondrially bound hexokinase is highly important in the control of the brain energy metabolism. Even though a solubilization of mitochondrially bound hexokinase means a reduction of brain energy metabolism, the solubilization effect of the pentobarbital anesthetic, methohexital, is due to a direct effect on the mitochondrial membrane and not mediated through a blockage of the metabolism.[58]

In the isolated rat brain, the intracellular brain glucose was linearly related to the glucose concentration in the perfusate.[52] Addition of thiopental to the perfusate in an anesthetic dose caused an increase in P-creatine and glucose, whereas the levels of ADP, AMP, lactate and pyruvate were diminshed. This is in accordance with what is seen in intact anesthetized animals. However, others have found a hyperbolical relation between the glucose uptake in the isolated rat brain and the perfusion rate.[123] In experiments using isolated dog brains, anerobic glucose metabolism did not appear to account for any substantial part of the glucose uptake.[41]

Calcium antagonists are known to improve postischemic brain energy metabolism. This has been confirmed by adding emopamile to the perfusate in the isolated rat brain after induced ischemia during thirty minutes.[10] In the early postischemic recovery phase, the creatine-phosphate, glucose, glucose-6-phosphate and fructose-6-phosphate were higher in the enopamile treated animals compared to controls, but the high-energy phosphate restoration turned out to depend on flow rates. Opposite changes in the same parameters could together with accumulations of creatine, AMP and lactate be seen postischemic in rats without drug addition.[67] Whereas the early postischemic restoration in high-energy phosphate seemed to be attributed to the vascular effect of the calcium entry blocker, the later decrease of lactate in the enopamile group was caused by a direct action on the brain parenchyma. In other experiments, nimodipine has been added to the isolated rat brain both before and after induced ischemia.[11] The cerebral flow rate was not changed by preischemic addition of nimodipine, but postischemic the nimodipine addition resulted in lowered levels of ADP, glucose-6-phosphate and lactate probably based on a direct improvement of mitrochaondrial function. This effect was also found for the calcium antagonists noremapamil and gallopamil.[108] The same protection of brain energy metabolism against ischemic damage was seen after addition of another calcium entry inhibitor, galopamil, to the perfusate during induced ischemia in an isolated rat perfusion model.[107] The isolated rat brain model has furthermore been used for studying the effect of calcium channel blocking drugs such as nicardipine and others on the rat brain muscarinic and α-adrenergic receptors.[104] The calcium entry blocking drug flunarizine has been demonstrated to improve postischemic flow and energy metabolism.[92]

Gasses

The oxygen consumption is considered a good measure for brain viability because the net cerebral metabolic rate for oxygen is related to intact mitochondrial function and preserved integrity of the cerebral neurons. However, to calculate the net oxygen flux from the differences in arterial and venous oxygen tension or content, one has to know the exact flow through the brain and this can be a problem in the set-ups where part of the infused blood is allowed to enter persisting soft tissue on the face and neck or the bones of the skull.

The isolated brain perfusion model enables studies of regional cerebral blood flow under different drug additions to the perfusate and to changing partial pressure of the ventilatory gasses. In the dog, anoxic anoxia quickly resulted in reduced blood flow to the cerebellum but an increase of more than 60% to the brain stem including the respiratoric entre.[20] In some models with isolated brain perfusion, the body of the animal is allowed to survive. In a canine model, the common carotids were shunted and when the vertebral arteries were ligated, the blood was forced into an extracorporeal oxygenator circulation.[102] By this model, it has been shown that hypercapnia in the brain circulation causes an increase in systemic arterial pressure, a decreased heart rate and an increased mean circulatory filling pressure in the head.

Cerebral oxygen metabolism in the intact animal is known to be influenced by the addition of nitrous oxide to the ventilatory air. However, this could not be shown in the isolated perfused canine brain,[26] but cerebral vascular resistance fell during exposure to nitrous oxide.

Electrophysiology

One of the parameters measured to evaluate the viability of the brain is the EEG, and as this parameter is dependent upon integrated functioning of large groups of cells, the EEG is considered a sensitive measure of brain function. In the dog it has been shown that the oxygen uptake changes parallel to the EEG activity expressed as amplitude and classes of frequency spectra.[25] Both energy and power spectral analyses of the EEG have been performed but besides these computer based methods, the visual evaluation, though less exact, is nevertheless important for the detection of particular waveforms or grouped patterns. This has been shown in quantitative analyses of the EEG effect produced by mipramine, desipramine and promazine given to the perfusate in the isolated rat brain.[68,90] These drugs produced EEG amplitude enlargement and slowing of the frequency as well as grouped spikes. Spontaneous electroencephalographic (EEG) activity in the rat was maintained for up to five hours when the brain was perfused in an open circuit and for up to two hours when a recirculation was applied.[2] Furthermore, addition of pentylenetetrazol (Metrazol) to the perfusate resulted in convulsive EEG patterns. Nitrous oxide added to the isolated canine changed the EEG pattern towards reduced amplitude and an increase in both low-voltage β activity and 3–5 Hz activity.[26]

Transport across the Blood Brain Barrier

The blood brain barrier seems to preserve its functionality after vascular isolation.[124] In contrast to brain slices or homogenates of brain cells, the isolated perfused brain utilizes pyruvate at a much lower rate compared to glucose indicating that the metabolism of the brain still relies on the glucose carrier situated in the blood brain barrier. Kinetic studies of glucose transport across the blood brain barrier indicate that the carrier mechanism is similar to that found in a cell membrane, e.g. the red blood cell. In the dog, it was found that the two structurally related compounds, phlorrizin and phlorretin, are competitive inhibitors of glucose transport indicating that the glucose carrier in the blood brain barrier comprises two binding loci, one for sugars and one for phenols.[8] The glucose transport across the blood brain barrier seems to be unaffected by changes in plasma insulin.[6,42,96] Furthermore, the glucose carrier mechanism seems maximally stimulated at normal plasma insulin levels as no changes in glucose uptake are seen after injection of a high dose of insulin. However, as insulin injected intracisternally affects glucose transport, the possibility remains that insulin has an effect on glucose transport from the interstitial fluid into the brain cell.[117] The kinetic constants for glucose transport across the blood brain barrier were found to be constant over a wide range of concentrations.[15] The net uptake of glucose from the blood stream is a result of two simultaneous solute fluxes, the undirectional uptake coupled to a glucose carrier and the glucose efflux. The difference between the two fluxes is, besides the glucose level in the blood stream, correlated to the glucose concentration in the brain tissue. In the dog, the glucose influx was increased after a short period of anoxia, probably due to a decrease in the glucose efflux. The subsequent decrease in net uptake was due to an impairment of the unidirectional glucose transport, indicating that the glucose uptaken was due to an impairment of the unidirectional glucose transport, indicating that the glucose uptake from the blood requires energy.[7] The post-hypoxia induced reduction in the undirectional transport seemed moreover to be related to the decrease in pH, which may change the glucose carrier's affinity for glucose.[9,45] However, others have not found any relation between the post-hypoxia unidirectional glucose transport across the blood brain barrier and pH.[121]

The synthesis of proteins, peptides and neurotransmitters such as dopamine, norepinephrine and serotonin in the brain also requires, besides the amino acids synthesized in the brain, supplies of at least twelve different neutral amino acids from the blood stream. The *in situ* brain perfusion technique has in the rat been used for studying the L-phenylalanine and the leucine transport to different brain regions.[47,78] The integrity of the blood brain barrier during extracorporeal perfusion was demonstrated by the lack of the free amino acids of brain tissue such as glutamate, aspartate, N-acetylaspartate and GABA in the perfusion fluid.[80] Furthermore, the unidirectional transport of leucine was unaffected by ten minutes of anoxia.[8] Even with a totally uncoupled oxidative phosphorylation, the blood brain barrier remains intact to diffusion of polar molecules for up to thirty minutes.[49]

The compound dimethylsulfoxide may improve the interpretation of the results due to its ability to selectively alter the permeability for drugs in the blood brain barrier.[38] The transport of primidone, an anticonvulsant agent, across the barrier has been compared to the transport rate for one of the metabolites of the prodrug to primidone, phenobarbital.[55,56] The isolated perfusion model allows for elimination of the substantial hepatic metabolism of most drugs and thus, an evaluation of the cerebral metabolic conversion.[66]

Drug Studies, Amino Acids, Transmitters and Hormones

The main question in using isolated brain preparations is whether the functionality of the brain can be compared to the conditions found *in vivo*. This has been evaluated for brain monoamines such as dopamine, norepinephrine, histamine, serotonin and their precursors tyrosine and tryptophan.[69] The level of these monoamines in the isolated rat perfusion model was equal to the results found in the intact animal. Furthermore, during two hours of perfusion, linear rates for the 5-hydroxytryptamine synthesis were demonstrated, and the rate was comparable to that found after L-tryptophan loading *in vivo*.[116] In an isolated mouse head model, the production of the primary metabolite of norepinephrine in the central nervous system, 3-methoxy-4-hydroxyphenethylenglycol, was similar to the turnover rate of norepinephrine in the intact mouse brain.[106]

The isolated guinea-pig brain has been used for demonstration of GABA immonoreactivity for up to twelve hours per perfusion.[16] Furthermore, by the olfactory signal processing in the guinea-pig piriform cortex has been evaluated.[12]

The acetylcholine in the perfusate of an isolated rat brain circulation was shown to increase after thirty minutes of perfusion compared to the levels found *in vivo*.[61] Physostigmine caused an acetylcholine rise in both the isolated brain circulation and *in vivo*. In contrast, oxotremorine, which is known to increase the acetylcholine level *in vivo*, did not result in any increase in the perfused model. The acetylcholinesterase enzyme is inhibited by the organophosphorus compound sarin and results in increased concentrations of acetylcholine in different organs.[94] In an isolated canine brain perfusion model, addition of either quinidine or physostigmine provided partial protection against the inhibition of acetylcholinesterase by sarin, provided that one of the two drugs was added prior to sarin.

The cerebral vessels are innervated by sympathetic fibres, but the response to physiological levels of circulating norepinephrine is minimal. This has been confirmed by an isolated brain circulation model in the dog, where the vasomotoric response to hormonal norepinephrine in cerebral vessels was compared to the responsiveness in the hind limb.[74]

Perfused isolated brain preparations of spontaneously hypertensive rats and of normotensive rats have been used for studying vascular resistance at different infusion flow rates and after addition of vasoconstrictors phenylephrine, prostaglandin $F_{2\alpha}$ and serotonin.[47] The hypertensive animals generally showed a higher cerebrovascular resistance, and the reactivity to both serotonin and phenylephrine was to a greater extent enhanced in this group.

An isolated mouse perfusion model has been used for studying the effect of nicotine on β-endorphin, α-MSH, and ACTH secretion from the pituitary gland.[75] The increase in these peptides after nicotine addition was shown to be due to a stimulatory effect on the brain, which resulted in increased pituitary secretion. The work was continued by studying changes in norepinephrine and dopamine and electrocortical activity after adding nicotine to the perfusate.[22] The nicotine induced desyncronization of the enhanced cortical electrical amplitude and, furthermore, increased the levels of norepinephrine and dopamine in various brain regions. The effect of injected morphine and methadone on the cerebral cortex of the isolated perfused rat brain was evaluated by measuring metabolic parameters.[82] In contrast to methadone, morphine resulted in a decrease in ATP and creatinine phosphate and an increase in lactate.

The relatively large amount of N-acetylaspartic acid in brain tissue compared to other organs indicates that this amino acid could be involved in the metabolism of the brain. However, results from the isolated perfused rat brain suggest that N-acetylaspartic acid is without metabolic significance.[81]

The uptake of estrogens in the isolated rat perfusion model has shown considerable sex related differences.[57] The uptake of estradiol-17β was higher in the female brain with the highest concentrations found in the hypophysis and in the hypothalamus. The metabolism of estrogens in the brain resulted in the creation of estrone and other metabolites important for the neuroendocrine regulation and for the control of the sex behavior. A model for vascular isolation of the hypothalamus-thalamus region in the rhesus monkey brain has been used to study the metabolism of progesterone, estradiol and androstenedione.[13] In the hypothalamus, the androstenedione was converted to estrone and progesterone to pregnanedione and hydroxypregneneone.

Neurotoxicants

Neurological disorders due to impairment of the hepatic function have been demonstrated to originate primarily from increased blood levels of the neurotoxicants ammonia and short-chain fatty acid. Ammonia and octanoic acid were added to the perfusate in a canine isolated brain model, and the effect was distended neuronal mitochondria and separation or even dispersion of the lamellae of the mitochondrial cristae[21] of such a magnitude that it could be compared to that seen in encephalopaties of hepatic origin seen in humans. Furthermore, it has been shown that cerebral detoxication of NH_4^+ during hyperammonemia syndromes is related to increased formation of glutamine and to a lesser extent to alanine.[5,29] The isolated rat brain perfusion model has furthermore been used in studying the neurotoxicity of cyanide and especially the means of displacing cyanide bound to cytrocrome aa_3.[59]

Conclusion

The isolated perfused brain models can be used for a large variety of studies of brain functions *per se* and of the importance of these for peripheral organ homeostasis. Though well-developed, the preparation technique may be improved. Especially the closed re-cycling systems imply accumulation of waste products, which may shorten the time where the extracorporeal perfusion can be carried out with preserved cerebral viability. This calls for dialyzing. Furthermore, important nutrients may be missing in the perfusate especially if artificial blood substitutes are used together with buffers.

The isolated brain preparation is most easily established in small animals like the rat, and a lot of valuable contributions to our understanding of metabolic and excretory functions in the brain and in the blood brain barrier as well originate from these experiments. However, larger animals like the dog and the pig may show neurophysiological and neuroendocrinological mechanisms which both during the healthy and the deceased state may be more equal to the conditions found in man.

References

1. Andersen, K., Waaben, J., Husum, B., Voldby, B., Bødker, A., Hansen, A. J. and Gjedde, A. (1985). Nonpulsatile cardiopulmonary bypass disrupts the flow-metabolism couple in the brain. *The Journal of Thoracic Cardiovascular Surgery*, **90**, 570–579.
2. Andjus, R. K., Suhara, K. and Sloviter, H. A. (1967). An isolated, perfused rat brain preparation, its spontaneous and stimulated activity. *Journal of Applied Physiology*, **22**, 1033–1039.
3. Andree, T. H. and Clarke, D. E. (1981). The isolated perfused rat brain preparation in the study of monoamine oxidase and benzylamine oxidase. *Biochemical Pharmacology*, **30**, 959–965.
4. Barrowcliffe, T. W., Johnson, E. A. and Thomas, D. (1978). Antithrombin III and heparin. *British Medical Bulletin*, **34**(2)., 143–150.
5. Benzi, G., Arrigoni, E., Strada, P., Pastoris, O., Villa, R. F. and Agnoldi, A. (1977). Metabolism and cerebral energy state: effect of acute hyperammoneia in beagle dog. *Biochemical Pharmacology*, **26**, 2397–2404.
6. Betz, A. L., Gilboe, D. D., Yudilevich, D. L. and Drewes, L. R. (1973). Kinetics of unidirectional glucose transport into the isolated dog brain. *The American Journal of Physiology*, **225**, 586–592.
7. Betz, A. L., Gilboe, D. D. and Drewes, L. R. (1974). Effects of anoxia on net uptake and unidirectional transport of glucose into the isolated dog brain. *Brain Research*, **67**, 307–316.
8. Betz, A. L., Gilboe, D. D. and Drewes, L. R. (1975). Kinetics of unidirectional leucine transport into brain: effects of isoleucine, valine and anoxia. *The American Journal of Physiology*, **228**, 895–900.
9. Betz, A. L., Drewes, L. R. and Gilboe, D. D. (1975). Inhibition of glucose transport into brain by phlorizin, phloretin and glucose analogues. *Biochimica et Biophysica Atca*, **406**, 505–515.
10. Bielenberg, G. W., Haubruck, H. and Krieglstein, J. (1987). Effects of calcium entry blocker emopamil on postischemic energy metabolism of the isolated perfused rat brain. *Journal of Cerebral Blood Flow Metabolism*, **7**, 489–496.
11. Bielenberg, G. W., Stierstorfer, H.-J., Weber, J. and Krieglstein, J. (1989). Nimodipine reduces postischemic lactate levels in the isolated perfused rat brain. *Biochemical Pharmacology*, **38**, 853–855.
12. Biella, G. and de Curtis, M. (1995). Associative synaptic potentials in the piriform cortex of the isolated guinea-pig brain *in vitro*. *European Journal of Neuroscience*, **7**, 54–64.
13. Billiar, R. B., Takaoka, Y., Johnson, W., White, R. J. and Little B. (1981). *In situ* subcellular distribution and metabolism of progesterone, estradiol and androstenedione in the vascularly separated and isolated hypothalamus of the female rhesus monkey. *Neuroendocrinology*, **32**, 355–363.
14. Biro, G. P. and Blais, P. (1987). Perfluorocarbon blood substitues. CRC. *Critical Reviews in Oncology/Hematology*, **6**, 311–374.
15. Christensen, O., Lund-Andersen, H., Betz, A. L. and Gilboe, D. D (1982). Transport of glucose across the blood-brain barrier: Reevaluation of the accelerative exchange diffusion. *Acta Physiologica Scandinavica*, **115**, 233–238.
16. Curtis, M. De, Arcelli, P., Spreafico, R. and Avanzini, G. (1994). Ultrastructural features of the isolated guinea-pig brain maintained *in vitro* by arterial perfusion. *Neuroscience*, **59**(3), 775–788.
17. Deutschmann, W., Wellhoner, H. H. and Erdmann, G. (1983). Perfusion of the cervical spinal cord *in situ* of adult rats using a perfluorocarbon emulsion. *Brain Research*, **280**, 239–249.
18. Dirks, B., Krieglstein, J., Lind, H. H., Rieger, H. and Schutz, H. (1980). Fluorocarbon perfusion medium applied to the isolated rat brain. *Journal of Pharmacological Methods*, **4**, 95–108.
19. Drewes, L. R. and Gilboe, D. D. (1973). Glycolysis and the permeation of glucose and lactate in the isolated, perfused dog brain during anoxia and postanoxic recovery. *Journal of Biological Chemistry*, **218**, 2489–2496.
20. Drewes, L. R., Frazin, L. and Levin, A. (1975). Blood flow in the isolated, perfused canine brain: Normal oxygenation vs. anoxic anoxia. *Physiologist*, **18**, 198.
21. Drewes, L. R. and Leino, R. L. (1985). Neuron-specific mitochondrial degeneration induced by hyperammonemia and octanoic acidemia. *Brain Research*, **340**, 211–218.
22. Erwin, V. G., Cornell, K. and Towell, J. F. (1986). Nicotine alters catecholamines and electrocortical activity in perfused mouse brain. *Pharmacological Biochemistry and Behavior*, **24**, 99–105.
23. Faithfull, N. S. (1987). Fluorocarbons. *Anaesthesia*, **42**, 234–242.
24. Faithfull, N. S., Fennema, M. and Erdmann, W. (1988). Myocardial oxygen tension during ischaemia in fluorocarbon diluted pigs. *Advanced Experimental Medical Biology*, **222**, 473–480.
25. Fitzpatrick, J. H., Gilboe, D. D., Drewes, L. R. and Betz, A. L. (1976). Relationship of cerebral oxygen uptake to EEG frequency in isolated canine brain. *The American Journal of Physiology*, **231**, 1840–1846.
26. Fitzpatrick, J. H. and Gilboe, D. D. (1982). Effects of nitrous oxide on the cerebrovascular tone, oxygen metabolism, and electroencephalogram of the isolated perfused canine brain. *Anesthesiology*, **57**, 480–484
27. Fitzpatrick, J. H. Jr. and Gilboe, D. D. (1982). Criteria of viability in isolated brain preparations. *Israel Journal of Medical Sciences*, **18**, 67–73.
28. Fleck, W., Krieglstein, J. and Reichmann, M. (1973). A two-circuit apparatus for the perfusion of the isolated rat brain. *Naunyn-Schmiedeberg's Archives of Pharmacology*, **278**, 319–322.

29. Flock, E. V., Block, A. M., Mann, F. C., Grinday, J. H. and Bollman, T. L. (1952). The effect of glucose on the amino acids of plasma after total hepatectomy. *Journal of Biological Chemistry*, **198**, 427.
30. Gaddis, M., MacAnespie, C. L. and Rothe, C. F. (1986). Vascular capacitance responses to hypercapnia of the vascularly isolated head. *The American Journal of Physiology*, **251**, H164–H170.
31. Geha, A. S., Malt, S. H., Nara, Y. and Baue, A. E. (1971). Effect of cardiopulmonary bypass on cerebral metabolism. *The Journal of Thoracic Cardiovascular Surgery*, **61**, 200–211.
32. Geha, A. S., Salaymeh, M. T., Abe, T. and Baue, A. E. (1972). Effect of pulsatile cardiopulmonary bypass on cerebral metabolism. *Journal of Surgical Research*, **12**(6)., 381–387.
33. Geiger, A. and Magnes, J. (1947). The isolation of the cerebral circulation and the perfusion of the brain in the living cat. *The American Journal of Physiology*, **149**, 517–536.
34. Geiger, A. (1958). Correlation of brain metabolism and function by the use of a brain perfusion method *in situ*. *Physiological Review*, **38**, 1–20.
35. Geyer, R. P. (1973). Fluorocarbon-polyol artificial blood substitutes. *New England Journal of Medicine*, **298**, 1077–1082.
36. Geyer, R. P. (1975). Bloodless rats through the use of artificial blood substitutes. *Federation Proceedings*, **34**, 1499–1505.
37. Ghosh, A. K., Mukherji, B. and Sloviter, H. A. (1972). Metabolism of isolated rat brain perfused with glucose or mannose as substrate. *Journal of Neurochemistry*, **19**, 1279–1285.
38. Ghosh, A. K., Ito, T., Ghosh, S. and Sloviter, H. A. (1976). Effects of dimethylsulfoxide on metabolism of isolated prefused rat brain. *Biochemical Pharmacology*, **25**, 1115–1117.
39. Gilboe, D. D., Cotanch, W. W. and Glover, M. B. (1964). Extracorporeal perfusion of the isolated head of a dog. *Nature*, **202**, 399–340.
40. Gilboe, D. D., Cotanch, W. W. and Glover, M. B. (1965). Isolation and mechanical maintenance of the dog brain. *Nature*, **206**, 94–96.
41. Gilboe, D. D., Glover, M. B. and Cotanch, W. W. (1967). Blood filtration and its effect on glucose metabolism by the isolated dog brain. *The American Journal of Physiology*, **213**, 11–15.
42. Gilboe, D. D., Andrews, R. L. and Dardenne, G. (1970). Factors affecting glucose uptake by the isolated dog brain. *The American Journal of Physiology*, **219**, 767–773.
43. Gilboe, D. D., Betz, A. L. and Langebartel, D. A. (1973). A guide for the isolation of the canine brain. *Journal of Applied Physiology*, **34**, 534–537.
44. Gilboe, D. D. and Betz, A. L. (1973). Oxygen uptake in the isolated canine brain. *The American Journal of Physiology*, **224**, 558–595.
45. Gilboe, D. D., Costello, D. and Fitzpatrick, J. H. (1980). The effect of severe hypoxia on cerebral glucose flux. *Advanced Experimental Medical Biology*, **131**, 279–293.
46. Gomes, M. M. R. and McGoon, D. C. (1970). Bleeding patterns after open-heart surgery. *The Journal of Thoracic and Cardiovascular Surgery*, **60**(1)., 87–96.
47. Gong, D., Uruno, T., Nagakura, Y., Matsuoka, Y. and Kubota, K. (1989). Effects of perfusion serotonin on isolated, perfused brains of spontaneously hypertensive rats. *Brain Research*, **482**, 122–128.
48. Gralnick, H. R. and Fischer, R. D. (1971). The hemostatic response to open-heart operations. *The Journal of Thoracic and Cardiovascular Surgery*, **61**(6)., 909–915.
49. Greenwoood, J., Luthert, P. J., Pratt, O. E. and Lantos, P. L. (1985). Maintenance of the integrity of the blood-brain barrier in the rat during an *in situ* saline-based perfusion. *Neuroscience Letters*, **56**, 223–227.
50. Grüner, J., Kriegelstein, J. and Rieger, H. (1973). Comparison of the effects of chloral hydrate and trichloroethanol on the EEG of the isolated perfused rat brain. *Naunyn-Schmiedeberg's Archives of Pharmacology*, **277**, 333–348.
51. Halley, M. M., Reemtsma, K. and Creech, O. (1958). Cerebral blood flow, metabolism, and brain volume in extracorporeal circulation. *The Journal of Thoracic Surgery*, **36**, 506–518.
52. Hein, H., Krieglstein, J. and Stock, R. (1975). The effects of increased glucose supply and thiopental anesthesia on energy metabolism of the isolated perfused rat brain. *Naunyn-Schmiedeberg's Archives of Pharmacology*, **289**, 399–407.
53. Hickey, P. R., Buckley, M. J. and Philbin, D. M. (1983). Pulsatile and nonpulsatile cardiopulmonary bypass: review of a counterproductive controversy. *Annals of Thoracic Surgery*, **36**, 720–737.
54. Hossmann, K. A. and Olsson, Y. (1970). Suppression and recovery of neuronal function in transient cerebral ischemia. *Brain Research*, **22**, 313–325.
55. Höller, M., Breuer, H. and Fleishhauer, K. (1983). An improved method for perfusion of the isolated brain of the rat – influence of perfusion conditions and application of analeptic and anticonvulsant drugs. *Journal of Pharmacological Methods*, **9**(1)., 19–32.
56. Höller, M. and Penin, H. (1984). Studies on the intracerebral metabolism of anticonvulsant drugs. Perfusion of primidone through the isolated brain of the rat. *Biochemistry and Pharmacology*, **33**, 1753–1756.
57. Höller, M., Dengler, K. and Breuer, H. (1984). Disposition of [4-^{14}C]oestradiol-17β in the isolated perfused brain of the rat. *Journal of Steroid Biochemistry*, **20**, 785–787.
58. Karcher, L., Rachman, A. and Krieglstein, J. (1984). Comparison of rat brain and heart mitochondrial hexokinase solubilized by methohexital. *Naunyn-Schmiedeberg's Archives of Pharmacology*, **326**, 80–82.
59. Kariman, K. and Burkhart, D. S. (1986). Reaction of cyanide with cytochrome aa3 in isolated perfused rat head *in situ*. *The Journal of Biochemistry*, **233**, 187–191.
60. Kariman, K., Chance, B., Burkhart, D. S. and Bolinger, L. A. (1986). Uncoupling effects of 2,4-dinitrophenol on electron transfer reactions and cell bioenergetics in rat brain *in situ*. *Brain Research*, **366**, 300–306
61. Kilbinger, H. and Krieglstein, J. (1974). Applicability of the isolated perfused rat brain for studying central cholinergic mechanisms. *Naunyn-Schmiedeberg's Archives of Pharmacology*, **285**, 407–411.
62. Kintner, D. B., Costello, D. J., Levin, A. B. and Gilboe, D. D. (1980). Brain metabolism after 30 minutes of

hypoxic or anoxic perfusion or ischemia. *The American Journal of Physiology*, **239**, E501–E509..

63. Kintner,D. B., Kao,J. L., Woodson,R. D. and Gilboe,D. D. (1986). Evaluation of artificial plasma for maintaining the isolated canine brain. *Journal of Cerebral Blood Flow Metabolism* **6**, 455–462..
64. Kintner, D. B., Kranner, P. W. and Gilboe, D. D. (1986). Cerebral vascular resistance following platelet and leukocyte removal from perfusate. *Journal of Cerebral Blood Flow Metabolism*, **6**, 52–58.
65. Kopp, S. J., Krieglstein, J., Freidank, A., Rachman, A., Seibert, A. and Cohen, M. M. (1984). P-31 nuclear magnetic resonance analysis of brain: II. Effects of oxygendeprivation on isolated perfused and nonperfused rat brain. *Journal of Neurochemistry*, **43**, 1716–1731.
66. Krieglstein, G., Krieglstein, J. and Urban, W. (1972). Long survival time of an isolated perfused rat brain. *Journal of Neurochemistry*, **19**, 885–886.
67. Krieglstein, G., Krieglstein, J. and Stock, R. (1972). Suitability of the isolated perfused rat brain for studying effects on cerebral metabolism. *Naunyn-Schmiedeberg's Archives of Pharmacology*, **275**, 124–134.
68. Krieglstein, J. and Stock, R. (1973). Influence of therapeutic and toxic doses of neurolepties and antidepressants on energy metabolism of the isolated perfused rat brain. *Naunyn-Schmiedebergs's Archives of Pharmacology*, **279**, 243–254.
69. Krieglstein, J. and Niemeyer, D. H. (1974). The isolated perfused rat brain as a model for brain monoamine research. *Journal of Neurochemistry*, **24**, 829–831
70. Krieglstein, J. and Stock, R. (1974). The isolated perfused rat brain as a model for studying drugs acting on the CNS. *Psychopharmacologica*, **35**, 169–177.
71. Krieglstein, J. (1984). An isolated rat brain preparation perfused with a fluorocarbon emulsion. *Advanced Experimental Medical Biology*, **180**, 119–129.
72. Lee, P. A., Sylvia, A. L. and Piantadosi, C. A. (1988). Effect of fluorocarbon-for-blood exchange on regional cerebral blood flow in rats. *The American Journal of Physiology*, **254**, H719–H726.
73. Llinas, R., Yarom, Y. and Sugimori, M. (1981). Isolated mammalian brain *in vitro*: new technique for analysis of electrical activity of neuronal circuit function. *Federal Proceedings*, **40**, 2240–2245.
74. Louie, J. A., Javid, M. J. and Gilboe, D. D. (1985). Vasomotor response to hormonal norepinephrine in canine cerebral and hindlimb vessels. *The American Journal of Physiology*, **248**, H232–H239.
75. Marty, M. A., Erwin, V. G., Cornell, K. and Zgombick, M. (1985). Effects of nicotine on β-endorfin, αMSH, and ACTH secretion by isolated perfused mouse brains and pituitary glands, *in vitro*. *Pharmacology Biochemistry and Behavior*, **22**, 317–325.
76. Mayevsky, A., Mizawa, I. and Sloviter, H. A. (1981). Surface fluorometry and electrical activity of the isolated rat brain with artificial blood. *Neurological Research*, **3**, 307–316.
77. Minder, R., Weder, H. J. and Bickel, M. H. (1970). Experimental errors resulting from uptake of lipohilic drugs by soft plastic materials. *Biochemical Pharmacology*, **19**, 2179–2180.
78. Momma, S., Aoyagi, M., Rapoport, S. I. and Smith, Q. R. (1987). Phenylalanine transport across the blood-brain barrier as studied in situ brain perfusion technique. *Journal of Neurochemistry*, **48**, 1291–1300.
79. Moss, G. (1964). Cerebral arterial isolation. The effects of differential pressure perfusion. *Journal of Surgical Research*, **4**, 170–177.
80. Mukherji, B., Turinsky, J. and Sloviter, H. A. (1971). Effects of perfusion without glucose on amino acids and glycogen of isolated rat brain. *Journal of Neurochemistry*, **18**, 1783–1785.
81. Mukherji, B. and Sloviter, H. A. (1973). Metabolism of acetate and N-acetylasparatate in isolated perfused rat brain. *Journal of Neurochemistry*, **20**, 633–636.
82. Mukherji, B., Suemaru, K., Sakai, N., Ghosh, A. K. and Sloviter, H. A. (1980). Effects of morphine and methadone on the isolated perfused rat brain. *Biochemical Pharmacology*, **29**, 1608–1611.
83. Mukherji, B. and Sloviter, H. A. (1987). Metabolism of an isolated brain perfused with perfluoro blood substitue. *Journal of Biomedical Science*, **11**, 23–33.
84. Mühlethaler, M., de Curtis, M., Walton, K. and Llinás, R. R. (1993). The *in vitro* isolated and perfused guinea pig brain. *Eur. J. Neurosci.*, **5**, 915–926.
85. Nara, Y., Geha, A. and Baue, A. (1971). Lactate production in blood circulating in a pump oxygenator: effect of carbondioxide. *Journal of Surgical Research*, **11**, 612–619.
86. Pakhotina, P. I. and Pakhonina, I. D. (1994). Preparation of isolated perfused ground squirrel brain. *Brain Research Bulletin*, **33**(6), 719–721.
87. Reed, C. C. and Clark, D. K. (1975). In *Cardiopulmonary Perfusion*, Texas Medical Press, Inc., Houston, Texas, 114–122.
88. Replagli, R. L., Kevy, S. V. and Gross, R. E. (1967). Some hematologic aspects of prolonged blood pumping. *The Journal of Thoracic and Cardiovascular Surgery*, **54**, 688.
89. Richerson, G. B. and Getting, P. A. (1987). Maintenance of complex neural function during perfusion of the mammalian brain. *Brain Research*, **409**, 128–132.
90. Rieger, H. and Krieglstein, J. (1974). Quantitative analysis of the EEG effects produced by imipramine, desipramine, promazine and monodesmethyl promazine in the isolated perfused rat brain. *Psychopharmacology*, **39**, 163–179.
91. Sanderson, J. M., Wright, G. and Sims, F. W. (1972). Brain damage in dogs immediately following pulsatile and non-pulsatile blood flows in extracorporeal circulation. *Thorax*, **27**, 275–286.
92. Scheller, D., Tegtmeier, F., Urenjak, J., Kolb, J., Peters, U., Bock, A. and Höller, M. (1989). Influence of flunarizine on postischemic flow and metabolism in the isolated rat brain. *Biomedical Biochemistry Acta*, **48**, S161–S165.
93. Shepard, R. B., Simpson, D. C. and Sharp, J. F. (1966). Energy equivalent pressure. *Archives of Surgery*, **93**, 730–740.
94. Singh, A. K., Zeleznikar, R. J. and Drewes, L. R. (1986). Protection from quinidien or physostigmine against *in vitro* inhibition by sarin of acetylcholinesterase activity. *Life Sciences*, **38**, 165–172.

95. Sloviter, H. A. and Kamimoto, T. (1967). Erythrocyte substitute for perfusion of brain. *Nature*, **216**, 458–460.
96. Sloviter, H. A. and Yamada, H. (1971). Absence of direct action of insulin on metabolism of the isolated perfused rat brain. *Journal of Neurochemistry*, **18**, 1269–1274..
97. Sloviter, H. A. (1982). Metabolic studies with an isolated, perfused rat brain perparation. *Israel Journal of Medical Sciences*, **18**, 59–66.
98. Smith, A. L. and Wollman, H. (1972). Cerebral blood flow and metabolism. Effects of anesthetic drugs and techniques. *Anesthesiology*, **36**, 378–400.
99. Smith, Q. R., Takasato, Y., Sweeney, D. J. and Rapoport, S. I. (1985). Regional cerebrovascular transport of leusine as measured by the *in situ* brain perfusion technique. *Journal of Cerebral Blood Flow Metabolism*, **5**, 300–311.
100. Stϕdkilde-Jϕrgensen, H., Frøkiaer, J., Kirkeby, H. J., Madsen, F. and Boye, N. (1986). Preparation of a cerebral perfusion model in the pig: technical aspects. Swine in *Biomedical Research*, In M. E. Tombleson (ed).. New York, Plenum Press, **2**, 727–734.
101. Stødkilde-Jørgensen, H., Frøkiaer, J., Kirkeby, H. J., Madsen, F. and Boye, N. (1986). Preparation of a cerebral perfusion model in the pig: anatomic considerations. Swine in *Biomedical Research*, In M. E. Tombleson (ed).. New York, Plenum Press, **2**, 719–725
102. Takasato, Y., Rapoport, S. I. and Smith, Q. R. (1984). An *in situ* brain perfusion technique to study cerebrovascular transport in the rat. *The American Journal of Physiology*, **247**, (Heart Circulation Physiology, 16)., H484–H493.
103. Tanaka, J., Shiki, K., Asou, T., Yasui, H. and Tokunaga, K. (1988). Cerebral autoregulation during deep hypothermic nonpulsatile cardiopulmonary bypass with selective cerebral perfusion in dogs. *The Journal of Thoracic Cardiovascular Surgery*, **95**, 124–132.
104. Thayer, S. A., Welcome, M., Chhabra, A. and Fairhurst, A. S. (1985). Effects of dihydropyridine calcium channel blocking drugs on rat brain muscarinic and α-adrenergic receptors. *Biochemical Pharmacology*, **34**, 175–180.
105. Thompson, A. M., Robertson, R. C. and Bauer, T. A. (1968). A rat head-perfusion technique developed for the study of brain uptake of materials. *The Journal of Applied Physiology*, **24**, 407–411.
106. Towell, J. F. and Erwin, V. G. (1981). Characterization of the isolated perfused mouse brain as a system for neourochemical studies. *Brain Research*, **209**, 476–481
107. Weber, J. and Krieglstein, J. (1985). Effect of gallopamil on energy metabolism of the isolated perfused rat brain in the postischemic period. *Naunyn-Schmiedeberg's Archives of Pharmacology*, **329**, 451–454.
108. Weber, J., Bielenberg, G. W. and Krieglstein, J. (1988). Effects of phenylalkylamine calcium entry blockers on postischemic energy metabolism in the isolated rat brain: Stereoselective action of emopamil. *Pharmacology*, **37**, 38–49.
109. White, R. J., Albin, M. S. and Verdura, J. (1963). Isolation of the monkey brain: *In vitro* preparation and maintenance. *Science*, **141**, 1060–1061.
110. White, R. J., Albin, M. S. and Verdura, J. (1964). Preservation of viability in the isolated monkey brain utilizing a mechanical extracorporeal circulation. *Nature*, **202**, 1082–1083.
111. White, R. J., Albin, M. S, Verdura, J. and Locke, G. E. (1967). The isolated monkey brain: Operative preparation and design of support systems. *Journal of Neurosurgery*, **27**, 216–225.
112. White, R. J., Albin, M. S., Davidson, E. and Yashon, D. (1968). Production of lactate and pyruvate by the brain: Accumulation of lactate and pyruvate in the recycling circulation of the isolated brain. *Surgery*, **64**, 60–67.
113. White, R. J., Wolin, L. R., Massopust, L. C., Taslitz, N. and Verdura, J. (1971). Primate cephalic transplanation: Neurogenic separation vascular association. *Transplantation Proceedings*, **3**, 602–604.
114. White, R. J. (1972). Preparation and mechanical perfusion of the isolated monkey brain. **S158**, 200–216.
115. Wilkens, H., Regelson, W. and Hoffmeister, F. S. (1962). The physiologic importance of pulsatile blood flow. *New England Journal of Medicine*, **267**, 443–446.
116. Woods, H. F., Graham, C. W., Green, A. R., Youdim, M. B. H., Grahame-Smith, D. G. and Huges, J. T. (1976). Some Histological and metabolic properties of an isolated perfused rat brain. *Acta Endocrinology*. preparation with special reference to monoamine metabolism. *Neurosciences*, **1**, 313–323.
117. Woods, S. C. and Porte, D. (1977). Relationship between plasma and cerebrospinal fluid insulin levels of dogs. *The American Journal of Physiology*, **233**, E331–E334.
118. Woods, H. F. and Youdim, M. B. H. (1978). *The isolated perfused rat brain preparation* - A Critical assessment. M. B. H. Youdim and W. E. Lovenberg (ed).: Essays in Neurochemistry and Neuropharmacology, **2**, J. Wiley, 49–69.
119. Woodson, R. D., Fitzpatrick, J. H., Costello, D. J. and Gilboe, D. D. (1982). Increased blood oxygen affinity decreases canine brain oxygen consumpiton. *Journal of Laboratory and Clinical Medicine*, **100**, 411–424.
120. Wright, G. and Sanderson, J. M. (1972). Brain damage and mortality in dogs following pulsatile and nonpulsatile blood flows in extracorporeal circulaton. *Thorax*, **27**, 738–747.
121. Yanushka, J. M., Costello, D. J. and Gilboe, D. D. (1983). Effect of blood pH and ischemia on kinetic constants for cerebral glucose transport. *The American Journal of Physiology*, **245**, E384–E390.
122. Zivin, J. A. and Snarr, J. F. (1972). Glucose and D(-).-3-hydroxybutyrate uptake by isolated perfused rat brain. *Journal of Applied Physiology*, **32**, 664–668.
123. Zivin, J. A. and Snarr, J. F. (1972). A stable perparation for rat brain perfusion: effect of flow rate on glucose uptake. *Journal of Applied Physiology*, **32**, 658–663.
124. Zlokovic, B. V., Begley, D. J., Djuricic, B. M. and Mitrovic, D. M. (1986). Measurement of solute transport across the blood-brain barrier in the perfused guinea pig brain: Method and application to N-methyl-α-aminoisobutyric acid. *Journal of Neurochemistry*, **46**, 1444–1451.
125. Zlokovic, B. V., Mackic, J. B., McComb, J. G., Weiss, M. H., Kaplowitz, N. and Kannan, R. (1994). Evidence for transcapillary transport of reduced glutathione in vascular perfused guinea-pig brain. *Biochemical and Biophysical Surgery*, **201**(1)., 402–8.

Index